T0418823

Mechanics of Fibrous Networks

Elsevier Series in Mechanics of Advanced Materials

Elsevier Series in Mechanics of Advanced Materials

Mechanics of Fibrous Networks

Edited by

Vadim V. Silberschmidt
Wolfson School of Mechanical, Electrical and Manufacturing Engineering, Loughborough University, Loughborough, Leicestershire, United Kingdom

ELSEVIER

Elsevier
Radarweg 29, PO Box 211, 1000 AE Amsterdam, Netherlands
The Boulevard, Langford Lane, Kidlington, Oxford OX5 1GB, United Kingdom
50 Hampshire Street, 5th Floor, Cambridge, MA 02139, United States

Library of Congress Cataloging-in-Publication Data
A catalog record for this book is available from the Library of Congress

British Library Cataloguing-in-Publication Data
A catalogue record for this book is available from the British Library

ISBN: 978-0-12-822207-2

For information on all Elsevier publications
visit our website at https://www.elsevier.com/books-and-journals

Publisher: Matthew Deans
Acquisitions Editor: Dennis McGonagle
Editorial Project Manager: Rafael G. Trombaco
Production Project Manager: Manju Thirumalaivasan
Cover Designer: Alan Studholme

Typeset by STRAIVE, India

Contents

Contributors

Mir Karim Razavi Aghjeh Department of Macromolecular Science and Engineering, Case Western Reserve University, Cleveland, OH, United States; Faculty of Polymer Engineering, Institute of Polymeric Materials, Sahand University of Technology, Tabriz, Iran

Kamel Berkache Ecole Supérieure des Sciences Appliquées d'Alger, ESSAA; Faculty of Physics, University of Science and Technology Houari Boumediene, Algiers, Algeria

Emanuela Bosco Department of the Built Environment, Eindhoven University of Technology, Eindhoven, The Netherlands

Christophe Bouvet Université de Toulouse, Institut Clément Ader, ISAE-SUPAERO—UPS—IMT Mines Albi—INSA, Toulouse, France

Fadhel Chatti Université de Toulouse, Institut Clément Ader, ISAE-SUPAERO—UPS—IMT Mines Albi—INSA, Toulouse, France

Vincenzo Cucumazzo Wolfson School of Mechanical, Electrical and Manufacturing Engineering, Loughborough University, Leicestershire, United Kingdom

Nik Dave Department of Mechanical Engineering, Eindhoven University of Technology, Eindhoven, The Netherlands

Jean-Francois Ganghoffer LEM3, Université de Lorraine, CNRS, Metz, France

Marc G.D. Geers Department of Mechanical Engineering, Eindhoven University of Technology, Eindhoven, The Netherlands

Carlos González Department of Materials Science, Technical University of Madrid, School of Civil Engineering; IMDEA Materials Institute, Getafe, Madrid, Spain

Alp Karakoç Department of Communications and Networking; Department of Bioproducts and Biosystems, Aalto University, Espoo, Finland

Artem Kulachenko Solid Mechanics, Department of Engineering Mechanics, KTH Royal Institute of Technology, Stockholm, Sweden

Javier Llorca Department of Materials Science, Technical University of Madrid, School of Civil Engineering; IMDEA Materials Institute, Getafe, Madrid, Spain

Horacio Lopez-Menendez Department of Physical-Chemistry, Complutense University of Madrid, Madrid, Spain

Yanhui Ma School of Engineering, Cardiff University, Cardiff, United Kingdom

Rami Mansour Solid Mechanics, Department of Engineering Mechanics, KTH Royal Institute of Technology, Stockholm, Sweden

Francisca Martínez-Hergueta School of Engineering, Institute for Infrastructure and Environment, The University of Edinburgh, William Rankine Building, Edinburgh, United Kingdom

Ron H.J. Peerlings Department of Mechanical Engineering, Eindhoven University of Technology, Eindhoven, The Netherlands

Dominique Poquillon CIRIMAT, Université de Toulouse, INP-ENSIACET, Toulouse, France

Amit Rawal Department of Textile and Fibre Engineering, Indian Institute of Technology Delhi, New Delhi, India

Mir Jalil Razavi Department of Mechanical Engineering, State University of New York at Binghamton, Binghamton, NY, United States

Hilal Reda LEM3, Université de Lorraine, CNRS, Metz, France; Faculty of Engineering, Section III, Lebanese University, Beirut, Lebanon

Alvaro Ridruejo Department of Materials Science, Technical University of Madrid, School of Civil Engineering, Madrid, Spain

Noud P.T. Schoenmakers Department of Mechanical Engineering, Eindhoven University of Technology, Eindhoven, The Netherlands

Vadim V. Silberschmidt Wolfson School of Mechanical, Electrical and Manufacturing Engineering, Loughborough University, Leicestershire, United Kingdom

Emrah Sozumert Wolfson School of Mechanical, Electrical and Manufacturing Engineering, Loughborough University, Leicestershire; College of Engineering, Swansea University, Swansea, United Kingdom

Tetsu Uesaka Department of Chemical Engineering, Mid Sweden University, Sundsvall, Sweden

Hanxing Zhu School of Engineering, Cardiff University, Cardiff, United Kingdom

Preface

From the point of view of mechanics of materials, fibrous networks represent a special class of materials. It contains dissimilar materials such as non-wovens, paper, hydrogels, and different biological systems and structures. One obvious common feature is their discontinuity, limiting the application of the formalism of mechanics of continuous media in their analysis. But they are not the only discontinuous materials, with porous ones being an obvious example. Still, the structure of fibrous networks makes them unique, allowing the development of common mechanical approaches and models. In these materials, a set of fibres (that can be made of different materials) is held together either by friction or by various types of forces at the intersections of fibres. Some speak about them as fibrous composites without a matrix. Such networks have various features of randomness: most demonstrate randomness in distribution, orientation, and curliness (crimp) of fibres that can be supplemented by randomness in their axial mechanical properties, spatial distribution, and/or mechanical characteristics of contacts between them. This can be complicated by the presence of various types of fibres in hybrid networks.

The pronounced discontinuity – together with specificity of mechanical behaviour of constituent fibres – defines the main mechanisms of deformation, damage, and fracture of fibrous networks that are principally different from those of traditional continuous structural materials. Let us consider some examples. The first case is of a fibrous network with a trivial structure – arrangement of two sets of orthogonal similar straight fibres connected at each intersection. As well known, uniaxial stretching of a traditional material results in its lateral contraction due to the Poisson's effect. In the considered example of the fibrous network, deformed along the direction of either set of fibres, there would be no deformation of another set: its fibres retain the initial length at all stages of deformation. Obviously, the introduction of fibres at various angles to the main orientations would change this.

The second example is of a fibrous network with an arbitrary orientation distribution of fibres, connected at some or all of their contacts/intersections. It shows a totally different realisation of deformation even in a case of a uniaxial stretching. In a conventional (homogeneous) material, all its parts are stretched in the same way in such a loading regime, with local deformations equal to those of the external ones (up to the onset of necking). By contrast, in the fibrous network of straight fibres, only fibres aligned with the stretching direction are deformed in this way (this can be affected by inter-fibre contacts), while others accommodate the external deformation by the rotation of their segments towards the direction of stretching. As a result, deformation of fibres is mostly less – sometimes significantly – than the external one. If the fibres are curly, this process is exacerbated by their straightening. Only after this deformation–accommodation mechanism is accomplished do the fibres start stretching axially. As a result, the initial stage of deformation is characterised by low stiffness that increases with the extent of stretching. Importantly, many networks are formed by polymeric fibres with high levels of axial deformation – more than 100%. This additionally complicates the mechanical analysis of deformations for such networks.

The third example is notch sensitivity. Continuous materials with sharp notches demonstrate considerable stress concentration in their vicinity, with a stress singularity near the tips for purely elastic materials. In fibrous networks, this process is governed by a specific load-transfer mechanism – along the axis of a (stretched) fibre's segment to the nearest contact (intersection) point. In this point, the load is split based on the number, orientation, and properties of joining fibres. Even the idea of the notch is different here – it is rather a sequence of cut (broken) neighbouring fibres. As a result, a local loss of load-bearing capacity can be delocalised to parts away from the 'notch tip' (as discussed, the definition of 'notch' is totally different here, not to speak of its 'tip'). This process is especially prominent in low-density networks, with a spatially sparse system of connections between fibres. It also affects the fracture-propagation mechanism: some fibres surviving just near the 'notch' effectively hinder its growth. Instead, local damage areas with a relatively small number of non-failed fibres are formed.

Both deformation and fracture mechanisms are affected by the nature of inter-fibre interaction that can be a mere friction contact, intersection allowing relative rotation or a full fixture of the contacting parts of the joining fibres. The levels of stiffness and strength of this contact can also vary, sometimes exceeding that of its constituents. In many cases they are also random, depending on the origin of the network. As a result, various damage/failure mechanisms can be realised: separation of fibres, their breakage at the contact or away from it. Such events change the local deformation mode, for example, by allowing its accommodation by rotation rather than stretching, also affecting the fracture evolution.

Apparently, all these complexities make any mechanical analysis of such networks a considerable challenge even in two-dimensional tension (since many examples of such materials have a rather small thickness and cannot bear any in-plane compression). Still, there are also different types of 3D fibrous networks.

This book covers different aspects of mechanics of such materials, starting from the discussion of relatively simple variants of their mechanical behaviour. This is followed by the analysis of more complex cases – elasto-plastic behaviour, hygromechanics, damage, time-dependant failure, and ballistic response. The effect of different types of inter-fibre contacts on different mechanical behaviours is studied thoroughly. Most of the networks have various random features, necessitating the development and application of dedicated numerical tools, which are covered in several contributions to the volume. Obviously, the main focus is on the specifics of mechanics of such networks, rooted in their microstructures; hence, micromechanics of such materials is presented prominently. Still, notwithstanding all the differences between the fibrous networks and traditional materials, a need to bridge the gap between discontinuous and continuous materials underpins development of methods of generalised continuum as well as numerical homogenisation. They are especially important for the design and optimisation of various products. The volume deals with both typical examples of fibrous materials such as nonwovens and papers and the more exotic ones, for example, F-actin and vimentin networks, and electrospun fibrous mats.

The volume is prepared by leading specialists in the area of mechanics of fibrous networks from the United Kingdom, India, France, Lebanon, Algeria, Sweden, Finland, the United States, The Netherlands, Spain, and Iran, and presents the recent developments in this emerging and challenging part of mechanics of materials.

Vadim V. Silberschmidt
Loughborough, United Kingdom

Mechanics of fibrous networks: Basic behaviour

1

Emrah Sozumert[a,b] *and Vadim V. Silberschmidt*[a]
[a]Wolfson School of Mechanical, Electrical and Manufacturing Engineering, Loughborough University, Leicestershire, United Kingdom, [b]College of Engineering, Swansea University, Swansea, United Kingdom

1.1 Introduction

This chapter investigates the mechanics of fibrous networks considering fundamental phenomena related to their microstructure and material properties, such as fibre alignments and curliness, cross links at fibre-to-fibre contacts, and material properties of individual fibres. In order to realise this aim, finite-element (FE) models of three unique fibrous networks are generated employing computer algorithms. Fibres represented by chains of beam and cross-link (beams in our case) elements are placed at fibre-to-fibre intersections. In addition to the microstructure of fibrous networks, the effect of boundary conditions, in particular, periodic and non-periodic ones, at network boundaries are of interest as well as the type of beam formulations (Timoshenko or Bernoulli). Deformation mechanisms of three configurations of fibrous networks are examined under tensile loading within the framework of a non-linear finite-element method. Mechanical response of a fibrous network depends on the material properties of individual fibres and their microstructural arrangements. For analysis of basic behaviours of fibrous networks, a material response of the constituent fibres is confined to elasticity. The load-transfer mechanisms between the fibres contacting at intersections are realised with cross-link elements of equal length. FE models of the fibrous networks with various densities of cross links are utilised to quantify and understand global (macroscopic) and local (microscopic) levels of deformational response. The global response of fibrous networks is tracked throughout the simulations and evolution of local stress–strain states are recorded for quantitative comparisons of the FE models with the investigated microstructures.

Microscopic images of three different fibrous networks are presented in Fig. 1.1: a bacterial-cellulose (BC) hydrogel, a collagen fibrous network from a starfish, and a soft collagenous membrane. Cellulose fibres of BC hydrogel, interacting with each other through entanglements, are produced by bacteria in an aqua environment [1]. In cases of the collagen network and the collagenous membrane, on the other hand, fibres are connected to each other with cross links [2,3]. The latter is composed of fibres highly curved at the microscale, which are combinations of fibrils in the nanoscale. These networks are examples of random fibrous networks with highly complex microstructures, which is the main source of non-linear deformation mechanisms at

Mechanics of Fibrous Networks. https://doi.org/10.1016/B978-0-12-822207-2.00005-2

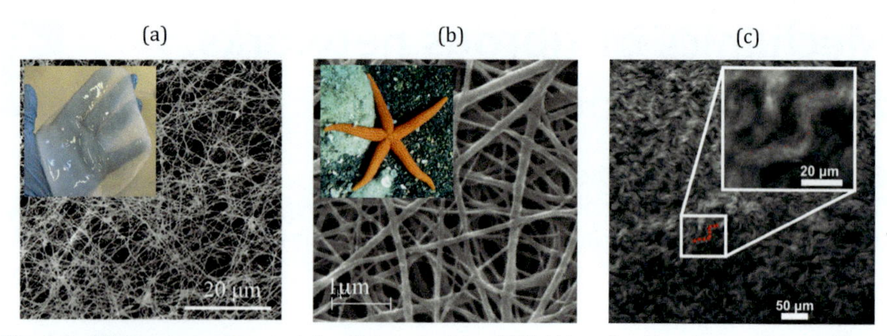

Fig. 1.1 SEM images of bacterial-cellulose hydrogel [1] (A) and echinoderm collagen fibrous network from starfish [2] (B); (C) multiphoton microscope image of the soft collagenous membrane [3].

various scales. Importantly, the deformability and strength of a fibrous network depends on the orientation of constituent fibres with respect to orthogonal vectors [4]. The orientation distributions of fibres in fibrous networks can be expressed with different distribution functions whilst their elastic properties can be estimated with numerical simulations of discrete fibrous networks [5].

To understand mechanical behaviours of complex, mostly random fibrous networks, it is easier to start with the analysis of more simple objects, for instance, eliminating such a factor as microstructural randomness. This, obviously, simplifies the orientational characteristics of the network. Still, its main feature – discreteness (as opposed to continuity of traditional structural materials) – affecting the load-transfer mechanism will be retained, albeit in a simplified form. So, the aim of this chapter is to investigate the mechanical behaviour of fibrous networks composed of fibres symmetrically and uniformly distributed at microstructural level, exposed to tensile loading (stretching). In order to realise this aim, fibre networks with three unique fibrous microstructures (i.e., square and triangular distributions, and collagen-like microstructure) were generated using Python scripts within an FE environment. A fully elastic constitutive behaviour of the constituent fibres in the analysed networks was assumed to exclude the contribution of material non-linearity, such as plastic hardening, at the fibre level.

Discrete FE models of fibrous networks with three different fibre alignments shown in Fig. 1.2 are considered in this study. Despite their resemblance (in the first two cases – Fig. 1.2A and B) to square and triangular lattice models used in the literature, the continuous fibres of our models meeting each other at fibre intersections are not welded; instead, cross links are introduced to transfer force from one fibre to another. Two spacings Δx and Δy along two main orthogonal directions between the fibres with a circular cross-section are kept constant. This assumption, however, results in different network densities (masses of fibres per unit area) for microstructures. This study does not aim to mimic the actual microstructure of the real fibrous network materials but rather elucidate the fundamental mechanical behaviours of fibrous networks with microstructures less complex than those of random networks. To eliminate (or minimise) the size effect on numerical investigations, at least 50

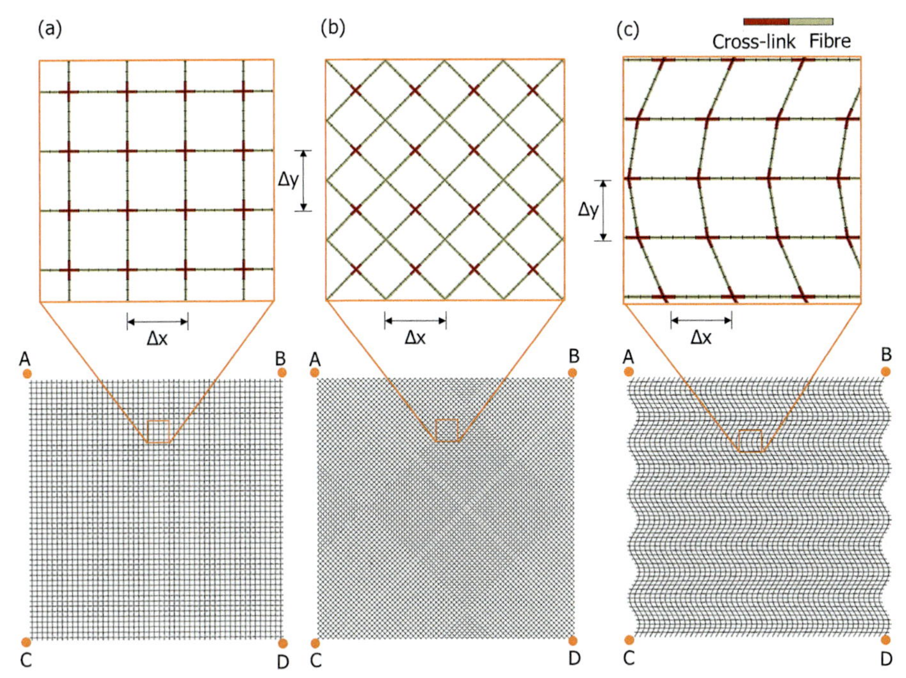

Fig. 1.2 FE models of fibrous network with square (A), triangular (B), and collagen-like (C) alignment of fibres and their corresponding zoomed-in views.

continuous fibres were placed at each domain boundary, meaning the total of 100 fibres in each model. The FE models utilised in the present study can be considered as gigantic representative volume elements (RVE) or stochastic fibrous networks, which are large enough to enable conduction of statistical analyses by varying geometrical and material properties at the fibre (microscopic) level. Taking advantage of this feature, the effect of inter-fibre cross-link density on the tensile deformation of fibrous networks is tested in this study. The effect of domain size on the mechanical properties of various fibrous networks can be found in Refs [6–8].

1.2 Numerical investigations

1.2.1 Finite-element models of fibrous networks

The individual fibre segments between adjacent fibre-crossings were modelled by an optimum number of Timoshenko beam elements in MSC Marc. The number was determined by conducting a mesh-sensitivity analysis. In some numerical simulations, FE models of fibrous networks were tested with Euler-Bernoulli beams to assess the effect of shear deformation, one of the dominant deformation mechanisms for the simulated fibrous networks under given tensile loading conditions. The quantitative analysis (in terms of force-displacement responses) and comparisons of deformation patterns for the developed FE models were carried out with both beam formulations. The difference in force-displacement curves and the patterns was found to be negligible. In our studies, beam elements were considered to have circular cross-sections with a fibre diameter of 1 μm. The mechanical behaviour of each beam was numerically modelled based on the linear-elasticity assumption. The modulus of elasticity and Poisson's ratio were assumed as 1 GPa and 0.35, respectively.

The type of cross links selected to model the force-transfer mechanisms between fibres controls the mechanical behaviour of fibrous networks. For simple cases, Picu in Ref. [9] classified the cross-link types for 2D fibrous networks as pin joint (a), rotating joint (b), and welded joint (c). The pin joint does not carry any moment from one fibre to another opposite to rotating joints. An example use of pint joints in a numerical study of deformation behaviour of the human amnion is available in Ref. [10]. Welded joints highly limit a relative motion of fibres by tying them at cross-link points (or surfaces). In contrast to 1D cross-link types, for instance, Deogekar and Picu [11] modelled inter-fibre bonds at cross links of 3D fibrous networks with a number of uniformly distributed 1D FE elements with large stiffness, transmitting both forces and moments. Yang et al. [12] mimicked the mechanical behaviour of binder particles at contacting surfaces of 3D glass fibres by bushing-like connector elements. The elastic and plastic parameters of 3D stochastic fibre networks were investigated in Ref. [13] with a 3D model with periodically distributed fibres composed of beam elements, with cross links represented by a single beam element at each intersection. Fig. 1.3A and B demonstrate the microstructure of a random fibrous network of paper and its corresponding FE model, formed from chains of Timoshenko beams (for fibre segments) and Bernoulli beams (for bond elements); their material behaviour was isotropic, elasto-plastic [14]. In contrast to a stated single-beam cross linker, four cross linkers were inserted between the fibrils of a collagen fibrous network in Ref. [2], also taken into account in this study.

A sample fibre intersection structure consisting basically of two fibres and two cross links used in the FE model of fibrous networks with square alignment of fibres is illustrated in Fig. 1.4. Whereas fibres 1 and 2 in Fig. 1.4A are chains of beams and

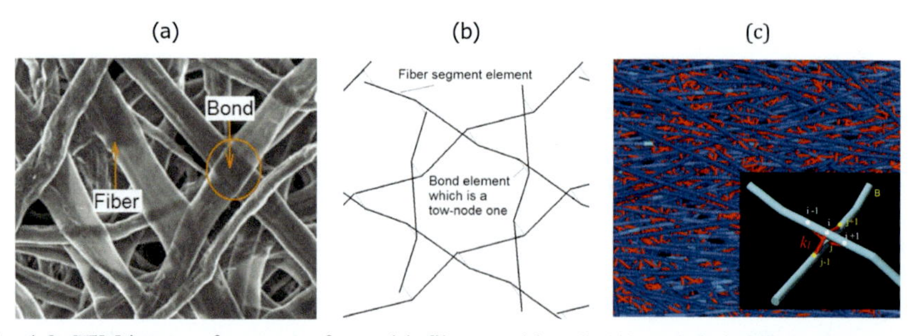

Fig. 1.3 SEM image of paper surface with fibres and bonds (A) and their FE model with 1D beam cross linkers [14] (B); (C) schematic of cross linking between two fibrils in a random fibrous network [2].

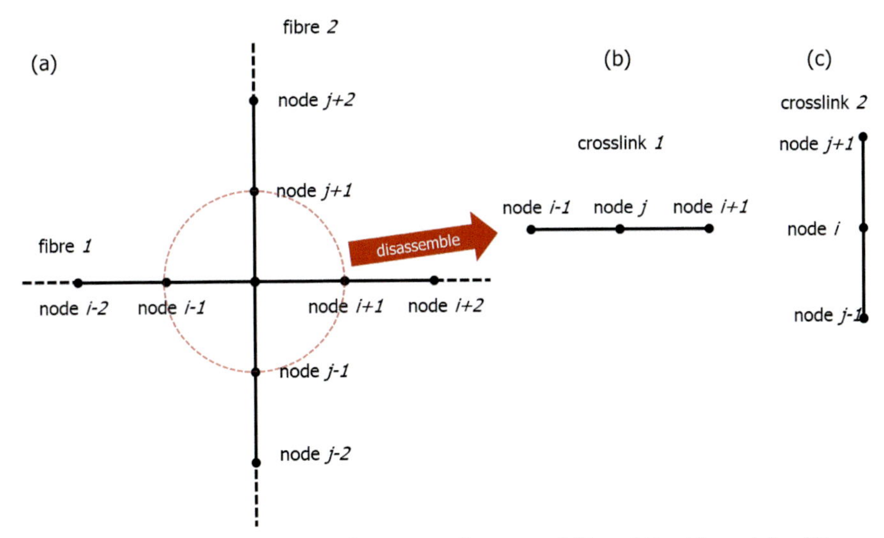

Fig. 1.4 Model of fibre intersection for square alignment of fibres (A) with models of its cross link 1 (B) and cross link 2 (C).

end nodes, they do not share any of them. The cross links (cross links 1 and 2 in Fig. 1.4A and B, respectively), however, share their end nodes with nodes of the fibres in order to facilitate the fibre-to-fibre load transfer.

1.2.2 Assumptions, boundary conditions, and solver

The corners of the domains simulated with the developed FE models were denoted with letters A to D (Fig. 1.2). The FE models of fibrous networks were subjected to a constant strain rate (constant velocity or linearly increasing displacement field) at the boundaries AB in the vertical direction in order to impose their stretch. The nodes at AB were able to rotate freely. Similarly, the nodes at the edge CD were free to rotate but they were fixed in the main orthogonal directions. Periodic boundary conditions were implemented on the edges AC and BD by taking the advantage of periodic fibre distributions and coupling (or linking) the nodes from both edges. The periodic boundary conditions satisfy the requirement that the summation of displacement and rotation vectors (u and θ) are equal to zero ($u_{AC} - u_{BD} = 0$ and $\theta_{AC} - \theta_{BD} = 0$).

The commercial finite-element software MSC Marc and programming language Python were used for pre- and pro-processing of simulations. Non-linear continuum equations governing physics of our current problem were solved with an implicit solver (Full Newton Solver) under quasi-static loading conditions. Large deformation formulations were implemented.

1.3 Results and discussion

The developed FE models of fibrous networks with three different fibrous microstructures and various cross-linking densities were subjected to tensile loading by applying prescribed and periodic boundary conditions. The cross links were initially placed on all the fibre intersections; subsequently, prescribed numbers of them were removed randomly by running an in-house Python script along with the commercial FE processor MSC Marc Mentat to implement the assigned cross-linking density.

1.3.1 Macroscale analysis of deformation

The deformation patterns and force-displacement data for the fibrous networks were obtained to quantify results for comparison of different scenarios. Deformed configurations of the fibrous networks with 100% cross-linked fibres at 15% stretch in the case of periodic boundary conditions for square, triangular, and collagen-like fibre distributions in microstructures are presented in Fig. 1.5. For the first type of fibre arrangement, when the periodic boundary conditions were removed, the deformation pattern and strain distribution were not altered significantly. The convergence for FE models of the third type of fibrous network – with collagen-like distribution – was severely affected because of the dangling fibre edges on sides AC and BD. As seen in Fig. 1.5B, a removal of periodic boundary conditions from the model of the fibrous network with triangular fibre distribution resulted in significant changes of deformation modes. The edge effect on local strain values was apparent, with a considerable fraction of fibre segments (beams) in the vicinity of these edges carrying no axial forces. A diffused necking behaviour was observed in the centre of this FE model. The effects of the FE model size and imposed boundary conditions (periodic or non-periodic) on elastic properties of random fibrous networks were investigated in detail in another study [15].

The global (macroscopic) axial-force data extracted from the FE simulations of the tested fibrous network geometries were normalised by a mass volume of constituent fibres, and the normalised force-effective (global/engineering) strain curves were plotted (Fig. 1.5C–F). It is apparent that the tensile response of the fibrous networks with a square fibre arrangement was linear as the material behaviour of fibres was confined to a linear elastic region. The tensile stiffness of the networks increased by different amounts with the cross-linking ratio increasing from 0 to 1.0. Since the material behaviour of individual fibres was linearly elastic in all the simulated fibrous networks, this variation is, therefore, mainly associated with the microstructural differences (orientation of fibres and fibre curliness). The trend in the graph of the normalised force-effective strain curve for the fibrous networks with triangular fibre alignment (Fig. 1.5D) was slightly non-linear in comparison to that for the square one. Even when the cross links were completely removed, this non-linearity was still observed in the former case due to the re-orientation of fibres by axial stretching.

The fibres in the case of collagen-like distribution (Fig. 1.5E) did not bear any significant load up to some level of stretching, corresponding to the effective strain of

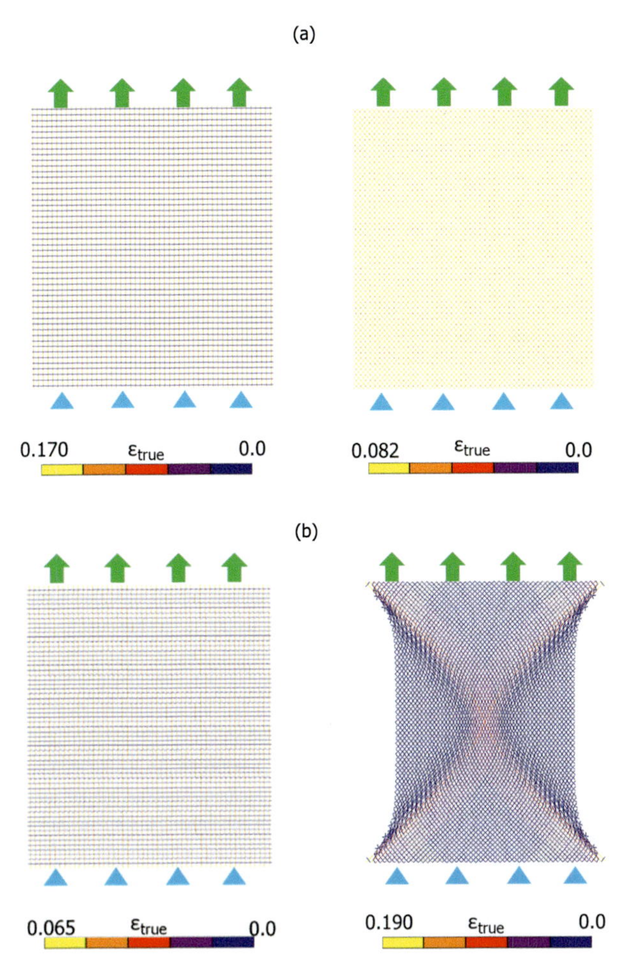

Fig. 1.5 Deformation patterns (levels of local fibre-level strains) for square (A) and triangular (B) fibrous networks at the effective external (global) strain of 15% for FE models with periodic boundary conditions *(left images)* and without them *(right images)*. Normalised force versus effective true strain for networks with square

(Continued)

0.06 in our case. This initial region was a result of the straightening of curved fibres, with the rest of the tensile mechanical response similar to those for the case of a square alignment. The magnitude of instantaneous elastic modulus of the studied fibrous networks were assessed from true stress-strain curves by post-processing force-displacement data for two levels of applied effective strain – 2.5% and 11.7% (Fig. 1.5F). Evidently, the fibrous networks with the square distribution of fibres demonstrated the stiffest elastic response with a propensity to increase with the growing cross-link density. Similar results for the effect of the cross-link number on the effective elastic modulus were reported in Ref. [2]. The case related to collagen-like fibre

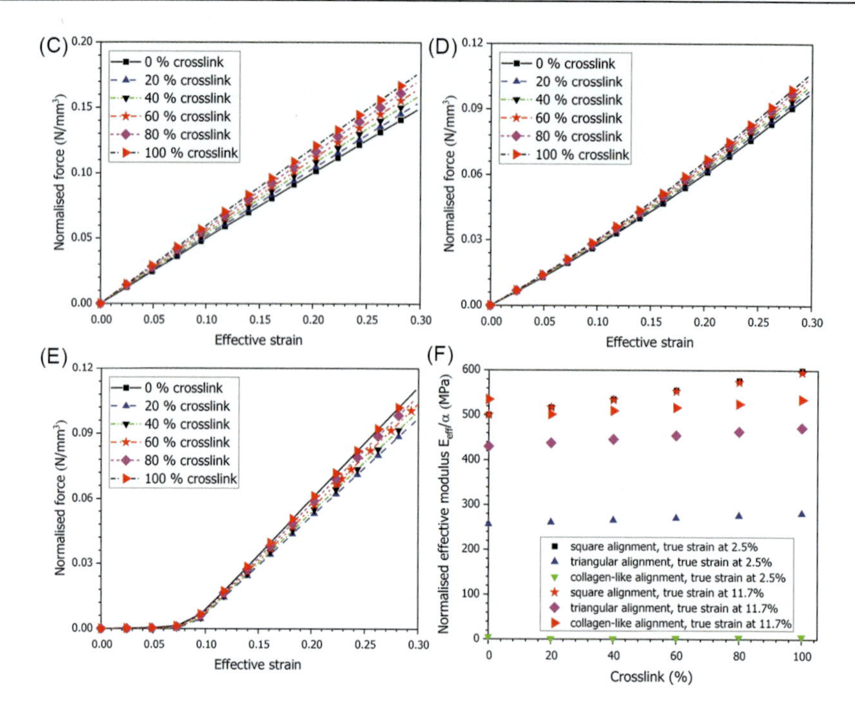

Fig 1.5, Cont'd (C), triangular (D), and collagen-like (E) fibre alignments. (F) Effect of cross-link fraction on the effective elastic modulus of three types of studied fibrous networks at two levels of stretching.

distribution at effective strain of 2.5% proved that fibre curvature was a more dominant factor in the elastic response of the analysed fibrous networks than the cross-linking density.

1.3.2 Microscale analysis of deformation

Analysing the deformed patterns of fibrous networks with a square fibre alignment, it was apparent that a major part of the total load was carried by fibres parallel to the loading direction and no changes in orientation of fibres were observed. A detailed study was based on the distribution of true strain in fibres and cross-link elements; zoomed-in views of the tested microstructures with 20% cross linking (i.e., cross-link density of 0.2) are presented in Fig. 1.6. Smaller local strains at fibre crossings were achieved due to their additional cross-link stiffness. The curved fibres in the collagen-like fibrous network were straightened whilst the straight connecting (horizontal) ones"buckled as a result, to accommodate the external deformation. Higher local strains were found in the models with square fibre alignment (Fig. 1.6A), whereas those in the case of triangular alignment were deformed less (Fig. 1.6B) with a part of external stretching accommodated by realigning (rotation) of fibres towards the

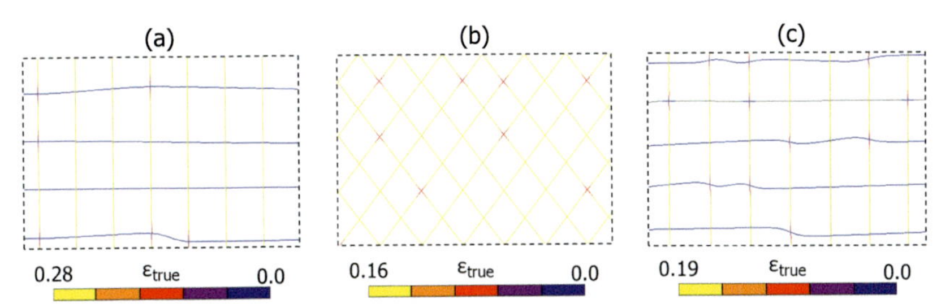

Fig. 1.6 Sample parts extracted from FE models of fibrous networks with square (A), triangular (B), and collagen-like (C) alignments under global strain of 0.3 (cross-link density 0.2).

loading direction. Obviously, in the square networks, fibres in parallel to the loading (vertical) direction were initially aligned with it and did not rotate with the external stretch. Interestingly, even in the case of a relatively low cross-link density in Fig. 1.6, the constituent fibres underwent almost a homogenised deformation over except for the cross-link regions, which can be easily detected.

In the detailed analysis, local strains in the fibres of the studied fibrous networks were computed by considering the coordinates of fibre segments (Timoshenko beams) at initial (at effective strain of 0) and deformed (at effective strains of 0.15–0.30) configurations. Due to the complexity of the problem and the number of beam elements to consider, an in-house Python script was developed to track the deformation of beams at selected stages of stretching. After normalising the obtained true strains with that applied to the network ($\varepsilon_{network}$), the corresponding results were summarised as histograms (Fig. 1.7) for various cross-link densities. Apparently, increased cross linking caused broader distributions of local strains in fibres. In most cases of cross linking, more uniform strain distributions (within a total range) appeared in the networks with triangular fibre alignment. As already discussed, the extent of fibre participation in the load transfer varied. Some of the fibres carried negligible amounts of load even at $\varepsilon_{network} = 0.3$.

In addition to the study on the influence of cross-link density, the effect of relative stiffness of cross links on an overall mechanical behaviour of the three analysed cases of fibrous-network microstructures was numerically examined. It is well known that in many real-life fibrous structures this parameter can vary significantly. In the initial investigations presented above, the cross links were assumed to have the same cross-section and material properties as the fibres. In order to assess the effect of their stiffness, a ratio E_c/E_f of elastic moduli of the cross-link and the fibre was introduced into the developed numerical models; it was varied in a range from 0.1 to 10.0. The obtained results (Fig. 1.8) demonstrated that in all cases, a large ratio E_c/E_f enhanced the tensile stiffness of fibrous networks. So, the elastic properties of cross links are as important in mechanics of fibrous networks as the type of cross links imitating the physical behaviour of bonds in typical fibrous networks such as collagens. In another research in Ref. [16], the strength of a stochastic fibrous network was also found to be linearly proportional to the cross-link density. An increase in the cross-link density improves the resistance of networks to deformation [17].

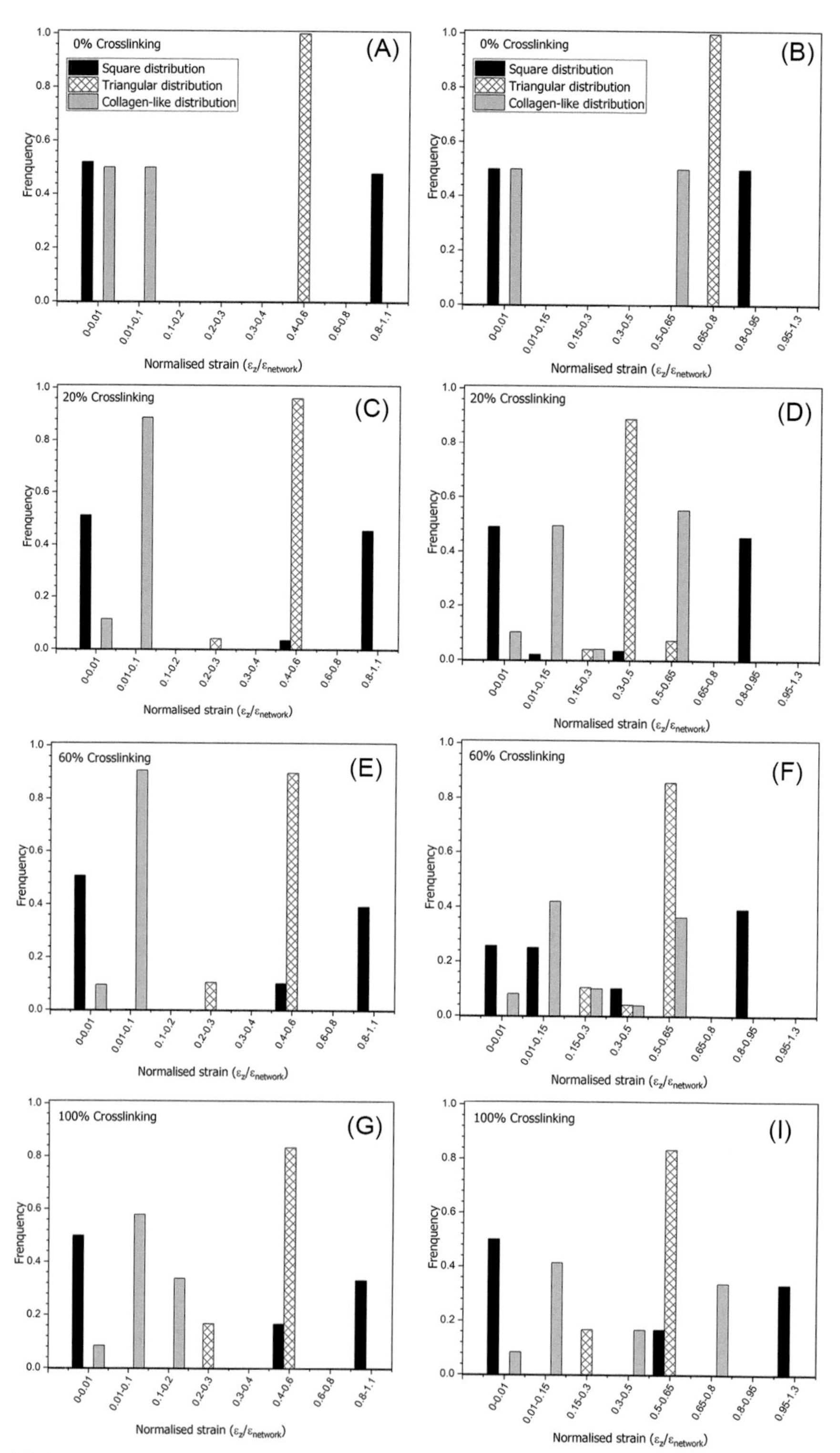

Fig. 1.7 Distributions of normalised strains in fibres for FE models with square, triangular, and collagen-like networks with various cross-link densities: 0% (A, B); 20% (C, D); 60% (E, F); and 100% (G, H) for different levels of external strain $\varepsilon_{network}$: 0.1 (A, C, E, G) and 0.3 (B, D, F, H).

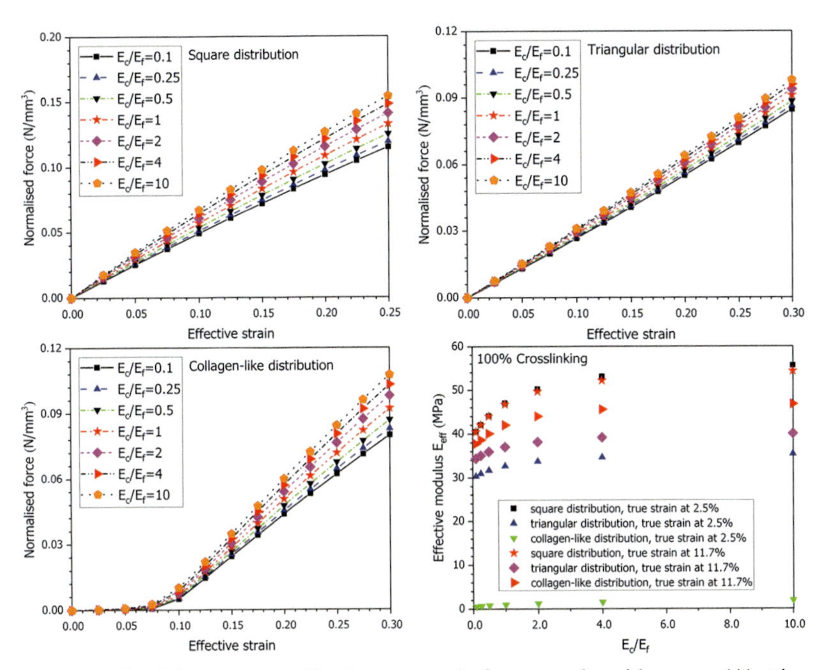

Fig. 1.8 Normalised force versus effective true strain for networks with square (A), triangular (B), and collagen-like (C) fibre alignments. (D) Effect of cross-link fraction on the effective elastic modulus of three types of studied fibrous networks at two levels of stretching.

1.4 Conclusions

Three different microstructural models of fibrous networks were investigated using FE analysis to study the basic deformation phenomena without the effect of complex random distributions of fibres in real-life systems. These models were formed by regularly distributed fibres with different patterns – square, triangular, and collagen-like. This study demonstrated the effect of orientation distribution of fibres on tensile deformation of fibrous networks at macro- and micro-scales. The FE models with various cross-link densities were generated and subjected to tensile stretch. Periodic and non-periodic boundary conditions were imposed in the developed numerical models of fibrous networks. It was established that features of constituent fibres (in particular, their spatial alignments and curvature) were the key parameters determining the mechanical response and material properties of these networks. The levels of spatial density and stiffness of cross links were found to be effective in improvement of the global stiffness of fibrous networks. Our investigations were, however, limited to elastic behaviours at macro- and micro-scales. Effects of plasticity and geometric irregularities in orientation and material properties of fibres in these networks on their mechanical behaviour and properties are worth examining in future studies.

Acknowledgement

This work has received funding in part from the EPSRC UK (grant number EP/R012091/1).

References

[1] X. Gao, P. Kuśmierczyk, Z. Shi, C. Liu, G. Yang, I. Sevostianov, V.V. Silberschmidt, Through-thickness stress relaxation in bacterial cellulose hydrogel, J. Mech. Behav. Biomed. Mater. 59 (2016) 90–98.

[2] M. Ovaska, Z. Bertalan, A. Miksic, M. Sugni, C. Di Benedetto, C. Ferrario, L. Leggio, L. Guidetti, M.J. Alava, C.A.M. La Porta, S. Zapperi, Deformation and fracture of echinoderm collagen networks, J. Mech. Behav. Biomed. Mater. 65 (2017) 42–52.

[3] K. Bircher, A.E. Ehret, E. Mazza, Microstructure based prediction of the deformation behavior of soft collagenous membranes, Soft Matter 13 (2017) 5107–5116.

[4] H.L. Cox, The elasticity and strength of paper and other fibrous materials, Br. J. Appl. Phys. 3 (1952) 72–79.

[5] Y. Lee, I. Jasiuk, Apparent elastic properties of random fiber networks, Comput. Mater. Sci. 79 (2013) 715–723.

[6] X. Gao, E. Sozumert, Z. Shi, G. Yang, V.V. Silberschmidt, Assessing stiffness of nanofibres in bacterial cellulose hydrogels: numerical-experimental framework, Mater. Sci. Eng. C 77 (2017) 9–18.

[7] S.B. Lindström, A. Kulachenko, L.M. Jawerth, D.A. Vader, Finite-strain, finite-size mechanics of rigidly cross-linked biopolymer networks, Soft Matter 9 (2013) 7302.

[8] S. Tyznik, Length scale dependent elasticity in random three-dimensional fiber networks, Mech. Mater. 12 (2019).

[9] R.C. Picu, Mechanics of random fiber networks—a review, Soft Matter 7 (2011) 6768.

[10] A. Mauri, R. Hopf, A.E. Ehret, C.R. Picu, E. Mazza, A discrete network model to represent the deformation behavior of human amnion, J. Mech. Behav. Biomed. Mater. 58 (2016) 45–56.

[11] S. Deogekar, On the strength of random fiber networks, J. Mech. Phys. Solids 116 (2018) 1–16.

[12] M. Yang, M. Ji, E. Taghipour, S. Soghrati, Cross-linked fiberglass packs: microstructure reconstruction and finite element analysis of the micromechanical behavior, Comput. Struct. 209 (2018) 182–196.

[13] Y.H. Ma, H.X. Zhu, B. Su, G.K. Hu, R. Perks, The elasto-plastic behaviour of three-dimensional stochastic fibre networks with cross-linkers, J. Mech. Phys. Solids 110 (2018) 155–172.

[14] J.X. Liu, Z.T. Chen, H. Wang, K.C. Li, Elasto-plastic analysis of influences of bond deformability on the mechanical behavior of fiber networks, Theor. Appl. Fract. Mech. 55 (2011) 131–139.

[15] A.S. Shahsavari, R.C. Picu, Size effect on mechanical behavior of random fiber networks, Int. J. Solids Struct. 50 (2013) 3332–3338.

[16] S. Deogekar, M.R. Islam, R.C. Picu, Parameters controlling the strength of stochastic fibrous materials, Int. J. Solids Struct. 168 (2019) 194–202.

[17] Y. Chen, M. Chen, E.A. Gaffney, C.P. Brown, Effect of crosslinking in cartilage-like collagen microstructures, J. Mech. Behav. Biomed. Mater. 66 (2017) 138–143.

Micromechanics of nonwoven materials

2

Amit Rawal
Department of Textile and Fibre Engineering, Indian Institute of Technology Delhi, New Delhi, India

2.1 Introduction

Over half a century ago, the term 'nonwoven' was coined primarily to omit the yarn spinning technology for the production of a new type of 'fabric' that extends beyond woven, knitted, or braided fabric. Nonwoven has been initially thought of as a cheap commodity material that would only serve low-end applications, but it has emerged as a 'wonder material' that spans its applications from lightweight masks to heavy geotextiles. Perhaps, the know-how of various fields, including polymer science, materials science, chemical engineering, and mechanical engineering, to name a few, have been embodied in the successful creation of nonwovens. Accordingly, the nonwoven industry has evolved from the developments in textile engineering in addition to the paper and polymer technologies. However, the nonwoven industry is reluctant to be associated with the traditional textile and paper industries. This has led to the establishment of nonwoven associations in Europe and the United States of America, i.e. EDANA (The European Disposables and Nonwovens Association) and INDA (North America's Association of the Nonwoven Fabrics Industry). Both of these associations have defined nonwovens in their own way.

EDANA defined nonwoven as "a manufactured sheet, web or batt of directionally or randomly orientated fibres, bonded by friction, and/or cohesion and/or adhesion," but excluded a number of materials from the definition, including paper, woven, knitted, tufted, or stitch-bonded or felted by wet-milling [1]. A distinction has also been made between the wet-laid nonwovens and wet-laid paper materials by stating the former with "more than 50% by mass of its fibrous content is made up of fibres (excluding chemically digested vegetable fibres) with a length to diameter ratio greater than 300." Similarly, INDA defined nonwoven as "sheet or web structures bonded together by entangling fibres or filaments, by various mechanical, thermal, and/or chemical processes. These are made directly from separate fibres or from molten plastic or plastic film." These definitions clearly indicate that the nonwoven fabrication is a two-step process, i.e. web formation (aligning the fibres in defined orientation characteristics) and bonding the fibres by mechanical, thermal, or chemical means to impart strength to the material [2]. Apparently, the classification of nonwovens stems from this two-step process, as discussed below.

Mechanics of Fibrous Networks. https://doi.org/10.1016/B978-0-12-822207-2.00008-8

2.1.1 Classification of nonwoven materials

Nonwoven materials can be classified on the basis of web formation and web bonding techniques, as illustrated in Fig. 2.1. In general, web formation is classified on the basis of dry laying, wet laying, and polymer laying methods. Specifically, the dry-laid materials originated from the field of textile technology; similarly, the wet-laid materials emerged from the papermaking process, and polymer science and engineering are the backbone of the polymer-laid nonwoven materials [1]. The web formation process is a cornerstone in aligning the fibres such that it dictates various physical, mechanical, geometrical, wetting, and other properties of nonwoven materials. Particularly, the micromechanics of nonwoven stems from the alignment of fibres. Therefore, it is important to understand some of the key characteristics of this web formation process. However, it is beyond the scope of this chapter to go into the depth of each of these processes, but the readers can refer to other references for further details of these processes [1,3].

Traditionally, carding is a web preparation method extensively used in the textile industry that allows the placement of short-staple fibres in the machine or production direction. Similarly, the cross-lapping method aligns the fibres in the cross-machine direction with the help of web transfer transportation systems. Generally, these methods of web formation lead to a highly anisotropic structure that can be easily characterised by determining the mechanical properties in the machine and cross-machine directions. A pseudo-random web can be obtained by placing the required number of layers of carded and cross-laid webs over each other to form a 'composite' web. On the other hand, the air-laid technique is a route to attain a transversely isotropic structure by placing the fibres in a random fashion primarily in the in-plane

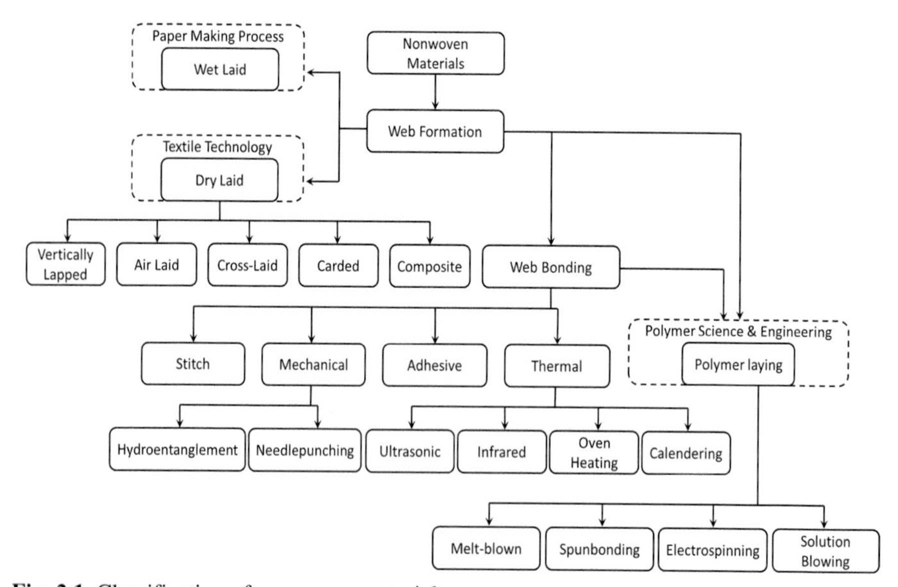

Fig. 2.1 Classification of nonwoven materials.

direction. Further, this technique can assist in producing a pseudo-three-dimensional (3D) structure, but realistically, the true 3D structure is obtained in a vertically lapped web formation method. The random structure is also realised in spunbonding and wet-laid methods. In a spunbonding process, the long and continuous filaments are produced similar to a typical melt spinning method employed in polymer processing, but the web is collected in a nearly random fashion by means of a conveyor belt. On the contrary, extremely short organic or inorganic fibres (0.3–10 mm) can be utilised for the production of the random web with the aid of the wet-laying process. The wet-laying process is a widely used method in the paper industry.

Web bonding is a method of conversion of a web of fibres to a realistic nonwoven material by creating bonds between them. As aforementioned, the bonding between the fibres can be realised by means of friction, and/or cohesion and/or adhesion. Accordingly, the web bonding formation is classified in terms of mechanical, thermal, chemical, stitch, and self-bonding methods. The mechanical web formation primarily utilises friction as a medium to hold the fibres together. The typical examples of this web formation technique are needlepunching and hydroentanglement. The needlepunching allows the utilisation of metal needles to transport some of the fibres from the surface to the through-thickness direction, which eventually leads to a 3D nonwoven structure. On the other hand, the hydroentanglement method uses high-pressure water jets to induce entanglements between the fibres. Similar to the mechanical bonding method, stitch bonding utilises the bonding of the web by means of stitching yarns.

Alternatively, the fused fibrous assemblies in the form of thermally bonded nonwoven can be obtained by applying the thermal or heat energy to the thermoplastic fibre. Here, the desired number of fused fibre–fibre contacts can be targeted on the basis of molten polymer flow with the aid of surface tension and capillary action. The thermally bonded nonwoven is broadly classified in terms of through-air bonding, calendaring, infrared, and ultrasonic bonding. The through-air bonding employs the passage of the web in a heated air chamber (oven), whereas calendaring involves the fibrous web to pass through the heated pair of rollers. In the case of the infrared bonding, the thermal radiation method is used for heating the fibres. Alternatively, an intense amount of heat can also be delivered via a high-frequency vibration horn in an ultrasonic bonding. On the other hand, a binder such as glue, rubber, casein, latex, cellulose derivative, or a synthetic resin is generally added to create adhesion between the long filaments or short fibres to form chemically bonded nonwoven materials.

In addition to the mechanical, thermal, and chemical bonding methods, there exists a 'self-bonding' method that essentially relies on the polymer laying method [1,4]. Apparently, spunbonding and melt-blown methods share the same technique of polymer melting and filament extrusion methods except for the fact that the latter system employs the stream of hot air for attenuating the extrudate to form extremely fine fibres. The application of high-voltage to polymer melts or solutions can also lead to a long and continuous length of random or anisotropic self-bonded submicron or even nanofibres using the electrospinning process. Recently, solution blowing has emerged as a nonwoven-producing technique that combines the scalability of melt-blowing with the versatility of electrospinning [5]. In the solution blowing method, the polymer solution is converted into fine fibres via high-velocity gas flow. These polymer laying methods offer a great advantage not only in terms of continuous filament length in comparison to staple fibres-based nonwovens that often result in a higher magnitude of mechanical properties, but the constituent filaments are 'self-adhering' in nature that essentially avoids an extra step of bonding. Occasionally, the bonding between the filaments can be improved by employing additional mechanical bonding. For example, heavy-weight spun-bonded nonwoven geotextiles can be needled via the needlepunching process in order to further enhance the mechanical properties. From the above discussion, it is of paramount importance to stratify the structure of nonwoven materials.

2.1.2 Structural characterisation of nonwoven materials

In general, the morphology of nonwoven material consists of fibre segments, fibre–fibre contacts, and irregularly shaped pores, as shown in Fig. 2.2. Now, we define some of the key structural parameters of nonwoven materials, and the details of these parameters are given in our recently published review [6].

- **Fibre orientation:** Fibre orientation is defined as the angle formed between the tangent to the fibre curl and one of the principal axes (e.g. machine or cross-machine direction). Fig. 2.2A shows the orientation of a typical fibre in two-dimensional (2D) space. Since the alignment of the fibre varies within the structure, the orientation distribution function (ODF) function, $\Omega(\varphi)$ can accordingly be defined as [7],

$$\Omega(\varphi) = \lim_{\Delta\varphi \to 0} \frac{1}{\Delta\varphi} \frac{\Delta L}{L} = \frac{dL}{Ld\varphi} \tag{2.1}$$

where L is the total length of fibre segments and ΔL is the sum of lengths of fibre segments with in-plane orientation angle that lies in the range of φ and $\varphi + \Delta\varphi$.

The following normalisation condition needs to be satisfied [8].

$$\int_0^{\pi} \Omega(\varphi) d\varphi = 1 \tag{2.2}$$

The orientation of a fibre segment in a 3D space is based on a spherical coordinate system, which can be defined in terms of in-plane (φ) and out-of-plane (θ) orientation angles, as illustrated in Fig. 2.3B. Further, the probability that fibre segment direction lies in the infinitesimal range of angles, θ and $\theta + d\theta$, and φ and $\varphi + d\varphi$ is given by $\Omega(\theta, \varphi) \sin \theta d\theta d\varphi$, where $\Omega(\theta, \varphi)$ is a probability density function. In this case, the following normalisation condition needs to be satisfied [8], i.e.,

$$\int_0^{\pi} \int_0^{\pi} \Omega(\theta, \varphi) \sin\theta d\theta d\varphi = 1 \tag{2.3}$$

Optical microscopy, image analysis, digital volumetric imaging (DVI), focused ion beam (FIB) sectioning, confocal laser scanning microscopy, magnetic resonance imaging or MRI, microtome sectioning, and X-ray micro-computed tomography (X-ray microCT) are some of the key techniques that are extensively used for the measurement of the ODF. The merits and demerits of these methods are given in Ref. [6].

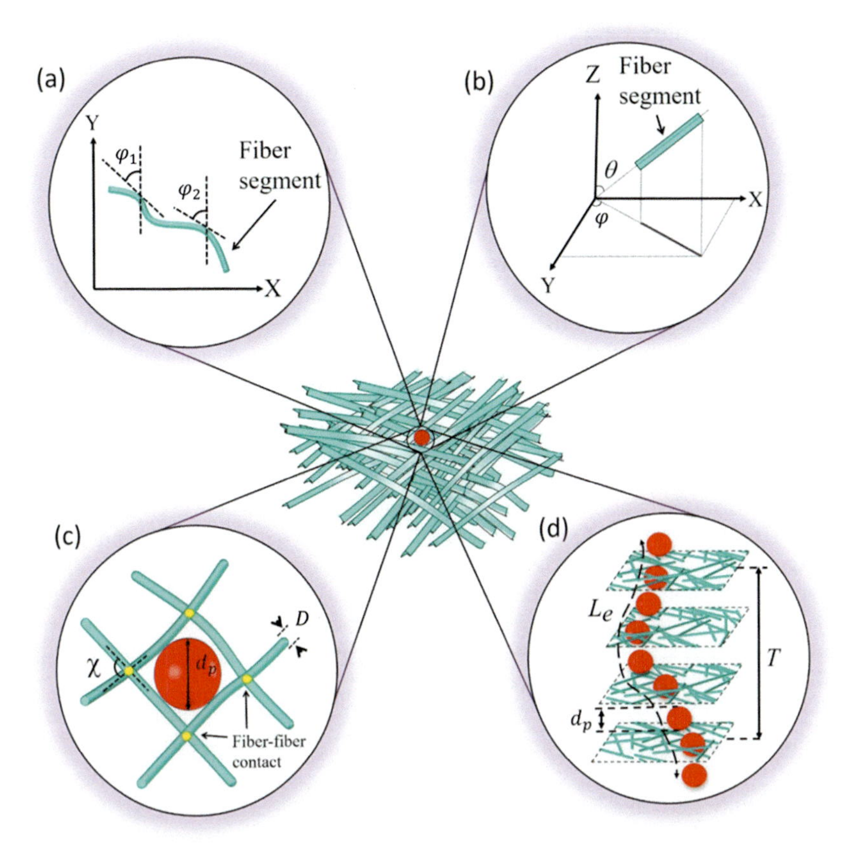

Fig. 2.2 A representative image of a nonwoven material (Originally referred to as absorptive glass mat (AGM) nonwoven separator in Ref. [6].) depicting key morphological parameters in two and three dimensions. Here, the magnified images show the determination of fibre orientation in (A) 2D and (B) 3D fibre networks. Fibre–fibre contacts form typical pore (shown with a *red colour circle*) with diameter d_p in (C) 2D and (D) 3D fibre networks. A typical spherical pore passing through the tortuous layers of fibre networks demonstrates the influence of tortuosity in nonwoven material. Here, D is the fibre diameter, φ is the in-plane orientation angle of the fibre, L_e is the actual length of the flow path, T is the straight length or thickness of the material, θ is the out-of-plane orientation angle formed with the Z-axis, and χ is the angle formed between the axes of two fibres [6].

- **Number of fibre–fibre contacts**: From a micro-mechanics viewpoint, the fibre segments in a nonwoven material are the load-bearing elements, whereas the fibre–fibre contacts are the load transferring elements. Whilst the fibre (yarn)–fibre (yarn) contacts in regular structures such as woven, knitted, and braided fabrics are easy to visualise, the number of fibre–fibre contacts in random fibrous structures such as nonwovens are strenuous to comprehend. Later, we will discuss in detail to predict the number of fibre–fibre contacts in a typical nonwoven material.

- **Type of fibre–fibre contact:** Typically, fibre–fibre contacts can be classified as follows:
  (a) Frictional contacts: These fibre–fibre contacts are formed due to the entanglement of fibres. For example, in needlepunched nonwoven materials, the fibres are made to contact with each other by means of friction.
  (b) Fused contacts: These fibre–fibre contacts are formed due to the melting or rubbery behaviour of the polymeric material. A typical example is a thermally bonded nonwoven material that comprises fused fibre networks.
  The other types of fibre–fibres contacts can be found elsewhere [9].
- **Distance between the fibre–fibre contacts:** This is the distance between the two successive contacts formed on a given fibre. In a regular structure, the distance between the fibre–fibre contacts is maintained uniformly, whereas, in the case of nonwoven materials, the distance between the fibre–fibre contacts varies within the structure.
- **Fibre volume fraction:** It is defined as the proportion of the volume occupied by fibres in a given volume of a nonwoven. Mathematically, the fibre volume fraction (V_f) is related to the mass per unit area (m), thickness (t), and fibre density (ρ_f), i.e. $V_f = \frac{m}{t\rho_f}$.
- **Porosity:** It is defined as the proportion of the volume occupied by pores in a given volume of a nonwoven material. Thus, the porosity (ϵ) of a fibrous assembly is given by $\epsilon = 1 - V_f$.
- **Pore size distribution:** As aforementioned, the nonwoven material comprises irregular arrays of fibres, and accordingly, the pores formed in a nonwoven material have dimensions that vary within the structure. The size of a pore in a nonwoven material is usually given by the diameter of the largest sphere that could be passed through a pore. The frequency distribution of pore sizes expressed in terms of their spherical diameters is called pore size distribution (see the magnified image in Fig. 2.2C).
- **Tortuosity:** It is a parameter that describes the ratio of the actual length of the flow path (L_e) to the straight length or thickness of the material (T), i.e. $\tau = \frac{L_e}{T}$ (see Fig. 2.2D).

As aforementioned, the fibre–fibre contacts are the load-bearing elements under defined mechanical loading conditions; thus, it is of paramount importance to quantify the fibre–fibre contacts.

2.2 Theory of fibre–fibre contacts

A fibre–fibre contact is formed when the region (area or volume) surrounding a fibre is occupied by a neighbouring fibre. In other words, the region around a fibre into which the centre of another fibre is being allowed to enter to form a fibre–fibre contact, which is contrary to the standard definition of excluded area or volume [10]. Based on the simple concept of the excluded area, van Wyk [11] considered a spherical particle that forms a contact with a cylindrical fibre for computing the number of fibre–fibre contacts in a 3D random fibrous assembly. Later, Komori and Makishima [8] generalised the concept of van Wyk [11] by formulating the theory of fibre–fibre contacts for anisotropic fibrous materials. Some of the key assumptions of Komori and Makishima's models [8] are given below.

(1) The fibres are considered to be cylinders of diameter D and segment length l.
(2) The distribution of the centres of mass of fibres is random.
(3) The effects of end-to-end or side-to-end fibre contacts are neglected.

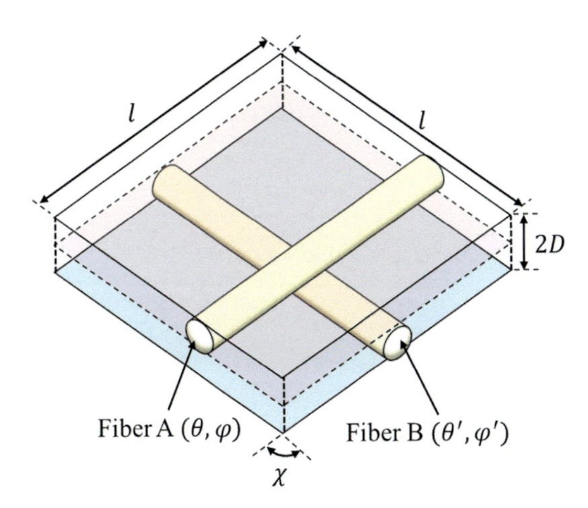

Fig. 2.3 A parallelepiped formed by fibre B sliding from one end of fibre A to another whilst making a single contact with the former. Here l and D are the fibre segment length and diameter, respectively.

Now consider two distinct fibres in a typical spherical coordinate system such that fibre A with orientation (θ, φ) is placed arbitrarily in a defined volume (V) and subsequently, another fibre B having an orientation (θ', φ') has been placed in the same volume such that two fibres contact each other (see Fig. 2.3). This fibre–fibre contact occurs as a result of the centre of their masses being brought in a defined region. Suppose fibre B slides over fibre A from one end to the other end keeping the direction and the contact point on B remain unchanged. The same procedure can be followed for fibre A. During these two movements, a parallelepiped is formed around the periphery of the axes of the fibres, as shown in Fig. 2.3. The volume (v) of this parallelepiped is given by,

$$v(\theta, \varphi; \theta', \varphi') = 2Dl^2 \sin\chi \tag{2.4}$$

where χ is the angle formed between the axes of the two fibres and is a function of $(\theta, \varphi; \theta', \varphi')$.

It is important to formulate a relationship between the angle between the axes of two fibres (χ) with the orientation of two fibres. As shown in Fig. 2.3, fibres can be represented using vectors $A\left(\sin\theta\cos\varphi\hat{i}, \sin\theta\sin\varphi\hat{j}, \cos\theta\hat{k}\right)$ and $\mathrm{B}\left(\sin\theta'\cos\varphi'\hat{i}, \sin\theta'\sin\varphi'\hat{j}, \cos\theta'\hat{k}\right)$ according to a typical spherical coordinate system. Therefore, the dot product of two vectors A and B is defined by,

$$\cos\chi = \frac{\vec{A} \cdot \vec{B}}{\left|\left|\vec{A}\right|\right| \cdot \left|\left|\vec{B}\right|\right|} \tag{2.5}$$

Here,

$$A \cdot B = \sin\theta \sin\theta' \cos\varphi \cos\varphi' + \sin\theta \sin\varphi \sin\theta' \sin\varphi' + \cos\theta \cos\theta' \quad (2.6)$$

and

$$\| \vec{A} \| = \sqrt{\sin^2\theta \cos^2\varphi + \sin^2\theta \sin^2\varphi + \cos^2\theta} = \sqrt{\sin^2\theta + \cos^2\theta} = 1 \quad (2.7)$$

Similarly,

$$\| \vec{B} \| = 1 \quad (2.8)$$

Thus combining the above Eqs. (2.5)–(2.8),

$$\cos\chi = \cos\theta \cos\theta' + \sin\theta \sin\theta' \cos(\varphi - \varphi') \quad (2.9)$$

Now, we are ready to compute the mean distance between the fibre–fibre contacts and the number of fibre–fibre contacts in anisotropic fibrous materials.

2.2.1 Fibre–fibre contacts in anisotropic materials

As aforementioned, a parallelepiped is formed around the periphery of fibres axes. Accordingly, the probability, $p(\theta,\varphi;\theta',\varphi')$, of formation of contact between the fibres A and B can be obtained using a simple geometrical probability approach, which is given by [8],

$$p(\theta, \varphi; \theta', \varphi') = \frac{v}{V} = \frac{2Dl^2}{V} \sin\chi \quad (2.10)$$

where V is the total volume of the fibrous assembly.

Suppose N fibres are contained in volume V, then the average number of fibre–fibre contacts formed on fibre A is given by [8],

$$\begin{aligned} n(\theta, \varphi) &= (N-1) \int_0^{\pi} \int_0^{\pi} p(\theta, \varphi; \theta', \varphi') \, \Omega(\theta', \varphi') \sin\theta' d\theta' d\varphi' \\ &= \frac{2D(N-1)l^2}{V} \int_0^{\pi} \int_0^{\pi} \Omega(\theta', \varphi') \sin\chi \sin\theta' d\theta' d\varphi' \end{aligned} \quad (2.11)$$

where l is the fibre segment length

For $N \gg 1$,

$$n(\theta, \varphi) = \frac{2DNl^2}{V} J(\theta, \varphi) \quad (2.12)$$

where

$$J(\theta,\varphi) = \int_0^{\pi}\int_0^{\pi}\Omega(\theta',\varphi')\sin\chi(\theta,\varphi;\theta',\varphi')\sin\theta' d\theta' d\varphi' \tag{2.13}$$

Averaging $n(\theta,\varphi)$ over the possible values of θ and φ gives the average number of fibre–fibre contacts on an arbitrary fibre [8],

$$\overline{n} = \int_0^{\pi}\int_0^{\pi} n(\theta,\varphi)\,\Omega(\theta,\varphi)\sin\theta d\theta d\varphi = \frac{2DNl^2}{V}I \tag{2.14}$$

where

$$I = \int_0^{\pi}\int_0^{\pi} J(\theta,\varphi)\,\Omega(\theta,\varphi)\sin\theta d\theta d\varphi \tag{2.15}$$

where I is the global orientation parameter.

Using the elementary definition of fibre volume fraction (V_f), i.e.,

$$V_f = \frac{\pi D^2 L}{4V} \tag{2.16}$$

where L ($=Nl$) is the total length of fibres defined in volume V.

Therefore, the expression of the number of fibre–fibre contacts per unit length of fibre is obtained by combining Eqs. (2.14)–(2.16),

$$\overline{n_l} = \frac{\overline{n}}{l} = \frac{2DNl}{V}I = \frac{2DL}{V}I = \frac{8V_f}{\pi D}I \tag{2.17}$$

Thus, the mean distance between fibre–fibre contacts is given by,

$$\overline{b} = \frac{1}{\overline{n_l}} = \frac{V}{2DLI} = \frac{\pi D}{8V_f I} \tag{2.18}$$

In addition, the total number of fibre–fibre contacts in the fibrous assembly of volume V, containing N fibres is given by (n_V) [8],

$$n_V = \frac{N}{2}\overline{n} = \frac{D(Nl)^2}{V}I = \frac{DL^2}{V}I \tag{2.19}$$

It should be noted that the number of fibre–fibre contacts has been divided by 2 to avoid their double counting. Now we take some special cases of nonwoven materials, i.e. 2D and 3D random structures.

2.2.2 2D random nonwoven materials

In a 2D random nonwoven material, the fibres lie parallel to the XY plane; therefore, $\theta = \frac{\pi}{2}$. Accordingly, the Dirac's delta function can be used to obtain the orientation density function as stated by Komori and Makishima [8],

$$\Omega(\theta, \varphi) = (\Omega_o)_{2D}\delta(\theta - \pi/2) \tag{2.20}$$

where $(\Omega_o)_{2D}$ is the orientation density function of 2D random fibrous material.

Using normalisation condition,

$$\int_0^{\pi}\int_0^{\pi} \Omega(\theta, \varphi)\sin\theta d\theta d\varphi = 1 \tag{2.21}$$

$$(\Omega_o)_{2D} = \left(\int_0^{\pi}\int_0^{\pi} \sin\theta\delta(\theta - \pi/2)d\theta d\varphi\right)^{-1} = \frac{1}{\pi} \tag{2.22}$$

$$\Longrightarrow \Omega(\theta, \varphi) = \frac{1}{\pi}\delta(\theta - \pi/2) \tag{2.23}$$

Using Eqs. (2.13), (2.15), and (2.22),

$$\Longrightarrow I = \int_0^{\pi}\int_0^{\pi} J(\theta, \varphi)\frac{1}{\pi}\delta(\theta - \pi/2)\,\sin\theta d\theta d\varphi = \frac{1}{\pi}\int_0^{\pi} J(\pi/2, \varphi)d\varphi \tag{2.24}$$

$$J(\pi/2, \varphi) = \int_0^{\pi}\int_0^{\pi} \frac{1}{\pi}\delta(\theta' - \pi/2)\,\sin\chi(\pi/2, \varphi; \pi/2, \varphi')\sin\theta' d\theta' d\varphi' \tag{2.25}$$

$$= \frac{1}{\pi}\int_0^{\pi} \sin\chi(\pi/2, \varphi; \pi/2, \varphi')d\varphi' = \frac{2}{\pi} \tag{2.26}$$

where

$$\cos\chi = \cos\theta\cos\theta' + \sin\theta\sin\theta'\cos(\varphi - \varphi') \tag{2.27}$$

Also,

$$\sin\chi(\pi/2, \varphi; \pi/2, \varphi') = |\sin(\varphi - \varphi')|$$

And we get the following value of global orientation parameter (I),

$$I = \left(\frac{1}{\pi}\right)^2 \int_0^{\pi}\int_0^{\pi} |\sin(\varphi - \varphi')|d\varphi d\varphi' = \frac{2}{\pi} \tag{2.28}$$

The above expression of I matches well with that of the orientation function (f) as defined by Toll and Manson [12]. Further, the total number of fibre–fibre contacts in a 2D random nonwoven material of volume V, can be calculated based on Eq. (2.19).

$$n_V = \frac{2DL^2}{\pi V} \tag{2.29}$$

2.2.3 3D random nonwoven materials

The fibres in a 3D random nonwoven material would be distributed in a manner that the orientation density function is independent of θ and φ. The orientation density function of a 3D random nonwoven material, $(\Omega_o)_{3D}$ can be computed on the basis of the normalisation condition, as defined in Eq. (2.3) [8],

$$(\Omega_o)_{3D} = \left(\int_0^{\pi}\int_0^{\pi} \sin\theta d\theta d\varphi\right)^{-1} = \frac{1}{2\pi} \tag{2.30}$$

Therefore, the average number of fibre–fibre contacts on an arbitrary fibre ($\overline{n}$) is given by,

$$\overline{n} = n(0,0) = \frac{2DNl^2}{V} J(0,0) \tag{2.31}$$

where

$$J(0,0) = \int_0^{\pi}\int_0^{\pi} \frac{1}{2\pi}\left[1-\cos^2\theta'\right]^{\frac{1}{2}} \sin\theta' d\theta' d\varphi' = \frac{\pi}{4} \tag{2.32}$$

$$I = \frac{1}{2\pi}\int_0^{\pi}\int_0^{\pi} J(0,0)\sin\theta d\theta d\varphi = \frac{\pi}{4} \tag{2.33}$$

The mean distance between fibre–fibre contacts ($\overline{b}$) can be calculated based on Eq. (2.18),

$$\overline{b} = \frac{D}{2V_f} \tag{2.34}$$

The above expression of $\overline{b}$ is the same as that being derived by van Wyk [11]. Moreover, the total number of fibre–fibre contacts in a 3D random nonwoven material of volume V, containing N fibres can be calculated based on Eq. (2.19).

$$n_V = \frac{\pi DL^2}{4V} \tag{2.35}$$

In order to make use of the calculation of number of fibre–fibre contacts for predicting various mechanical properties of nonwoven materials, it is necessary to define the contacts in a mesodomain. This is an intermediate length scale that comprises a statistical ensemble of microelements [13]. Generally, the physical and geometrical parameters of the mesodomain are predicted statistically based on the properties of microelements (lowest length scale element). The mesodomain is a representative element of the network, on which the continuum mechanics is applicable [14]. Therefore, it is pertinent to derive an expression of the number of fibre–fibre contacts in the mesodomain.

2.2.4 Number of fibre–fibre contacts in mesodomain

Considering a mesodomain of volume (dV) defined by two planes of unit cross-sectional area, $A_c = 1$ and comprising two successive contact points with a fibre element sandwiched between them [2,11,13], as depicted in Fig. 2.4. This mesodomain has a thickness of mean distance between the contacts, which is projected on the thickness or z-axis ($\overline{b_j}$) [2,13]. Therefore, the total number of fibre–fibre contacts (n_{dV}) in this mesodomain can be computed from the total number of fibre–fibre contacts, n_V, and assuming that the fibres are distributed uniformly in the space thus [2],

$$n_{dV} = n_V \frac{\overline{b_j} A_c}{V} \tag{2.36}$$

The component, $\overline{b_j}$, can be deduced from $\overline{b}$, i.e.,

$$\overline{b_j} = \overline{b} K_j = \frac{V}{2DLI} K_j = \frac{\pi D K_j}{8 V_f I} \tag{2.37}$$

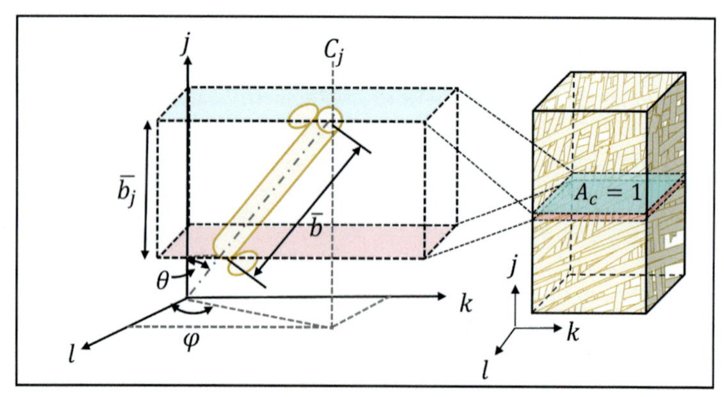

Fig. 2.4 A magnified image of a fibre segment between two successive contact points in a mesodomain, dV, sectioned by two planes of unit cross-sectional area, $A_c = 1$ with a thickness of $\overline{b_j}$.

where K_j is the directional parameter representing the projection of the fibre segment between the two fibre–fibre contacts on a given (j) direction [2,13].

$$K_j = \int_0^{\pi}\int_0^{\pi} |\sin\theta\cos\theta|\Omega(\theta,\varphi)\, d\theta d\varphi \tag{2.38}$$

The number of fibre–fibre contacts in a mesodomain can be calculated by combining Eqs. (2.36) and (2.37).

$$n_{dV} = \frac{2V_f}{\pi D^2}K_j \tag{2.39}$$

The detailed derivation and the description can be obtained from Ref. [2]. For a 3D random nonwoven material, we know the value of $(\Omega_o)_{3D}$ is $\frac{1}{2\pi}$, therefore, the value of the directional parameter in three dimensions, $(K_j)_{3D}$ can be computed as

$$(K_j)_{3D} = \frac{1}{2\pi}\int_0^{\pi}\int_0^{\pi} |\sin\theta\cos\theta|\, d\theta d\varphi = \frac{1}{2} \tag{2.40}$$

In the case of 2D nonwoven material, the value of the directional parameter in two dimensions, $(K_j)_{2D}$ can be obtained by combining Eqs. (2.20) and (2.38), i.e.,

$$(K_j)_{2D} = \int_0^{\pi}\int_0^{\pi} \cos\varphi\frac{1}{\pi}\delta(\theta-\pi/2)\sin\theta d\theta d\varphi = \frac{2}{\pi} \tag{2.41}$$

2.3 Tensile properties of nonwoven materials

Nonwoven materials have innumerable applications such as geotextiles, scaffolds, composites, ballistic protection, etc. All of the above-mentioned applications and many others demand that the nonwoven structure should maintain its structural integrity and should not disintegrate into constituent fibres/filaments, especially at the on-site performance. Thus, the tensile property is extremely pertinent – a key parameter to determine the performance of nonwoven material [15]. Theoretical prediction of the tensile properties of a nonwoven material requires a deep understanding of the distinct structural characteristics of fibres and the nature of bonds.

In general, various analytical approaches and numerical modelling techniques have been employed to predict the tensile properties of various types of nonwoven materials. Although it is beyond the scope of this chapter to discuss all these approaches and techniques in detail, a key micromechanical approach that can be universally applicable to all types of nonwoven materials would be described later in this section. Nevertheless, the analytical prediction of tensile properties has classically evolved from the shear-lag models, orthotropic, and fibre network theories [16–23]. Particularly, Cox [16] published a seminal work on predicting the elastic behaviour of fibrous materials and their composites by using *shear-lag* theory. In this theory, it is invariably assumed that the load transfer from fibre to matrix occurs as a result of shear stresses only. The concept of *shear-lag* theory has been successfully applied for the prediction of tensile properties of paper-based materials [24–26]. However, the usefulness of *shear-lag* theory has been disputed for simulated random fibre networks as the dominating mode of load transfer in such materials was found to be axial stress at the fibre intersections [27].

Another approach pioneered by Backer and Petterson [17] for predicting the elastic properties of nonwoven materials was employed by using a well-known orthotropic theory. Here, the fibre stretching was considered as a main mode of deformation under uniaxial tensile loading. Surprisingly, the predictive in-plane Poisson's ratio obtained via Backer and Petterson's work has resulted in the vicinity of 0.5, which is the upper bound of Poisson's ratio based on the classical theory of elasticity. In reality, the in-plane Poisson's ratio of various nonwoven materials has been found to be extremely large [15,28–31]. Nevertheless, the work of Backer and Petterson [17] has been extensively modified by Hearle and his colleagues by utilising fibre network theory [18,32,33]. The fibre network theory utilises the incremental deformation principle. Here, the nonwoven material is divided into numerous unit cells that experience the same strain as that of a nonwoven itself. Further, the average strain developed in each fibre is presumably equivalent to the strain experienced by the fibre, which is bonded at the boundaries of the unit cell [18]. The theory has been validated with a variety of nonwoven materials. However, a poor agreement between the theory and experiments was obtained for some of the nonwoven materials [18]. In contrast, the nonwoven has been modelled as a layered structure known as *laminate*, and each layer is considered as a *lamina* [19]. The modelling scheme allowed the incorporation of nonuniform fibre orientation distribution, and the fibres in each layer were oriented in the same direction. Subsequently, these layers were stacked over the top of each other in a way that the overall composition resulted in anisotropic nonwoven material. The theory was primarily validated for spunbonded nonwoven materials; however, the effect of heterogeneity of these nonwoven materials was clearly observed during the measurement of Poisson's ratio.

Another class of model that accounted for the realistic microstructure on the fibre network structure is defined on the basis of computational modelling techniques. Some of these models are computationally costly, but they allow a realistic approximation of the governing micromechanisms involved during deformation [34]. These models can be divided into well-defined classes. The first class of these models has been developed on the basis of the constituent of nonwoven materials, and they are regarded as 'indestructible' [35–38]. The second class is focused on the motion of the fibre–fibre contacts and their evolution [39–41]. In addition, the finite element models have been developed to investigate various aspects of mechanical behaviour of nonwoven materials, including stress-strain behaviour, notch-sensitive behaviour, and mechanisms involved in deformation and damage [34,42–60]. A finite element model incorporating all these mechanisms and, thus, predicting the macroscopic response of the nonwoven material whilst keeping it computationally efficient is an extremely challenging task [61]. Accordingly, various strategies have been adopted in the literature to simulate nonwoven materials and predict their behaviour [62]. The reader is referred to an excellent review by Farukh et al. [61] for the critical analysis of key strategies to predict the mechanical behaviour of nonwoven materials using finite element modelling. Next, we discuss a key micromechanical approach that can be universally applicable to a variety of nonwoven materials.

2.3.1 A 'generalised' initial tensile model of nonwoven materials

Pan and his coworkers [13] have published a 'generalised' micromechanical model for predicting the initial tensile behaviour of planar fibrous materials, which is also applicable for a range of nonwovens. Considering the external tensile load T_j, which is exerted on a fibre network in direction j ($j = 1$ or 2). Intuitively, the transmission of the tensile load has been carried out by means of fibre–fibre contacts. The fibre orientation and fibre–fibre contact region vary within the nonwoven structure, which leads to a complex situation to determine the actual direction and magnitude of force C acting on each fibre–fibre contact. In this scenario, a cross-section of the nonwoven is obtained in a manner such that all the forces that act on the contact points be C_j and the resultant of these forces counterbalance the applied tensile load T_j, as shown in Fig. 2.5. Hence [13],

$$C_j = \frac{T_j}{n_{dV}} \tag{2.42}$$

where n_{dV} is the total number of fibre–fibre contacts in the mesodomain as defined in Section 2.2.4.

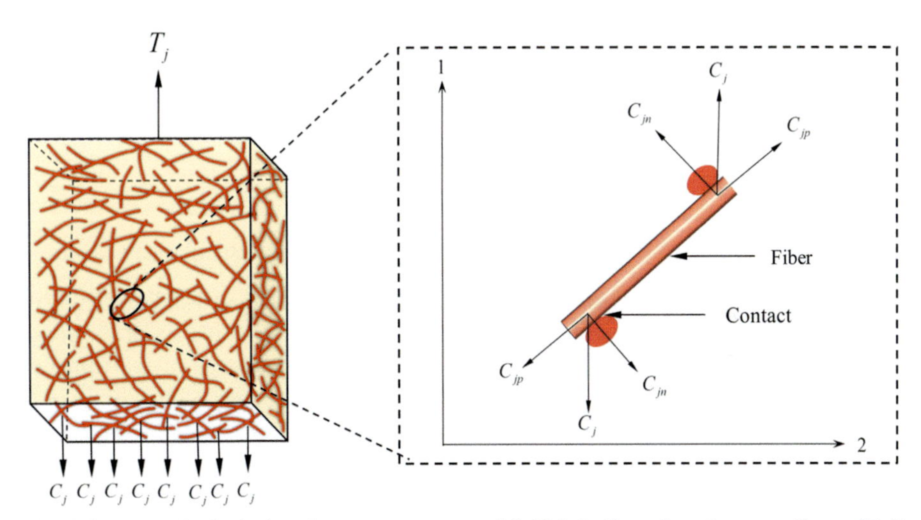

Fig. 2.5 A sketch displaying the nonwoven material (Originally referred to as collagen fibril network in Ref. [63].) being exerted with uniaxial tensile load, T_j, in direction, j, and the network is in a state of equilibrium through the contacts exhibiting load, C_j. The magnified image illustrates the resolution of forces at the contact points [63].

Now consider a fibre–fibre contact formed on an arbitrary fibre having a pair of the orientation angle (θ, φ) in the spherical coordinate system. The force C_j acting on this contact point can be bifurcated into the tangential component C_{jp} that act in the direction of the fibre axis and the normal component C_{jn}, which act perpendicular to the fibre axis (see Fig. 2.5). Here, the C_{jn} create a torque couple on the contact points that will bend the fibre segment and C_{jp} would stretch the contact points and fibre segment along the direction of the fibre axis, as shown in Fig. 2.5. These components of forces for a planar nonwoven material can be easily deduced on the basis of simple vector analysis, as shown below [13],

$$C_{1p} = C_1 \sin\theta\cos\varphi(\sin\theta\cos\varphi i_1 + \sin\theta\sin\varphi i_2) \tag{2.43}$$

$$C_{2p} = C_2 \sin\theta\sin\varphi(\sin\theta\cos\varphi i_1 + \sin\theta\sin\varphi i_2) \tag{2.44}$$

$$C_{1n} = C_1\left[\left(1 - \sin^2\theta\cos^2\varphi\right)i_1 - \left(\sin^2\theta\sin\varphi\cos\varphi\right)i_2\right] \tag{2.45}$$

$$C_{2n} = C_2\left[-\sin^2\theta\sin\varphi\cos\varphi i_1 + \left(1 - \sin^2\theta\sin^2\varphi\right)i_2\right] \tag{2.46}$$

Assuming that the fibre segments are straight before applying tensile load and consider the nonwoven consists of only one type of fibre. Further, the forces at the contact points are distributed based on the fibre orientation density function. Therefore, the statistical mean of the deformation of all microelements $\overline{\delta}_{jr}$ within the mesodomain in direction r due to the external load T_j is given by [13],

$$\delta_{jr} = \int_0^{\pi} d\theta \int_0^{\pi} d\varphi \delta\Omega(\theta,\varphi) \tag{2.47}$$

$$\overline{\delta}_{jr} = \left[+\overline{\delta}_1 M'_{jr} \pm \overline{\delta}_2 M''_{jr} \pm \overline{\delta}_3 M''_{jr}\right] \quad (\pm : +j = r,\ -j \neq r) \tag{2.48}$$

where $\overline{\delta}_1\left(=\frac{C_j\left(m_l\overline{b}\right)^3}{12EI_d}\right)$,[a] $\overline{\delta}_2\left(=\frac{C_j m_l \overline{b}}{AE}\right)$ and $\delta_3\left(=\frac{C_j n_l \overline{b}}{AE}\right)$ are the statistical mean deformations in the mesodomain due to bending, stretching of fibre segments, and stretching of fibre–fibre contacts, respectively, I_d is the diametric moment of inertia of the fibre, E is the elastic modulus of the fibre, A is the cross-sectional area of the fibre, m_l and n_l being the proportion of the free length of fibre and contact with respect to $\overline{b}$, respectively, M_{jr}', and M_{jr}'' are the geometric coefficients representing the averaging effects of fibre and load orientations [13].

[a] Here, we have considered the fibre segment between the contacts as a fixed-guided beam under the action of a normal force.

For computing the overall modulus (E_{jj}) of the nonwoven material, we can divide it into three components representing the contributions from the free fibre segments and those of contacts by considering the whole system made up of a series of three elastic components [13]. Thus,

$$\frac{1}{E_{jj}} = \frac{1}{E'_{jj}} + \frac{1}{E''_{jj}} + \frac{1}{E'''_{jj}} \tag{2.49}$$

where $E_{jj}{}' \left(= \frac{T_j \bar{b}_j}{\bar{\delta}_1 M'_{jj}} \right)$, $E_{jj}{}'' \left(= \frac{T_j \bar{b}_j}{\bar{\delta}_2 M''_{jj}} \right)$, and $E_{jj}{}''' \left(= \frac{T_j \bar{b}_j}{\bar{\delta}_3 M''_{jj}} \right)$ are the elastic moduli due to free fibre segment bending, stretching of fibre segments, and stretching of fibre–fibre contacts, respectively.

Similarly, the Poisson's ratio of the nonwoven material (ν_{jr}) can be obtained, as [13],

$$\nu_{jr} = \frac{\left[\bar{\delta}_1 M'_{jr} + \bar{\delta}_2 M''_{jr} + \bar{\delta}_3 M''_{jr} \right] \bar{b}_j}{\left[\bar{\delta}_1 M'_{jj} + \bar{\delta}_2 M''_{jj} + \bar{\delta}_3 M''_{jj} \right] \bar{b}_r} \tag{2.50}$$

Or,

$$\nu_{jr} = \nu'_{jr} + \nu''_{jr} + \nu'''_{jr} \tag{2.51}$$

where
$\nu'_{jr} \left(= \frac{\bar{\delta}_1 M'_{jr} \bar{b}_j}{\left[\bar{\delta}_1 M'_{jj} + \bar{\delta}_2 M''_{jj} + \bar{\delta}_3 M''_{jj} \right] \bar{b}_r} \right)$, $\nu''_{jr} \left(= \frac{\bar{\delta}_2 M''_{jr} \bar{b}_j}{\left[\bar{\delta}_1 M'_{jj} + \bar{\delta}_2 M''_{jj} + \bar{\delta}_3 M''_{jj} \right] \bar{b}_r} \right)$ and $\nu'''_{jr} \left(= \frac{\bar{\delta}_3 M''_{jr} \bar{b}_j}{\left[\bar{\delta}_1 M'_{jj} + \bar{\delta}_2 M''_{jj} + \bar{\delta}_3 M''_{jj} \right] \bar{b}_r} \right)$ are the Poisson's ratios that correspond to the bending deformation, fibre extension, and the deformation of fibre–fibre contact.

In general, ν'_{jr} is significantly larger than ν''_{jr} and ν'''_{jr} or in other words, the bending deformation of fibres dominates in the nonwoven materials under tensile loading [13]. Recently, we have revealed that the nature of deformation in needlepunched nonwovens is attributed to the fibre bending under uniaxial tensile loading conditions [31]. However, there are cases that ν'''_{jr} is comparatively larger than the other two components when the total fibre volume fraction approaches its maximum allowable value [13]. It should also be noted that the bending stiffness of constituent fibres of nonwoven materials is a few orders of magnitude lower than their corresponding axial stiffness [54]. Under tensile loading, the fibres tend to bend easily rather than being stretched. Alternatively, it is well established that a parameter with units of length $l_b \left(= \sqrt{\frac{(EI)_f}{(EA)_f}} \right)$, is significantly important in determining the bending-dominated or stretch-dominated regime, where $(EA)_f$, and $(EI)_f$ are the axial and bending stiffnesses, respectively [64]. When $l_b \gg \bar{l}$, where $\bar{l}$ is the mean fibre segment length, the nonwoven modulus would tend to follow stretch-dominated regime whereas if $l_b \ll \bar{l}$ then the nonwoven modulus follows the bending-dominated regime. Perhaps, this is a reason that the stretch-dominated regime is expected to occur in electrospun nonwoven materials [65].

2.4 Compression properties of nonwoven materials

Nonwoven materials experience compression forces during their end-use performance, including thermal insulation, geotextiles, sound insulation, fluid filtration, seats, beddings, and many more. Over the years, the compression properties of nonwoven materials have been extensively investigated, but the theoretical explanation of the observed compression-recovery behaviour has largely remained ambiguous [66]. The compression properties of nonwoven materials have been predicted primarily on the basis of phenomenological and micromechanics approaches [67,68]. In general, the phenomenological models fit a mathematical equation to the experimental stress-strain curve without revealing the *exact* relationship between the fibre and nonwoven properties. For example, the compression behaviour of nonwoven materials can be empirically modelled by the 'rheological' model, whereby the nonwoven can be considered as an elastic/viscoplastic material [69]. On the other hand, the micromechanical models overcome the ambiguity posed by phenomenological models by adapting a continuum mechanics approach that considers two-phase (fibre-pore) material as a homogenous discrete continuum [67].

More than seven decades back, van Wyk [11] pioneered the work on the compressibility of random fibrous assemblies, which is also applicable to nonwoven materials. Here, the compression behaviour of 3D random fibrous assemblies was expressed purely in terms of bending strains in the constituent fibres by neglecting the shearing, twisting, slippage, and fibre extension. The fibrous assembly was considered as a 'layered system'; each layer has been defined by straight and bending units of fibre segments. Under compression loading, the deflections exhibited by the fibre segments between the two successive fibre–fibre contacts caused the changes in the thickness of each layer. The prediction of the 'pressure-volume' curve made use of some of the key approaches, including the continuum mechanics, stereological, geometrical probability, least square method, and excluded area concept. However, the model was applicable to the random fibrous assemblies and did not account for the changes in the orientation of fibres during compression. Stearn [70] predicted the number of fibre–fibre contacts per unit volume of anisotropic fibrous assemblies by updating the fibre orientation during compression. Later, the isotropic assumption of van Wyk [11] was circumvented by Lee and Lee [71], who introduced the generalised concepts of fibre orientation distribution, as derived by Komori and Makishima [8]. Initial compressional modulus and Poisson's ratio of an anisotropic fibrous assembly were predicted. However, there was a poor agreement between the theoretical and experimental values of initial compressional modulus vs cube of the fibre volume fraction, which was partly ascribed to the negligence of fibre slippage [71].

Carnaby and Pan [66] included the effects of fibre slippage by classifying all the fibre–fibre contacts as slipping or nonslipping and predicted the compression behaviour of fibrous assemblies. The theory was not only validated with the experimental results but also revealed a good agreement between the initial compressional modulus vs cube of the fibre volume fraction. During compression loading, some of the updated orientation parameters, i.e. I [as defined in Eq. (2.15)], exceeded the maximum allowable magnitude for a 3D random fibrous material. Perhaps, this has resulted in an unusual trend and magnitudes of Poisson's ratio, with the level of the compression strain [72]. Nevertheless, one of the notable contributions of Carnaby and Pan's work was to predict the recovery behaviour of fibrous assemblies, which was scarcely available in the literature at that time. In a series of papers, Carnaby and colleagues [73,74] revisited the prediction of compression properties by considering the joint density function of length and orientation of each fibre segment and subsequently minimised the total energy of the bending, straightening, and slipping fibre segments. Apparently, the whole idea of the joint density function of length and orientation was essentially to capture the crimp information of the fibrous assemblies. Although a good agreement between the theory and experiments was observed, the inclusion of fibre crimp did not provide such evidence [74].

Komori and colleagues [72,75,76] not only predicted the compression behaviour of fibrous assembly under large levels of strains but also individual bending behaviours are combined into the overall response of the mass. Here, the length of the bending element depended on the respective orientation, and the mechanical relationships were derived based on the energy method [72]. This theory was not only applicable to a simple case of uniaxial compression but has been extended to consider the cases of isotropic compression and laterally confined compression. However, the theory was not validated with the required set of experiments, except in Komori et al. [75], whereby the theoretical predictions overestimated the experimental data by 10–120 times. Subsequently, Neckar [77] also published a semiempirical model of compressibility of fibrous assemblies under uniaxial and bi-axial deformations. Further, an attempt was also made to predict the compressibility of thermally bonded nonwoven materials under a wide range of stresses based on the bending of free fibre segments followed by the transverse compression of fibre contacts through the classical Hertzian contact mechanics approach [78]. A good agreement was observed between the theory and experiments.

A key takeaway from the above discussion is that the bending and slippage of fibre segments are the main deformations that occur in fibrous assemblies that also include nonwovens. However, Narter et al. [67] argued that the tensile (axial) deformation of fibres primarily occurs for the nonwovens with a high coefficient of friction or self-bonded (e.g. spun-bonded, melt-blown, etc.) materials, but no experimental validation of their proposed deformation scheme was offered. Recently, we have extended the work of Carnaby and Pan [66] by incorporating the internal friction losses, and subsequently, the compression-recovery behaviour of non-wovens prepared from glass fibres with a high coefficient of friction was validated [14]. More importantly, we used the 3D data of fibre orientation from X-ray micro-computed tomography for predicting the compression-recovery behaviour. Next, we present some of the key sets of equations that were used to model the compression-recovery behaviour of nonwoven materials [14].

2.4.1 Compression-recovery model of nonwoven materials

Based on the approach of Carnaby and Pan [66], compressive stress (P) has been applied to the mesodomain of a nonwoven in the j-direction [14]. As aforementioned, a mean force (C_j) acts at the fibre–fibre contact points and can be resolved into the normal (C_{jn}) and tangential (C_{jp}) components. Now considering the friction forces between the fibres; accordingly, a *slipping* or a *nonslipping* fibre–fibre contact has been formed, as shown in Fig. 2.6. The fibres tend to *slip* if the following criteria have been satisfied [66], i.e.,

$$C_{jp} \geq \mu C_{jn} + Wf_o\overline{b} \tag{2.52}$$

where μ is the coefficient of friction between the fibres, and Wf_o is the mean withdrawal force per unit fibre length.

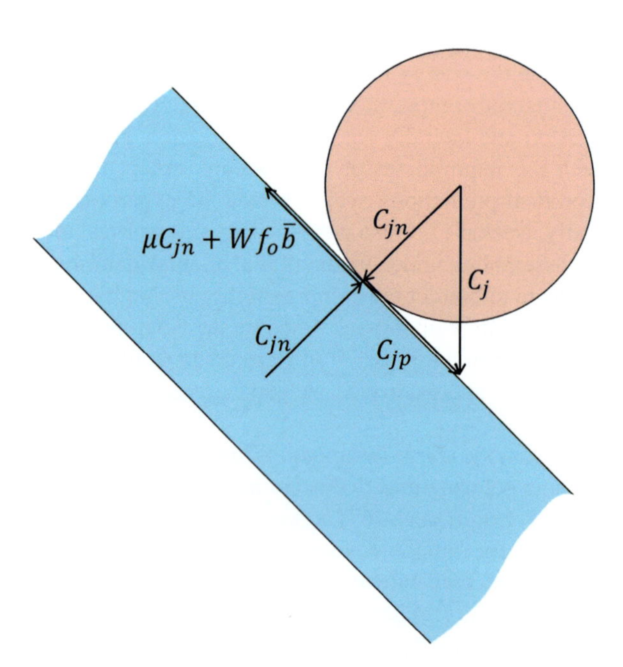

Fig. 2.6 A sketch depicting the force equilibrium between the two fibres.
Adapted from G.A. Carnaby, N. Pan, Theory of the compression hysteresis of fibrous assemblies, Text. Res. J. 59 (1989) 275–284.

Under uniaxial compression loading, the forces act primarily in the out-of-plane direction, the fibre slippage criteria in terms of critical polar angle (θ_c) can be defined as shown below[66],

$$C_j \cos\theta \geq \mu C_j \sin\theta + Wf_o \bar{b} \tag{2.53}$$

Thus, the threshold criteria for fibre slippage is given by [66],

$$\sin\theta_c = \frac{-\mu\omega \pm \left\{\mu^2\omega^2 - (1+\mu^2)(\omega^2-1)\right\}^{1/2}}{1+\mu^2} \tag{2.54}$$

where

$$\omega = \frac{Wf_o \bar{b}}{C_j};\ C_j = \frac{P}{n_{dV}}$$

For, $\theta \geq \theta_c$, there is no slippage between the fibres, whereas the slippage between the fibres occurs when $\theta < \theta_c$. A majority of the parameters given in Eq. (2.54) are readily available except Wf_o. Based on the analysis of Grosberg [79], the expression for withdrawal force per unit fibre length was modified for obtaining its mean value via orientation averaging approach [14], as shown below.

$$Wf_o = \int_0^{\pi} d\theta \int_0^{\pi} d\varphi \frac{\pi P D \sin\theta \Omega(\theta,\varphi)}{2V_f(\cos\theta \pm \mu \sin\theta)} \tag{2.55}$$

It should be noted that the '+' and '−' signs imply compression and recovery behaviour of nonwoven material, respectively. Further θ_c has been included in the structural parameters given in Eqs. (2.13), (2.15), (2.18), (2.38) for computing the effective proportion of nonslipping contacts and the mean distance between the nonslipping contact points, as depicted below.

$$\bar{b}^C = \frac{\pi D}{8V_f I^c};\ I^c = 2\int_{\theta_c}^{\pi/2} d\theta \int_0^{\pi} J^c(\theta,\varphi)\sin\theta\Omega(\theta,\varphi)d\varphi \tag{2.56}$$

$$J^c(\theta,\varphi) = 2\int_{\theta_c}^{\pi/2} d\theta' \int \sin\chi(\theta,\varphi,\theta',\varphi')\Omega(\theta',\varphi')\sin\theta' d\varphi' \tag{2.57}$$

$$K_j^c = 2\int_{\theta_c}^{\pi/2} d\theta \int_0^{\pi} d\varphi \sin\theta\cos\theta\Omega(\theta,\varphi) \tag{2.58}$$

For brevity, we have used the same symbols of $J^c(\theta,\varphi)$ in this chapter. In the above equation, the superscript 'c' refers to the inclusion of the θ_c in computing the structural parameters. Moreover, the compressive stress or pressure (P) that has been applied to the mesodomain of volume (dV) would be distributed between slipping and non-slipping contacts [66], as illustrated below,

$$P = f_s n_{dV} C_{sj} + (1 - f_s) n_{dV} C_{nsj} \tag{2.59}$$

$$C_{nsj} = \frac{P - f_s n_{dV} C_{sj}}{(1 - f_s) n_{dV}} \tag{2.60}$$

$$f_s = 2\int_0^{\theta_c} d\theta \int_0^{\pi} \Omega(\theta,\varphi)\sin\theta d\varphi;\ C_{sj} = 2\int_0^{\theta_c} d\theta \int_0^{\pi} d\varphi \frac{W f_o \bar{b} \sin\theta\Omega(\theta,\varphi)}{\cos\theta - \mu\sin\theta} \tag{2.61}$$

where f_s is the fraction of slipping contact points within the mesodomain, C_{nsj} and C_{sj} are the mean forces per nonslipping and slipping contact points, respectively.

As aforementioned, C_{nsj} creates bending in the fibre segments at the fibre–fibre contact points. In this direction, the fibre segment between the contact points can be treated as an elastic beam. Using the orientation averaging approach, the statistical mean of the deformation of various fibre segments in the direction j ($\bar{\delta}_{jj}$) present within the mesodomain is given by [14,66],

$$\bar{\delta}_{jj} = \pm \frac{C_{nsj}\left(\bar{b}^c\right)^3}{\eta B} M_{jj}^c \tag{2.62}$$

where

$$M_{jj}^c = 2\int_{\theta_c}^{\pi/2} d\theta \int_0^{\pi} d\varphi \sin^3\theta\Omega(\theta,\varphi) \tag{2.63}$$

where B is the bending modulus of the fibre, η is the beam constant that depends on the region of application of the force, and M_{jj}^c is the geometric coefficient that represents the combined effects of projection of force and deflection for all possible alignment of fibres in the direction of compressive stress (j) [71].

The principle of continuum mechanics in the mesodomain has then been applied, which defines the strain experienced by the fibre segment in the mesodomain ($\Delta\varepsilon$) causes a proportional change in the thickness of the nonwoven material. Thus,

$$\frac{T_{(i-1)} - T_{(i)}}{T_{(i-1)}} = \frac{\bar{\delta}_{jj}}{\bar{b}^c K_j^c} = \Delta\varepsilon_i \tag{2.64}$$

where T is the thickness of the nonwoven, indices 'i − 1' and 'i' refer to the preceding and current structural level being analysed, respectively.

Further, the out-of-plane orientation angle of the fibre has been updated based on the application of compressive stresses [66,71,74]. Subsequently, the orientation density function, Poisson's ratio and fibre volume fractions have been updated. By combining Eqs. (2.13), (2.15), (2.18), (2.38), (2.39), (2.52)–(2.64) would yield the value of the thickness of nonwoven material under the defined amount of compressive stress.

Further, the recovery stage releases the compressive load gradually, i.e. in a succession of reduction in the magnitude of C_j in the mesodomain of volume (dV). Accordingly, the stored bending energy in the fibres is the primary cause of the recovery of the nonwoven material. During the recovery of nonwoven material, a defined value of C_{sj} is directed in regression by overcoming the resistance, $Wf_o\overline{b}$. Thus [66],

$$C_j \cos\theta = Wf_o\overline{b} \tag{2.65}$$

where

$$C_{sj} = 2\int_0^{\theta_c} d\theta \int_0^{\pi} d\varphi Wf_o\overline{b}\tan\theta\Omega(\theta,\varphi) \tag{2.66}$$

Next, the remnant pressure (P_r) exerted by the nonwoven material during the recovery stage was computed. In general, the energy due to the applied stresses is expended for the pressure to bend the fibres (P_b), pressure loss due to slippage of contacts (P_s), and pressure losses due to the internal friction that accounted for the withdrawal forces ($\Delta P_{Wf\overline{b}}$) [14]. Therefore, P_r can be calculated as,

$$P_r = P_b - \left(P_s + \Delta P_{Wf\overline{b}}\right) \tag{2.67}$$

where

$$P_b = (1 - f_s)n_{dV}C_{nsj}, \Delta P_{Wf\overline{b}} = P_{Wf\overline{b}(i)} - P_{Wf\overline{b}(0)}, P_{Wf\overline{b}(0)} = Wf_o\overline{b}n_{dV} \text{ and } P_s = f_sC_{sj}n_{dV}$$

In the above equation, the indices '0' and 'i' refer to the initial and current structural level being analysed. As recovery proceeds, the cumulative pressure losses during the successive iterations were accounted for by modifying Eq. (2.67) that resulted in the recovery pressure, as shown below.

$$P_{r(i)} = \left[P_{b(i)} - \left(P_{s(i)} + \Delta P_{Wf\overline{b}(i)}\right)\right]\frac{P_{r(i-1)}}{P_{b(i-1)}} \tag{2.68}$$

Using this analysis, an algorithm to compute the compression-recovery behaviour of nonwoven material has been developed as the magnitude of geometrical and structural parameters was stored during the compression stage and reused at the recovery stage [14].

With the aid of parametric analysis, the nonwoven material can be designed and tailored for a desired compression-recovery curve [14]. Fig. 2.7 depicts the effect of the coefficient of friction between the fibres on the compression hysteresis of nonwoven material. The compression hysteresis of the nonwoven material can be minimised by increasing the coefficient of friction between the fibres. Similarly, the hysteresis of compression-recovery behaviour can be reduced by varying the structural parameters, namely, fibre volume fraction and ODF, as illustrated in Figs. 2.8 and 2.9, respectively. The readers are referred to the work of Rao et al. [14] for obtaining further details of this parametric analysis.

In general, the micromechanics approach presumes that the forces on all the fibre-fibre contacts are equal, and the direction matches that of compression forces that act on the fibrous assembly [71]. The numerical simulations carried out by Beil and Roberts [80,81] did not support this assumption. Nevertheless, the nonwoven material that possesses a small volume fraction of fibres and is randomly aligned would be least affected by this assumption.

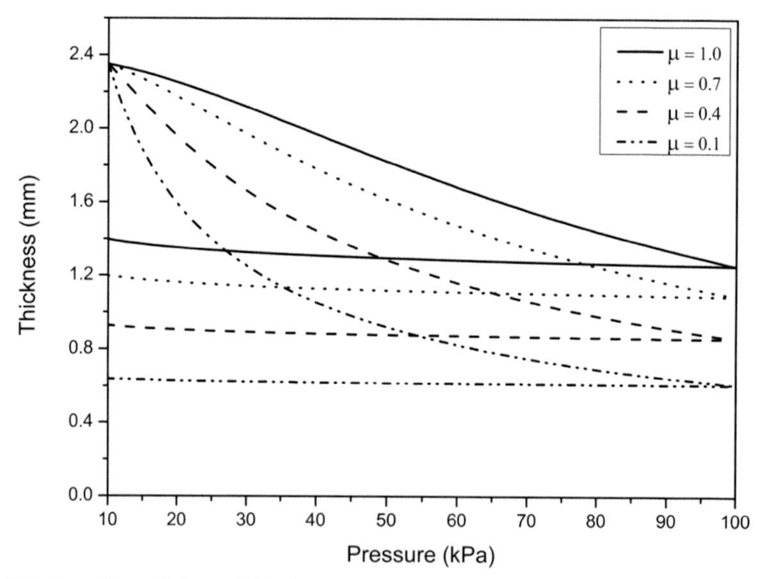

Fig. 2.7 Effect of coefficient of friction (μ) of glass fibres on the compression-recovery behaviour of nonwoven materials [14].

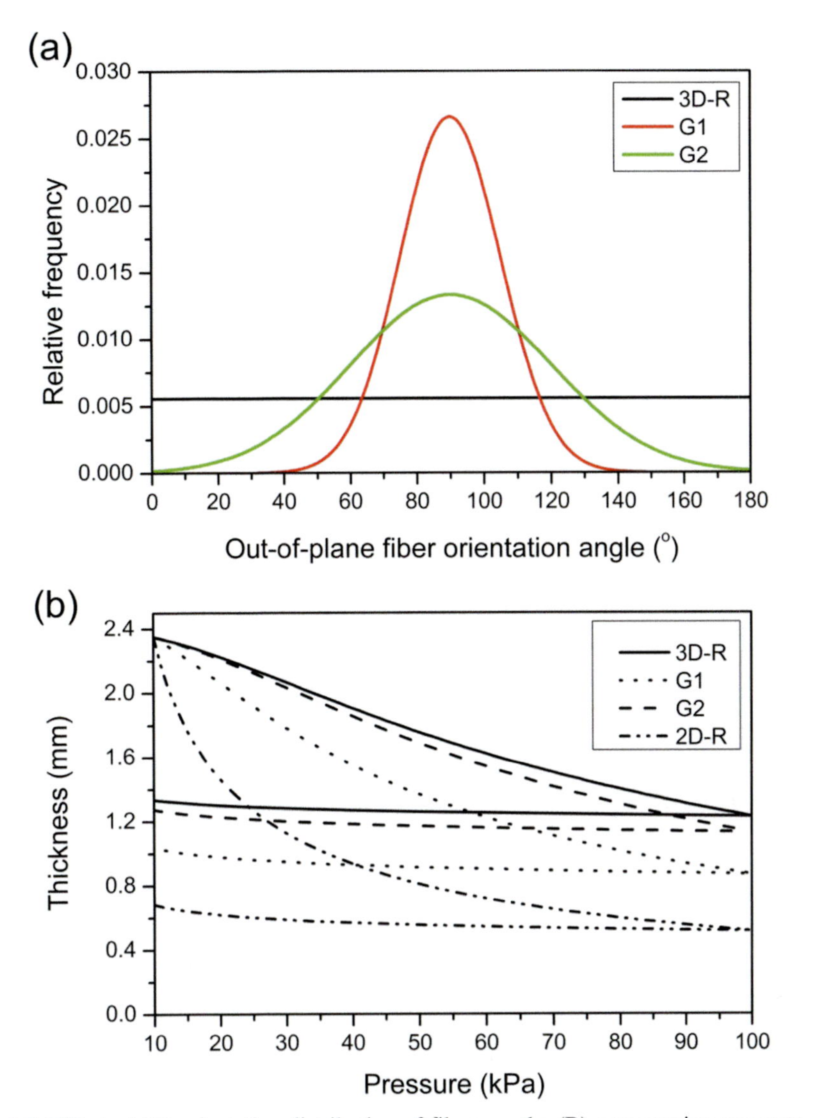

Fig. 2.8 Effect of (A) orientation distribution of fibres on the (B) compression-recovery behaviour of nonwoven material comprising glass fibres. Here, the '3D-R' refers to uniform three-dimensional random material; '2D-R' refers to planar random material; G1 and G2 are the Gaussian distributions with a mean of 90 degrees and standard deviations of 15 and 30 degrees, respectively, for out-of-plane fibre orientation angle and uniform random distribution was considered for in-plane fibre orientation angle [14].

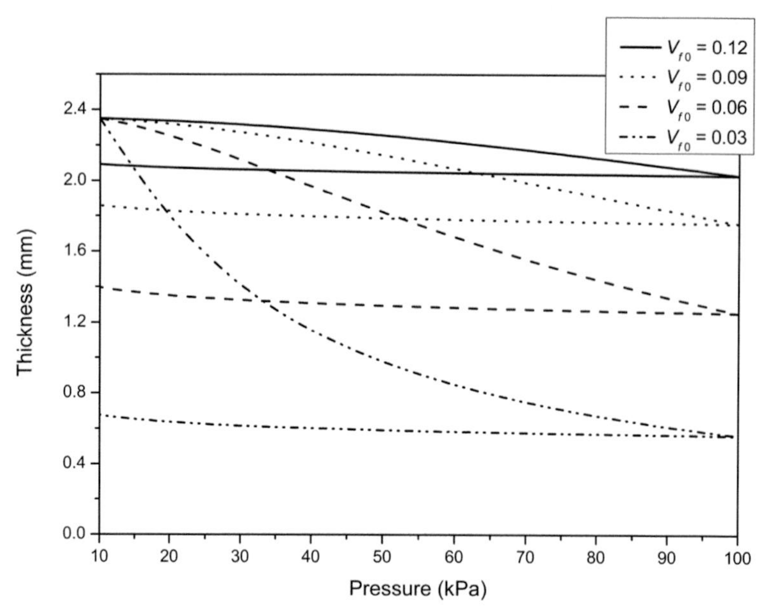

Fig. 2.9 Effect of initial fibre volume fraction (V_{f0}) on the compression-recovery behaviour of nonwoven material comprising glass fibres [14].

2.5 Shear properties of nonwoven materials

Nonwoven materials have an extensive range of end-use applications, but they are particularly restricted in apparel applications. One of the key reasons is that nonwoven materials possess a papery appearance due to poor drapeability [82]. Even in the manufacture of components of complex-shaped composites, the reinforcement material is expected to drape onto a mould surface [83]. In general, the shearing of fabrics is a key mode of deformation during draping [84]. Fibre–fibre contacts are one of the main constituents that are responsible for high shear characteristics or poor drapeability. Stevenson and his coworkers [82,85] pointed out that some of the key characteristics of nonwovens that can improve the drapeability characteristics are: firstly, for chemically bonded nonwovens, the fabric stiffness is less sensitive to the binder stiffness than the fibre stiffness. Secondly, a nearly threefold increase in the weight reduces the drape coefficient to within the acceptable textile region. Thirdly, the direct approach would be to modify the nonwoven structure itself, which can be easily attained by the alignment of fibres, porosity, and the proportion of fibre–fibre contacts. Considering the viewpoint of 3D fibre orientation and fibre–fibre contacts, Cheng and Duckett [86] computed the energy losses of fibrous materials *whilst* they are being subjected to an oscillating shear. One of the key contributions of this work was to include the slippage between the fibres on the basis of critical polar angle. Although the 'off-the-shelf' predictive model of the shear properties of nonwoven materials is scarcely available in the literature, the model of the initial shear response of fibrous materials [87] can easily be adapted for nonwoven materials.

Pan and Carnaby [87] implemented the modelling strategy that was used for predicting the compression behaviour of fibrous materials [66]. Accordingly, it was presumed that the imposed shear deformation would either cause slippage at the fibre–fibre contacts or result in the bending of fibre segments between the contact points. Similar to Eq. (2.52), an equivalent condition for the general loading case was proposed [87].

$$C_1 \sin\theta\cos\varphi + C_2 \sin\theta\sin\varphi + C_3\cos\theta \geq \mu\left\{C_1\left(1-\sin^2\theta\cos^2\varphi\right)^{\frac{1}{2}} + C_2\left(1-\sin^2\theta\sin^2\varphi\right)^{\frac{1}{2}} + C_3\sin\theta\right\} + Wf_o\bar{b} \tag{2.69}$$

where C_1, C_2, and C_3 be the net forces per contact in the 1, 2, and 3 directions, respectively.

The above equation defines the slippage criterion that leads to the modifications in the orientation parameters and the mean distance between the fibre–fibre contacts projected in the 'j' direction similar to the compression case, as defined in Eqs (2.56), (2.57). Here, the critical azimuthal angle (φ_c) can define the fibre slippage, as shown below [87],

$$\bar{b}'_j = \frac{\pi D K_j}{8V_f I'};\ I' = 4\int_0^{\pi/2} d\theta \int_{\varphi_c}^{\pi/2} J'(\theta,\varphi)\sin\theta\,\Omega(\theta,\varphi)\,d\varphi \tag{2.70}$$

$$J'(\theta,\varphi) = 4\int_0^{\pi/2} d\theta' \int_{\varphi_c}^{\pi/2} \sin\chi(\theta,\varphi,\theta',\varphi')\,\Omega(\theta',\varphi')\sin\theta'\,d\varphi' \tag{2.71}$$

In a simple shear, it can be assumed that there are two stresses, P_j and P_k acting on a unit area[b] of the same magnitude in the orthogonal directions for maintaining the equilibrium, as illustrated in Fig. 2.10.

Considering a simple case as defined by surface tractions P_j ($j=3$) and P_k ($k=1$), which results in net point contact forces as C_j and C_k at any arbitrary fibre–fibre contact in j and k directions. Thus, the slippage criterion, as defined in Eq. (2.69), can be modified, as shown below [87],

$$C_3\cos\theta + C_1\sin\theta\cos\varphi \geq \mu\left\{C_3\sin\theta + C_1\left(1-\sin^2\theta\cos^2\varphi\right)^{\frac{1}{2}}\right\} + Wf_o\bar{b} \tag{2.72}$$

[b] Pan and Carnaby [87] assumed the area of cube as $V^{2/3}$, where V is the volume of the cube. For simplicity and maintaining brevity, we have used the unit cross-sectional area of the mesodomain.

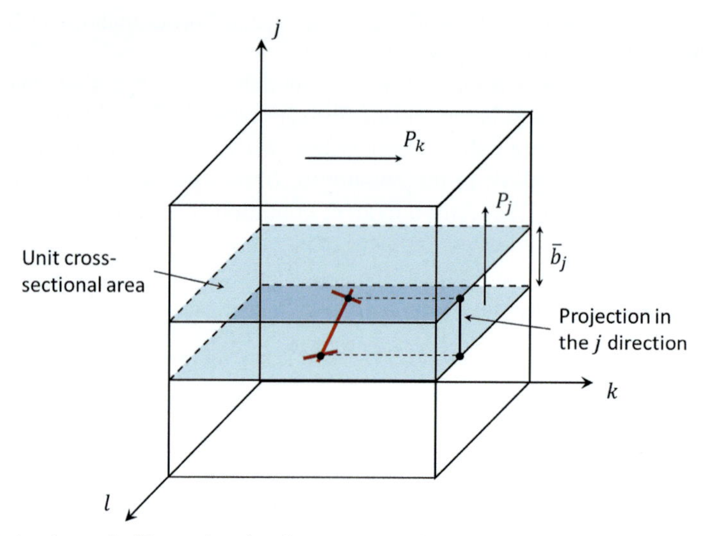

Fig. 2.10 A schematic illustrating the shear stresses, P_j and P_k acting on the surface of mesodomain.
Adapted from N. Pan, G.A. Carnaby, Theory of the shear deformation of fibrous assemblies, Text. Res. J. 59 (1989) 285–292.

Analogous to the compression case, the P_j and P_k that has been applied to the mesodomain of volume (dV) would be distributed between slipping and nonslipping contacts [87], i.e.,

$$P_j = f_s'(n_{dV})_k C_{sj}' + \left(1 - f_s'\right)(n_{dV})_k C_{nsj}' \tag{2.73}$$

$$P_k = f_s'(n_{dV})_j C_{sk}' + \left(1 - f_s'\right)(n_{dV})_j C_{nsk}' \tag{2.74}$$

where

$$f_s' = 4\int_0^{\frac{\pi}{2}} d\theta \int_0^{\varphi_c} \Omega(\theta, \varphi)\sin\theta d\varphi \tag{2.75}$$

$$C_{sj}' = 4\int_0^{\pi/2} d\theta \int_0^{\varphi_c} \Omega(\theta\varphi)\sin\theta \frac{Wf_o\overline{b} - C_k\left\{\sin\theta\cos\varphi - \mu(1-\sin^2\theta\cos^2\varphi)^{\frac{1}{2}}\right\}}{\cos\theta - \mu\sin\theta} d\varphi \tag{2.76}$$

$$C'_{sk} = 4\int_{0}^{\pi/2} d\theta \int_{0}^{\varphi_c} \Omega(\theta,\varphi)\sin\theta \frac{Wf_o\overline{b} - C_j\{\cos\theta - \mu\sin\theta\}}{\sin\theta\cos\varphi - \mu(1-\sin^2\theta\cos^2\varphi)^{\frac{1}{2}}} d\varphi \tag{2.77}$$

$$C'_{nsj} = \frac{-f'_s}{1-f'_s} C'_{sj} + \frac{f'_s}{\left(1-f'_s\right)} \frac{K_j}{K_k} C'_{sk} + \frac{K_j}{K_k} C'_{nsk} \tag{2.78}$$

where $(n_{dV})_j$ and $(n_{dV})_k$ are the number of fibre–fibre contacts in a mesodomain of the unit cross-sectional area with a thickness defined as the mean distance between the two successive fibre–fibre contacts projected in j and k directions, respectively, K_j and K_k are the directional parameters when the mean distance between the two successive fibre–fibre contacts projected in j and k directions, respectively.

Assuming α and β are the angular deformations of the face in the Z-Y plane with respect to the Z and Y axes, respectively caused by P_j and P_k then the following definition of shear strains would be satisfied.

$$\alpha \sim \tan\alpha = \frac{\overline{\delta}_{kk} + \overline{\delta}_{jk}}{\overline{b}'_j} \tag{2.79}$$

$$\beta \sim \tan\beta = \frac{\overline{\delta}_{jj} + \overline{\delta}_{kj}}{\overline{b}'_k} \tag{2.80}$$

where

$$\overline{\delta}_{jk} = \pm \frac{C'_{nsj}\left(\overline{b}'\right)^3}{\eta B} m_{jk} \tag{2.81}$$

$$\begin{aligned}
m_{11} &= \int_0^{\pi} d\theta \int_0^{\pi} d\varphi \left(1 - \sin^2\theta\cos^2\varphi\right) \Omega(\theta,\varphi)\sin\theta \\
m_{22} &= \int_0^{\pi} d\theta \int_0^{\pi} d\varphi \left(1 - \sin^2\theta\sin^2\varphi\right) \Omega(\theta,\varphi)\sin\theta \\
m_{33} &= \int_0^{\pi} d\theta \int_0^{\pi} d\varphi \sin^2\theta\, \Omega(\theta,\varphi)\sin\theta \\
m_{12} = m_{21} &= \int_0^{\pi} d\theta \int_0^{\pi} d\varphi \sin^2\theta\sin\varphi\cos\varphi\, \Omega(\theta,\varphi)\sin\theta \\
m_{23} = m_{32} &= \int_0^{\pi} d\theta \int_0^{\pi} d\varphi \sin\theta\cos\theta\sin\varphi\, \Omega(\theta,\varphi)\sin\theta \\
m_{31} = m_{13} &= \int_0^{\pi} d\theta \int_0^{\pi} d\varphi \sin\theta\cos\theta\cos\varphi\, \Omega(\theta,\varphi)\sin\theta
\end{aligned} \tag{2.82}$$

Therefore, the shear strain due to P_j and P_k is defined as,

$$\gamma_{jk} = \frac{\overline{\delta}_{kk} + \overline{\delta}_{jk}}{\overline{b}_j'} + \frac{\overline{\delta}_{jj} + \overline{\delta}_{kj}}{\overline{b}_k'} \tag{2.83}$$

Combining Eqs. (2.18), (2.37), (2.38), and (2.79)–(2.83),

$$\gamma_{jk} = \frac{C'_{nsk} V^2}{4\eta B D^2 L^2 I'^2} S(j,k) \tag{2.84}$$

where

$$S(j,k) = \left[\frac{K_k m_{kk} + K_j m_{kj} + K_k \dfrac{C'_{nsj}}{C'_{nsk}} m_{jk} + K_j \dfrac{C'_{nsj}}{C'_{nsk}} m_{jj}}{K_j K_k} \right] \tag{2.85}$$

Using the basic definition of shear modulus G_{jk},

$$G_{jk} = \frac{P_k}{\gamma_{jk}} = \frac{P_j}{\gamma_{kj}} \tag{2.86}$$

Combining Eqs. (2.73), and (2.83)–(2.86),

$$G_{jk} = \frac{2\eta E_f V_f^3 K_j I'^2}{\pi^2 S(j,k)} \left\{ (1 - f_s') + f_s' \frac{C_{sk}}{C'_{nsk}} \right\} \tag{2.87}$$

Therefore, Eq. (2.87) provides the shear modulus, which is expressed in terms of fibre orientation distribution, type of beam considered, fibre modulus, proportion of contacts that slip, average forces per (non) slipping contact, and fibre volume fraction. Ostensibly, the above model of shear modulus was not validated with the required experimental results. However, Pan and Carnaby [87] have simplified the above model for random fibrous assemblies and compared their results with the well-known relationship of shear modulus and tensile modulus for isotropic materials.

2.6 Summary and future outlook

The brief discussion of 3D micromechanical models of nonwoven materials shows the continuing and ever-increasing interest in this field. In general, micromechanics requires the use of a continuum mechanics approach and presumes affine deformations in the homogenous fibre network. A combinatorial design strategy of micromechanics involving stereological, stochastic, excluded volume, and geometrical probability approaches has successfully predicted various mechanical properties of nonwoven materials. Micromechanics can assist in designing nonwoven materials with a unique and innovative set of properties. A typical example is out-of-plane 'auxetic' nonwovens that tend to expand in the through-thickness direction on stretching, and the negative Poisson's ratio can be tailored by micromechanics approach [28,30,31]. In other words, the porosity of out-of-plane auxetic nonwoven materials would rise on stretching, unlike conventional nonwovens, which would be extremely useful in numerous applications, such as filtration, sound absorption, scaffolds, etc.

Though micromechanics presents a long list of merits, there are challenges associated with this approach. Firstly, the theory is scarcely validated with suitable experiments, as the determination of input parameters such as 3D fibre orientation distribution and porosity poses a serious conundrum. However, state-of-the-art techniques such as the X-microcomputed tomography technique provide the measurement of such input parameters with reasonable precision. Secondly, the micromechanics approach invariably presumes equal distribution of forces on fibre-fibre contact points without accounting for the local alignment of fibres. Perhaps, the numerical simulations such as the finite element approach can be combined together with the micromechanical approaches to compute the realistic distribution of forces on fibre-fibre contact points. Thirdly, the visco-elastic behaviour of constituent fibres in nonwovens has not been considered whilst predicting the mechanical properties. Perhaps, a semi-empirical approach similar to van Wyk's work [11] involving a 'constant' emerging from the appropriate set of springs and dashpots based on Maxwell's approach can be encapsulated in predicting the micromechanical behaviour of nonwovens. Lastly, the generalised theory of fibre–fibre contacts [8], which forms a cornerstone in the micromechanics approach, needs to be revisited. Particularly, the assumption of Komori and Makishima's work [8], whereby the effects of end-to-end or side-to-end fibre contacts have been neglected [88]. Despite these challenges, micromechanics is an enormously resourceful approach that not only unravels the structure-property relationship of nonwoven materials but paves a way to unfold various properties of a family of disordered arrays of fibre networks that also comprises conventional paper, collagen, stainless steel networks, spider silk, and many others [89].

References

[1] S.J. Russell, Handbook of nonwovens, Woodhead Publishing, 2006.

[2] A. Rawal, S. Lomov, T. Ngo, I. Verpoest, J. Vankerrebrouck, Mechanical behavior of thru-air bonded nonwoven structures, Text. Res. J. 77 (2007) 417–431.

[3] W. Albrecht, H. Fuchs, W. Kittelmann, Nonwoven Fabrics: Raw Materials, Manufacture, Applications, Characteristics, Testing Processes, John Wiley & Sons, 2006.

[4] S.M. Lee, A.S. Argon, The mechanics of the bending of non-woven fabrics part I: spunbonded fabric (cerex), J. Text. Inst. 74 (1983) 1–11.

[5] T.Y. Nilsson, M.A. Trojer, A solution blown superporous nonwoven hydrogel based on hydroxypropyl cellulose, Soft Matter 16 (2020) 6850–6861.

[6] A. Rawal, P.K. Rao, V. Kumar, A. Kukovecz, A critical review on the absorptive glass mat (AGM) separators synergistically designed via fiber and structural parameters, J. Power Sources 430 (2019) 175–192.

[7] B.S. Jeon, Theoretical orientation density function of spunbonded nonwoven fabric, Text. Res. J. 71 (2001) 509–513.
[8] T. Komori, K. Makishima, Numbers of fiber-to-fiber contacts in general fiber assemblies, Text. Res. J. 47 (1977) 13–17.
[9] L. Berhan, A.M. Sastry, On modeling bonds in fused, porous networks: 3d simulations of fibrous–particulate joints, J. Compos. Mater. 37 (2003) 715–740.
[10] I. Balberg, N. Binenbaum, N. Wagner, Percolation thresholds in the three-dimensional sticks system, Phys. Rev. Lett. 52 (1984) 1465.
[11] C.M. Van Wyk, Note on the compressibility of wool, J. Text. Inst. Trans. 37 (1946) T285–T292.
[12] S. Toll, J.-A. Manson, Elastic compression of a fiber network, J. Appl. Mech. 62 (1995) 223–226.
[13] N. Pan, J. Chen, M. Seo, S. Backer, Micromechanics of a planar hybrid fibrous network, Text. Res. J. 67 (1997) 907–925.
[14] P.K. Rao, A. Rawal, V. Kumar, K.G. Rajput, Compression-recovery model of absorptive glass mat (AGM) separator guided by X-ray micro-computed tomography analysis, J. Power Sources 365 (2017) 389–398.
[15] A. Rawal, A. Priyadarshi, N. Kumar, S.V. Lomov, I. Verpoest, Tensile behaviour of nonwoven structures: comparison with experimental results, J. Mater. Sci. 45 (2010) 6643–6652.
[16] H.L. Cox, The elasticity and strength of paper and other fibrous materials, Br. J. Appl. Phys. 3 (1952) 72.
[17] S. Backer, D.R. Petterson, Some principles of nonwoven fabrics, Text. Res. J. 30 (1960) 704–711.
[18] J.W.S. Hearle, P.J. Stevenson, Studies in nonwoven fabrics. Part IV: prediction of tensile properties, Text. Res. J. 34 (1964) 181–191.
[19] S. Bais-Singh, B.C. Goswami, Theoretical determination of the mechanical response of spun-bonded nonwovens, J. Text. Inst. 86 (1995) 271–288.
[20] H.S. Kim, Orthotropic theory for the prediction of mechanical performance in thermally point-bonded nonwovens, Fibers Polym. 5 (2004) 139–144.
[21] A. Rawal, A. Kochhar, A. Gupta, Biaxial tensile behavior of spunbonded nonwoven geotextiles, Geotext. Geomembr. 29 (2011) 596–599.
[22] A. Rawal, A modified micromechanical model for the prediction of tensile behavior of nonwoven structures, J. Ind. Text. 36 (2006) 133–149.
[23] A. Rawal, A. Priyadarshi, S.V. Lomov, I. Verpoest, J. Vankerrebrouck, Tensile behaviour of thermally bonded nonwoven structures: model description, J. Mater. Sci. 45 (2010) 2274–2284.
[24] O.J. Kallmes, M. Perez, A new theory for the load/elongation properties of paper, in: Consolidation of the Paper Web, Trans. IIIrd Fund. Res. Symp. Cambridge, 1965, pp. 779–800.
[25] L.A. Carlsson, T. Lindstrom, A shear-lag approach to the tensile strength of paper, Combust. Sci. Technol. 65 (2005) 183–189.
[26] M.K. Ramasubramanian, Y. Wang, A computational micromechanics constitutive model for the unloading behavior of paper, Int. J. Solids Struct. 44 (2007) 7615–7632.
[27] V.I. Räisänen, M.J. Alava, K.J. Niskanen, R.M. Nieminen, Does the shear-lag model apply to random fiber networks? J. Mater. Res. 12 (1997) 2725–2732.
[28] A. Rawal, V. Kumar, H. Saraswat, D. Weerasinghe, K. Wild, D. Hietel, M. Dauner, Creating three-dimensional (3D) fiber networks with out-of-plane auxetic behavior over large deformations, J. Mater. Sci. 52 (2017) 2534–2548.

[29] J.P. Giroud, Poisson's ratio of unreinforced geomembranes and nonwoven geotextiles subjected to large strains, Geotext. Geomembr. 22 (2004) 297–305.
[30] A. Rawal, S. Sharma, D. Singh, N.K. Jangir, H. Saraswat, D. Sebők, A. Kukovecz, D. Hietel, M. Dauner, L. Onal, Out-of-plane auxetic nonwoven as a designer metabiomaterial, J. Mech. Behav. Biomed. Mater. 112 (2020) 104069.
[31] A. Rawal, S. Sharma, V. Kumar, P.K. Rao, H. Saraswat, N.K. Jangir, R. Kumar, D. Hietel, M. Dauner, Micromechanical analysis of nonwoven materials with tunable out-of-plane auxetic behavior, Mech. Mater. 129 (2019) 236–245.
[32] J.W.S. Hearle, V. Ozsanlav, Studies of adhesive-bonded non-woven fabrics part I: a theoretical model of tensile response incorporating binder deformation, J. Text. Inst. 70 (1979) 19–28.
[33] J.W.S. Hearle, A. Newton, Nonwoven fabric studies: part XIII: the influence of the binder on the tensile properties of nonwovens, Text. Res. J. 37 (1967) 495–503.
[34] A. Ridruejo, C. González, J. LLorca, A constitutive model for the in-plane mechanical behavior of nonwoven fabrics, Int. J. Solids Struct. 49 (2012) 2215–2229.
[35] P.N. Britton, A.J. Sampson, C.F. Elliott Jr., H.W. Graben, W.E. Gettys, Computer simulation of the mechanical properties of nonwoven fabrics: part I: the method, Text. Res. J. 53 (1983) 363–368.
[36] P.N. Britton, A.J. Sampson, W.E. Gettys, Computer simulation of the mechanical properties of nonwoven fabrics: part II: bond breaking, Text. Res. J. 54 (1984) 1–5.
[37] P.N. Britton, A.J. Sampson, W.E. Gettys, Computer simulation of the mechanical properties of nonwoven fabrics: part III: fabric failure, Text. Res. J. 54 (1984) 425–428.
[38] X.-F. Wu, Y.A. Dzenis, Elasticity of planar fiber networks, J. Appl. Phys. 98 (2005), 093501.
[39] T.H. Grindstaff, S.M. Hansen, Computer model for predicting point-bonded nonwoven fabric strength, part I, Text. Res. J. 56 (1986) 383–388.
[40] O. Jirsák, D. Lukás, Computer modelling of geotextiles related to mechanical properties evaluated by micromechanoscopy, Geotext. Geomembr. 10 (1991) 115–124.
[41] O. Jirsák, D. Lukáš, R. Charvát, A two-dimensional model of the mechanical properties of textiles, J. Text. Inst. 84 (1993) 1–15.
[42] E. Demirci, M. Acar, B. Pourdeyhimi, V.V. Silberschmidt, Computation of mechanical anisotropy in thermally bonded bicomponent fibre nonwovens, Comput. Mater. Sci. 52 (2012) 157–163.
[43] E. Demirci, M. Acar, B. Pourdeyhimi, V.V. Silberschmidt, Finite element modelling of thermally bonded bicomponent fibre nonwovens: tensile behaviour, Comput. Mater. Sci. 50 (2011) 1286–1291.
[44] F. Farukh, E. Demirci, B. Sabuncuoğlu, M. Acar, B. Pourdeyhimi, V.V. Silberschmidt, Characterisation and numerical modelling of complex deformation behaviour in thermally bonded nonwovens, Comput. Mater. Sci. 71 (2013) 165–171.
[45] F. Farukh, E. Demirci, M. Acar, B. Pourdeyhimi, V.V. Silberschmidt, Meso-scale deformation and damage in thermally bonded nonwovens, J. Mater. Sci. 48 (2013) 2334–2345.
[46] F. Farukh, E. Demirci, B. Sabuncuoglu, M. Acar, B. Pourdeyhimi, V.V. Silberschmidt, Mechanical behaviour of nonwovens: analysis of effect of manufacturing parameters with parametric computational model, Comput. Mater. Sci. 94 (2014) 8–16.
[47] F. Farukh, E. Demirci, B. Sabuncuoglu, M. Acar, B. Pourdeyhimi, V.V. Silberschmidt, Numerical analysis of progressive damage in nonwoven fibrous networks under tension, Int. J. Solids Struct. 51 (2014) 1670–1685.
[48] F. Farukh, E. Demirci, B. Sabuncuoglu, M. Acar, B. Pourdeyhimi, V.V. Silberschmidt, Numerical modelling of damage initiation in low-density thermally bonded nonwovens, Comput. Mater. Sci. 64 (2012) 112–115.

[49] F. Farukh, E. Demirci, B. Sabuncuoglu, M. Acar, B. Pourdeyhimi, V.V. Silberschmidt, Mechanical analysis of bi-component-fibre nonwovens: finite-element strategy, Compos. Part B Eng. 68 (2015) 327–335.
[50] X. Hou, M. Acar, V.V. Silberschmidt, 2D finite element analysis of thermally bonded nonwoven materials: continuous and discontinuous models, Comput. Mater. Sci. 46 (2009) 700–707.
[51] X. Hou, M. Acar, V.V. Silberschmidt, Finite element simulation of low-density thermally bonded nonwoven materials: effects of orientation distribution function and arrangement of bond points, Comput. Mater. Sci. 50 (2011) 1292–1298.
[52] B. Sabuncuoglu, M. Acar, V.V. Silberschmidt, A parametric finite element analysis method for low-density thermally bonded nonwovens, Comput. Mater. Sci. 52 (2012) 164–170.
[53] M.N. Silberstein, C.-L. Pai, G.C. Rutledge, M.C. Boyce, Elastic–plastic behavior of nonwoven fibrous mats, J. Mech. Phys. Solids 60 (2012) 295–318.
[54] R.C. Picu, Mechanics of random fiber networks—a review, Soft Matter 7 (2011) 6768–6785.
[55] S. Deogekar, R.C. Picu, On the strength of random fiber networks, J. Mech. Phys. Solids 116 (2018) 1–16.
[56] H. Hatami-Marbini, R.C. Picu, Scaling of nonaffine deformation in random semiflexible fiber networks, Phys. Rev. E 77 (2008), 062103.
[57] V. Negi, R.C. Picu, Mechanical behavior of cross-linked random fiber networks with inter-fiber adhesion, J. Mech. Phys. Solids 122 (2019) 418–434.
[58] T. Liao, S. Adanur, J.-Y. Drean, Predicting the mechanical properties of nonwoven geotextiles with the finite element method, Text. Res. J. 67 (1997) 753–760.
[59] U.H. Erdogan, N. Erdem, Modeling the tensile behaviour of needle punched nonwoven geotextiles, Int. J. Cloth. Sci. Technol. 23 (2011) 258–268.
[60] A. Ridruejo, R. Jubera, C. González, J. LLorca, Inverse notch sensitivity: cracks can make nonwoven fabrics stronger, J. Mech. Phys. Solids 77 (2015) 61–69.
[61] F. Farukh, E. Demirci, H. Ali, M. Acar, B. Pourdeyhimi, V.V. Silberschmidt, Nonwovens modelling: a review of finite-element strategies, J. Text. Inst. 107 (2016) 225–232.
[62] P. Jearanaisilawong, A Continuum Model for Needlepunched Nonwoven Fabrics, PhD Thesis, Massachusetts Institute of Technology, 2008.
[63] A. Rawal, V. Kumar, D. Hietel, M. Dauner, Modulating the Poisson's ratio of articular cartilage via collagen fibril alignment, Mater. Lett. 194 (2017) 45–48.
[64] E. Ban, V.H. Barocas, M.S. Shephard, C.R. Picu, Effect of fiber crimp on the elasticity of random fiber networks with and without embedding matrices, J. Appl. Mech. 83 (2016), 041008.
[65] V. Kumar, A. Rawal, Elastic moduli of electrospun mats: importance of fiber curvature and specimen dimensions, J. Mech. Behav. Biomed. Mater. 72 (2017) 6–13.
[66] G.A. Carnaby, N. Pan, Theory of the compression hysteresis of fibrous assemblies, Text. Res. J. 59 (1989) 275–284.
[67] M.A. Narter, S.K. Batra, D.R. Buchanan, Micromechanics of three-dimensional fibrewebs: constitutive equations, in: Proceedings of the Royal Society of London A: Mathematical, Physical and Engineering Sciences, The Royal Society, 1999, pp. 3543–3563.
[68] A. Rawal, Application of theory of compression to thermal bonded non-woven structures, J. Text. Inst. 100 (2009) 28–34.
[69] I. Krucinska, I. Jalmuzna, W. Zurek, Modified rheological model for analysis of compression of nonwoven fabrics, Text. Res. J. 74 (2004) 127–133.

[70] A.E. Stearn, The effect of anisotropy in the randomness of fibre orientation on fibre-to-fibre contacts, J. Text. Inst. 62 (1971) 353–360.
[71] D.H. Lee, J.K. Lee, Initial compressional behaviour of fibre assembly, in: Objective Measurements: Application to Product Design and Process Control, The Textile Machinery Society of Japan, Osaka, 1985, pp. 613–622.
[72] T. Komori, M. Itoh, A new approach to the theory of the compression of fiber assemblies, Text. Res. J. 61 (1991) 420–428.
[73] D.H. Lee, G.A. Carnaby, Compressional energy of the random fiber assembly: part I: theory, Text. Res. J. 62 (1992) 185–191.
[74] D.H. Lee, G.A. Carnaby, S.K. Tandon, Compressional energy of the random fiber assembly part II: evaluation, Text. Res. J. 62 (1992) 258–265.
[75] T. Komori, M. Itoh, A. Takaku, A model analysis of the compressibility of fiber assemblies, Text. Res. J. 62 (1992) 567–574.
[76] T. Komori, M. Itoh, Theory of the general deformation of fiber assemblies, Text. Res. J. 61 (1991) 588–594.
[77] B. Neckář, Compression and packing density of fibrous assemblies, Text. Res. J. 67 (1997) 123–130.
[78] V. Kumar, A. Rawal, A model for predicting uniaxial compression behavior of fused fibrous networks, Mech. Mater. 78 (2014) 66–73.
[79] P. Grosberg, The strength of twistless slivers, J. Text. Inst. Trans. 54 (1963) T223–T233.
[80] N.B. Beil, W.W. Roberts Jr., Modeling and computer simulation of the compressional behavior of fiber assemblies: part I: comparison to Van Wyk's theory, Text. Res. J. 72 (2002) 341–351.
[81] N.B. Beil, W.W. Roberts Jr., Modeling and computer simulation of the compressional behavior of fiber assemblies: part II: hysteresis, crimp, and orientation effects, Text. Res. J. 72 (2002) 375–382.
[82] R.I.C. Michie, P.J. Stevenson, Nonwoven fabric studies: part VII: the effect of stretching on the mechanical properties of nonwoven fabrics, Text. Res. J. 36 (1966) 494–501.
[83] U. Mohammed, C. Lekakou, L. Dong, M.G. Bader, Shear deformation and micromechanics of woven fabrics, Compos. A: Appl. Sci. Manuf. 31 (2000) 299–308.
[84] K.D. Potter, The influence of accurate stretch data for reinforcements on the production of complex structural mouldings: Part 1. Deformation of aligned sheets and fabrics, Composites 10 (1979) 161–167.
[85] J.W.S. Hearle, R.I.C. Michie, P.J. Stevenson, Nonwoven fabric studies: part V: the dependence of fabric stiffness and drape on the properties of the constituent fiber and binder, Text. Res. J. 34 (1964) 275–282.
[86] C.C. Cheng, K.E. Duckett, Energy losses within sheared fiber assemblies, Text. Res. J. 42 (1972) 51–60.
[87] N. Pan, G.A. Carnaby, Theory of the shear deformation of fibrous assemblies, Text. Res. J. 59 (1989) 285–292.
[88] A. Rawal, Excluded volume and its relation to the theory of fiber-fiber contacts, Mech. Mater. 160 (2021), 103901.
[89] A. Rawal, D. Singh, A. Rastogi, S. Sharma, On the origin of negative Poisson's ratio in buckypapers, Extreme Mech. Lett. 42 (2021) 101059.

Generalised continuum mechanics of random fibrous media

Jean-Francois Ganghoffer[a], *Hilal Reda*[a,b], *and Kamel Berkache*[c,d]
[a]LEM3, Université de Lorraine, CNRS, Metz, France, [b]Faculty of Engineering, Section III, Lebanese University, Beirut, Lebanon, [c]Ecole Supérieure des Sciences Appliquées d'Alger, ESSAA, Algiers, Algeria, [d]Faculty of Physics, University of Science and Technology Houari Boumediene, Algiers, Algeria

3.1 Introduction

A large variety of materials presenting a stochastic fibrous microstructure and constituting the elementary bricks of complex and multifunctional parts exist in nature. Protein is recurrently found in nature in a fibrous form, especially the collagen which is the most abundant in mammals and constitutes the major part of tendons, ligaments, and most of the organic matrix in bone and dentin. It provides mechanical stability, strength, and toughness to these tissues [1]. The structural characteristics of random fibrous networks depend upon the properties of the fibres; thus, their modelling is necessary in order to understand the mechanism of deformation and failure on a system sub-scale, due to the difficulty of measuring the in situ deformation mechanisms of the fibrous microstructure. The tradeoff between local axial stretching and bending deformations of the fibres has an important impact on the overall mechanical response; especially, the response of the network to imposed deformations is likely to be non-affine, the degree of non-affinity being controlled by the fibre bending length, a scalar quantity which quantifies the relative importance of the bending to the stretching stiffnesses, as pointed out in Refs [2–4], who evidenced that the network shifts from the non-affinely deforming structure to an affinely deforming one by increasing the fibre bending length. In Refs [5, 6], the authors further concluded that the degree of heterogeneity decreases by increasing the network density; one of the most important result as to scale effects is that the heterogeneity leads to a strong dependency of the apparent moduli on the size of the probed network domain; this has been modelled in Ref. [3] by evaluating the correlation functions of the tensile modulus versus window size.

As described in Ref. [7], models in the literature developed to simulate the mechanical behaviour of fibrous networks fall into two main categories, namely phenomenological models and micromechanical models. Micromechanical models overcome the shortcomings of the phenomenological models, which very often are not able to capture the relation of the fibre properties to the model parameters. The primary focus of micromechanically based constitutive models of non-woven fibrous networks is the elastic behaviour. Cox (1952) in Ref. [8] was one of the first authors to propose a

Mechanics of Fibrous Networks. https://doi.org/10.1016/B978-0-12-822207-2.00003-9

model for the elastic modulus of paper based on the mechanics of the fibre network, with all fibres extending from one end of the mat to the other, and assuming stretching of the fibres as the dominant deformation mechanism. However, since fibres have a relatively low stiffness and are randomly oriented, bending is an important feature, particularly in the absence of a supporting medium [9,10]. As mentioned in recent contributions [11], the dynamics of periodic fibre networks has recently raised a lot of interest, whereas wave propagation in random fibrous networks has not deserved yet the attention of researchers. Although fibre networks are often subjected to dynamical loadings like vibrations, most of the works have been indeed devoted to the analysis of their static behaviour. The contribution [11] is a pioneering work on the wave propagation in random fibrous networks in the small strains range. The authors show via frequency-domain FE computations performed at the microscopic scale of the entire network that the response is non-dispersive at long wavelengths, whilst it becomes dispersive at intermediate and short wavelengths. In this last situation, the Bloch modes are highly non-affine (the degree of non-affinity is increasing when decreasing the wavelength), most of the deformation localising within the longest network fibres.

Fibrous materials prove to be efficient in noise reduction, and the main mechanisms of sound attenuation in such fibrous networks are the friction between propagating waves and fibre assemblies, leading to a viscous and thermal damping at the interface between fibres and the existing porosities at the microscopic level. Tortuosity is also one important geometrical characteristic responsible for sound attenuation, especially for random fibrous assemblies. The relationship between noise attenuation and various macro- and microstructural parameters has been investigated in recent works [12,13], from both an empirical and theoretical point of view. The parameters influencing wave propagation at the macroscopic scale are principally the bulk density, thickness, porosity, and fibre size. The relations between bulk density and other measured parameters such as porosity, resistivity, tortuosity, viscous and thermal characteristic lengths, and static thermal permeability have been established in the literature. The models developed to evaluate the acoustic damping behaviour of fibrous networks based on microstructural approaches attempt to predict the impact of microscopic physical characteristics, including porosity, tortuosity, fibre density, viscoelasticity, thermal loss, and air motion within the pores.

In this work, we will use generalised continuum theories at an intermediate mesoscopic level in order to address the issue of size effects occurring in the dynamic response of random fibrous microstructures; those theories have been extensively used to explain size effects for a wide class of materials, but not for random fibrous networks, to our knowledge. This constitutes the main originality advocated in the present contribution. Real materials such as biological membranes and tissues often exhibit a number of important length scales, which must be included in any realistic model. We will not analyse noise reduction per se, but we shall provide a quantitative assessment of the impact of RFN (this abbreviation for random fibrous networks will be used here and in the sequel) micro-parameters on the acoustic properties of these materials at the intermediate level of a constructed effective substitution continuum. Thereby, the microstructural parameters will scale up to new internal parameters at the mesoscopic level, including the mean fibre length, the network density, and a characteristic bending length quantifying the importance of bending effects. We shall restrict our analysis of acoustic wave propagation to a single-phase effective continuum (discarding the air and air-fibre network acoustic interactions) endowed with purely elastic effective properties, determined in a first step from a static homogenisation of the mechanical response of the fibrous assembly. From this basis, the properties of the band diagrams will be analysed; especially, the impact of the new parameters of generalised continuum models identified at the continuum level will be assessed.

Many advanced theories and models have been proposed to study wave propagation problems accounting for non-locality and microstructural effects in materials. One category of approaches is multiscale homogenisation techniques [14], which aims at computing asymptotically the solutions of the wave equations involving multiple spatial and temporal scales, and to capture the long-term response of the homogenised response. The mechanics of generalised continua accounts for the non-locality of the elastic fields due to microstructural effects in a macroscopic manner. An overview of the literature on the advanced theories and models proposed to study wave propagation problems accounting for non-locality and microstructural effects in materials is provided in Ref. [15, 16], in which the authors analyse the dynamic properties of periodic textile structures. The two main classes of generalised continuum theories are the higher-grade theories, in which the gradients of strains or the higher-order gradients of the displacement are incorporated; the higher-order theories incorporate additional degrees of freedom and the microrotation gradient is the source of an extra internal energy, and constitute a second class. The reader is referred to Cihan and Onck [17], who proposed a historical overview of generalised continuum theories.

The development of the micropolar theory of elasticity also coined Cosserat theory of elasticity traces back to the seminal work of the Cosserat brothers in 1909 that did, however, not get the attention it deserved for a long time. At the beginning of the 1960s, prominent authors became interested in Cosserat theories [18,19], and a special case of the Cosserat continuum theory was investigated by Koiter [20], in which the rotation of the rigid Cosserat triad is defined in terms of the displacement gradients, deserving the name of couple-stress theory. In a recent work [21], the authors identify the couple-stress moduli of 2D random fibrous networks; this was followed by a more recent contribution of the same authors [22] devoted to the identification of second gradient models for stochastic materials.

We will use in the present work successively the couple-stress and strain gradient theories as a possible modelling framework a priori, in order to analyse size effects of the acoustic properties of random fibrous networks at the level of windows of analysis. Both strategies for enhancing Cauchy elasticity will be compared; especially, the criterion of accounting for the observed dispersive nature of propagating waves will be essential to discriminate between them.

The chapter is organised as follows: the generation of a random set of fibres in a 2D context within windows of analysis is explained in Section 3.2, together with the method used for the identification of the couple-stress moduli based on the strain energy equivalence. The influence of important parameters characterising the network (fibre bending length, network density) on the effective mechanical moduli of the couple-stress substitution medium is studied in Section 3.3. The wave propagation properties of the second gradient homogenised substitution medium are exposed in Section 3.4. The acoustic properties of random fibrous networks are next analysed based on strain gradient models as an alternative framework in Section 3.4, providing a comparative analysis of the acoustic properties obtained by the couple-stress theory. Finally, a summary of the work is exposed in Section 3.5 together with perspectives for future work.

3.2 Model

The reconstruction of the real microstructure and understanding the architecture of random fibrous networks occurring in nature is considered as the important step for their modelisation; this step is essential in evaluation and estimation of the effect of different microstructural parameters in order to construct adequate and predictable structures. The microstructural information of the fibrous networks can be obtained by using different imaging techniques in order to estimate the mechanical properties. In biomechanics, for hard and porous tissues such as bone, we can cite optical coherence tomography and magnetic resonance imaging to obtain 3D images [23–25]. In contrast, for soft tissues the structural images can be obtained from electron, confocal, or multiphoton microscopy [26–29]. Therefore, from these images the architecture and microstructural parameters are then drawn such as fibre diameter, fibre orientation, cross-link nature, and network morphology in general. Detailed information on the geometrical parameters of the network is required if the local response of the system is of interest; otherwise, if we focus on the global average response of the system, models which represent average geometric properties are sufficient.

For biological networks which present approximately a distinct and periodic unit cell such as hexagonal geometry in trabecular bone [30,31] or a triangular one in erythrocyte cytoskeleton [32], models with repeat unit cells may be used and those most encountered to represent the structure of cellular networks are the Voronoi tessellation and Delaunay triangulation models. Other biological networks exhibit stochastic structures, and the random ones are considered as a special case of these structures. With the aim of generating random networks, two approaches can be exploited; the first one is by growing straight fibres with random orientations from randomly distributed seed points generated by the Monte Carlo method integrated in Mathematica kernel; fibre grows with a constant rate and stops growing when it punches another growing fibres or a domain boundary (Fig. 3.1).

The second approach relies on the Mikado model in which the generation of two-dimensional random fibrous networks is made by placing randomly a set of fibres of length, L_0, in a square region of dimension L called the window of analysis (abbreviated WOA in the sequel), so that the random network consists of fibres with uniformly distributed orientation and each passing through a point distributed according to a Poisson point process in a plane [33]. The characterising parameters of the random geometry that have been identified to influence the mechanical behaviour of the

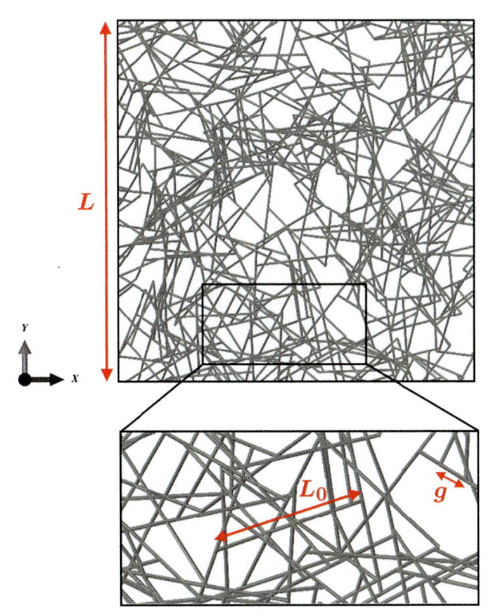

Fig. 3.1 Generation of 2D random fibrous network using the Mikado model.

network are the nature and density of cross links, the distribution of ligament length and orientation, the coordination at cross links, and the fibre density, which is defined as $D = NL_0$, where N is the fibre number density, i.e. the number of fibre centres per unit area. The number of crossings per unit area depends only on the expected fibre density (total fibre length per unit area); also, the expected number of crossings per fibre is directly proportional to this parameter and to the length of fibres [34]. The distribution of ligament lengths in random fibrous network generated by the Mikado model is exponential with mean $E(g) = \overline{g}$ and variance $V(g) = \overline{g}^2$ and given by

$$p(g) = \frac{1}{\overline{g}} \cdot e^{-\frac{g}{\overline{g}}} \tag{3.1}$$

This continuous probability density is obtained from a discrete Poisson probability distribution describing the incidence of cross links in the window of analysis 'WOA in short', and is considered as relatively broad, which indicates that the structure is multiscale.

The coordination plays an important role in network stability and connectivity and networks become more convoluted in the case of a spatial variation of the coordination of the cross links. In biological membranes, the hexagonal, square, and triangular networks with coordination number $z = 3$, $z = 4$, and $z = 6$, respectively, are the most encountered ones. The peptidoglycan network of the bacterial cell is a well-known example of a network with a threefold coordination number and T-shaped junctions, resulting from the assembly of two non-equivalent chains [35]. Note that the

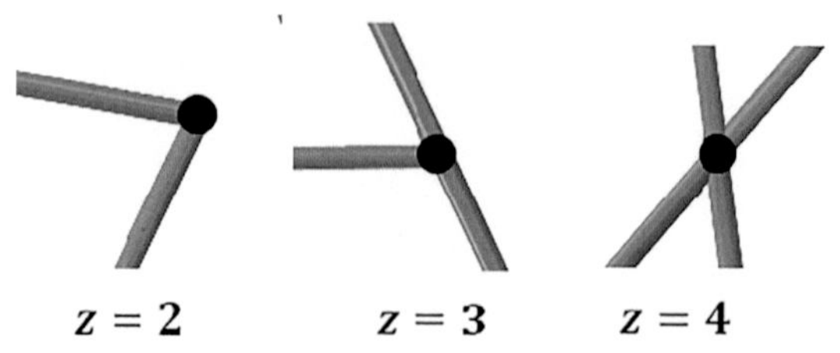

Fig. 3.2 Different coordination numbers for all possible cases in the Mikado model.

knowledge of the mechanical properties of peptidoglycan is also of importance for understanding bacterial growth and form. In our case, cross links are introduced at all points where fibres intersect and for these intersection points the coordination number is $z = 4$. The fibre dangling ends are eliminated as they do not store elastic energy during deformation. The cross links at the ends of fibres have a coordination number $z = 2$ or 3 (Fig. 3.2).

In order to identify a constitutive law that describes efficiently the mechanical behaviour of the biological network and in addition of the morphological data given by previous techniques concerning the architecture and network geometry, one needs to determine the mechanical properties of individual fibres, which constitute the network. The material which constitutes the fibre material is considered as linear elastic with Young's modulus E_f and shear modulus G_f. The ligaments between each pair of cross links are modelled as beam elements and are characterised by stretching, bending, and shear rigidities; $\eta = E_fA$, $\kappa = E_fI$, and $\psi = \gamma G_fA$ ($\gamma = 0.88$ in the case of beams with a circular cross section). We use the Timoshenko beam model because of the occurrence of a large number of short ligaments in the random generation process used to describe the actual structure of biological networks.

The total strain energy stored in the window of analysis is computed numerically as the sum of all strain energies associated with stretching, bending, and shear deformation modes of fibres in the WOA, viz.,

$$U = \frac{1}{2}\sum_{fibres}\int_{0}^{L_0}\left[\eta\left(\frac{du}{ds}\right)^2 + \kappa\left(\frac{d\xi}{ds}\right)^2 + \psi\left(\frac{dv}{ds} - \xi\right)^2\right]ds \tag{3.2}$$

In expression (3.2), the derivative $\frac{du}{ds}$ represents the axial strain at position s along the beam, $\frac{dv}{ds}$ is the rotation of the fibre cross section, and variable ξ represents the rotation of the section perpendicular to the neutral axis of the beam whilst $\frac{dv}{ds} - \xi$ represents the beam shear deformation. The main characteristic lengths that play a role within our window of analysis are the fibre length L_0, the window size L, the fibre bending length which quantifies the importance of the bending to the stretching stiffness $l_b = \sqrt{\kappa/\eta}$, and the mean ligament length $\overline{g}$ which is related directly to the fibre number density N by $\overline{g} = \frac{\pi}{2NL_0} = \frac{\pi}{2D}$. It is interesting to observe that the mean ligament length scales inversely with the density.

3.3 Identification of 2D continuum equivalent moduli based on couple-stress and second gradient theories

3.3.1 Couple-stress substitution continuum

The non-affine deformation character present in the random fibrous networks is due to the energy storage predominantly in the bending deformation mode of fibres [2–4]. The degree of non-affinity increases rapidly with decreasing bending stiffness of the fibres, the importance of which being quantified by the internal bending length l_b, a parameter elaborated from the ratio of the fibre bending modulus to its axial stiffness (E_f is the fibre tensile modulus, I the quadratic moment of inertia, and A the fibre section area). Increasing the level of heterogeneity leads to more pronounced size effects, which have deserved previous works [2]. Size effects can be captured by generalised continuum theories, as it has been demonstrated for a wide class of materials [5,36]. The rotation field inside the WOA of an RFN subjected to pure bending (by finite elements) has been reconstructed from the discrete fibre rotations map, as illustrated in Fig. 3.3 (in the small strains regime), in both affine and non-affine regimes. Abbreviations ADR and NADR are used here and below in place of affine deformation regime and non-affine deformation regime, respectively. The rotation field has been amplified by a factor of 10 for visualisation purposes. The fibre rotation shows (Fig. 3.2) a nearly uniform gradient in the affine deformation regime (left view in Fig. 3.3), since it is controlled by the flexion applied to the boundary of the WOA, whereas the rotation field clearly does not follow the kinematics of the boundary for the non-affine situation (right view in Fig. 3.3). The existence of pronounced microrotations and their gradients therefore motivates the identification of a couple stress substitution continuum for the initial discrete random fibrous network.

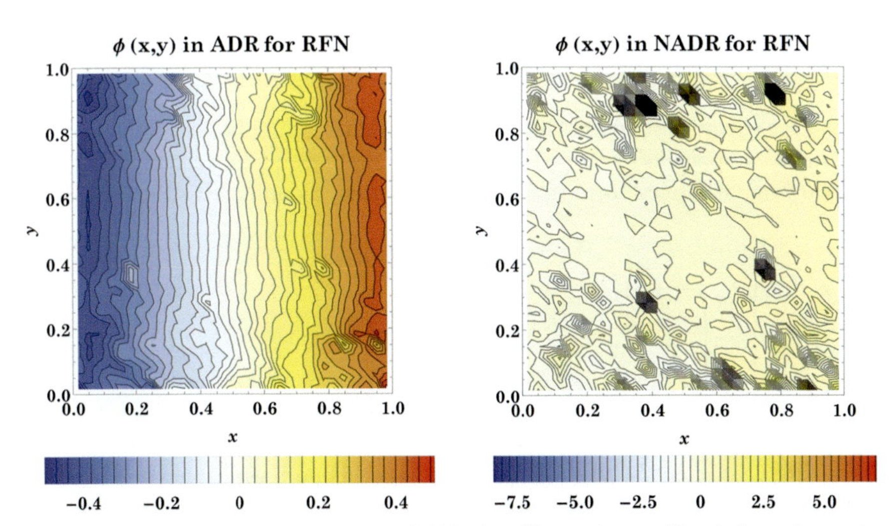

Fig. 3.3 Local distribution of the rotation field in the affine and non-affine deformation regime inside the WOA of a random fibrous network (RFN) subjected to a bending test.

In the micropolar theory, the deformation is described by the displacement vector $\boldsymbol{u}$ and an independent rotation vector $\boldsymbol{\varphi}$, whereas in the couple-stress theory, the rotation vector $\boldsymbol{\varphi}$ is not independent from the displacement vector, since it is given as the anti-symmetric part of the displacement gradient, leading in the present 2D context to the expression of the microrotation around the z out-of-plane axis as

$$\varphi = \frac{1}{2}\left(\frac{\partial v}{\partial x} - \frac{\partial u}{\partial y}\right) \tag{3.3}$$

In the frame of 2D plane stress situation of couple-stress theory, the stress tensor presents four independent components σ_{xx}, σ_{yy}, σ_{xy}, and σ_{yx}, and the couple-stress tensor has two components m_{xz} and m_{yz}. The four independent deformation components and the two independent micro-curvature components κ_{xz} and κ_{yz} express versus the displacement gradients and the micro-rotation as

$$\varepsilon_{xx} = \frac{\partial u}{\partial x},\ \ \varepsilon_{yy} = \frac{\partial v}{\partial y},\ \ \varepsilon_{xy} = \frac{\partial v}{\partial x} - \varphi,\ \ \varepsilon_{yx} = \frac{\partial u}{\partial y} + \varphi,\ \ \kappa_{xz} = \frac{\partial \varphi}{\partial x},\ \ \kappa_{yz} = \frac{\partial \varphi}{\partial y} \tag{3.4}$$

As a result of the kinematic coupling (3.3) in the couple-stress theory, the strain tensor ε_{ij} is symmetrical with components defined as

$$\varepsilon_{xy} = \varepsilon_{yx} = \frac{1}{2}\left(\frac{\partial v}{\partial x} + \frac{\partial u}{\partial y}\right)$$

Ignoring body forces and body moments, the dynamical equilibrium in translation and rotation, is written as the set of two equations

$$\frac{\partial \sigma_{xx}}{\partial x} + \frac{\partial \sigma_{xy}}{\partial y} = 0,\ \ \frac{\partial m_{xz}}{\partial x} + \frac{\partial m_{yz}}{\partial x} + \sigma_{xy} - \sigma_{yx} = 0 \tag{3.5}$$

The balance equation of internal bending momentum (3.5) implies the equality of both shear stress components $\sigma_{xy} = \sigma_{yx}$. Thus, the constitutive equation can be expressed in the following uncoupled form (for a centrally symmetric unit cell structure) as

$$\begin{Bmatrix} \sigma_{xx} \\ \sigma_{yy} \\ \sigma_{xy} \\ m_{xz} \\ m_{yz} \end{Bmatrix} = \begin{pmatrix} A_{11} & A_{12} & 0 & 0 & 0 \\ A_{12} & A_{22} & 0 & 0 & 0 \\ 0 & 0 & A_{33} & 0 & 0 \\ 0 & 0 & 0 & D_{11} & 0 \\ 0 & 0 & 0 & 0 & D_{22} \end{pmatrix} \begin{Bmatrix} \varepsilon_{xx} \\ \varepsilon_{yy} \\ \varepsilon_{xy} \\ \kappa_{xz} \\ \kappa_{yz} \end{Bmatrix} \tag{3.6}$$

wherein the coefficients A_{ij} are the classical Cauchy moduli, whilst the coefficients D_{ij} are the micropolar moduli relating the two independent non-zero couple-stress components to the corresponding curvatures. The effective Young's moduli and Poisson ratios can be expressed versus coefficients A_{ij} as:

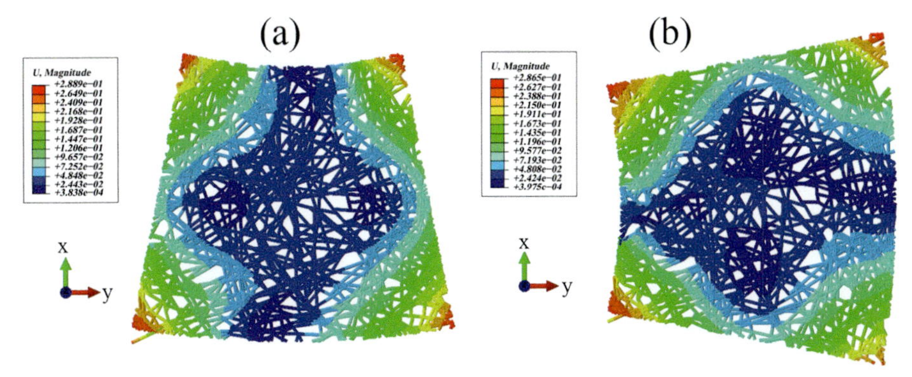

Fig. 3.4 Displacement fields corresponding to boundary conditions applied to identify the effective constitutive coefficients (A) D_{11} and (B) D_{22}.

$$E_x = A_{11} - \frac{{A_{12}}^2}{A_{22}}, \quad E_y = A_{22} - \frac{{A_{12}}^2}{A_{11}}, \quad \nu_{xy} = \frac{A_{12}}{A_{22}}, \quad \nu_{yx} = \frac{A_{12}}{A_{11}}$$

We determine the effective moduli of the couple-stress continuum from the response of random fibrous networks within so-called windows of analysis (WOA) of different sizes. We design different boundary conditions for the determination of the independent components of the constitutive constants over a domain Ω with boundary $\partial\Omega$. An example of the deformed shape of the WOA under a given macroscopic kinematic load triggering a specific couple stress rigidity coefficient is plotted in Fig. 3.4. In each case, we force the WOA to bear a set of specific deformation, as described in Ref. [21], and compute numerically the total elastic strain energy U_{WOA} stored in the WOA under the corresponding boundary conditions.

The numerical procedure used here is similar to the one used in Ref. [21]. The total strain energy stored in the WOA is equated to the energy of an equivalent homogeneous couple-stress continuum based on the extended Hill macrohomogeneity condition; thus, it holds the identity

$$U_{WOA} = U_{CS} = \frac{V}{2}\left(\varepsilon_{ij} A_{ijkl} \varepsilon_{kl} + \kappa_{ij} D_{ijkl} \kappa_{kl}\right) \tag{3.7}$$

In Eq. (3.7), V represents the volume of the WOA. The strain energy stored in the effective homogeneous couple-stress continuum can be obtained by the prescribed displacement fields. The fibre bending length is defined as the ratio between the axial stiffness to the bending stiffness, $l_b = \sqrt{E_f I / E_f A}$. In Fig. 3.5, we plot the evolution of the classical and couple-stress moduli versus the fibre bending length l_b in logarithmic axes, for a constant network density. Low values of l_b/L_0 enhance local rotations of the fibres, which do not follow the imposed deformation over the boundary of the window of analysis; thus, the network responds essentially in a non-affine manner; opposite to this, high values of l_b lead to a rather affine response, whilst intermediate l_b values correspond to the transition regime. The variation of the four classical moduli with

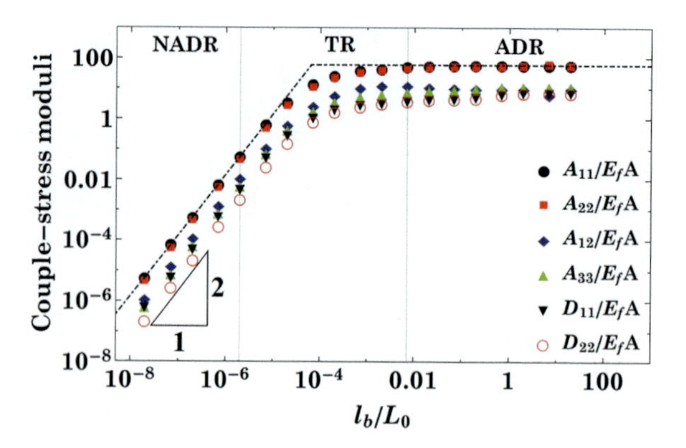

Fig. 3.5 Variation of classical elastic moduli with normalised fibre bending length l_b/L_0 for a constant normalised network density $DL_0 = 75$.

l_b is shown in Fig. 3.4; the vertical axis is normalised with the tensile rigidity of the fibres, quantity E_fA, and the horizontal axis is normalised by the fibre length L_0. In these computations, the network density is kept constant at $kL_0 = 75$. Note that the variable in the horizontal axis is proportional to the aspect ratio of fibres. The acronyms ADR, NADR, and TR in Fig. 3.4 stand for the affine deformation regime, the non-affine deformation regime, and the transition regime, respectively. The transition from affine to non-affine regimes is controlled by a number of parameters: l_b, the density (itself inversely proportional to the distance between cross links), and the coordination number [6].

For large l_b/L_0 values the classical moduli are proportional to E_fA; thus, the strain energy is stored predominantly in the axial deformation mode of fibres and the deformation field is approximately affine. At small l_b/L_0 ratio, the classical moduli are proportional to the mechanical parameter $E_fA\left(\frac{l_b}{L_0}\right)^2 \propto E_fI$, and accordingly the strain energy is stored predominantly in the bending deformation mode of fibres and the deformation is non-affine. Interestingly, the non-classical moduli D_{11} and D_{22} exhibit the same behaviour as the classical moduli. Furthermore, the transition from ADR to NADR takes place in the same range of values of l_b/L_0. The way the local deformation mechanisms in each of these two regimes will affect the overall dynamic properties of the network is one essential issue analysed in this contribution. In order to investigate the anisotropic dynamic behaviour of the networks, we quantify in Fig. 3.6 the state of anisotropy for the classical and couple-stress coefficients. The relative variation of the classical tensile moduli, quantity $\eta = \frac{A_{11}-A_{22}}{A_{22}}$, is defined as the anisotropy measure, to quantify the degree of anisotropy for the classical part of the constitutive law, as pictured in Fig. 3.6A.

Fig. 3.6 shows that the network is nearly isotropic, since the previously defined anisotropic measure η does not exceed 12% for the classical modulus and the deviation from isotropy is 5% for the couple-stress moduli. We can thus conclude that the generated fibrous networks are nearly isotropic (up to statistical fluctuations) for all large enough WOAs. The dynamical response of such fibrous networks can be evaluated either based on computations done at the microscopic level [11], or at the scale of

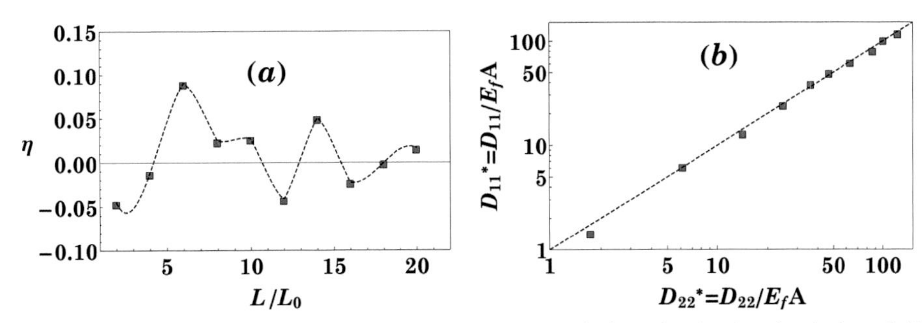

Fig. 3.6 (A) Variation of the anisotropy measure versus window size for the classical moduli and (B) for the couple-stress moduli.

the homogenised continuum. The first approach however involves huge computations due to the large number of fibres (a few thousands); so, we shall instead follow the second route in this contribution, and aim at capturing the influence of the microstructure (which controls the dynamic response of the system) by a suitable enhancement of Cauchy elasticity, considering successively higher-order and higher-grade effective continua as two possible enrichment of the Cauchy continuum.

3.3.2 Second gradient substitution continuum

In classical continuum mechanics, only the first displacement gradient is of importance and all the higher-order displacement gradients are neglected in measuring the deformations of a body. In this case, the stress tensor at a material point is linked to the strain tensor through the classical elasticity tensor. The second gradient elasticity is a kinematic enhancement of classical elasticity taking into account the second gradient of the displacement field. The strain gradient theory is next exploited in terms of the strain (first displacement gradient) and second gradient of the displacement field which are, respectively, the second- and third-order tensors:

$$\begin{aligned} \boldsymbol{\varepsilon} &= \frac{1}{2}\left(\boldsymbol{u}\otimes\nabla + \boldsymbol{u}\otimes\nabla^T\right) \Leftrightarrow \varepsilon_{ij} = \frac{1}{2}\left(u_{i,j} + u_{j,i}\right) \\ \boldsymbol{K} &= \boldsymbol{u}\otimes\nabla\otimes\nabla \Leftrightarrow K_{ijk} = u_{i,jk} \end{aligned} \tag{3.8}$$

The classical infinitesimal strain tensor $\boldsymbol{\varepsilon}$ therein is the symmetric part of the displacement gradient, with three independent components in the present 2D situation. The second gradient of the displacement field, tensor $\boldsymbol{K}$, is symmetric in the last two indices and it has thus six independent components, due to the symmetries $\varepsilon_{ij} = \varepsilon_{ji}$ and $K_{ijk} = K_{ikj}$. In 2D, the displacement field is the vector $\boldsymbol{u} = [u, v]^T$, which entails the following strain tensor written here in vector format

$$\boldsymbol{\varepsilon} = \left[\frac{\partial u}{\partial x}, \frac{\partial v}{\partial y}, \frac{1}{2}\left(\frac{\partial u}{\partial y} + \frac{\partial v}{\partial x}\right)\right]^T \tag{3.9}$$

Similarly, the second displacement gradient tensor can be written in vector format from the second gradient of the displacement field as

$$\boldsymbol{K} = \left[\frac{\partial^2 u}{\partial x^2}, \frac{\partial^2 v}{\partial y^2}, \frac{\partial^2 u}{\partial y^2}, \frac{\partial^2 v}{\partial x^2}, \frac{\partial^2 u}{\partial x \partial y}, \frac{\partial^2 v}{\partial x \partial y}\right] \tag{3.10}$$

In the strain-gradient theory of linear elasticity, the constitutive law involves the symmetric Cauchy stress tensor $\boldsymbol{\sigma}$ and the hyper-stress (or double stress) tensor $\boldsymbol{\sigma}^S$. The tensors $\boldsymbol{\sigma}$ and $\boldsymbol{\sigma}^S$ are related to the strain tensor $\boldsymbol{\varepsilon}$ and the second displacement gradient tensor $\boldsymbol{K}$ through the following general constitutive law for a homogeneous anisotropic second-order grade continuum:

$$\begin{aligned} \sigma_{ij} &= A_{ijlm}\varepsilon_{lm} + M_{ijlmn}K_{lmn} \\ \sigma^S_{ijk} &= M_{ijklm}\varepsilon_{lm} + D_{ijklmn}K_{lmn} \end{aligned} \tag{3.11}$$

with A_{ijlm} therein the classical fourth-order elastic tensor, D_{ijklmn} the sixth-order tensor of elastic moduli, and M_{ijlmn} the fifth-order coupling tensor between the first and second-order elastic responses which does not vanish for non-centrosymmetric microstructures. For microstructures exhibiting central symmetry – which we assume here and in the sequel – the fifth-order coupling elastic stiffness tensor M vanishes, so that the previous constitutive law takes the simplified form

$$\begin{aligned} \sigma_{ij} &= A_{ijlm}\varepsilon_{lm} \\ \sigma^S_{ijk} &= D_{ijklmn}K_{lmn} \end{aligned} \tag{3.12}$$

The stress and hyper-stress for the effective 2D second-order grade continuum can be defined in vector format as

$$\begin{aligned} \boldsymbol{\sigma} &= \left[\sigma_{xx}, \sigma_{yy}, \sigma_{xy}\right]^T \\ \boldsymbol{\sigma}^S &= \left[\sigma^S_{xxx}, \sigma^S_{yyy}, \sigma^S_{xyy}, \sigma^S_{yxx}, \sigma^S_{xxy}, \sigma^S_{yxy}\right]^T \end{aligned} \tag{3.13}$$

The main purpose of this section is to determine the effective constitutive coefficients of the strain-gradient continuum from the response of random fibre networks; the abbreviation SG is conveniently used in the sequel to denote the employed effective second gradient continuum. We next design a set of boundary conditions for the sequential determination of the components of the constitutive (rigidity) constants over various 2D domains Ω with boundary $\partial\Omega$. In each case, we force the WOA to bear a set of specific deformation and deformation gradients, as detailed in Ref. [22], and compute numerically the total elastic strain energy U_{WOA} stored in the window of analysis under the corresponding boundary conditions. An example of determination of the effective constitutive corresponding to the hyperstress coefficient is plotted in Fig. 3.7.

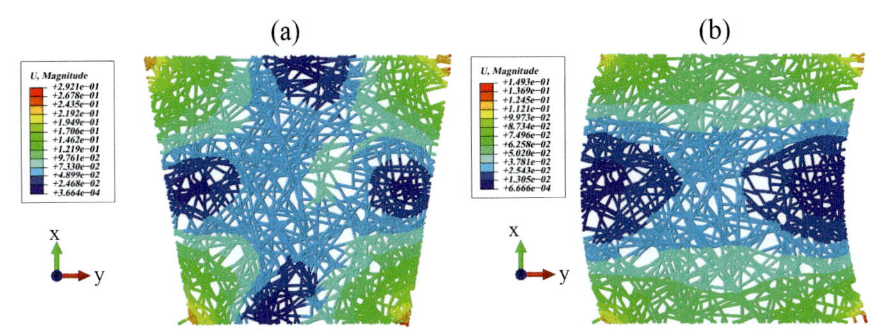

Fig. 3.7 Displacement fields corresponding to boundary conditions applied to identify the effective constitutive coefficients (A) D_{112} and (B) D_{122}.

Similar to the identification procedure for the couple-stress continuum, the first- and second-order gradient elastic moduli are computed by equating the total strain energy stored in the RVE with the energy of an equivalent homogeneous strain-gradient continuum, viz.,

$$U_{WOA} = U_{SG} = \frac{V}{2}\left[\varepsilon_{ij}A_{ijkl}\varepsilon_{kl} + K_{ijk}D_{ijklmn}K_{lmn}\right] \tag{3.14}$$

The left-hand side in (3.14) is the total elastic strain energy stored in the window of analysis, whilst the right-hand side is the expression of the energy of the postulated effective strain-gradient continuum. The characteristic lengths are essential parameters for the second gradient continua; we generalise the definition of these parameters to an anisotropic continuum in terms of the engineering constants. In 3D, the six internal lengths associated to the independent classical moduli A_{ij} can be identified by the expressions [22]:

$$l_{\alpha\beta} = \left(\sum_{r=1}^{3} \frac{D_{\alpha r\beta r}}{A_{\alpha\beta}}\right)^{\frac{1}{2}} \tag{3.15}$$

The previous equation can be simplified in the present 2D context, resulting in three internal lengths in tension and shear associated to the independent classical and second-order elastic moduli defined as follows:

$$l_{xx} = \left(\frac{D_{1111} + D_{1212}}{A_{11}}\right)^{\frac{1}{2}}, l_{yy} = \left(\frac{D_{2121} + D_{2222}}{A_{22}}\right)^{\frac{1}{2}}, l_{xy} = \left(\frac{D_{1121} + D_{1222}}{A_{33}}\right)^{\frac{1}{2}} \tag{3.16}$$

The variation of these internal lengths with normalised window size L/L_0 is shown in Fig. 3.8, corresponding to ADR and NADR. For $l_b/L_0 = 3.5 \times 10^{-2}$ (Fig. 3.8A), the normalised internal lengths become model size independent once $L/L_0 > 2$. Similarly, in the non-affine regime, for $l_b/L_0 = 2 \times 10^{-7}$, the internal length become model size independent once $L/L_0 > 12.5$. In the second part of this chapter, the impact of the identified higher-order mechanical properties on wave propagation features within RFNs will be assessed.

3.4 Wave propagation analysis

3.4.1 Equivalent couple stress continuum

The dynamical equilibrium equations for the identified effective couple-stress medium corresponding to the two independent degrees of freedom u, v in the present planar situation are written successively as the two independent equations (using the balance equation of momentum to express the following derivative of the shear stress components $\sigma_{xy,x}$, $\sigma_{yx,y}$ which are then inserted back into the balance of linear momentum):

$$\begin{aligned} \sigma_{xx,x} + \sigma_{yx,y} - m_{xz,xy} - m_{yz,yy} &= \rho^* \ddot{u} \\ \sigma_{xy,x} + \sigma_{yy,y} - m_{xz,xx} - m_{yz,yx} &= \rho^* \ddot{v} \end{aligned} \tag{3.17}$$

with $\rho^* = \frac{M_1}{A_{WOA}}$ is the effective density, M_1 is the mass of the fibrous microstructure, and A_{WOA} the area of the WOA. Note that we have reduced the set of initially three dynamical equations to a set of two truly independent dynamical equations in (3.17). Inserting the constitutive law, and adopting a planar harmonic wave Ansatz for the solution of the dynamical equilibrium equation, the wave motion equations are further written as

$$\begin{pmatrix} A_{11}k_1^2 + A_{33}k_2^2/2 - D_{11}k_1^2k_2^2/2 - D_{22}k_2^4/2 - \rho^*\omega^2 & k_1k_2\left(2A_{12} + A_{33} + D_{22}k_2^2 + D_{11}k_1^2\right)/2 \\ k_1k_2\left(2A_{12} + A_{33} + D_{22}k_2^2 + D_{11}k_1^2\right)/2 & A_{22}k_2^2 + A_{33}k_1^2/2 - D_{22}k_1^2k_2^2/2 - D_{11}k_1^4/2 - \rho^*\omega^2 \end{pmatrix} \begin{pmatrix} \hat{u} \\ \hat{v} \end{pmatrix} = \begin{pmatrix} 0 \\ 0 \end{pmatrix} \tag{3.18}$$

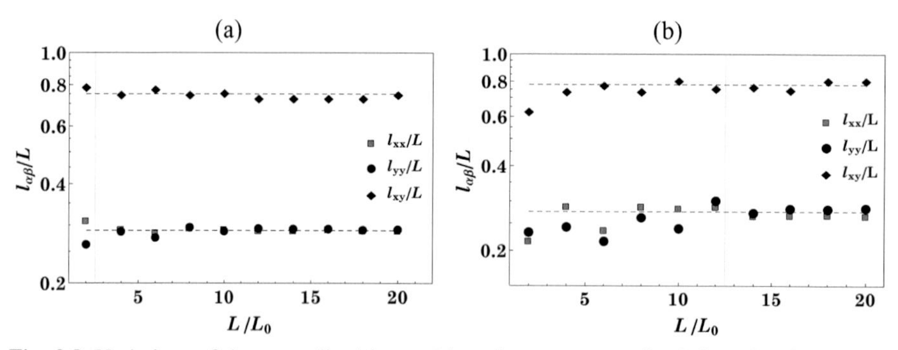

Fig. 3.8 Variations of the normalised internal length versus normalised size of WOA in (A) affine deformation regime and (B) non-affine deformation regime. The vertical bar shows the lower limit of model size which insures model size independence of the classical moduli.

The determinant $\Delta(\omega, k_1, k_2)$ of the matrix in the left-hand side of Eq. (3.18) must be zero to have non-trivial solutions; the obtained positive roots characterise the dispersion relations for planar wave propagation. Two modes of wave propagation exist, namely the longitudinal mode (designated by L), and the shear mode (labelled S). The structure of the coefficients of the wave motion matrix in (3.18) shows different powers of the wave vector components; by comparison, for the Cauchy medium, the frequency and the wave number only have the same quadratic powers in the wave motion equation; thus, the medium is non-dispersive. The phase and group velocities can be expressed as follows

$$c^p = \frac{\omega}{|k|}, c^g = \left(\frac{\partial \omega}{\partial k_1}, \frac{\partial \omega}{\partial k_2}\right) \tag{3.19}$$

In the case of wave propagation in the longitudinal direction (x direction in the Cartesian basis), the dispersive longitudinal and transverse modes have frequencies expressing versus the wave number and effective moduli as

$$\omega_1 = \left(\frac{(2A_{11}+A_{33})k_1^2 - D_{11}k_1^4 + \left(D_{11}^2k_1^8 + 4D_{11}A_{11}k_1^6 - 2D_{11}A_{33}k_1^6 + 4A_{11}^2k_1^4 - 4A_{11}A_{33}k_1^4 + A_{33}^2k_1^4\right)^{\frac{1}{2}}}{\rho^*}\right)^{\frac{1}{2}}$$
$$\omega_2 = \left(\frac{(2A_{11}+A_{33})k_1^2 - D_{11}k_1^4 - \left(D_{11}^2k_1^8 + 4D_{11}A_{11}k_1^6 - 2D_{11}A_{33}k_1^6 + 4A_{11}^2k_1^4 - 4A_{11}A_{33}k_1^4 + A_{33}^2k_1^4\right)^{\frac{1}{2}}}{\rho^*}\right)^{\frac{1}{2}} \tag{3.20}$$

Expressions (3.20) show that the two propagation modes are controlled by the components of the rigidity matrix in traction along the x direction, the shear components along the y direction, and the couple-stress components along the x direction. Nonlinear relations are obtained between frequency and wave number, and the frequency reaches a plateau for large values of the wave number. For in-plane wave propagation ($k_1 \neq 0, k_2 \neq 0$), the presence of the shear strain components in the two tensile stress components σ_{xx}, σ_{yy} introduces a coupling between the shear and tensile effective moduli, which in turn impact wave propagation for the longitudinal and shear modes. In the sequel, we shall analyse the influence of fibre bending length l_b, network density D, and window size L on wave propagation within the network. We introduce the following non-dimensional parameters:

The dimensionless wave number	kL_0
The dimensionless frequency	$\frac{\omega L_0}{\sqrt{E/\rho}}$
The dimensionless phase and group velocities, respectively	$\frac{c^p}{\sqrt{E/\rho}}, \frac{c^g}{\sqrt{E/\rho}}$
The dimensionless fibre bending length	$\frac{l_b}{L_0}$

E, ρ, and L_0 are the Young's modulus, density of fibres and fibre length, respectively.

The frequency band structure can be obtained based on the previous methodology, relying on the computed effective moduli of the homogenised couple-stress continuum. In all subsequent plots, the continuous line corresponds to wave propagation in the longitudinal mode, whilst the dashed line corresponds to the shear mode. Note that all frequency band structures presented in this work focus on the direction of propagation $\pi/4$.

3.4.1.1 Influence of fibre bending length on the dispersion relation and on phase and group velocities

We shall first investigate the effect of the fibre bending length l_b on the dispersion relation, and evaluate the phase velocity; recall that l_b is defined as the ratio of the bending stiffness to the axial fibre stiffness. Fig. 3.9 displays the band structure of the random fibrous medium with a normalised network density $DL_0 = 50$, for different values of l_b. Obviously, an increase of the partial band gap with frequency occurs between the two modes, until it becomes constant when moving from the non-affine deformation regime (at small l_b) to the affine deformation regime (at large l_b, when the macroscopic deformation becomes very close to the microscopic deformation).

We also observe in Fig. 3.9 the stabilisation of the shear mode beyond the value $kL_0 = 0.8$ in the affine regime (when $l_b/L_0 = 20$), noting that the other modes tend to an asymptote beyond certain values of k, based on previous expressions (3.9). The same evolutions of the frequency are obtained in Ref. [11], in which the authors perform finite element analysis of the dispersion relations at the fibre level. Fig. 3.10 illustrates the evolution of the modulus of the phase velocities in the form of polar plots, for three different values of the fibre bending length l_b in order to highlight the effect of this parameter on the anisotropic dynamic behaviour and the dispersive

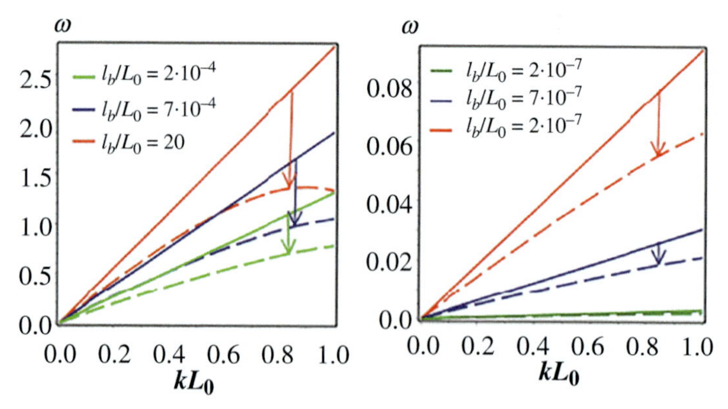

Fig. 3.9 Frequency band structure for the random fibrous medium versus wave number and normalised fibre bending length for a normalised network density $DL_0 = 50$ and for the longitudinal and shear modes. *Continuous line*: longitudinal mode. *Dashed line*: shear mode.

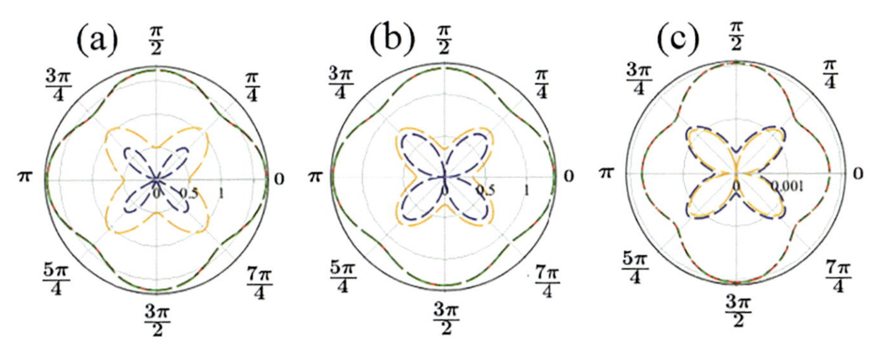

Fig. 3.10 Modulus of the phase velocity for three values of normalised fibre bending length and for normalised network density $DL_0 = 50$, for the two modes of propagation, (A) $l_b/L_0 = 20$, (B) $l_b/L_0 = 2 \times 10^{-3}$ and (C) $l_b/L_0 = 2 \times 10^{-7}$. *Red* and *green lines*, respectively: longitudinal mode for $kL_0 = 1.5$ and $kL_0 = 0.5$. *Orange* and *blue lines*, respectively: shear mode for $kL_0 = 1.5$ and $kL_0 = 0.5$.

characteristics of the network. Two different values of the wave number are selected in Fig. 3.10 in the analysis of the dispersive behaviour of the couple-stress medium.

The anisotropic nature of wave propagation for the longitudinal and shear modes is evidenced by the corresponding irregular shape of the phase velocity plot (Fig. 3.10), despite the isotropy of the static mechanical behaviour of the effective couple-stress medium previously shown in Fig. 3.6. Works from the literature [37–39] show that anisotropy of wave propagation changes with frequency, and is not in conflict with the static isotropy of the network. Furthermore, Fig. 3.10 shows that the state of anisotropy is not affected by the fibre bending length and wave number. In the longitudinal mode, the medium is non-dispersive as shown in Fig. 3.10 (there is no change in the phase velocity when wave number increases), whilst the shear mode is dispersive (the phase velocity increases with wave number). This behaviour can be explained by the additional rotational degree of freedom ϕ of the couple-stress theory which leads to second gradient terms in flexion and shear and thus affecting the shear modes, but not the longitudinal mode (which is independent from φ). The dispersive behaviour of the shear modes in the couple-stress theory is the basic difference in comparison with Cauchy medium which is non-dispersive for both longitudinal and shear modes.

3.4.1.2 Influence of the network density on the dispersion relation and on the phase velocity

It is well recognised that the dispersion relation and phase velocity depend on the static properties of the fibres. In order to assess the effect of the network density on the dispersion diagram, we represent in Fig. 3.11 the frequency band structure versus wave number for different network density values in both affine and non-affine regimes.

In the non-affine regime, an increase in frequency and partial band gap width occurs; in contrast to this, for affine deformations, the frequency and the width of

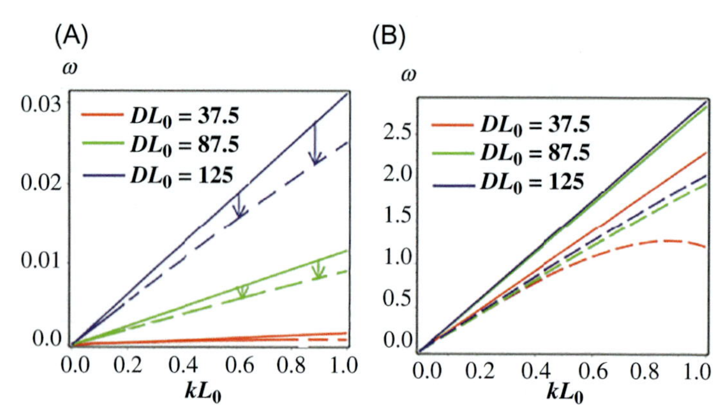

Fig. 3.11 Frequency band structure versus wave number for different network densities. (A) Non-affine regime and (B) affine regime.

the partial band gap remain constant, and all data converge to a horizontal asymptote (stabilisation effect), so that the long wavelength speeds of the networks become independent of network density.

The anisotropic behaviour for the longitudinal and shear modes is highlighted by the polar plot of the modulus of the phase velocity for different values of the network density (Figs 3.12 and 3.13), in the affine and non-affine regimes, for two values of the wave number. We conclude from these plots that the anisotropic behaviour is the same in both regimes, and it is furthermore independent from network density. The degree of anisotropy becomes higher for the longitudinal and shear modes when moving from the affine to the non-affine regime, whatever the value of network density. In the non-affine regime, the medium remains non-dispersive for the longitudinal modes and

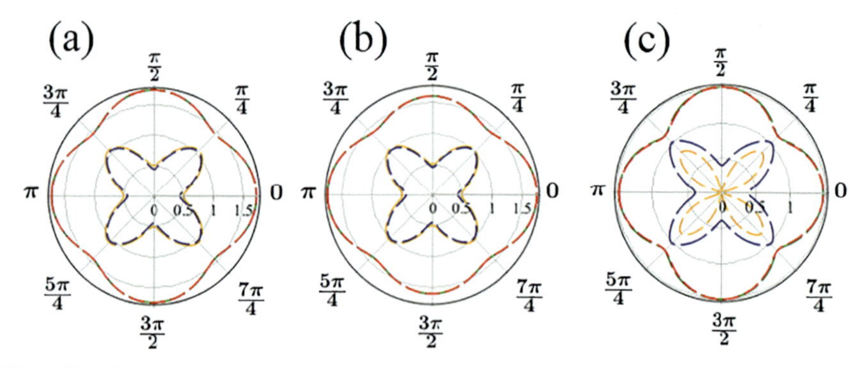

Fig. 3.12 Modulus of phase velocities for three values of normalised network density in the affine regime ($l_b/L_0 = 0.02$) for the two modes of propagation, (A) $L_0 = 125$, (B) $DL_0 = 87.5$, and (C) $DL_0 = 37.5$. *Red* and *green lines*, respectively: longitudinal mode for $kL_0 = 1.5$ and $kL_0 = 0.5$. *Orange* and *blue lines*, respectively: shear mode for $kL_0 = 1.5$ and $kL_0 = 0.5$.

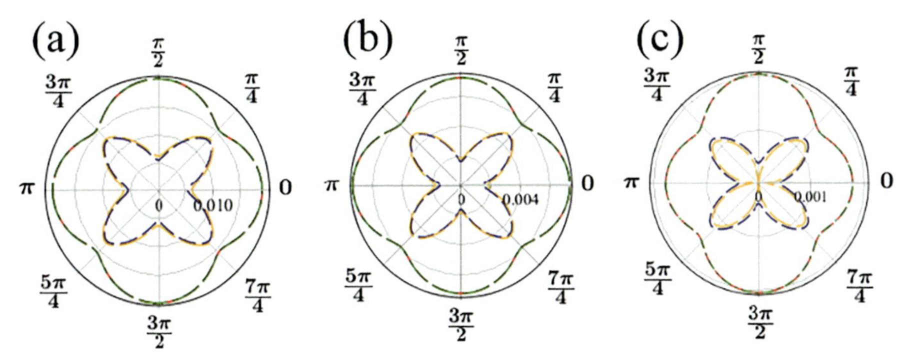

Fig. 3.13 Modulus of phase velocities for three values of normalised network density in the affine regime ($l_b/L_0 = 2 \times 10^{-7}$) for the two modes of propagation, (A) $L_0 = 125$, (B) $DL_0 = 87.5$, and (C) $DL_0 = 37.5$. *Red* and *green lines*, respectively: longitudinal mode for $kL_0 = 1.5$ and $kL_0 = 0.5$. *Orange* and *blue lines*, respectively: shear mode for $kL_0 = 1.5$ and $kL_0 = 0.5$.

dispersive for the shear mode when network density increases. In the affine regime, the medium moves from a dispersive to a non-dispersive response as the network density increases: this can be explained by the fact that in the affine regime and for increasing network density, the macroscopic deformation is close to the microscopic deformation, which entails that the medium behaves as an effective Cauchy medium showing no influence of the microstructure (no dispersion).

In both affine and non-affine regimes, shear waves are dispersive whereas longitudinal waves are not dispersive for the reasons mentioned before. It shall be pointed out that there is in fact no unique choice for the effective substitution continuum of the initial random fibrous network, since computations performed at the microscopic level of the lattice reveal both important local rotations of the individual fibres, as well as noticeable internal strain gradients. Models based on couple-stress type theory do however not give realistic predictions of the effective medium properties, such as the dispersion relation (since the longitudinal mode is non-dispersive). It has however been proven by experiments that most waves are dispersive [40]. Accordingly, we explore in the sequel an alternative and more general modelling strategy, considering an enhancement of the Cauchy continuum by higher strain gradients.

3.4.2 Equivalent second gradient continuum

The equations of motion for an effective second gradient continuum are easily obtained; they can be written in components form as the two following differential equations along the x and y directions of the Cartesian coordinates system:

$$\frac{\partial \sigma_{xx}}{\partial x} + \frac{\partial \sigma_{xy}}{\partial y} - \frac{\partial^2 \sigma^S_{xxx}}{\partial x^2} - \frac{\partial^2 \sigma^S_{xxy}}{\partial x \partial y} - \frac{\partial^2 \sigma^S_{xyx}}{\partial x \partial y} - \frac{\partial^2 \sigma^S_{xyy}}{\partial y^2} = \rho^* \ddot{u} \tag{3.21}$$

$$\frac{\partial \sigma_{yx}}{\partial x}+\frac{\partial \sigma_{yy}}{\partial y}-\frac{\partial^2 \sigma^S_{yxx}}{\partial x^2}-\frac{\partial^2 \sigma^S_{yxy}}{\partial x \partial y}-\frac{\partial^2 \sigma^S_{yyx}}{\partial x \partial y}-\frac{\partial^2 \sigma^S_{yyy}}{\partial y^2}=\rho^* \ddot{v} \tag{3.22}$$

The second-order time derivatives $\ddot{u}$ and $\ddot{v}$ therein are the horizontal and vertical components of the acceleration vector. Relying on the plane harmonic wave solutions of the dynamical equations leads to a wave motion equation describing the propagation of longitudinal and shear waves, written in compact form as

$$\begin{pmatrix} A_{11}k_1^2+A_{33}k_2^2/2+D_{111}k_1^4+D_{122}k_2^4+D_{112}k_2^2k_1^2-\rho^*\omega^2 & k_1k_2(2A_{12}+A_{33})/2 \\ k_1k_2(2A_{12}+A_{33})/2 & A_{22}k_2^2+A_{33}k_1^2/2+D_{211}k_1^4+D_{222}k_2^4+D_{212}k_2^2k_1^2-\rho^*\omega^2 \end{pmatrix}\begin{pmatrix}\hat{u}\\ \hat{v}\end{pmatrix}=\begin{pmatrix}0\\0\end{pmatrix} \tag{3.23}$$

Nontrivial solutions of the last equation exist if the determinant $\Delta(\omega, k_1, k_2)$ of the matrix on the left-hand side of Eq. (3.23) vanishes, which shall provide the dispersion relations. We shall next investigate the effect of fibre bending length l_b, fibre network density and WOA size on the dispersion relations, and compare them with those obtained for the couple-stress effective medium.

3.4.2.1 Dispersion relations and phase velocity for the second-order effective continuum versus fibre bending length l_b. Comparison with the couple-stress effective medium

We evaluate in Fig. 3.14 the frequency for the homogenised second gradient medium versus the normalised fibre bending length, and compare with the frequency evaluated for the couple-stress effective medium. Inspection of Fig. 3.14 shows that the influence of fibre bending length is the same for both couple-stress and second gradient theories, the partial band gap increasing with l_b for both models, whilst the existing small shifts for the L and S modes increase with l_b.

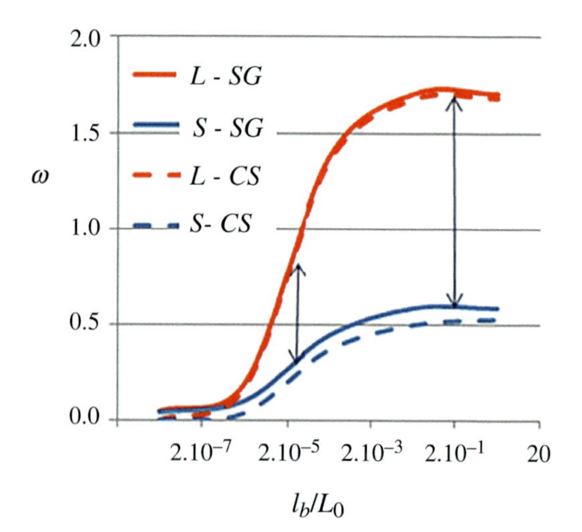

Fig. 3.14 Frequency versus normalised fibre bending length for $kL_0 = 0.5$. Comparison between second gradient and couple-stress theories.

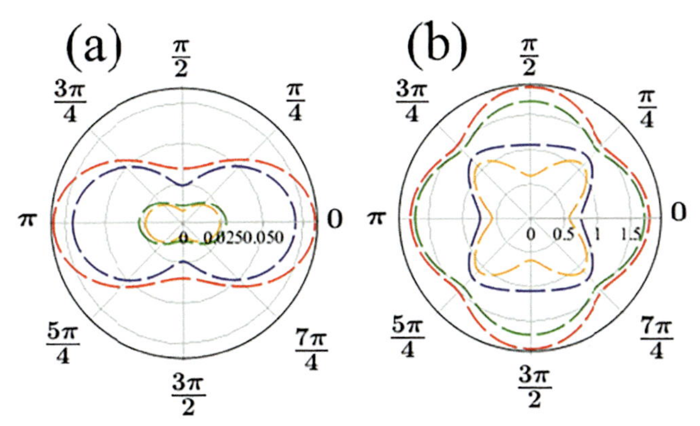

Fig. 3.15 Modulus of the phase velocity for random fibrous medium for two different values of the wave number in the SG medium, (A) non-affine regime and (B) affine regime. *Red* and *green lines*, respectively: longitudinal mode for $kL_0 = 1.5$ and $kL_0 = 0.5$. *Blue* and *orange lines*, respectively: shear mode for $kL_0 = 1.5$ and $kL_0 = 0.5$.

The main difference between the couple-stress theory and the second-order gradient substitution media is clearly visible looking at the polar plot of the modulus of the phase velocity (Fig. 3.15). The dispersive behaviour of the second gradient medium in the non-affine regime is evidenced by the modification of the shape of the plot when changing the wave number k. In the affine regime, when the macroscopic deformation coincides with the microscopic deformation, the influence of the second gradient disappears and no significant effect in the phase velocity is observed when increasing k, and the medium behaves as a Cauchy continuum.

3.4.2.2 Effect of network density and window size on the dispersion relation, phase velocity for the second gradient medium

We next investigate the effect of the network density and window size on the frequency band structure for the second gradient medium, and make a comparison with the couple-stress theory in both affine and non-affine regimes.

Inspection of Fig. 3.16 shows the same influence of network density and window size on the dispersion relation for the second gradient and couple-stress models in both regimes; a significant increase in the frequency for the L and S modes occurs when the network density or the window size increases when moving from the couple-stress to the second gradient effective continuum. The dispersive behaviour of the second gradient medium and the anisotropic characteristic of the random fibrous medium do not change when varying the network density or the size of the window of analysis. It is important to note that the second gradient effective medium is dispersive for the longitudinal mode, which is not the case for the effective couple-stress model.

The same effect of the network density and WOA size on wave dispersion can also be seen in Fig. 3.17; the anisotropy of wave propagation does not change when

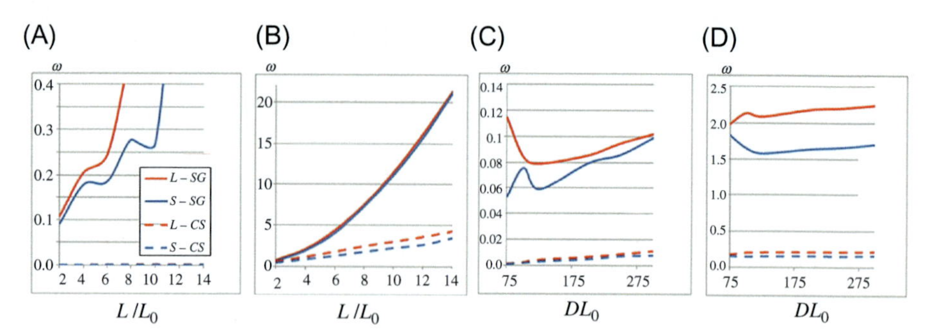

Fig. 3.16 Effect of the window size and network density on the frequency band structure for $kL_0 = 0.5$ (A) and (C) non-affine regime (B) and (D) affine regime.

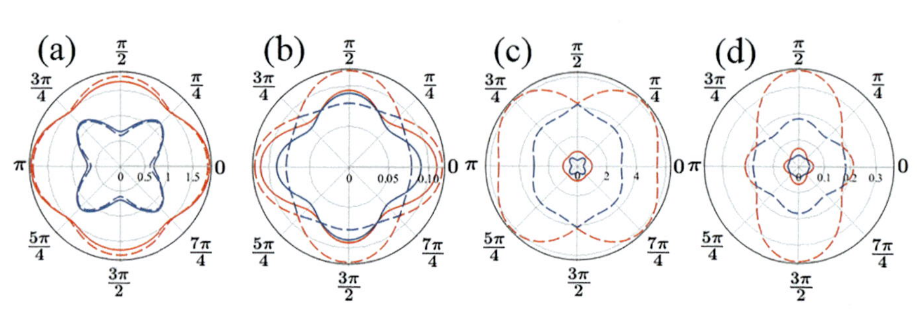

Fig. 3.17 Polar plot of the modulus of the phase velocity for random fibrous media with a wave number $kL_0 = 0.5$. (A) Affine regime, (B) non-affine regime, *continuous* and *dashed lines*: phase velocity for $DL_0 = 100$ and $DL_0 = 150$, respectively. (C) Affine regime, and (D) non-affine regime. *Continuous* and *dashed lines*: phase velocity for $L/L_0 = 4$ and $L/L_0 = 14$, respectively.

changing network density or WOA size. The shape of the phase velocity plot changes with the wave number k, thereby reflecting the dispersive behaviour of the second gradient medium, for any values of window size, network density and in both affine and non-affine regimes. This highlights the need for a second gradient effective continuum rather than couple stress substitutions, which is not able to capture the dispersive nature of propagating waves within RFNs.

3.5 Conclusion

The dynamic analysis of random fibrous networks (RFN) is a rather new topic, which has many applications in materials science or in the biomedical area, since these networks are very often subjected to dynamical loading conditions. In order to bypass the complexity of performing dynamic computations at the microscopic scale of the full network, we have identified couple-stress and second gradient models as effective continua at the mesoscopic level of windows of analysis of different sizes, with the objective to investigate the size effects of such networks on their dynamic properties.

The static mechanical properties which are at the root of the dynamical analysis are computed thanks to FE simulations performed over windows of analysis subjected to mixed boundary conditions allowing us to capture the classical and non-classical effective moduli. The acoustic properties of these networks are captured by the dispersion diagrams and the phase and group velocities plots; we have analysed the influence on the dynamic properties of the fibre bending length, the size of the window of analysis, and the fibre network density as three main parameters of interest. The impact of these parameters has been successively assessed for the couple-stress and strain gradient substitution continua.

One important result from these analyses is the need for a second gradient effective continuum rather than a couple stress substitution continuum since the last effective medium is not able to capture the dispersive nature of propagating waves within RFNs.

The influence of fibre bending length is the same in both effective theories, the partial band gap increasing when increasing the fibre bending length l_b. Network density and window size have the same influence on the dispersion relation for both models and in both affine and non-affine regimes. A significant increase in frequency for the longitudinal and shear modes occurs when network density or window size increases when moving from couple-stress to second gradient medium. The same effect of network density and size of the window of analysis on the dispersion diagram is also obtained. The anisotropic feature of wave propagation does not change when changing the network density or the size of the window of analysis.

The main benefit of using random fibrous networks for wave propagation is the possible control of band gap by different parameters, like fibre network density, window size, and fibre bending length. In the affine regime, when the macroscopic deformation is very close to the microscopic deformation, the influence of the microstructure disappears and no significant effect of the microstructure on the phase velocity occurs; thus, Cauchy elasticity is sufficient.

Although the replacement of the initial random network by effective generalised continua is able to capture the local deformation mechanisms of the fibrous microstructure better than Cauchy elasticity does and is more tractable from a numerical point of view in comparison to full network simulations, it loses some of the details of the microstructure response when waves propagate. Accordingly, such mesoscopic models shall be complemented by fully resolved local analyses at the scale of the fibres using Bloch's theorem or suitable extensions of it in order to better isolate individual phenomena associated with the mechanical response of the fibres themselves.

The influence of large strains developed by random fibrous networks on wave propagation is an important aspect encountered in real situations that shall be investigated in future contributions.

References

[1] P. Fratzl, Collagen: structure and mechanics, an introduction, in: Collagen: Structure and Mechanics, Springer, 2008. Department of Biomaterials 14424 Potsdam Germany.

[2] R.C. Picu, Mechanics of random fiber networks—a review, Soft Matter 7 (15) (2012) 6768–6785.

[3] H. Hatami-Marbini, R.C. Picu, An eigenstrain formulation for the prediction of elastic moduli of defective fiber networks, Eur. J. Mech. A. Solids 28 (2009) 305–316.

[4] H. Hatami-Marbini, R.C. Picu, Scaling of nonaffine deformation in random semiflexible fiber networks, Phys. Rev. E 77 (2008), 062103.

[5] A. Shahsavari, R.C. Picu, Model selection for athermal cross-linked fiber networks, Phys. Rev. E 86 (2012), 011923.

[6] A. Shahsavari, R.C. Picu, Size effect on mechanical behavior of random fiber networks, Int. J. Solids Struct. 50 (2013) 20–21.

[7] Y. Lee, I. Jasiuk, Apparent elastic properties of random fiber networks, Comput. Mater. Sci. 79 (1995) 715–723.

[8] D.R. Cox, The inter-fiber pressure in slivers confined in rectangular grooves, J. Text. Inst. 43 (1952) 87–94.

[9] D.R. Petterson, Mechanics of Nonwoven Fabrics, Ind. Eng. Chem. 51 (8) (1959) 902–903.

[10] X.F. Wu, Y.A. Dzenis, Elasticity of planar fiber networks, J. Appl. Phys. 98 (093501) (2005) 1–9.

[11] S. Babaee, A.S. Shahsavari, P. Wang, R.C. Picu, K. Bertoldi, Wave propagation in cross-linked random fiber networks, Appl. Phys. Lett. 107 (2015), 211904.

[12] M.E. Nute, K. Slater, A study of some factors affecting sound reduction by carpeting, J. Text. Inst. 64 (11) (1973) 645–651.

[13] M.E. Nute, K. Slater, The effect of fabric parameters on sound-transmission loss, J. Text. Inst. 64 (11) (1973) 652–658.

[14] R.D. Mindlin, H.F. Tiersten, Effects of couple-stresses in linear elasticity, Arch. Ration. Mech. Anal. 11 (1962) 415–448.

[15] H. Reda, Y. Rahali, J.F. Ganghoffer, H. Lakiss, Wave propagation in 3D viscoelastic auxetic and textile materials by homogenized continuum micropolar models, Compos. Struct. 141 (2016) 328–345.

[16] H. Reda, Y. Rahali, J.F. Ganghoffer, H. Lakiss, Analysis of dispersive waves in repetitive lattices based on homogenized second-gradient continuum models, Compos. Struct. 152 (2016) 712–728.

[17] T. Cihan, P.R. Onck, Size effects in two-dimensional Voronoi foams: a comparison between generalized continua and discrete models, J. Mech. Phys. Solids 56 (2008) 3541–3564.

[18] A.C. Eringen, Microcontinuum Field Theories, vol. 325, Springer, 1999.

[19] E. Cosserat, F. Cosserat, Théorie des Corps Déformables, Hermann, Paris, 1909.

[20] W.T. Koiter, Effects of couple stresses in linear elasticity, Proc. K. Ned. Akad. Wet. B 67 (1964) 17–44.

[21] K. Berkache, S. Deogekar, I. Goda, R.C. Picu, J.F. Ganghoffer, Identification of equivalent couple-stress continuum models for planar random fibrous media, Contin. Mech. Thermodyn. 31 (2019) 1035–1050.

[22] K. Berkache, S. Deogekar, I. Goda, R.C. Picu, J.F. Ganghoffer, Construction of second gradient continuum models for random fibrous networks and analysis of size effects, Compos. Struct. 181 (2017) 347–357.

[23] B. Van Rietbergen, S. Majumdar, W. Pistoia, D.C. Newitt, M. Kothari, A. Laib, P. Rüegsegger, Assessment of cancellous bone mechanical properties from micro-FE models based on micro-CT pQCT and MR images, Technol. Health Care 6 (1998) 413–420.

[24] C. Hitzenberger, E. Gotzinger, M. Sticker, M. Pircher, A.F. Fercher, Measurement and imaging of birefringence and optic axis orientation by phase resolved polarization sensitive optical coherence tomography, Opt. Express 9 (2001) 780–790.

[25] J. Rogowska, N.A. Patel, J.C. Fujimoto, M.E. Brezinski, Optical coherence tomographic elastography technique for measuring deformation and strain of atherosclerotic tissues, Heart 90 (2004) 556–562.
[26] P. Friedl, E. Brocker, Biological confocal reflection microscopy: reconstruction of three-dimensional extracellular matrix, cell migration, and matrix reorganization, in: D. Hader (Ed.), Image Analysis: Methods and Applications, second ed., CRC Press, Boca Raton, 2001, pp. 9–22.
[27] J. Liu, G.H. Koenderink, K.E. Kasza, F.C. Mackintosh, D.A. Weitz, Visualizing the strain field in semiflexible polymer networks: strain fluctuations and nonlinear rheology of F-actin gels, Phys. Rev. Lett. 98 (2007), 198304.
[28] P.P. Provenzano, R. Vanderby, Collagen fibril morphology and organization: implications for force transmission in ligament and tendon, Matrix Biol. 25 (2006) 71–84.
[29] A. D'Amore, J.A. Stella, W.R. Wagner, M.S. Sacks, Characterization of the complete fiber network topology of planar fibrous tissues and scaffolds, Biomaterials 31 (2010) 5345–5354.
[30] A. Yoo, I. Jasiuk, Couple-stress moduli of a trabecular bone idealized as a 3D periodic cellular network, J. Biomech. 39 (2006) 2241–2252.
[31] P. Fratzl, Statistical model of the habit and arrangement of mineral crystals in the collagen of bone, J. Stat. Phys. 77 (1994) 125–143.
[32] J. Li, M. Dao, C.T. Lim, S. Suresh, Spectrin-level modeling of the cytoskeleton and optical tweezers stretching of the erythrocyte, Biophys. J. 88 (2005) 3707–3719.
[33] O. Kallmes, H. Corte, The structure of paper. I. The statistical geometry of an ideal two dimensional fiber network, TAPPI J. 43 (1960) 737–752.
[34] W.W. Sampson, Modelling Stochastic Fibrous Materials with Mathematica®, Springer, 2009.
[35] A.L. Koch, S.W. Woeste, The elasticity of the sacculus of Escherichia coli, J. Bacteriol. 174 (1992) 4811–4819.
[36] A.J.M. Spencer, K.P. Soldatos, Finite deformations of fibre-reinforced elastic solids with fibre bending stiffness, Int. J. Non Linear Mech. 42 (2007) 355–368.
[37] G. Rosi, N. Auffray, Anisotropic and dispersive wave propagation within strain-gradient framework, Wave Motion 63 (2016) 120–134.
[38] A. Spadoni, M. Ruzzene, S. Gonella, F. Scarpa, Phononic properties of hexagonal chiral lattices, Wave Motion 46 (2009) 435–450.
[39] H. Reda, J.F. Ganghoffer, H. Lakiss, Micropolar dissipative models for the analysis of 2D dispersive waves in periodic lattices, J. Sound Vib. 392 (2017) 325–345.
[40] V.I. Erofeyev, Wave Processes in Solids with Microstructure, World Scientific, Singapore, 2003.

Stochastic constitutive model of thin fibre networks

Rami Mansour and Artem Kulachenko
Solid Mechanics, Department of Engineering Mechanics, KTH Royal Institute of Technology, Stockholm, Sweden

4.1 Introduction

Fibre networks are truly abundant structures that can be found in both man-made products such as packaging and composites as well as in nature, for example, in biological tissues, cells, and bones. Fibre-based materials offer a range of unique properties, such as high stiffness to mass ratio. At the same time, they can be made from renewable resources at a relatively small cost with paper and packaging products being an excellent example of that.

Thin fibre networks are anisotropic and have drastically different mechanical properties in-plane and out-of-plane. Owing to anisotropy, thin networks have a remarkable feature, namely, the ability to withstand creasing and folding without significant deterioration of the in-plane properties. As with the majority of materials, the stiffness and strength of fibre-based materials are among the most important properties in a number of applications. The mechanisms that control these properties originate from the structure at the microscale, where the following contributing factors play a crucial role: mechanical properties of fibres and fibre bonds, fibre morphology and fibre alignment inside the network, which includes fibre orientation, number of interfibre contacts, bonding properties, and disordered nature of the fibre network. Therefore, it is natural to tackle the questions related to the mechanics of the fibre networks at a scale where all the essential components can be taken into consideration.

Due to the disordered nature of thin fibre networks, fibrous materials are characterised by a certain degree of randomness in their mechanical properties [1]. Such variations can be the cause of unexplained occasional failures that cannot always be predicted by deterministic material models [2–4]. Naturally, the question of predicting such stochastic failures is a part of the engineering process and it needs to be addressed. It is, therefore, crucial to develop a sound stochastic approach in studying the randomness of thin fibre networks of arbitrary size. The rapid development of characterisation tools enables the quantification of randomness at different scales and the construction of random realisations of a fibre network. The mechanical behaviour of the reconstructed networks can be investigated using detailed direct micromechanical simulations [5–8]. However, although such simulations can capture the complicated mechanisms of failure and strain localisation, they cannot yet be employed for product development due to the overwhelming computational costs

Mechanics of Fibrous Networks. https://doi.org/10.1016/B978-0-12-822207-2.00014-3

required to capture the relevant product sizes. Therefore, the straightforward use of direct micromechanical simulations for predicting the mechanical response of large structures of complex geometry will remain impossible in the foreseeable future.

In this chapter, we will demonstrate how to connect the state of the art in micromechanical simulations of thin fibre networks with efficient stochastic continuum modelling of large fibre networks. This connection enables considering products of relevant sizes and yet including the stochastic aspects of failure.

4.2 Micromechanical simulation of thin random networks

The mechanical properties of the fibre network depend on three main components that need to be taken into account during modelling:

(i) mechanical properties of fibres and fibre bonds;
(ii) fibre morphology; and
(iii) fibre alignment inside the network.

We will briefly discuss how these properties can be measured and therefore used as inputs.

4.2.1 Mechanical properties of fibre and fibre bonds

The traditional way to experimentally determine the mechanical properties of fibres and fibre bonds has been using micromechanical stages, which require rather elaborate and time-consuming assembly and mounting procedures that can damage the fibres [9,10]. Emerging technologies in the area of micro- and nanorobotics offer a novel approach for characterising fibres and fibre bonds. In particular, micromechanical manipulation devices can be used to characterise the constitutive relations of individual fibre as well as fibre segments under both tension and compression.

The main principle behind the testing of the fibres using micromanipulators with load cells and motion sensors is shown in Fig. 4.1A. The fibres are handled by grippers (Fig. 4.1B), which are operated remotely by either a haptic three-dimensional (3D)

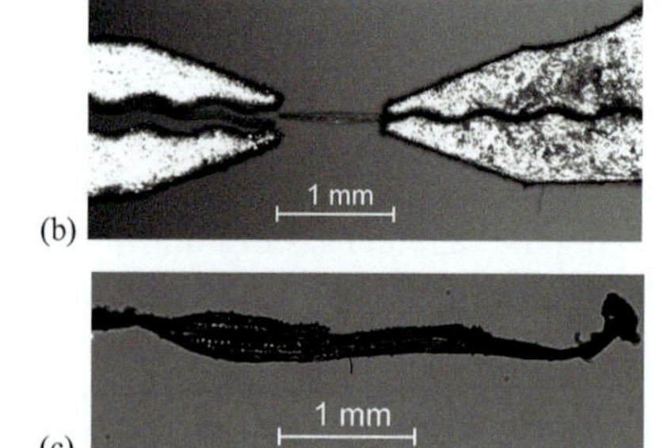

Fig. 4.1 Operational principles of fibre and fibre bond testing using micromanipulators (courtesy of Tampere University): (A) the layout of the device; (B) the gripping principles using the microgrippers with integrated force sensing; and (C) a softwood fibre tested.

controller or by automation. An important piece of information to capture besides the force response is the geometry of the cross-sectional area of the fibres. This can be done by optical tomography or by performing cross-sectional scans after the test [11].

Measuring fibre bonds is a far more challenging task that is not completely solved. The fibre bond composition and properties depend on the process of making the network even to a greater degree than the fibres. Obviously, the bonds cannot be tested inside the network and therefore the testing is often done on artificially created bonds [10,12]. Another complication is that during testing, the gripping cannot be done close to the bond, which means that the tested force response includes deformation of the fibres and not only the bond. The fracture mode at bond separation cannot be easily controlled and therefore it is often a mix of peeling and shearing.

To single out the contribution of each factor, one should employ inverse analysis. It includes reconstructing the fibre shape during the deformation process, making a virtual twin of the tests. By following the reconstructed geometry of the bond during testing, the problem of separation can be addressed by back calculating the forces acting in the bond interfaces [13].

Despite the progress, the measurements of the fibres and fibres bonds are still difficult and not robust. Accumulating the statistics on this level requires a lot of measurements. The current progress in this field involves automation of the test procedures using machine learning [14].

4.2.2 Fibre morphology

Fibre morphology, which includes length, curl or tortuosity, and cross-sectional properties, greatly affects the mechanical properties. A simple example: it is difficult to tear a network consisting of long fibres compared to the one with short fibres at the same network density. There are tools that can characterise the fibre morphology automatically and extract geometry data [15]. However, they often require fibres to be diluted in liquid. As the fibre cross-section of many natural fibres changes significantly upon pressing and drying inside the network, the cross-sectional data acquired in the wet state cannot be trusted.

One way of solving this problem is using X-ray tomography of the formed sheets to correct the data. Let us illustrate this by using the sheets made of the kraft pulp scanned by microcomputed tomography (μCT) methods [16,17]. First, each fibre needs to be segmented or at least a centreline needs to be identified (Fig. 4.2). The cross-section of the fibre can be extracted from the image by taking a slice orthogonal to the centreline. Examples of cross-sections are shown in Fig. 4.3. Along with the extracted cross-section, it is possible to fit one of the standard cross-sections, which can be used in the modelling. In the fitting, it is important to preserve the cross-sectional area, the orientation of the cross-section, and the geometrical moments of inertia.

4.2.3 Network geometry

Network geometry, which incorporates the through-thickness fibre orientation, anisotropy, and density profiles, is a vital component affecting to a large extent the mechanical properties of the network (see Fig. 4.4). Currently, the availability of 3D imaging

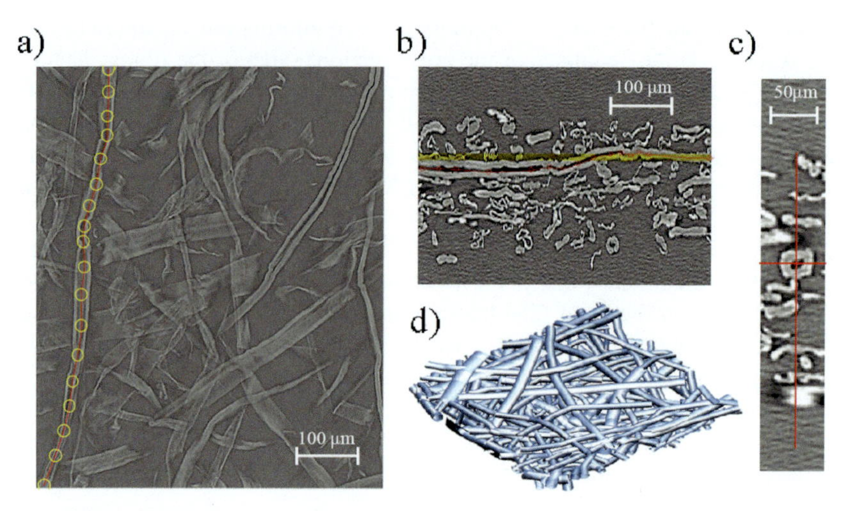

Fig. 4.2 Schematic diagram [18] of how the microCT data is processed by following the fibres in different planes (A–C) to extract the centrelines (D).

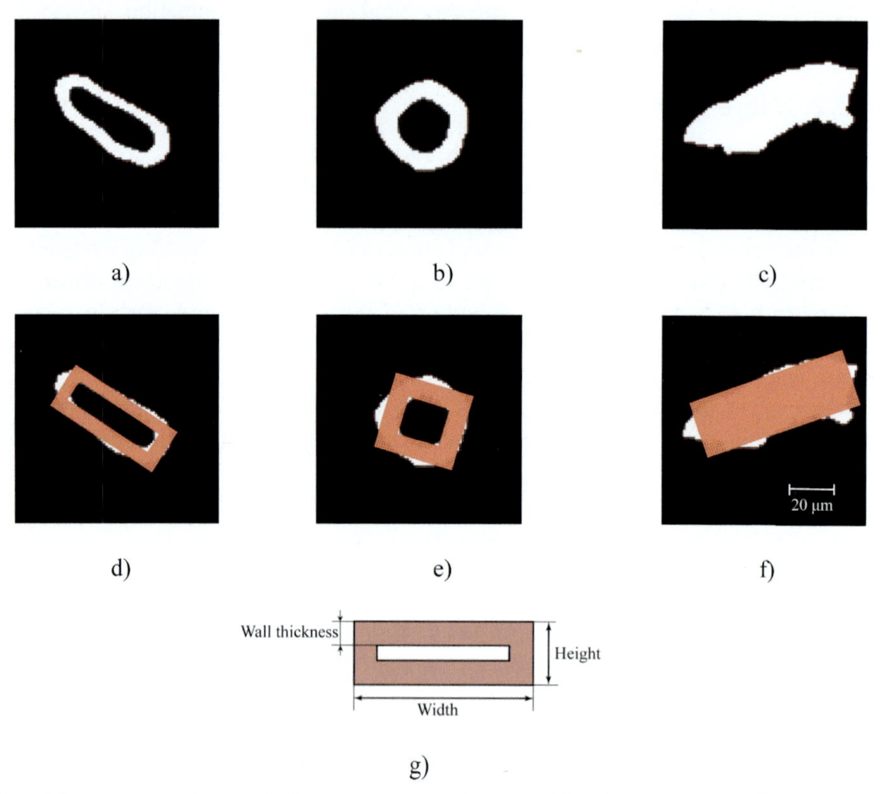

Fig. 4.3 Images of the actual fibre cross-section in paper [6]: (A, B) with and (C) without a lumen, computed from X-ray tomograms. Geometrical representation of the same cross-sections defined by a rectangle with hollow (D, E) and solid (F) cross-sections. (G) Model definition of the fibre geometrical characteristics.

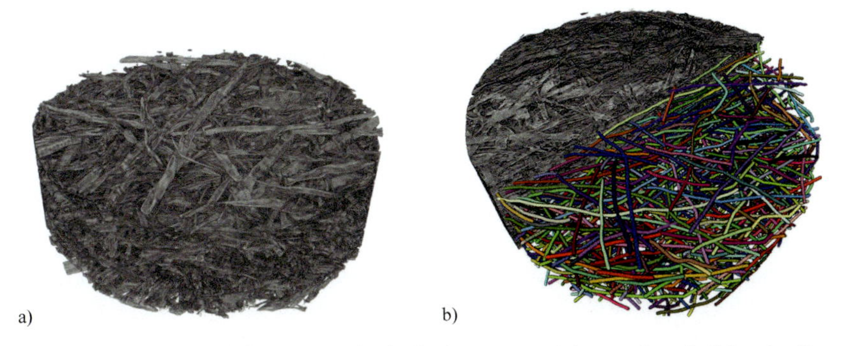

Fig. 4.4 (A) Reconstructed fibre network of a 3-ply commercial paper board; (B) orthoslices of the reconstructed network. The sparse middle ply contains bulky fibres with hollow, uncollapsed fibres. The dense top and bottom ply contain slender collapsed fibres [19].

is rapidly increasing, and it is possible to attain resolutions in the nanometre range, although often at the expense of long scanning times and large volumes of generated data. Extracting the through-thickness fibre orientation and connectivity properties of the network is yet another challenge, as the individual fibres have to be distinguished, and the contact area should be accurately resolved. The complexity of this task is increased by irregular fibre shapes, which can be curved with either hollow or compressed cross-sections varying along their length.

Using synchrotron facilities, methods for automated or semi-automated fibre detection have been developed [16,20]. Although the latter method requires a certain degree of manual processing, it has proved to be robust and applicable to both dense and sparse networks [20].

The disadvantage of the tomography-based method for detecting fibre orientation and anisotropy profiles is the limited size of the specimen, which may not always be representative of a given material. The alternative methods are based on sheet-splitting [21]. With this technique, the fibre network is delaminated in thin layers and each layer is analysed using optical methods through identifying the fibres in planar images. The disadvantage of this method is the inability to detect the orientation in ZD or thickness direction, which is important for certain applications, in particular for the compressive strength [22].

4.2.4 Fibre network simulation

4.2.4.1 Random generation with target properties

Given the geometry of the fibres that constitute a fibre network, acquired using a fibre image analyser and a characterisation methodology outlined in Ref. [6], random networks can be simulated with target characteristics. These characteristics include average and standard deviation of fibre length, width, shape factor, wall thickness, and width-to-height ratio. The random generation is based on a deposition technique in which the fibres are sequentially deposited on a flat surface from two sides. The deposition algorithm can be outlined as follows and the code for generating the network along with the documentation is available as supplementary material in Ref. [19].

1. The fibre geometries are taken from the fibre characterisation data containing length, width, height, wall thickness, and curvature. The curvature is represented through an arc of constant curvature located in a single plane parallel to the deposition plane. The cross-sectional data are corrected [6] using microtomography scans (see Section 4.2.2).
2. The fibre orientation distribution (FOD) is computed from the fibre characterisation data. The random deposition of the fibres in the generated network is performed following this FOD. For the generation of isotropic fibre networks, the FOD follows a uniform distribution. An example of a randomly generated fibre network with corresponding FOD is shown in Fig. 4.5A and B.

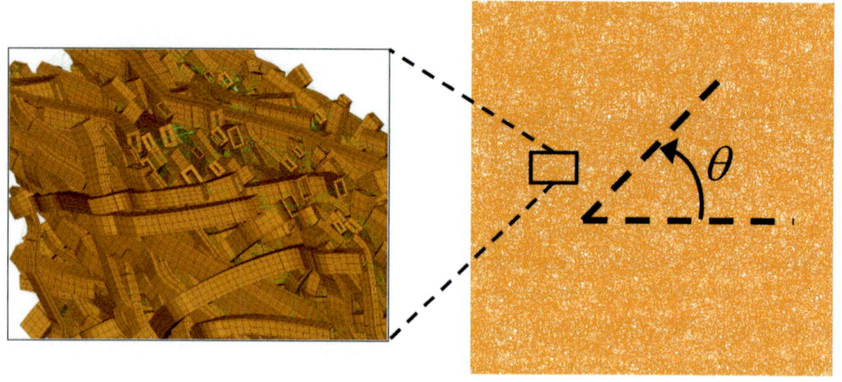

(a) 18 mm × 18 mm randomly generated fiber network with main fiber orientation θ

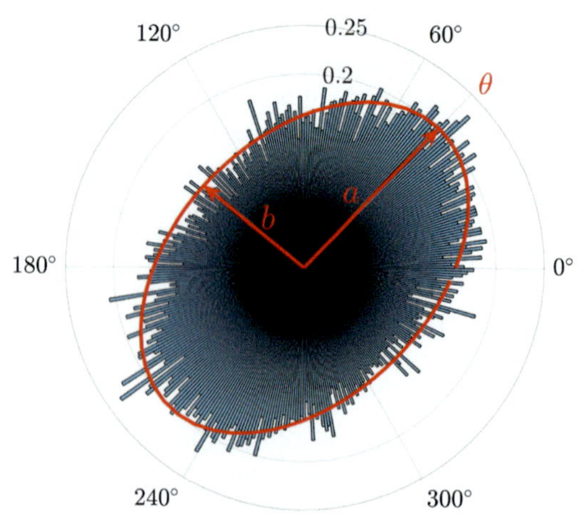

(b) Fiber orientation distribution

Fig. 4.5 Randomly generated fibre network with corresponding fibre orientation distribution (FOD) [23].

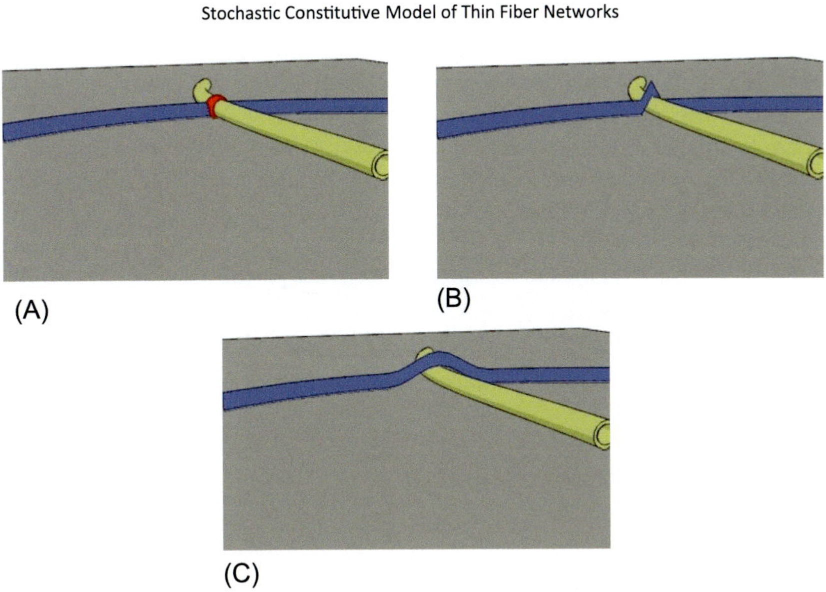

Fig. 4.6 Steps in deposition procedure [19]: (A) finding intersections with previously deposited fibres, (B) lifting the intersection point, and (C) smoothing the fibre.

3. The fibre position before deposition onto the domain is chosen randomly.
4. The first fibres are deposited on the flat plane consecutively from above or below. For the subsequent fibres, we first find the intersection between them and the previously deposited fibres in the plane (Fig. 4.6A).
5. The found intersection points are lifted discretely to exclude penetration (Fig. 4.6B). The contact search diameter depicted in the figure corresponds to the height of the fibre, which is normally smaller than or equal to the width of the fibre.
6. The fibre geometry is smoothed to remove discontinuities caused by the previous step (Fig. 4.6C). During the smoothing, we control the maximum angle the fibres can form.
7. When the grammage (the basis weight or the weight per unit area) of the network has reached a prescribed value, the deposition procedure is stopped.
8. The thickness of the network is evaluated and measured using the procedure described in Ref. [6]. Thereafter, the thickness is brought to the target value by uniform scaling of the coordinates in the thickness direction with respect to the centre plane of the network. The scaling may result in interpenetration, which are zeroed out during the subsequent computations.

4.2.4.2 Finite-element model

After the generation of the network, the fibres are represented with a fine mesh of curved segments and imported into a custom finite-element code. The fibres are converted into cubic splines and a full 3D network finite element model [6,7,24] is used, in which fibres are resolved as a chain of 3D quadratic Timoshenko/Reissner beam elements [25] with six degrees of freedom at each node. The implicit time integration is used. During contact detection, cross-sections of the fibres are treated as circular and rigid, with the diameter equal to the mean of the values of fibre width and height. The mechanical behaviour of the bonds is described with traction-separation laws with a cohesive zone model based on the contact forces [26]. The constitutive response of the fibre is modelled using bilinear plasticity with an isotropic hardening law. The flow stress is given by $\sigma_{s,\,f} + E_{tan,\,f}\varepsilon_{e,\,pl}$ where $\sigma_{s,\,f}$ is the initial yield stress of the fibre, $E_{tan,\,f}$ is the tangent modulus, and $\varepsilon_{e,\,pl}$ is the equivalent plastic strain. The latter is incrementally computed from the plastic stain increments $\mathbf{d\varepsilon}_{pl}$ as $d\varepsilon_{e,pl} = \left(\frac{2}{3}\mathbf{d\varepsilon}_{pl}{}^{T}\mathbf{d\varepsilon}_{pl}\right)^{1/2}$, which has three components (one normal and two transverse shear strains) in the used Timoshenko beam element.

4.2.4.3 Finite-element simulation

In the following, the considered network represents paper materials produced using softwood kraft pulp. The grammage used is 28 g/m^2, which is relatively low but corresponds to the set of handsheets used to calibrate the measurements in Ref. [6]. The target network thickness is 68 μm, which also corresponds to the measured value used in the calibration. The details of the fibres used in the analysis are summarised in Table 4.1. The material parameters in the constitutive response of the fibres, described by a bilinear plasticity model (see Section 4.2.4.2), are listed in Table 4.2.

The contact conditions at the fibre bonds are governed by a bilinear cohesive traction-separation law. It requires the definitions of a bond's stiffness, a bond's strength, and separation (see Table 4.3). Since the beam's cross-section is rigid against the local normal forces and shear forces, the physical compliance of the fibre at the bonding sites is represented solely by the stiffness of the penalty-based contact

Table 4.1 Fibre geometrical data used in the network simulation, based on the direct measurement on wet pulp (fibre morphology analyser, FMA), on dry sheets (μCT), and numerical parameters in terms of length-weighted mean and standard deviation (SD) values.

	Mean	**SD**	**Source of data**
Fibre length, mm	2.34	0.90	FMA
Fibre width, μm	23.83	7.09	FMA
Fibre wall thickness, μm	3.96	1.90	μCT
Width-to-height ratio, (–)	2.9	1.72	μCT
Fibre shape factor, (–)	0.945	0.015	FMA
Maximum interface angle, °	5	–	μCT
Radius swelling factor[a], (–)	0.78	0.68	μCT
Wall thickness swelling factor, (–)	0.528	0.31	μCT

[a] The ratio between the dry and wet measured radius and wall thickness, respectively.

Table 4.2 Fibre material parameters used in the network simulation.

Elastic modulus (GPa)	Tangent modulus (GPa)	Yield stress (MPa)
30	10	150

Table 4.3 Characteristics of bonds used in the network simulation.

	Tangential direction	Normal direction
Bond strength, mN	11.00	2.75
Bond stiffness, 10^9 N/m	8.90	8.00
Separation distance, μm	1.56	0.35

element. The selection of the parameters listed in Tables 4.2 and 4.3 are based on the experimental calibration in the tensile test described in Ref. [6].

Fig. 4.7 shows a 24 mm × 24 mm fibre network subjected to uniaxial loading in the x-direction. The load is applied through prescribed displacement in a finite-element model. In the plot, the strain field is mapped onto the network with colour plots showing non-uniformity. The strain field was determined by calculating the first partial derivative of the axial displacement field with respect to the axial coordinate x i.e. the coordinate in the direction of loading. Fig. 4.7A shows the strain field distribution for a very small loading. As can be seen, the non-uniformity of the strain field is already apparent. In Fig. 4.7B, the development of localisation is observed as seen by the presence of some nucleation. However, the propagation path is not yet formed. In Fig. 4.7C, there is a clear localisation of failure across the network at this point and in this region most of the slippage between the fibres occurs.

Although these simulations offer a powerful tool for studying and tailoring fibre properties to achieve the desired performance, they cannot be directly used for product development due to prohibitive computational cost. Therefore, fibre-network simulations at product scale are impossible in the near future. Instead, equivalent continuum realisation of fibre networks and continuum finite-element simulation are necessary, with principles and mathematical preliminaries described in the remainder of this chapter.

4.3 Mathematical theory of random spatial fields

In multiscale modelling of fibre networks, stochasticity stems from a number of parameters that can be modelled as random variables. These random variables can represent material properties that generally vary with spatial location. In the following, we present necessary mathematical preliminaries related to spatial field representation and simulation.

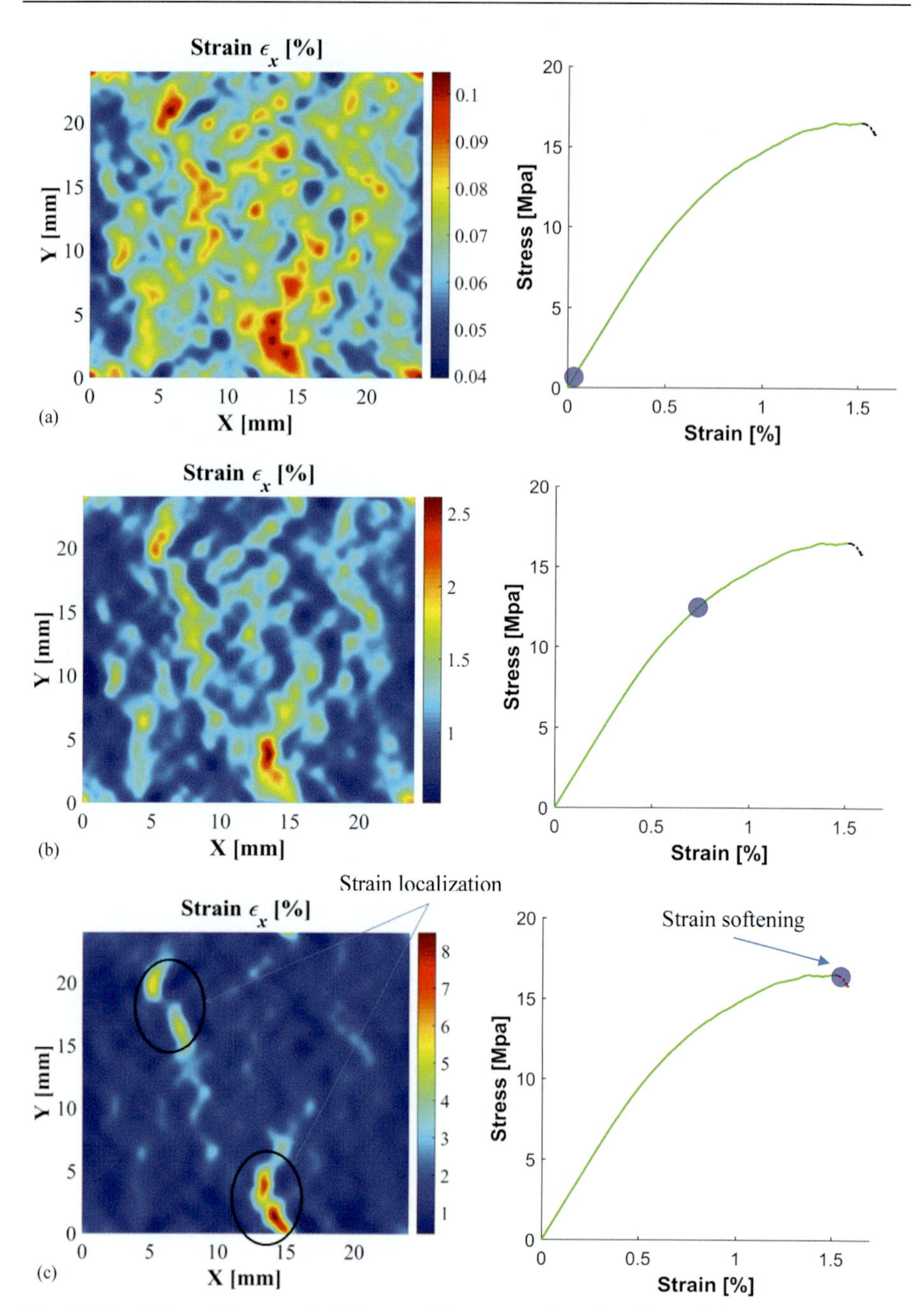

Fig. 4.7 Evolution of damage in 24 mm × 24 mm sample using direct finite-element simulation of a random realisation of fibre network: (A) very small loading, (B) development of localization, and (C) clear localization of failure and strain softening.

4.3.1 Random variables

A random property such as a material parameter or geometrical quantity can be represented by a random variable. This random variable, denoted by M_1, is defined as a function that maps elements $\omega \in \Omega$ of a sample space Ω to the set of real numbers i.e. $M_1 : \Omega \rightarrow \mathbb{R}$. Denoting an individual realisation of M_1 by m_1, the cumulative distribution function (CDF) $F_{M_1}(m_1)$ is defined as

$$F_{M_1}(m_1) = \Pr(M_1 < m_1), \tag{4.1}$$

where $\Pr(M_1 < m_1)$ is the probability of the occurrence of the event $\{M_1 < m_1\}$. The random variable is completely defined by its CDF, which is a strictly increasing function with the properties $\lim_{m_1 \rightarrow -\infty} F_{M_1}(m_1) = 0$ and $\lim_{m_1 \rightarrow \infty} F_{M_1}(m_1) = 1$. The probability density function (PDF) is defined as $f_{M_1}(m_1) = \lim_{dm_1 \rightarrow 0} \Pr(m_1 < M_1 \leq m_1 + dm_1)/dm_1$. Hence, the PDF can be obtained by differentiation of the CDF according to

$$f_{M_1}(m_1) = \frac{dF_{M_1}(m_1)}{dm_1}. \tag{4.2}$$

The CDF can therefore be defined as

$$F_{M_1}(m_1) = \int_{-\infty}^{m_1} f_{M_1}(v)dv. \tag{4.3}$$

It is noted that the integral of the PDF and not the function itself expresses a meaningful probability measure. For the PDF, the normalisation rule reads $\int_{-\infty}^{\infty} f_{M1}(m_1)dm_1 = 1$.

The random variable M_1 is completely defined by its PDF or CDF. Its n-th moment is given by

$$\mathbb{E}[M_1^n] = \int_{-\infty}^{\infty} m_1^n f_{M_1}(m_1)dm_1 \tag{4.4}$$

and its n-th non-central moment by

$$\mathbb{E}[(M_1 - \mathbb{E}[M_1])^n] = \int_{-\infty}^{\infty} (m_1 - \mathbb{E}[M_1])^n f_{M_1}(m_1)dm_1, \tag{4.5}$$

where $\mathbb{E}[\cdot]$ is the expectation operator. These moments are used to characterise the distribution as well as fit empirical or common distribution functions [27,28]. The mean value is given by the first moment

$$\mu_{M_1} = \mathbb{E}[M_1]. \tag{4.6}$$

The variance is given by the second non-central moment according to

$$\mathrm{Var}[M_1] = \mathbb{E}\left[\left(M_1 - \mu_{M_1}\right)^2\right], \tag{4.7}$$

which is often used as a measure of the dispersion of the PDF. The skewness coefficient is defined as normalised third central moment according to

$$\gamma_{M_1} = \frac{\mathbb{E}\left[\left(M_1 - \mu_{M_1}\right)^3\right]}{\mathrm{Var}[M_1]^{3/2}}. \tag{4.8}$$

If $\gamma_{M_1} = 0$, then the PDF of M_1 is symmetric about the mean. If $\gamma_{M_1} < 0$, then the left tail is longer, and if $\gamma_{M_1} > 0$ then the right tail is longer. The normalised fourth central moment is the kurtosis coefficient defined as

$$\kappa_{M_1} = \frac{\mathbb{E}\left[\left(M_1 - \mu_{M_1}\right)^4\right]}{\mathrm{Var}[M_1]^2}. \tag{4.9}$$

The kurtosis coefficient is a measure of the flatness of a PDF.

4.3.2 Univariate stationary random spatial field

The random material parameter M_1 may vary with spatial location. In this case, the random spatial field is written as $M_1(\boldsymbol{r})$, where $\mathbf{r} = [x, y, z]^T \in \mathcal{D}$ is the spatial location vector and $\mathcal{D}$ is the spatial domain. A realisation (outcome) of this random spatial field at a location $\mathbf{r}$ is denoted by $m_1(\mathbf{r})$ (see Fig. 4.8).

Given different spatial locations $\mathbf{r}_1, \mathbf{r}_2, \ldots, \mathbf{r}_p$, a random vector $\mathbf{M}_1 = [M_1(\mathbf{r}_1), M_1(\mathbf{r}_2), \ldots, M_1(\mathbf{r}_p)]^T$ can be defined such as $\mathbf{M}_1 : \Omega \to \mathbb{R}^p$. A random realisation of the vector $\mathbf{M}_1$ is denoted by $\mathbf{m}_1 = [m_1(\mathbf{r}_1), m_1(\mathbf{r}_2), \ldots, m_1(\mathbf{r}_p)]^T$. For stationary spatial fields, all random variables $M_1(\mathbf{r}_1)$, $M_1(\mathbf{r}_2)$, ..., $M_1(\mathbf{r}_p)$ follow the same marginal

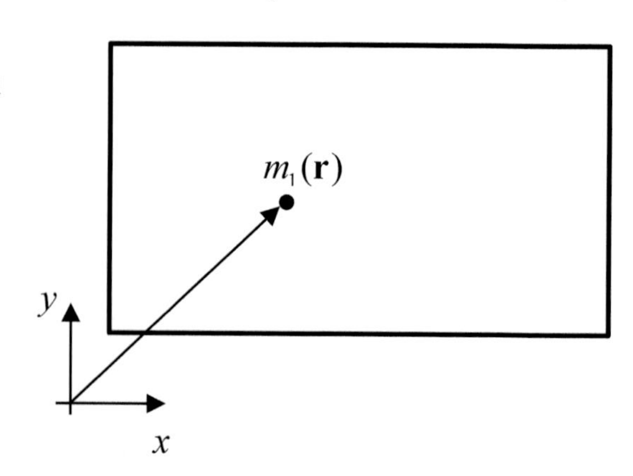

Fig. 4.8 A realisation m_1 of the random spatial field M_1 at a spatial location $\mathbf{r}$.

distribution $F_{M_1}(m_1)$ according to Eq. (4.1) independently of spatial location. The covariance of two component random variables $M_1(\mathbf{r}_i)$ and $M_1(\mathbf{r}_j)$ of the random vector $\mathbf{M}_1$ is defined as

$$\mathrm{Cov}\left[M_1(\mathbf{r}_i), M_1(\mathbf{r}_j)\right] = \mathbb{E}\left[\left(M_1(\mathbf{r}_i) - \mu_{M_1}\right)\left(M_1(\mathbf{r}_j) - \mu_{M_1}\right)\right]. \tag{4.10}$$

It is noted that, if $i = j$, Eq. (4.10) can be written as $\mathrm{Cov}[M_1(\mathbf{r}_i), M_1(\mathbf{r}_i)] = \mathrm{Var}[M_1]$ (see Eq. 4.7). The covariance is a measure of the linear dependence between $M_1(\mathbf{r}_i)$ and $M_1(\mathbf{r}_j)$, where $\mathrm{Cov}[M_1(\mathbf{r}_i), M_1(\mathbf{r}_j)] = 0$ implies statistical uncorrelation. Considering all p discrete spatial locations, the covariance matrix $\sum_{\mathbf{M}_1\mathbf{M}_1} \in \mathbb{R}^{p\times p}$ is defined as

$$\sum\nolimits_{\mathbf{M}_1\mathbf{M}_1} = \begin{bmatrix} \mathrm{Var}[M_1] & \ldots & \mathrm{Cov}\left[M_1(\mathbf{r}_1), M_1(\mathbf{r}_p)\right] \\ \vdots & \ddots & \vdots \\ \mathrm{Cov}\left[M_1(\mathbf{r}_p), M_1(\mathbf{r}_1)\right] & \ldots & \mathrm{Var}[M_1] \end{bmatrix}, \tag{4.11}$$

which is a square symmetric and positive-definite matrix.

4.3.3 Simulation of stationary univariate random spatial fields

A random realisation $m_1(\mathbf{r})$ of the arbitrarily distributed spatial field $M_1(\mathbf{r})$ can be simulated by first simulating random realisations $g_1(\mathbf{r})$ of a standard Gaussian spatial field $G_1(\mathbf{r}) \sim \mathcal{N}(0,1)$ followed by a probabilistic transformation.

To simulate realisations $g_1(\mathbf{r})$ of $G_1(\mathbf{r})$, the standard Gaussian spatial field is expressed using the Karhunen-Loève (KL) expansion [29],

$$G_1(\mathbf{r}) = \sum_{k=1}^{\infty} h_k(\mathbf{r})\sqrt{\lambda_k}\xi_k, \tag{4.12}$$

where ξ_k are independent standard normal variables, $h_k(\mathbf{r})$ and λ_k are eigenfunctions and eigenvalues of the covariance function $\mathrm{Cov}[G_1(\mathbf{r}), G_1(\mathbf{r}')]$, respectively, where $\mathbf{r}$ and $\mathbf{r}'$ are two spatial locations. This implies that the covariance function has the spectral decomposition [29]

$$\mathrm{Cov}\left[G_1(\mathbf{r}), G_1(\mathbf{r}')\right] = \sum_{k=1}^{\infty} \lambda_k h_k(\mathbf{r}) h_k(\mathbf{r}'), \tag{4.13}$$

where λ_k and $h_k(\mathbf{r})$ satisfy the Fredholm integral equation,

$$\int_{\mathcal{D}} \mathrm{Cov}[G_1(\mathbf{r}), G_1(\mathbf{r}')] h_k(\mathbf{r}) d\mathbf{r} = \lambda_k h_k(\mathbf{r}'). \tag{4.14}$$

and the eigenfunctions satisfy the orthogonality condition $\int_{\mathcal{D}} h_k(\mathbf{r}) h_l(\mathbf{r}) d\mathbf{r} = \delta_{kl}$. It can be shown [29] that of all possible choices of expansion basis, the KL expansion is the best approximation of the original random field in the sense that it minimises $\int_{\mathcal{D}} e_p(\mathbf{r})^2 d\mathbf{r}$, where $e_p(\mathbf{r}) = \sum_{k=p+1}^{\infty} \sqrt{\lambda_k} h_k(\mathbf{r}) \xi_k$ is the error resulting from its truncation at the pth term. It should be noted that the underlying covariance function Cov[$G_1(\mathbf{r})$, $G_1(\mathbf{r}')$] is generally not known, and an appropriate model needs to be chosen based on a set of observations.

In stochastic constitutive modelling applied in a finite-element framework, the simulation of discrete spatial fields is of interest. The spatial domain $\mathcal{D}$ is therefore discretised at p spatial points $\mathbf{r}_1$, $\mathbf{r}_2$, ..., $\mathbf{r}_p$. A random realisation of this discretised spatial field can be written as

$$\mathbf{g}_1 = \left[g_1(\mathbf{r}_1)\, g_1(\mathbf{r}_2) \cdots g_1(\mathbf{r}_p) \right]^T. \tag{4.15}$$

This set of observations can be regarded as one sample from a multivariate Gaussian distribution denoted by

$$\mathbf{G}_1 \sim \mathcal{N}\left(\mathbf{0}, \sum\nolimits_{\mathbf{G}_1\mathbf{G}_1}\right), \tag{4.16}$$

where $\mathbf{0} \in \mathbb{R}^p$ is the zero mean vector and

$$\sum\nolimits_{\mathbf{G}_1\mathbf{G}_1} = \begin{bmatrix} 1 & \dots & \mathrm{Cov}\left[G_1(\mathbf{r}_1), G_1(\mathbf{r}_p)\right] \\ \vdots & \ddots & \vdots \\ \mathrm{Cov}\left[G_1(\mathbf{r}_p), G_1(\mathbf{r}_1)\right] & \dots & 1 \end{bmatrix} \tag{4.17}$$

is the $p \times p$ covariance matrix. The diagonal terms in $\sum_{\mathbf{G}1\mathbf{G}1}$ are equal to unity since $G_1 \sim \mathcal{N}(0, 1)$. The covariance matrix is therefore composed of the covariance of all pairs of the Gaussian spatial field values at the different spatial locations in the discretised domain. For this discrete case, the Fredholm integral equation (Eq. 4.14) can be rewritten as

$$\sum\nolimits_{\mathbf{G}_1\mathbf{G}_1} \mathbf{h}_k = \lambda_k \mathbf{h}_k, \tag{4.18}$$

where λ_1, ...,λ_p and $\mathbf{h}_1$, ...,$\mathbf{h}_p$ are eigenvalues and eigenvectors, respectively, of the covariance matrix $\sum_{\mathbf{G}1\mathbf{G}1}$. That is, when the random field is discretised, operations on functions are transformed into operations on matrices. Similarly, the KL expansion (Eq. 4.12) of the discrete random field can be written as [30–33].

$$\mathbf{G}_1 = \sum_{k=1}^{p} \sqrt{\lambda_k} \mathbf{h}_k \xi_k. \tag{4.19}$$

Therefore, given the covariance matrix according to Eq. (4.17), a random realisation $\mathbf{g}_1$ according to Eq. (4.15) of the discretised spatial field $\mathbf{G}_1$ can be simulated using Eq. (4.19) simply by simulating random realisations of the standard independent normal variables $\xi_1, \ldots, \xi_p$. The corresponding realisation $\mathbf{m}_1$ of the original spatial field is thereafter computed using the isoprobabilistic transformation [34–36]:

$$F_{M_1}(m_1(\mathbf{r})) = \Phi(g_1(\mathbf{r})), \tag{4.20}$$

where $\Phi(\cdot)$ is the standard normal CDF and $F_{M_1}(m_1)$ is the marginal CDF of M_1 defined in Eq. (4.1). This results in

$$\mathbf{m_1} = \left[F_{M_1}{}^{-1}\{\Phi(g_1(\mathbf{r_1}))\}\, F_{M_1}{}^{-1}\{\Phi(g_1(\mathbf{r_2}))\}\cdots F_{M_1}{}^{-1}\left\{\Phi\left(g_1\left(\mathbf{r}_p\right)\right)\right\}\right]^T. \tag{4.21}$$

It should be emphasised that this random vector represents a realisation of the same random spatial field M_1 at different spatial locations. It is also noted that $F_{M_1}(m_1)$ is generally not known and assumptions need to be made based on a set of observations.

From Eqs (4.12)–(4.14) as well as Eqs (4.18)–(4.21), it can be seen that the spatial distribution of $G_1(\mathbf{r})$ is determined by the covariance function $\mathrm{Cov}[G_1(\mathbf{r}), G_1(\mathbf{r}')]$. Since the underlying covariance function is generally not known, a proper choice of model is crucial in that it directly determines the properties of the Gaussian random field. A covariance function applicable to isotropic random fields that is indefinitely differentiable is the squared exponential kernel given by [37]

$$\mathrm{Cov}[G_1(\mathbf{r}), G_1(\mathbf{r}')] = \exp\left(\frac{|\mathbf{r}-\mathbf{r}'|^2}{2\ell_1^2}\right), \tag{4.22}$$

where ℓ_1 is a characteristic length scale determining the spatial correlation. From Eq. (4.22), it can be seen that this covariance function only depends on the spatial distance $|\mathbf{r} - \mathbf{r}'|$ and the characteristic length ℓ_1. Therefore, ℓ_1 affects how far two spatial locations need to be before realisations of the spatial field can vary significantly. In Fig. 4.9, two realisations, $g_1(x)$ and $g_2(x)$, of two different one-dimensional spatial fields, $G_1(x)$ and $G_2(x)$, with characteristic length $\ell_1 = 0.3$ and $\ell_2 = 0.6$ are shown. An important characteristic of the Gaussian spatial fields is the distance between two zero level upcrossings, defined by $G_1(x) = 0$ and $\partial G_1(x)/\partial x > 0$ [37]. If this distance is described by a random variable, it can be shown that its meanμ_{upcr1} is given by

$$\mu_{\mathrm{upcr}\,1} = 2\pi\ell_1, \tag{4.23}$$

when $\mathrm{Cov}[G_1(x), G_1(x')]$ is given by Eq. (4.22) (see Fig. 4.9). The characteristic length $\ell_1 = \mu_{\mathrm{upcr1}}/2\pi$ can therefore be determined by counting the number of upcrossings and by computing the mean distance between them. This is performed based on a realisation $g_1(\mathbf{r})$ computed from an observed realisation $m_1(\mathbf{r})$ using Eq. (4.20). It is noted that, for isotropic covariance functions, a realisation along one coordinate direction (Fig. 4.9) is sufficient to determine ℓ_1. It should be emphasised that the simple relation $\ell_1 = \mu_{\mathrm{upcr1}}/2\pi$ is not necessarily applicable in the presence of multiple correlated spatial fields.

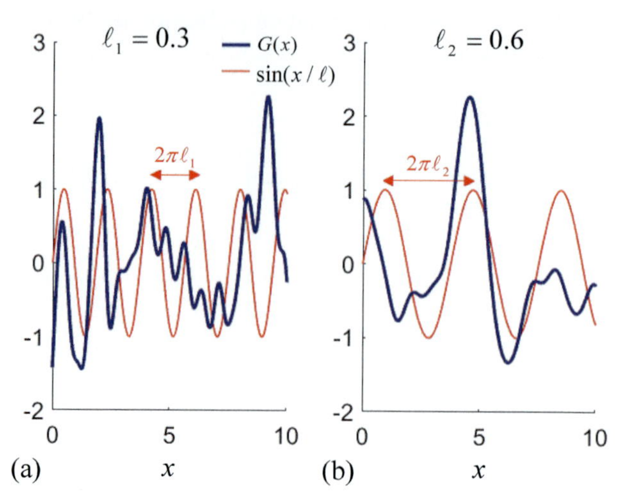

Fig. 4.9 One-dimensional spatial field realisations from two different Gaussian spatial fields with characteristic length (A) $\ell_1 = 0.3$ and (B) $\ell_2 = 0.6$.

4.3.4 Multivariate stationary random fields

In stochastic constitutive modelling, a number of properties are represented by random variables denoted by $M_1, M_2, \ldots, M_N$. Each random variable M_i, $i = 1, \ldots, N$, is defined as a function that maps elements $\omega \in \Omega$ of a sample space Ω to the set of real numbers i.e. $M_i : \Omega \rightarrow \mathbb{R}$. In addition, they vary randomly with spatial location such as $M_i(\mathbf{r})$ is a stationary spatial field. The correlation between the random variables is quantified using the correlation coefficient $\rho_{M_iM_j} \in [-1, 1]$, given by

$$\rho^{M_iM_j} = \frac{\mathrm{Cov}[M_i, M_j]}{\sqrt{\mathrm{Var}[M_i]\,\mathrm{Var}[M_j]}} = \frac{\mathbb{E}\left[\left(M_i - \mu_{M_i}\right)\left(M_j - \mu_{M_j}\right)\right]}{\sqrt{\mathrm{Var}[M_i]\,\mathrm{Var}[M_j]}}. \tag{4.24}$$

Since the considered random variables are stationary spatial fields, a realisation $m_i(\mathbf{r})$ of $M_i(\mathbf{r})$ follows the marginal CDF $F_{M_i}(m_i)$ regardless of spatial location.

For the discretised spatial fields at spatial points $\mathbf{r}_1, \mathbf{r}_2, \ldots, \mathbf{r}_p$, the covariance matrix $\sum_{\mathbf{M}_i\mathbf{M}_i}$ defined in Eq. (4.11) is named the *auto-covariance matrix* since it describes the spatial correlation for the same spatial field $M_i(\mathbf{r})$. Similarly, $\mathrm{Cov}[M_i(\mathbf{r}), M_i(\mathbf{r}')]$ is named the *auto-covariance function*. For multiple correlated spatial fields, the *cross-covariance matrix* $\sum_{\mathbf{M}_i\mathbf{M}_j}$describing the correlation between two different spatial fields $M_i(\mathbf{r})$ and $M_j(\mathbf{r})$ needs to be computed based on the *cross-covariance* function $\mathrm{Cov}[M_i(\mathbf{r}), M_j(\mathbf{r}')]$, such as

$$\sum\nolimits_{\mathbf{M}_i\mathbf{M}_j} = \begin{bmatrix} \mathrm{Cov}\left[M_i(\mathbf{r}_1), M_j(\mathbf{r}_1)\right] & \cdots & \mathrm{Cov}\left[M_i(\mathbf{r}_1), M_j(\mathbf{r}_p)\right] \\ \vdots & \ddots & \vdots \\ \mathrm{Cov}\left[M_i(\mathbf{r}_p), M_j(\mathbf{r}_1)\right] & \cdots & \mathrm{Cov}\left[M_i(\mathbf{r}_p), M_j(\mathbf{r}_p)\right] \end{bmatrix}. \tag{4.25}$$

The multivariate covariance matrix is composed of auto-covariance matrices and cross-covariance matrices as

$$\sum_{\mathbf{MM}} = \begin{bmatrix} \sum_{\mathrm{M}_1\mathrm{M}_1} & \cdots & \sum_{\mathrm{M}_1\mathrm{M}_N} \\ \vdots & \ddots & \vdots \\ \sum_{\mathrm{M}_N\mathrm{M}_1} & \cdots & \sum_{\mathrm{M}_N\mathrm{M}_N} \end{bmatrix}. \tag{4.26}$$

4.3.5 Simulation of multivariate stationary random fields

Similar to the univariate case presented in Section 4.3.2, multivariate arbitrarily distributed spatial fields $M_1(\mathbf{r})$, $M_2(\mathbf{r})$, …, $M_N(\mathbf{r})$ can be simulated by first simulating multivariate Gaussian random spatial fields $G_1(\mathbf{r})$, $G_2(\mathbf{r})$, …, $G_N(\mathbf{r})$.

A random realisation of the discretised multivariate Gaussian spatial field can be written as a N-variate random vector of size pN according to

$$\mathbf{g} = \left[\mathbf{g}_1^T\ \mathbf{g}_2^T\ \cdots\ \mathbf{g}_N^T\right]^T \in \mathbb{R}^{pN}, \tag{4.27}$$

where $\mathbf{g}_i \in \mathbb{R}^p$ is given by Eq. (4.15). This set of observations can be regarded as one sample from a multivariate Gaussian distribution denoted by

$$G \sim \mathcal{N}\left(0, \sum_{\mathbf{GG}}\right), \tag{4.28}$$

where $0 \in \mathbb{R}^{pN}$ is the zero mean vector and

$$\sum_{\mathbf{GG}} = \begin{bmatrix} \sum_{\mathbf{G}_1\mathbf{G}_1} & \cdots & \sum_{\mathbf{G}_1\mathbf{G}_N} \\ \vdots & \ddots & \vdots \\ \sum_{\mathbf{G}_N\mathbf{G}_1} & \cdots & \sum_{\mathbf{G}_N\mathbf{G}_N} \end{bmatrix} \tag{4.29}$$

is the $pN \times pN$ covariance matrix. The auto-covariance matrices $\sum_{\mathbf{G}i\mathbf{G}i}$ are given by Eq. (4.17) and the cross-covariance matrices $\sum_{\mathbf{G}i\mathbf{G}j}$ are constructed from the cross-covariance function $\mathrm{Cov}[\mathrm{G}_i(\mathbf{r}), G_j(\mathbf{r}')]$. The underlying auto- and cross-covariance functions are generally not known and model assumptions need to be made.

The KL expansion of the discrete N-variate Gaussian random field (see Eq. 4.18 for the univariate case) can be written as

$$\mathbf{G} = \sum_{k=1}^{pN} \sqrt{\lambda_k}\mathbf{h}_k\xi_k, \tag{4.30}$$

where $\lambda_1, \ldots, \lambda_{pN}$ and $\mathbf{h}_1, \ldots, \mathbf{h}_{pN}$ are eigenvalues and eigenvectors, respectively, of the covariance matrix $\sum_{\mathbf{GG}}$ and $\mathbf{G} = \left[\mathbf{G}_1{}^T\ \mathbf{G}_2{}^T\ \cdots\ \mathbf{G}_N{}^T\right]^T$ is the N-variate random vector. Therefore, given the covariance matrix according to Eq. (4.29), a random realisation

$\mathbf{g}$ according to Eq. (4.27) of the discretised spatial field $\mathbf{G}$ can be simulated using Eq. (4.30), by simulating random realisations of the standard independent normal variables $\xi_1, \ldots, \xi_{PN}$. The corresponding realisation of the original N-variate spatial field can thereafter be computed by isoprobabilistic transformation, such as

$$\mathbf{m} = \left[\mathbf{m}_1^T, \ \mathbf{m}_2^T, \ \cdots \ , \mathbf{m}_N^T\right]^T, \tag{4.31}$$

where $\mathbf{m}_i$ are given by Eq. (4.21).

It should be noted that the correlation coefficient ρ_{MiMj} is different from the correlation coefficient in the standard Gaussian variable space ρ_{GiGj}due to the non-linear transformation according to Eq. (4.20). The relation is given by

$$\rho_{M_iM_j} = \int_{-\infty}^{\infty}\int_{-\infty}^{\infty} \left(\frac{m_i - \mu_{M_i}}{\sqrt{\mathrm{Var}(M_i)}}\right)\left(\frac{m_i - \mu_{M_j}}{\sqrt{\mathrm{Var}(M_j)}}\right)\varphi_2\left(g_i, g_j, \rho_{G_iG_j}\right) dg_i dg_j, \tag{4.32}$$

where $\varphi_2(g_i, g_j, \rho_{G_iG_j})$ is the bivariate standard normal distribution.

The underlying covariance function, necessary to compute the covariance matrix $\sum_{\mathbf{GG}}$ and any random realisation $\mathbf{g} = [\mathbf{g}_1{}^T \ \mathbf{g}_2{}^T \ \cdots \ \mathbf{g}_N{}^T]^T$ from the KL expansion in Eq. (4.30), is generally not known. Therefore, model covariance functions are used with a number of unknown parameters. Since any random realisation $\mathbf{g}$ is a sample of the multivariate Gaussian distribution $\mathcal{N}(0, \sum_{\mathbf{GG}})$, the unknown parameters can be determined by the maximisation of the log-marginal likelihood function (MLII) [37]

$$\log \mathcal{N}\left(\mathbf{0}, \sum\nolimits_{\mathbf{GG}}\right) = -\frac{1}{2}\left(\mathbf{g}^T \sum\nolimits_{\mathbf{GG}}{}^{-1} \mathbf{g} + \log\left|\sum\nolimits_{\mathbf{GG}}\right| + 2p\log 2\pi\right) \tag{4.33}$$

The maximisation is generally performed with respect to the unknown parameters in the covariance matrix for a given known observed realisation $\mathbf{g}$.

4.4 Stochastic characterisation and continuum realisation of fibre network

Direct random fibre network simulation is computationally prohibitive. Therefore, the random nature of fibre networks must be assessed using stochastic continuum modelling. In this section, a continuum model developed to simulate isotropic fibre networks is described. The continuum representation of the random fibre network includes three main steps:

(i) characterisation of stochasticity in random fibre networks
(ii) continuum random generation
(iii) continuum mechanical simulation

In this section, these steps are thoroughly described based on the technical background presented in Section 4.3. At the end of this chapter, the applicability of the proposed stochastic multiscale model is demonstrated by studying random failures in paper-making machines.

4.4.1 Characterisation of stochasticity in random fibre networks

Random realisations of 3D fibre networks can be modelled by statistically equivalent two-dimensional (2D) random continuum realisations (see Fig. 4.10).

This is performed using spatial field representation of material properties. For the isotropic fibre network, two spatial fields have been shown to be sufficient for the modelling [19], namely the strain to failure $\varepsilon_f(\mathbf{r},\omega)$ and strength $\sigma_f(\mathbf{r},\omega)$, where $\mathbf{r} = (x, y)^T$ is the spatial vector and $\omega \in \Omega$ denotes an element of the sample space indicating that the involved quantity is random. These two spatial fields are characterised based on direct mechanical simulation on stochastic volume elements (SVE) as is described in the following.

4.4.1.1 Stochastic volume elements

Multiscale mechanical modelling of thin fibre networks is a powerful technique aimed at predicting the mechanical response at the macro-mechanical level, based on properties at the meso- and microscale. The upscaling methodology in multiscale methods generally relies on either representative volume element (RVEs) [38] or SVEs [39].

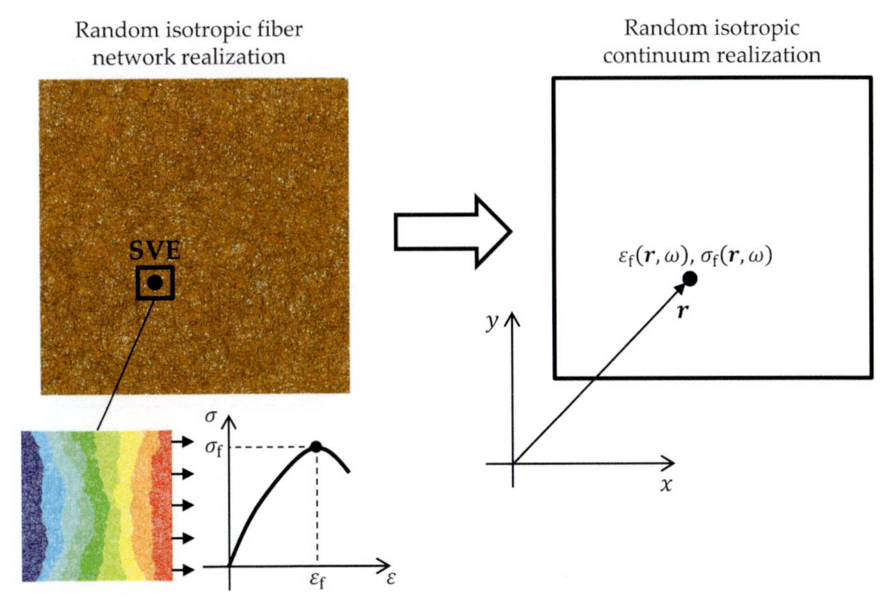

Fig. 4.10 A random realisation of a fibre network using a deposition technique and an equivalent continuum realisation based on random spatial fields of strain and strength to failure [19].

An RVE is defined as the smallest volume over which statistical representation can be made for the considered material property. It is used to determine the corresponding effective properties for a homogenised macroscopic model. According to the micro–meso–macro principle [40], the RVE should be sufficiently large to contain representative information about the microstructure, but its size should be small compared to the studied macroscopic body.

An SVE is smaller than the RVE. Its mechanical response generally faces two sources of uncertainties: one contribution resulting from the applied boundary conditions and the other one from the uncertainties in the microstructure. If a large enough SVE is chosen, the influence of boundary conditions is small [41]. However, uncertainties resulting from the microstructure randomness remain. The use of SVE-based approaches is beneficial in that local variability and strain localisation patterns may be captured accurately. In the following, a stochastic multiscale approach based on SVEs is presented.

4.4.1.2 Sampling of spatial fields

The first step in the stochastic multiscale approach is the sampling of the spatial fields. Since the fibre network is isotropic, it is sufficient to sample the spatial fields along one coordinate direction (see Fig. 4.11). A randomly generated fibre network of length L is therefore discretised using SVEs of size $L_{\mathrm{SVE}} \times L_{\mathrm{SVE}}$. The discretisation distance ΔL is the distance between the centres of two neighbouring SVEs corresponding to material points in the continuum model. All SVEs are cut from the fibre network and a numerical uniaxial test is performed. The resulting one-dimensional spatial fields of the strain to failure $\varepsilon_{\mathrm{f}}(x)$ and strength $\sigma_{\mathrm{f}}(x)$ are shown in Fig. 4.12. The latter

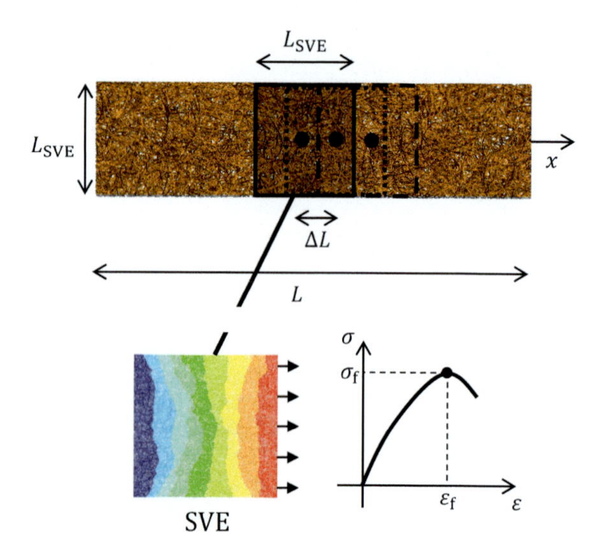

Fig. 4.11 Samples used to characterise the spatial random fields of ε_{f} and σ_{f} with discretisation distance ΔL.

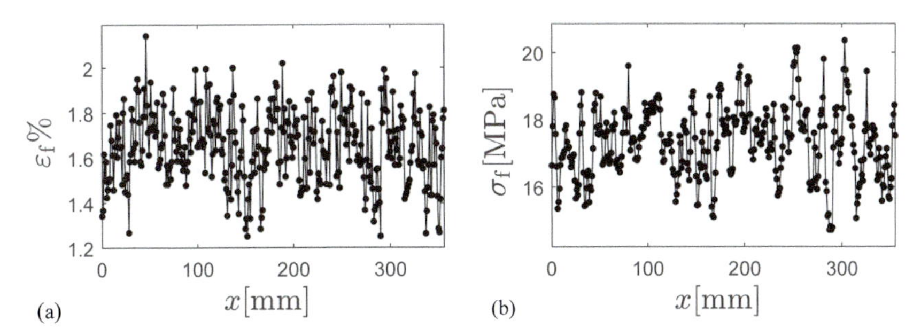

Fig. 4.12 Sampled realisation of the spatial random fields of (A) ε_f and (B) σ_f along the x-coordinate.

is defined as the ratio of the applied force on the SVE and the average thickness of the whole fibre network.

4.4.1.3 Marginal probability distributions

The next step is to fit a probability density function for the random variables $\varepsilon_f(\omega)$ and $\sigma_f(\omega)$. In Fig. 4.13, histograms of m random realisations of ε_{fi} and σ_{fi} are shown. These are generated by cutting SVEs from random spatial locations in a large fibre network. The corresponding fitted probability density functions (PDF), $f_{\varepsilon_f}(\varepsilon_f)$ and $f_{\sigma_f}(\sigma_f)$, are approximated based on the kernel density estimators

$$\begin{cases} f_{\varepsilon_f}(\varepsilon_f) = \dfrac{1}{m h_{\varepsilon_f}} \displaystyle\sum_{i=1}^{m} \varphi\left(\dfrac{\varepsilon_{fi} - \varepsilon_f}{h_{\varepsilon_f}}\right) \\ f_{\sigma_f}(\sigma_f) = \dfrac{1}{m h_{\sigma_f}} \displaystyle\sum_{i=1}^{m} \varphi\left(\dfrac{\sigma_{fi} - \sigma_f}{h_{\sigma_f}}\right) \end{cases}, \tag{4.34}$$

where $\varphi(\cdot)$ is the standard normal density function. The fitted smoothing bandwidths are given by $h_{\varepsilon_f} = 0.0693$ and $h_{\sigma_f} = 0.389$.

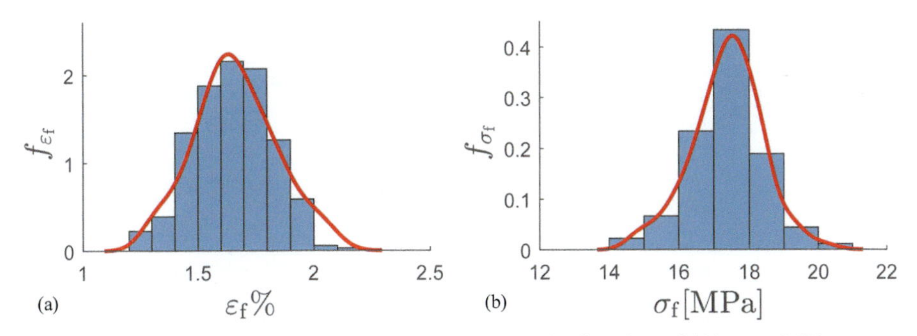

Fig. 4.13 Kernel approximation of the probability density function of (A) ε_f and (B) σ_f.

4.4.1.4 Transformation to Gaussian spatial fields

The third step is to transform the sampled spatial field realisations of $\varepsilon_f(x)$ and $\sigma_f(x)$ shown in Fig. 4.12 to Gaussian random field realisations, $g_1(x)$ and $g_2(x)$. It should be emphasised that this step is performed to facilitate the modelling of the covariance function and that simulated random Gaussian fields realisations (Section 4.4.2) are later transformed back to the original spatial fields. The transformation to the Gaussian random realisations is performed using the isoprobabilistic transformation,

$$\begin{cases} g_1(x,\omega) = \Phi^{-1}\{F_{\varepsilon_f}[\varepsilon_f(x,\omega)]\} \\ g_2(x,\omega) = \Phi^{-1}\{F_{\sigma_f}[\sigma_f(x,\omega)]\} \end{cases}, \tag{4.35}$$

where $F_{\varepsilon_f}(\varepsilon_f)$ and $F_{\sigma_f}(\sigma_f)$ are the cumulative density functions (CDFs) found by integration of the kernel density estimators according to Eq. (4.34), and $\Phi(\cdot)$ is the standard normal CDF. Applying the transformation yields the results shown in Fig. 4.14.

4.4.1.5 Correlation coefficient

An important characteristic of the Gaussian spatial fields is their correlation coefficient. It is computed from the one-dimensional realisations in Fig. 4.14 using Eq. (4.24) according to

$$\rho_{12} = \frac{1}{N}\sum_{i=1}^{N} g_1(x_i)g_2(x_i), \tag{4.36}$$

resulting in $\rho_{12} = 0.55$. This is a target value that needs to be satisfied when simulating new random realisations $g_1(\mathbf{r},\omega)$ and $g_2(\mathbf{r},\omega)$. It is noted that the correlation coefficient in the transformed Gaussian random variable space is generally different from that in the original random variable space, refer to Section 4.3.5.

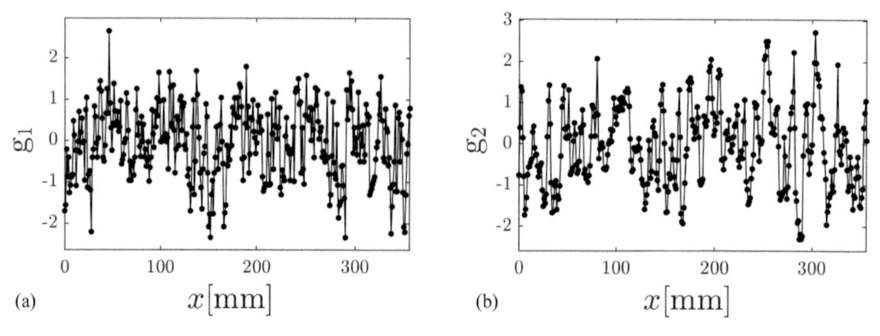

Fig. 4.14 Realisation of correlated Gaussian random spatial fields given by isoprobabilistic transformation of sampled spatial fields of (A) $\varepsilon_f(x)$ and (B) $\sigma_f(x)$.

4.4.1.6 Zero-level upcrossings

An important characteristic of each of the Gaussian spatial fields is the average distance between two zero level upcrossings denoted by μ_{upcr1} and μ_{upcr2}, respectively. These are computed based on the number of upcrossings in the interval $[0, L]$, where L is the network length defined in Fig. 4.11, according to

$$N_{\text{upcr}i} \triangleq \# \left\{ x \in [0, L] : g_i(x) = 0, \frac{dg_i}{dx} > 0 \right\}, \quad i = 1, 2 \tag{4.37}$$

where $g_1(x)$ and $g_2(x)$ are the sampled realisations in Fig. 4.14. The average distance between two zero level upcrossings is therefore given by $\mu_{\text{upcr}i} = \frac{L}{N_{\text{upcr}i} - 1}$. From Fig. 4.15, it can be seen that a sampling length of $L = 250$ mm is sufficiently large for convergence to approximately stable values of $\mu_{\text{upcr1}} = 6.3$ mm and $\mu_{\text{upcr2}} = 11.9$ mm. These are target values that need to be satisfied when simulating new realisations $g_1(\mathbf{r}, \omega)$ and $g_2(\mathbf{r}, \omega)$. A summary of the parameters that characterise the random fibre network is shown in Table 4.4.

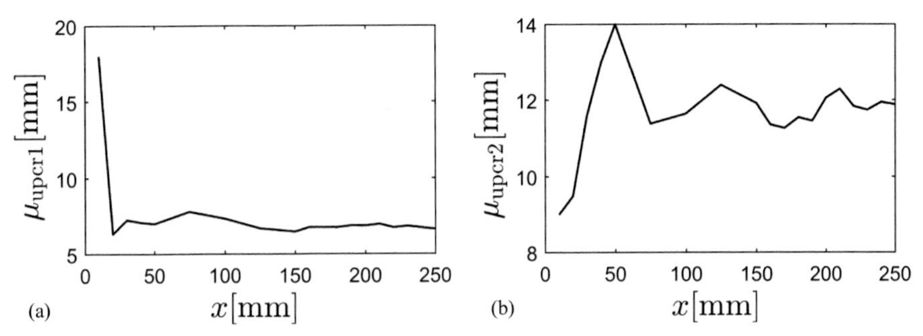

Fig. 4.15 Average distance between two zero-level upcrossings for the sample realisation (A) g_1 and (B) g_2 as a function of specimen length.

Table 4.4 Summary of parameters characterising the randomness in the studied fibre network.

Target Gaussian random field properties			**PDF kernel density estimator bandwidths for $f_{\varepsilon_f}(\varepsilon_f)$ and $f_{\sigma_f}(\sigma_f)$**	
ρ_{12}	μ_{upcr1}	μ_{upcr2}	h_{ε_f}	h_{σ_f}
0.55	6.3 mm	11.9 mm	0.0693	0.389

4.4.2 Random generation

4.4.2.1 Modelling of auto-covariance and cross-covariance functions

The square exponential multivariate covariance function

$$\mathrm{Cov}\left(g_i(\mathbf{r},\omega), g_j(\mathbf{r}',\omega)\right) = \rho_{ij}\exp\left(-\frac{|\mathbf{r}-\mathbf{r}'|^2}{2l_{ij}^2}\right), \quad \rho_{ii}=1, \tag{4.38}$$

is used to model the unknown underlying covariance. In Eq. (4.38), $\mathbf{r}$ and $\mathbf{r}'$ are two spatial location vectors, ρ_{ij} is the correlation coefficient, and ℓ_{ij} are characteristic lengths to be determined. As can be seen, the auto-covariance function $\exp\left(\frac{|\mathbf{r}-\mathbf{r}'|^2}{2\ell_{ii}^2}\right)$ is the well-known square exponential kernel where it is noted that $\rho_{ii}=1$. The cross-covariance function $\rho_{ij}\exp\left(\frac{|\mathbf{r}-\mathbf{r}'|^2}{2\ell_{ij}^2}\right)$, $i \neq j$, is a multivariate extension of the univariate case. As noted in Ref. [42], a characterisation of parameters for the validity of this multivariate extension is still ongoing research in the field of mathematical statistics. For the studied bivariate spatial field, the correlation coefficient is $\rho_{12}=0.55$ (Table 4.4) and the unknown parameters are ℓ_{11}, ℓ_{22}, and ℓ_{12}. These are determined by maximisation of the log-marginal likelihood function $\log\mathcal{N}\left(0, \sum_{\mathbf{GG}}\right)$ where $\sum_{\mathbf{GG}}$ is the bivariate covariance matrix (see Section 4.3.5), subject to constraints $\mu_{\mathrm{upcr1}}=6.3$ mm and $\mu_{\mathrm{upcr2}}=11.9$ mm (Table 4.4). This maximisation is equivalent to the minimisation problem

$$\begin{cases} \min\limits_{l_{11},l_{22},l_{12}} \mathbf{g}(\omega)^T {\sum_{\mathrm{GG}}}^{-1}\mathbf{g}(\omega) + \log\left|\sum_{\mathrm{GG}}\right| \\ \mu_{\mathrm{upcr1}} = 6.3\mathrm{mm} \\ \mu_{\mathrm{upcr2}} = 11.9\mathrm{mm} \end{cases} \tag{4.39}$$

where

$$\mathbf{g}(\omega) = \left[\mathbf{g}_1(\omega)^T\ \mathbf{g}_2(\omega)^T\right], \tag{4.40}$$

is a random realisation of the discretised bivariate Gaussian spatial field. If the spatial fields are discretised at p spatial locations $\mathbf{r}_1, \mathbf{r}_2, \ldots, \mathbf{r}_p$, the component random vectors can be written as $\mathbf{g}_1(\omega) = [\mathrm{g}_1(\mathbf{r}_1,\omega)\ \cdots \mathrm{g}_1(\mathbf{r}_p,\omega)]^T$ and $\mathbf{g}_2(\omega) = [\mathrm{g}_2(\mathbf{r}_1,\omega)\ \cdots\ \mathrm{g}_2(\mathbf{r}_p,\omega)]^T$. The multivariate covariance matrix is composed of auto-covariance matrices and the cross-covariance matrix such as

$$\sum_{\mathbf{GG}} = \begin{bmatrix} \sum_{\mathrm{G_1G_1}} & \sum_{\mathrm{G_1G_2}} \\ \sum_{\mathrm{G_2G_1}} & \sum_{\mathrm{G_2G_2}} \end{bmatrix}.$$

The auto-covariance matrices $\sum_{\mathbf{G}i\mathbf{G}i} \in \mathbb{R}^{p\times p}$ are computed using Eq. (4.38) as

$$\sum\nolimits_{\mathbf{G}_1\mathbf{G}_1} = \begin{bmatrix} 1 & \cdots & \exp\left(-\dfrac{|\mathbf{r}_1-\mathbf{r}_p|^2}{2l_{11}{}^2}\right) \\ \vdots & \ddots & \vdots \\ \exp\left(-\dfrac{|\mathbf{r}_1-\mathbf{r}_p|^2}{2l_{11}{}^2}\right) & \cdots & 1 \end{bmatrix} \tag{4.41}$$

and

$$\sum\nolimits_{\mathbf{G}_2\mathbf{G}_2} = \begin{bmatrix} 1 & \cdots & \exp\left(-\dfrac{|\mathbf{r}_1-\mathbf{r}_p|^2}{2l_{22}{}^2}\right) \\ \vdots & \ddots & \vdots \\ \exp\left(-\dfrac{|\mathbf{r}_p-\mathbf{r}_I|^2}{2l_{22}{}^2}\right) & \cdots & 1 \end{bmatrix}. \tag{4.42}$$

Similarly, from Eq. (4.38), the cross-covariance matrix $\sum_{\mathbf{G}1\mathbf{G}2} \in \mathbb{R}^{p\times p}$ is given by

$$\sum\nolimits_{\mathbf{G}_1\mathbf{G}_2} = \begin{bmatrix} \rho_{12} & \cdots & \rho_{12}\exp\left(-\dfrac{|\mathbf{r}_1-\mathbf{r}_p|^2}{2l_{12}{}^2}\right) \\ \vdots & \ddots & \vdots \\ \rho_{12}\exp\left(-\dfrac{|\mathbf{r}_1-\mathbf{r}_p|^2}{2l_{12}{}^2}\right) & \cdots & \rho_{12} \end{bmatrix}. \tag{4.43}$$

The random realisations can be simulated using the KL expansion (Section 4.3.5) according to

$$\mathbf{g}(\omega) = \sum_{k=1}^{pN} \sqrt{\lambda_k}\mathbf{h}_k\xi_k(\omega), \tag{4.44}$$

where $\lambda_1, \ldots, \lambda_{2p}$ and $\mathbf{h}_1, \ldots, \mathbf{h}_{2p}$ are eigenvalues and eigenvectors, respectively, of the covariance matrix $\sum_{\mathbf{GG}}$ and $\xi_1, \ldots, \xi_{2p}$ are standard independent normal variables.

It is noted that both $\mathbf{g}(\omega)$ and $\sum_{\mathbf{GG}}$ in the objective function in Eq. (4.39) are functions of the unknown parameters ℓ_{11}, ℓ_{22}, and ℓ_{12}. The average distance between two zero-level upcrossings, μ_{upcr1} and μ_{upcr2}, are computed from the random realisations $\mathbf{g}(\omega)$ and are therefore also functions of the unknown parameters. Since the covariance function is isotropic, random realisations along one coordinate direction are sufficient for the determination of the unknown parameters (Table 4.5).

Table 4.5 Parameters in the auto- and cross-covariance function of the Gaussian random fields.

ρ_{12}	ℓ_{11}	ℓ_{22}	ℓ_{12}
0.55	0.85	1.95	1.25

4.4.2.2 Simulation of random spatial fields of strength and strain to failure

To generate random realisations of the 2D correlated non-Gaussian spatial fields of strain to failure $\varepsilon_f(\mathbf{r},\omega)$and strength $\sigma_f(\mathbf{r},\omega)$, simulation of correlated Gaussian random fields $g_1(\mathbf{r},\omega)$ and $g_2(\mathbf{r},\omega)$ is first performed. Given the covariance function in Eq. (4.38) with parameters in Table 4.4, random Gaussian realisations over a discretised 2D spatial domain can be generated using the KL expansion according to Eq. (4.44). A transformation is thereafter performed according to

$$\begin{cases} \varepsilon_f(\mathbf{r},\omega) = F_{\varepsilon_f}{}^{-1}\{\Phi[g_1(\mathbf{r},\omega)]\} \\ \sigma_f(\mathbf{r},\omega) = F_{\sigma_f}{}^{-1}\{\Phi[g_2(\mathbf{r},\omega)]\} \end{cases}. \tag{4.45}$$

An example of random generated spatial fields of strength and strain to failure corresponding to an 18 mm × 18 mm fibre network is shown in Fig. 4.16.

4.4.3 Continuum mechanical simulation

In the following, a constitutive model of the 6 mm × 6 mm SVE response is presented. The model is a function of the SVE strength σ_f and strain to failure ε_f.

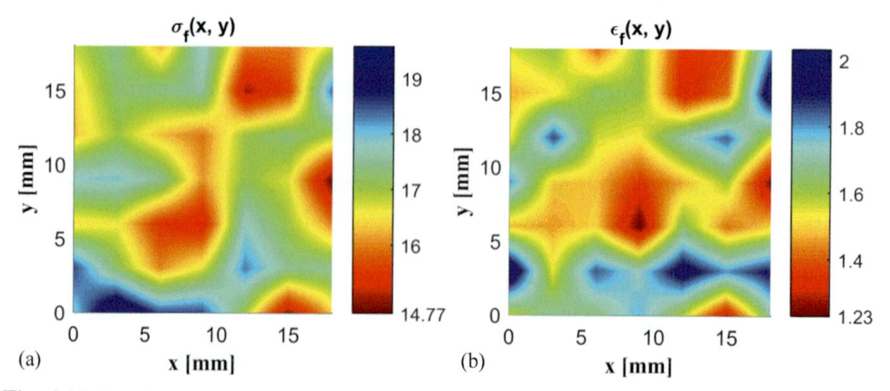

Fig. 4.16 Randomly simulated spatial field of (A) strength σ_f (MPa) and (B) strain to failure ε_f (%).

4.4.3.1 SVE-based constitutive model

A constitutive model can be written in rate form as $\dot{\boldsymbol{\sigma}} = \mathbf{D}_{\text{tan}}\,\dot{\boldsymbol{\varepsilon}}$ where $\boldsymbol{\sigma} = [\sigma_x, \sigma_y, \tau_{xy}]^T$ is the stress vector, $\boldsymbol{\varepsilon} = [\varepsilon_x, \varepsilon_y, \gamma_{xy}]^T$ is the strain vector and $\mathbf{D}_{\text{tan}} = \mathbf{D}_{\text{el}} - \mathbf{D}_{\text{pl}}$ is the local tangent stiffness matrix computed from the elastic and elastic-plastic stiffness matrix $\mathbf{D}_{\text{el}}$ and $\mathbf{D}_{\text{pl}}$, respectively. For the isotropic case, the elastic stiffness matrix with a plane stress condition can be written as

$$\mathbf{D}_{\text{el}} = \begin{bmatrix} \frac{E}{1-\nu^2} & \frac{\nu E}{1-\nu^2} & 0 \\ \frac{\nu E}{1-\nu^2} & \frac{E}{1-\nu^2} & 0 \\ 0 & 0 & G \end{bmatrix}, \tag{4.46}$$

where E is the elastic modulus, $G = \frac{E}{2(\nu+1)}$ is the shear modulus, and $\nu = 0.293$ [43] is Poisson's ratio. An isotropic hardening is assumed, resulting in the following expression for the elastic-plastic stiffness matrix

$$\mathbf{D}_{\text{pl}} = \frac{\mathbf{D}_{\text{el}} \frac{\partial f}{\partial \boldsymbol{\sigma}} \frac{\partial f}{\partial \boldsymbol{\sigma}}^T \mathbf{D}_{\text{el}}}{\frac{\partial \sigma_{\text{flow}}}{\partial \varepsilon_{\text{e,pl}}} + \frac{\partial f}{\partial \boldsymbol{\sigma}}^T \mathbf{D}_{\text{el}} \frac{\partial f}{\partial \boldsymbol{\sigma}}}, \tag{4.47}$$

where $f(\boldsymbol{\sigma})$ and σ_{flow} are the yield surface and flow stress, respectively. The plastic behaviour is assumed to follow von Mises 2D yield criterion,

$$f(\boldsymbol{\sigma}) = \sqrt{\boldsymbol{\sigma}^T \mathbf{H} \boldsymbol{\sigma}} - \sigma_{\text{flow}} = 0, \tag{4.48}$$

where

$$\mathbf{H} = \begin{bmatrix} 1 & -\frac{1}{2} & 0 \\ -\frac{1}{2} & 1 & 0 \\ 0 & 0 & 3 \end{bmatrix}. \tag{4.49}$$

The flow stress follows a variant of a Bammann-Chiesa-Johnson (BCJ) model [44] according to

$$\frac{\sigma_{\text{flow}}}{\sigma_{\text{f}}} = \left(1 - f_1(\kappa, n)\left(\frac{\varepsilon_{\text{e,pl}}}{\varepsilon_{\text{f}}}\right)^n\right) f_2(\kappa, n) \tanh\left(\kappa \frac{\varepsilon_{\text{e,pl}}}{\varepsilon_{\text{f}}}\right), \quad \varepsilon_{\text{e,pl}} \leq \varepsilon_{\text{f}} \tag{4.50}$$

where the softening branch is modelled by

$$\frac{\sigma_{\text{flow}}}{\sigma_{\text{f}}} = -\xi\left(\frac{\varepsilon_{\text{e,pl}}}{\varepsilon_{\text{f}}} - 1\right)^2 + 1, \; \varepsilon_{\text{e,pl}} \geq \varepsilon_{\text{f}}. \tag{4.51}$$

In Eq. (4.51), $\varepsilon_{\mathrm{e,pl}} = \int \sqrt{\frac{2}{3} d\varepsilon_{ij} d\varepsilon_{ij,\mathrm{pl}}}$ is the equivalent plastic strain computed from the plastic strain increments $d\varepsilon_{ij,\ \mathrm{pl}}$, $\xi > 0$ is a parameter controlling the softening rate and

$$\begin{cases} f_1(\kappa, n) = \dfrac{\kappa}{n \sinh(\kappa) \cosh(\kappa) + \kappa} \\[2ex] f_2(\kappa, n) = \dfrac{n \sinh(\kappa) \cosh(\kappa) + \kappa}{n \sinh^2(\kappa)}, \end{cases} \tag{4.52}$$

where $\kappa > 0$ and the damage exponent $n > 0$ are constants to be determined. The relations in Eqs (4.50) and (4.51) are shown in Fig. 4.17A for different parameter combinations. As can be seen, the flow stress satisfies $\sigma_{\mathrm{flow}}(\varepsilon_{\mathrm{e,\ pl}} = \varepsilon_{\mathrm{f}}) = \sigma_{\mathrm{f}}$ and $\frac{\partial \sigma_{\mathrm{flow}}}{\partial \varepsilon_{\mathrm{e,pl}}}\Big|_{\varepsilon_{\mathrm{e,pl}} = \varepsilon_{\mathrm{f}}} = 0$. This is true for all combinations of κ and n, as is seen by inserting Eq. (4.52) in Eq. (4.50). It is also noted that Eq. (4.50) implies instantaneous yielding i.e. the yield stress is given by $\sigma_s = 0$. The elastic modulus used in Eq. (4.46) is given by differentiation of Eq. (4.50) such as $E = \frac{\partial \sigma_{\mathrm{flow}}}{\partial \varepsilon_{\mathrm{e,pl}}}\Big|_{\varepsilon_{\mathrm{e,pl}} = 0}$ i.e.

$$E = \frac{n \sinh(\kappa) \cosh(\kappa) + \kappa}{n \sinh^2(\kappa)} \frac{\sigma_{\mathrm{f}}}{\varepsilon_{\mathrm{f}}}. \tag{4.53}$$

For a given damage exponent n, statistical analysis reveals a strong correlation between the parameter κ and the strain to failure ε_{f}. A linear relation of the form

$$\kappa = c_1 + c_2 \varepsilon_{\mathrm{f}} + e, \tag{4.54}$$

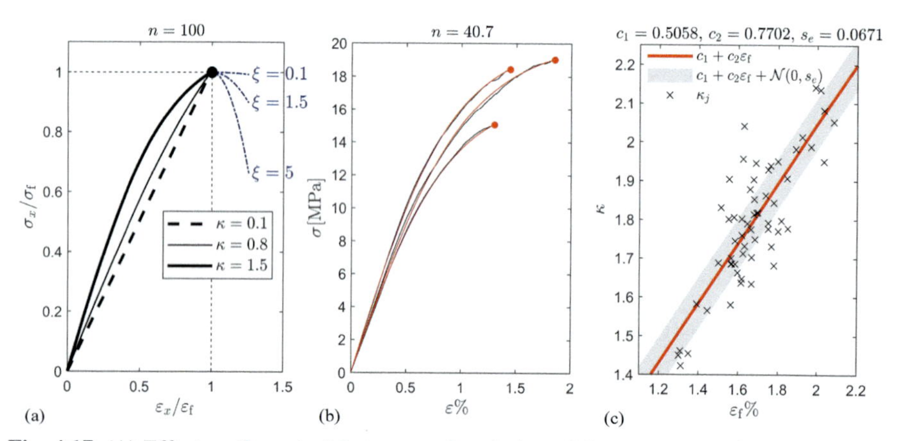

Fig. 4.17 (A) Effect on the uniaxial stress-strain relation of the parameter κ for a constant damage parameter $n = 100$. (B) Response from direct simulation and fitted models for three 6 mm × 6 mm SVEs. (C) Fitting of Eq. (4.54) to direct fibre network simulation.

Table 4.6 Constitutive model parameters for 6 mm × 6 mm SVEs.

n	c_1	c_2	s_e	ξ	ν
40.7	0.5058	0.7702	0.0671	0.15	0.293

is therefore assumed where e is the fitting error. The three fitting parameters in the constitutive model, n, c_1, and c_2, are found by minimising the least-square error between model prediction and a large number of 6 mm × 6 mm (SVE) fibre network uniaxial responses. The fitted parameter values are shown in Table 4.6. In Fig. 4.17B, both model and fibre network responses are shown for three different SVEs.

For each of the N_{SVE} SVEs used to fit the linear relation in Eq. (4.54), an error $e_j = \kappa_j - (c_1 + c_2\varepsilon_{\mathrm{f},j}), j = 1, \ldots, N_{\mathrm{SVE}}$ can be computed. These errors can be regarded as random realisations of the epistemic uncertainty [45] variable $e \sim \mathcal{N}(0, s_e^2)$ where $s_e^2 = \frac{1}{N_{\mathrm{SVE}}} \sum_{j=1}^{N_{\mathrm{SVE}}} e_j^2$. In Fig. 4.17C, the linear fit is shown together with the epistemic uncertainty corresponding to 1 standard deviation (grey-shaded region). It is noted that, for a given ε_{f} in Eq. (4.54), the normally distributed uncertainty parameter e results in a normally distributed $\kappa \sim \mathcal{N}(c_1 + c_2\varepsilon_{\mathrm{f}}, s_e^2)$.

The softening part of the SVE response is modelled by Eq. (4.51) where $\xi > 0$ is a parameter controlling the softening rate of the SVE response (see Fig. 4.17A). A small softening parameter $\xi = 0.15$ is chosen to assure numerical stability during mechanical simulation of a large sample composed of many SVEs, but large enough for strain localisation to onset in the larger sample (see Sections 4.4.3.2 and 4.4.4.1). It is noted that the response after strain localisation is not of interest in this study since failure is defined as the onset of strain localisation. Furthermore, for continuum FE models, the behaviour at the softening branch of the response is generally mesh dependent.

4.4.3.2 Finite-element implementation

The determined parameters n, c_1, c_2, and s_e and the spatial fields realisations, $\varepsilon_{\mathrm{f}}(\mathbf{r})$ and $\sigma_{\mathrm{f}}(\mathbf{r})$, are used to specify the elemental constitutive material properties in a finite-element (FE) simulation. The randomly generated spatial fields are applied on the integration points and extrapolated to the nodes. This ensures a continuous material property spatial field.

Consider the 18 mm ×18 mm simulated spatial fields of ε_{f} and σ_{f} in Fig. 4.16. In Fig. 4.18, the corresponding mechanical response based on these spatial fields, using the constitutive model in Section 4.4.3.1, is shown. It should be emphasised that the randomness in the model is not only determined by ε_{f} and σ_{f} but also depends on s_e. This can be seen from the expression of the stiffness according to Eq. (4.54), which is a function of the random parameter $\kappa \sim \mathcal{N}(c_1 + c_2\varepsilon_{\mathrm{f}}, s_e^2)$. In Fig. 4.18B, the uniaxial strain defined as the spatial derivative of the displacement field, is computed for the maximum applied stress marked in red in Fig. 4.18A.

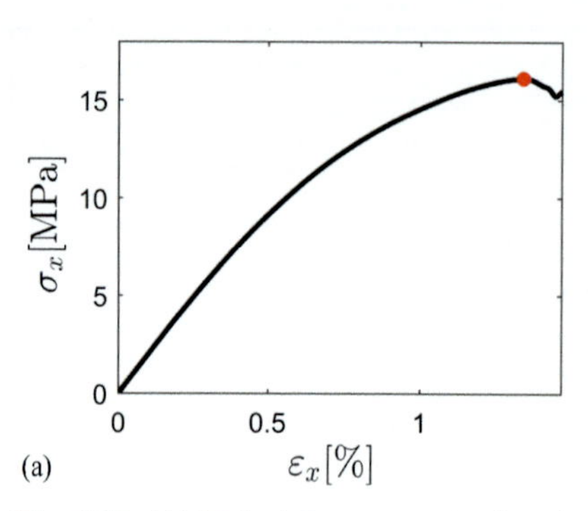

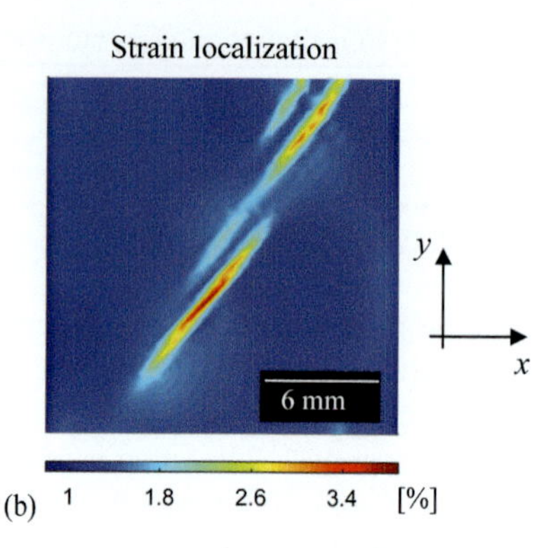

Fig. 4.18 (A) Uniaxial response (x-direction) of an 18 mm × 18 mm sample based on the stochastic constitutive model and (B) corresponding spatial distribution of uniaxial strain (%) at the maximum applied stress (marked by a red point in (A)).

A 1 mm × 1 mm finite-element size is used in the simulation i.e. each 6 mm × 6 mm SVE is composed of 36 finite elements. The choice of element size is based on the strain localisation bands. From direct fibre network simulations (see Section 4.4.4.1), it is observed that the width of the strain localisation bands is roughly 1 mm. Although the width of the strain localisation bands depends on the size of the mesh, the mechanical response up to the maximum stress, corresponding to the onset of strain localisation, is mesh independent.

4.4.4 Method applicability

4.4.4.1 Validation

Validation of the methodology was performed by comparing results from uniaxial tensile tests on the fibre network and the SVE-based continuum model. The boundary conditions applied to the fibre network are shown in Fig. 4.19A. For the proposed continuum model, the same boundary conditions would result in stress concentration and erroneous strain localisation initiation. Therefore, the clamped ends in the fibre network were modelled in the continuum model using a contact boundary condition (see Fig. 4.19B). Very low contact penalty stiffness was used to allow free contraction at the boundaries. To avoid uneven deformation at the ends due to the low contact stiffness, the boundary nodes were coupled with respect to their axial deformation u_x.

A uniaxial tensile test was performed on a specimen of size 18 mm × 18 mm using direct fibre network simulation and the continuum model based on the 6 mm × 6 mm SVE size. The constitutive model parameters used are according to Table 4.6. The spatial field sampling distance (see Fig. 4.11) i.e. the distance between the centres of the SVEs, is half of the SVE length (3 mm) in both the length and width direction. The resulting spatial distributions of stress σ_f and strain to failure ε_f are presented in Fig. 4.20. As can be seen, the weakest material point is at the lower right part of the specimen and has a strength value of 14.86 MPa.

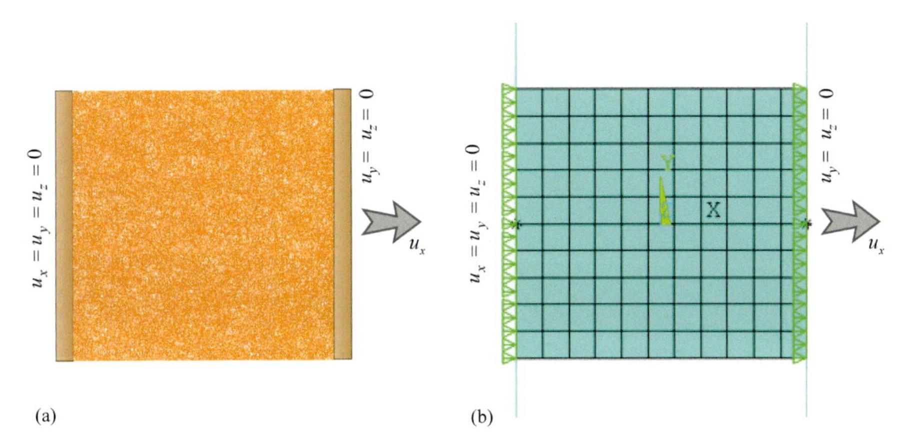

Fig. 4.19 Finite element model and boundary conditions of a uniaxial test using (A) direct fibre simulation and (B) continuum FE model.

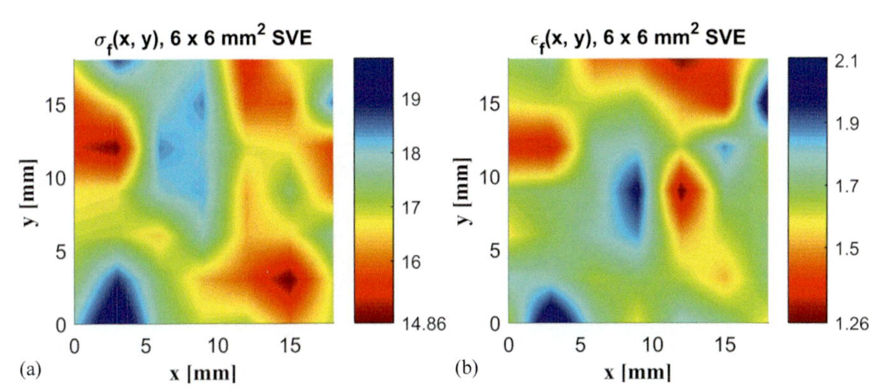

Fig. 4.20 Spatial distribution of (A) strength σ_{f} (MPa) and (B) strain to failure ε_{f} (%) using SVEs of size 6 mm $\times$ 6 mm and a spatial field sampling distance of 3 mm.

In Fig. 4.21A, the uniaxial response in the x-direction using the continuum model is shown together with the response from the direct simulation. The results are in good agreement with direct simulation, with a relative error in strength and strain to failure of -2% and -6%, respectively. In Fig. 4.21B, the responses are also compared for an applied load in the transverse y-direction, with overall performance similar to the x-direction i.e. 1% and 7% error in strength and strain to failure.

The spatial distribution of uniaxial strain in the loading (x-direction) is shown for direct fibre network simulation as well as the stochastic continuum model in Fig. 4.22A and B, respectively. By comparing to Fig. 4.20A, it is clear that strain localisation was initiated at weak material points having low local strength. It is also seen that the strain localisation pattern from direct simulation matches the ones from the continuum models relatively well.

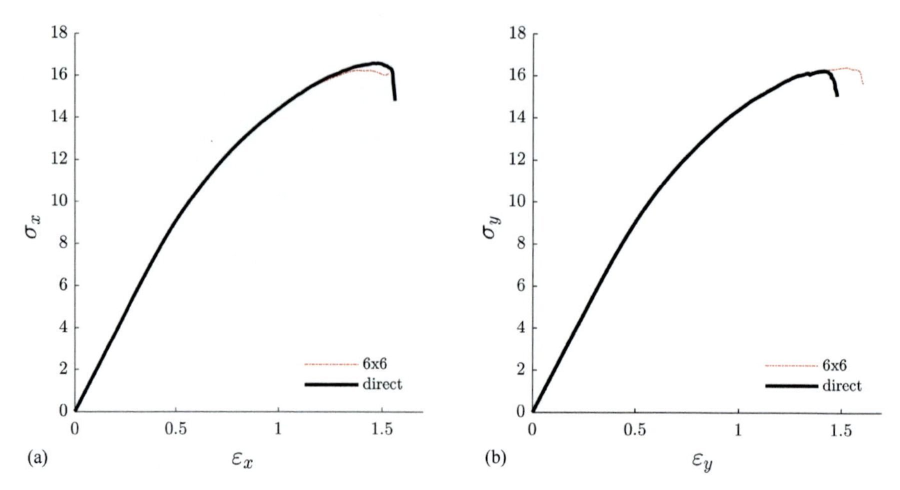

Fig. 4.21 Uniaxial response in the (A) x-direction and (B) y-direction using an SVE size of 6 mm × 6 mm.

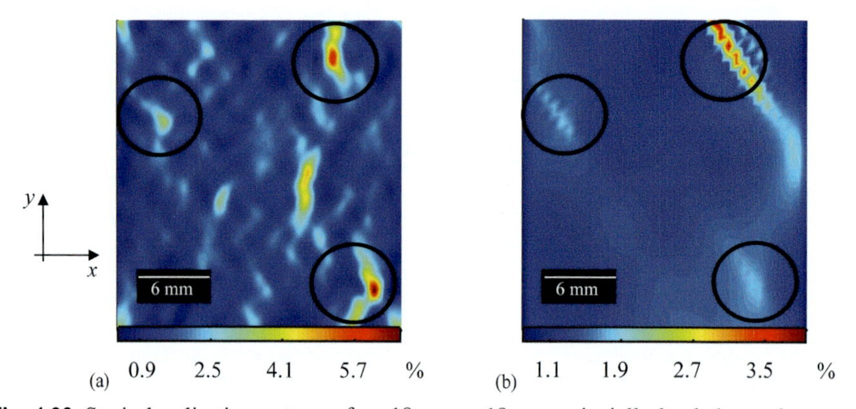

Fig. 4.22 Strain localisation pattern of an 18 mm × 18 mm uniaxially loaded sample (x-direction) using (A) direct simulation and (B) continuum model based on 6 mm × 6 mm SVEs.

4.4.4.2 Influence of SVE size

In the following, we will study the influence of decreasing the SVE size on both strain localisation pattern and uniaxial response. An SVE size of 4 mm × 4 mm is now used to analyse the same fibre network studied in Section 4.4.4.1. The spatial distribution of stress σ_{f} and strain to failure ε_{f} is shown in Fig. 4.23. As can be seen, the weakest material point using the 4 mm × 4 mm SVE is at the lower right part of the specimen and has a value of 13.90 MPa. The corresponding value using the 6 mm × 6 mm SVEs is 14.86 MPa at approximately the same location (see Fig. 4.20). For the smaller SVE

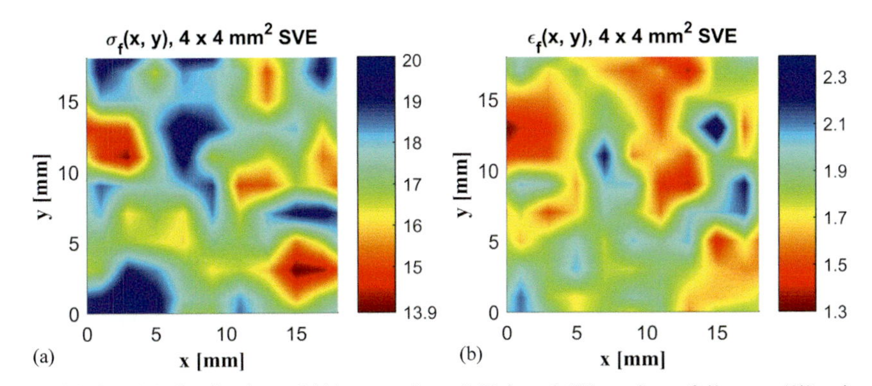

Fig. 4.23 Spatial distribution of (A) strength σ_f (MPa) and (B) strain to failure ε_f (%) using SVEs of size 4 mm × 4 mm and a spatial field sampling distance of 2 mm.

Table 4.7 Constitutive model parameters for 4 mm × 4 mm SVEs.

n	c_1	c_2	s_e	ξ	ν
34.6	0.4591	0.8017	0.0977	0.15	0.293

size, a higher variation in strength and strain to failure values can be observed within the specimen.

It is noted that the material parameters in Table 4.6 are not applicable for spatial fields based on the 4 mm × 4 mm SVE. Instead, the fitting procedure needs to be repeated for smaller SVE size. The corresponding new parameters are shown in Table 4.7.

In Fig. 4.24A and B the uniaxial response in the x- and y-direction is shown together with the response from the direct simulation. Overall, the performance is slightly better than the 6 mm × 6 mm SVE-based constitutive model (Fig. 4.21). The strain localisation pattern in the loading (x-direction) is shown for direct fibre network simulation as well as the stochastic continuum model in Fig. 4.25A and B. When compared to Fig. 4.22, it can be clearly seen that the use of 4 mm × 4 mm SVE-based constitutive model is superior in terms of predicting failure locations.

4.4.4.3 Random failure simulation of large fibre networks (paper machines)

The major potential of the stochastic continuum model is demonstrated by the construction of larger specimens for which results cannot be validated using direct numerical simulation. As an example of the applicability of the method, consider the random occurrence of breaks from paper-making machines. Every break cost around 6000€ and some paper machines can experience up to six breaks a day. With the speed of the machine reaching 2000 m/min, the break can, therefore, occur once per 480 km

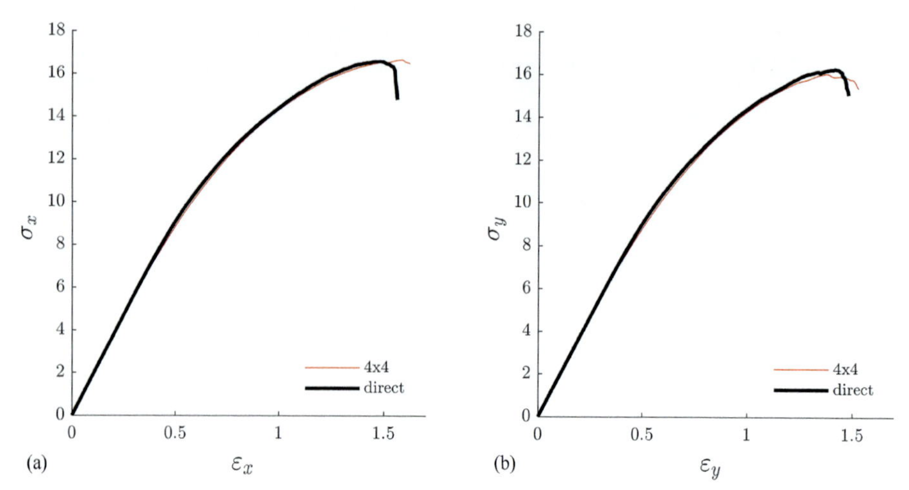

Fig. 4.24 Uniaxial response in the (A) x-direction and (B) y-direction using an SVE size of 4 mm × 4 mm.

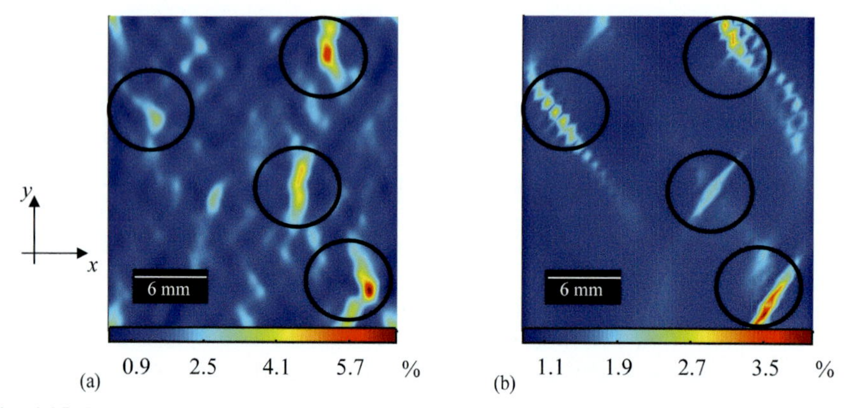

Fig. 4.25 Strain localisation pattern of an 18 mm × 18 mm uniaxially loaded sample (x-direction) using (A) direct simulation and (B) continuum model based on 4 mm × 4 mm SVE.

of produced paper, which makes it an extremely stochastic event. We take an example of a paper-making machine and the case where the breaks occur in an open draw in the section where paper is dry. The length of the open draw is 1 m [39] (see Fig. 4.26A).

Using the proposed method, random realisations of 1 m × 8 m samples can be simulated. Three random realisations of the spatial fields $\sigma_f(\mathbf{r}, \omega)$ and $\varepsilon_f(\mathbf{r}, \omega)$ are first generated using the method presented in Section 4.4.2. The uniaxial finite-element response for each of the three realisations (see Fig. 4.26B), is thereafter simulated using the 6 mm × 6 mm SVE-based constitutive model (see Section 4.4.3). Similar to Section 4.4.4.1, a finite-element size of 1 mm × 1 mm is used. A relatively small variation in strength can be observed. The variation in corresponding strain to failure

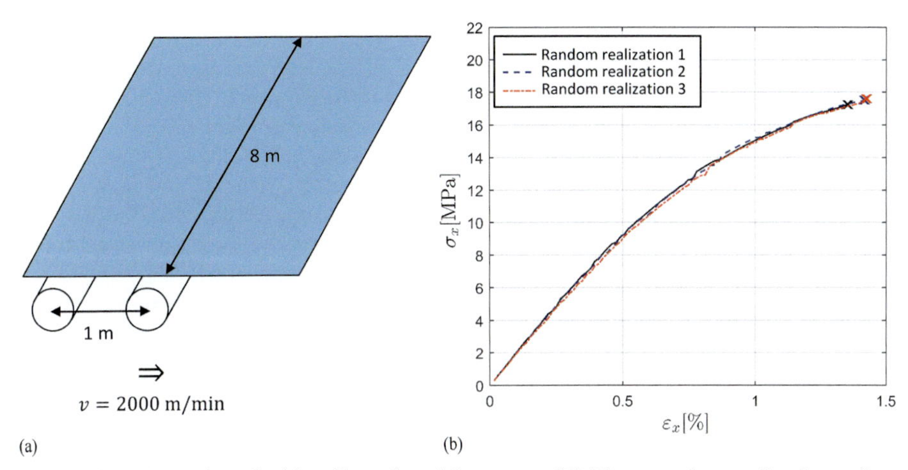

Fig. 4.26 (A) Paperboard with rollers placed 1 m apart. (B) Three random realisations of 1 m × 8 m dry samples and corresponding uniaxial response.

is, however, larger. This statistical variation in material response demonstrated using the proposed model, together with the variation in the applied load, can be seen as a major reason for the random occurrence of breaks in paper machines. From this example, it can be concluded that the disordered nature of the fibre network contributes to the statistical variation in mechanical response, even for such large specimens.

4.5 Summary

Stochastic multiscale modelling of thin fibre networks considering failure at the product scale is a challenging field of research. Herein, we have introduced a computationally efficient approach and demonstrated its applicability. The important steps including fibre network characterisation, micromechanical simulation, and random multivariate spatial field modelling have been described.

The presented approach enables the propagation of uncertainties at the microscale to the macroscale. This paves the way to a better understanding of the origin of size dependency in material properties, such as strength scaling and weakest-link assumptions [2,46]. Furthermore, by modelling and efficiently propagating the variability at different scales, future research challenges may be directed into tailoring the fibre properties to reduce the variability in the end-product performance.

References

[1] D.W. Coffin, P.-J. Gustafsson, R. Hägglund, A. Kulachenko, P. Mäkelä, M. Nygards, S. Östlund, T. Uesaka, K. Niskanen, L. Berglund, L.A. Carlsson, T. Uesaka, 8. Statistical aspects of failure of paper products, in: Mechanics of Paper Products, De Gruyter, 2012, https://doi.org/10.1515/9783110254631.139.

[2] D.T. Hristopulos, T. Uesaka, Structural disorder effects on the tensile strength distribution of heterogeneous brittle materials with emphasis on fiber networks, Phys. Rev. B Condens. Matter Mater. Phys. 70 (2004), https://doi.org/10.1103/PhysRevB.70.064108.
[3] T. Uesaka, J. Juntunen, Time-dependent, stochastic failure of paper and box, Nord. Pulp Pap. Res. J. 27 (2012) 370–374, https://doi.org/10.3183/NPPRJ-2012-27-02-p370-374.
[4] A. Mattsson, T. Uesaka, Characterisation of time-dependent, statistical failure of cellulose fibre networks, Cellulose 25 (2018) 2817–2828, https://doi.org/10.1007/s10570-018-1776-5.
[5] A. Kulachenko, T. Uesaka, Direct simulations of fiber network deformation and failure, Mech. Mater. 51 (2012) 1–14, https://doi.org/10.1016/j.mechmat.2012.03.010.
[6] S. Borodulina, H.R. Motamedian, A. Kulachenko, Effect of fiber and bond strength variations on the tensile stiffness and strength of fiber networks, Int. J. Solids Struct. 154 (2018) 19–32, https://doi.org/10.1016/j.ijsolstr.2016.12.013.
[7] S. Borodulina, A. Kulachenko, D.D. Tjahjanto, Constitutive modeling of a paper fiber in cyclic loading applications, Comput. Mater. Sci. 110 (2015) 227–240, https://doi.org/10.1016/j.commatsci.2015.08.039.
[8] S. Borodulina, A. Kulachenko, S. Galland, M. Nygårds, Stress-strain curve of paper revisited, Nord. Pulp Pap. Res. J. 27 (2012) 318–328, https://doi.org/10.3183/NPPRJ-2012-27-02-p318-328.
[9] A. Miletzky, W.J. Fischer, C. Czibula, C. Teichert, W. Bauer, R. Schennach, How xylan effects the breaking load of individual fiber–fiber joints and the single fiber tensile strength, Cellulose 22 (2015) 849–859, https://doi.org/10.1007/s10570-014-0532-8.
[10] A. Torgnysdotter, A. Kulachenko, P. Gradin, L. Wågberg, Fiber/fiber crosses: finite element modeling and comparison with experiment, J. Compos. Mater. 41 (2007) 1603–1618, https://doi.org/10.1177/0021998306069873.
[11] C. Lorbach, U. Hirn, J. Kritzinger, W. Bauer, Automated 3D measurement of fiber cross section morphology in handsheets, Nord. Pulp Pap. Res. J. 27 (2012) 264–269, https://doi.org/10.3183/NPPRJ-2012-27-02-p264-269.
[12] M.S. Magnusson, X. Zhang, S. Östlund, Experimental evaluation of the interfibre joint strength of papermaking fibres in terms of manufacturing parameters and in two different loading directions, Exp. Mech. 53 (2013) 1621–1634, https://doi.org/10.1007/s11340-013-9757-y.
[13] M.S. Magnusson, Interfibre Joint Strength under Mixed Modes of Loading, PhD thesis, KTH Royal Institute of Technology, 2014.
[14] Y. Xiong, M. Von Essen, J. Hirvonen, P. Kallio, Vision based 3D calibration of micromanipulator in microrobotic fiber characterization platform, in: 2012 International Conference on Manipulation, Manufacturing and Measurement on the Nanoscale (3M-NANO), 2012, pp. 154–159, https://doi.org/10.1109/3M-NANO.2012.6472968.
[15] U. Hirn, W. Bauer, A review of image analysis based methods to evaluate fiber properties, Lenzing. Ber. 86 (2006) 96–105.
[16] E.L.G. Wernersson, S. Borodulina, A. Kulachenko, G. Borgefors, Characterisations of fibre networks in paper using micro computed tomography images, Nord. Pulp Pap. Res. J. 29 (2014) 468–475, https://doi.org/10.3183/npprj-2014-29-03-p468-475.
[17] S. Borodulina, E.L.G. Wernersson, A. Kulachenko, C.L. Hendriks Luengo, Extracting fiber and network connectivity data using microtomography images of paper, Nord. Pulp Pap. Res. J. 31 (2016) 469–478.
[18] G. Urstöger, A. Kulachenko, R. Schennach, U. Hirn, Microstructure and mechanical properties of free and restrained dried paper: a comprehensive investigation, Cellulose 27 (2020) 8567–8583, https://doi.org/10.1007/s10570-020-03367-4.

[19] R. Mansour, A. Kulachenko, W. Chen, M. Olsson, Stochastic constitutive model of isotropic thin fiber networks based on stochastic volume elements, Materials (Basel) 12 (2019) 538, https://doi.org/10.3390/ma12030538.
[20] J. Viguié, P. Latil, L. Orgéas, P.J.J. Dumont, S. Rolland du Roscoat, J.F. Bloch, C. Marulier, O. Guiraud, Finding fibres and their contacts within 3D images of disordered fibrous media, Compos. Sci. Technol. 89 (2013) 202–210, https://doi.org/10.1016/j.compscitech.2013.09.023.
[21] U. Hirn, W. Bauer, Evaluating an improved method to determine layered fibre orientation by sheet splitting, in: 61st Appita Annual Conference and Exhibition, Gold Coast, Australia, 6–9 May 2007: Proceedings, 2007, pp. 71–79.
[22] A. Brandberg, A. Kulachenko, Compression failure in dense non-woven fiber networks, Cellulose 27 (2020) 6065–6082, https://doi.org/10.1007/s10570-020-03153-2.
[23] M. Alzweighi, R. Mansour, J. Lahti, U. Hirn, A. Kulachenko, The influence of structural variations on the constitutive response and strain variations in thin fibrous materials, Acta Mater. 203 (2021) 116460, https://doi.org/10.1016/j.actamat.2020.11.003.
[24] H. Reza Motamedian, A. Kulachenko, Rotational constraint between beams in 3-D space, Mech. Sci. 9 (2018) 373–387, https://doi.org/10.5194/ms-9-373-2018.
[25] A. Ibrahimbegović, On finite element implementation of geometrically nonlinear Reissner's beam theory: three-dimensional curved beam elements, Comput. Methods Appl. Mech. Eng. 122 (1995) 11–26, https://doi.org/10.1016/0045-7825(95)00724-F.
[26] H.R. Motamedian, A.E. Halilovic, A. Kulachenko, Mechanisms of strength and stiffness improvement of paper after PFI refining with a focus on the effect of fines, Cellulose 26 (2019) 4099–4124, https://doi.org/10.1007/s10570-019-02349-5.
[27] R. Mansour, M. Olsson, A closed-form second-order reliability method using noncentral chi-squared distributions, J. Mech. Des. Trans. ASME 136 (2014), https://doi.org/10.1115/1.4027982.
[28] R. Mansour, J. Zhu, M. Edgren, Z. Barsoum, A probabilistic model of weld penetration depth based on process parameters, Int. J. Adv. Manuf. Technol. 105 (2019) 499–514,-https://doi.org/10.1007/s00170-019-04110-5.
[29] R.G. Ghanem, P.D. Spanos, Stochastic Finite Elements: A Spectral Approach, Springer, New York, 1991, https://doi.org/10.1007/978-1-4612-3094-6.
[30] W. Chen, X. Yin, S. Lee, W.K. Liu, A multiscale design methodology for hierarchical systems with random field uncertainty, J. Mech. Des. Trans. ASME (2010) 0410061–04100611, https://doi.org/10.1115/1.4001210.
[31] X. Yin, S. Lee, W. Chen, W.K. Liu, M.F. Horstemeyer, Efficient random field uncertainty propagation in design using multiscale analysis, J. Mech. Des. Trans. ASME 131 (2009) 0210061–02100610, https://doi.org/10.1115/1.3042159.
[32] R.G. Ghanem, A. Doostan, On the construction and analysis of stochastic models: characterization and propagation of the errors associated with limited data, J. Comput. Phys. 217 (2006) 63–81, https://doi.org/10.1016/j.jcp.2006.01.037.
[33] A. Clément, C. Soize, J. Yvonnet, Uncertainty quantification in computational stochastic multiscale analysis of nonlinear elastic materials, Comput. Methods Appl. Mech. Eng. 254 (2013) 61–82, https://doi.org/10.1016/j.cma.2012.10.016.
[34] R. Popescu, G. Deodatis, J.H. Prevost, Simulation of homogeneous non Gaussian stochastic vector fields, Probab. Eng. Mech. 13 (1998) 1–13, https://doi.org/10.1016/s0266-8920(97)00001-5.
[35] R. Mansour, M. Olsson, Efficient reliability assessment with the conditional probability method, J. Mech. Des. Trans. ASME 140 (2018), https://doi.org/10.1115/1.4040170.

[36] Z. Hu, R. Mansour, M. Olsson, X. Du, Second-order reliability methods: a review and comparative study, Struct. Multidiscip. Optim. 64 (2021) 3233–3263, https://doi.org/10.1007/s00158-021-03013-y.

[37] C.E. Rasmussen, C.K.I. Williams, Gaussian Processes for Machine Learning, The MIT Press, 2018, https://doi.org/10.7551/mitpress/3206.001.0001.

[38] I.M. Gitman, H. Askes, L.J. Sluys, Representative volume: existence and size determination, Eng. Fract. Mech. 74 (2007) 2518–2534, https://doi.org/10.1016/j.engfracmech.2006.12.021.

[39] X. Yin, W. Chen, A. To, C. McVeigh, W.K. Liu, Statistical volume element method for predicting microstructure-constitutive property relations, Comput. Methods Appl. Mech. Eng. 197 (2008) 3516–3529, https://doi.org/10.1016/j.cma.2008.01.008.

[40] Z. Hashin, Analysis of composite materials: a survey, J. Appl. Mech. 50 (1983) 481–505,- https://doi.org/10.1115/1.3167081.

[41] V. Lucas, J.C. Golinval, S. Paquay, V.D. Nguyen, L. Noels, L. Wu, A stochastic computational multiscale approach; application to MEMS resonators, Comput. Methods Appl. Mech. Eng. 294 (2015) 141–167, https://doi.org/10.1016/j.cma.2015.05.019.

[42] M.G. Genton, W. Kleiber, Cross-covariance functions for multivariate geostatistics, Stat. Sci. 30 (2015) 147–163, https://doi.org/10.1214/14-STS487.

[43] G.A. Baum, C.C. Habeger, E.H. Fleischman, Measurement of the Orthotropic Elastic Constants of Paper, The Institute, Appleton, Wisconsin, 1981.

[44] M. Steven Greene, Y. Liu, W. Chen, W.K. Liu, Computational uncertainty analysis in multiresolution materials via stochastic constitutive theory, Comput. Methods Appl. Mech. Eng. 200 (2011) 309–325, https://doi.org/10.1016/j.cma.2010.08.013.

[45] D. Sandberg, R. Mansour, M. Olsson, Fatigue probability assessment including aleatory and epistemic uncertainty with application to gas turbine compressor blades, Int. J. Fatigue 95 (2017) 132–142, https://doi.org/10.1016/j.ijfatigue.2016.10.001.

[46] G. Hultgren, R. Mansour, Z. Barsoum, M. Olsson, Fatigue probability model for AWJ-cut steel including surface roughness and residual stress, J. Constr. Steel Res. 179 (2021), https://doi.org/10.1016/j.jcsr.2021.106537, 106537.

Numerical models of random fibrous networks

Emrah Sozumert[a,b] *and Vadim V. Silberschmidt*[a]
[a]Wolfson School of Mechanical, Electrical and Manufacturing Engineering, Loughborough University, Leicestershire, United Kingdom, [b]College of Engineering, Swansea University, Swansea, United Kingdom

5.1 Introduction

Random fibrous networks are commonly found in nature, for instance, in collagens [1] and tissues of marine animals [2], but can be also produced, e.g. in the form of synthesised materials such as bacterial hydrogels [3] and artificial tendon-like tissues [4] as well as electrospun networks [5,6] and non-wovens [7]. Some fibrous networks form a highly efficient environment for drug-delivery purposes [8] or can be used for treatment, or replacement, of damaged tissues and organs of a human body [9]. Also, two fibrous networks can be combined to produce a composite network with higher mechanical strength and toughness than its constituent networks [10]. So, understanding their complex mechanical behaviours is crucial for the design of new fibrous networks and the improvement of the existing ones. However, their complex nature defined by various anisotropies – randomness in both orientation and spatial distributions as well as curliness of their fibres – makes their mechanical characterisation highly challenging, preventing quantitative analytical approaches in most cases. This chapter, therefore, focuses on various numerical models or modelling techniques used for random fibrous networks such as collagens, non-wovens, etc. to ease the difficulty in the characterisation of these materials. For discussion of mechanics of fibrous networks, the key fundamental concepts of their microstructure and mechanical behaviour are essential to be mentioned before giving insights into the relevant numerical schemes and models.

5.2 Fundamental concepts of fibrous networks

Challenges in understanding and modelling the mechanical behaviour of random fibrous networks stem from four fundamental sources of non-linearity: (i) distribution of fibres; (ii) non-linear material behaviour of fibres; (iii) fibre curvature; and (iv) fibre-to-fibre interactions. In this section, before introducing the numerical models of fibrous networks suggested in the literature, fundamental concepts playing important roles in the deformation mechanisms of fibrous networks are discussed.

Mechanics of Fibrous Networks. https://doi.org/10.1016/B978-0-12-822207-2.00012-X

5.2.1 Fibre orientation distribution and randomness

Mechanical properties, such as failure strength and toughness, of fibrous networks are sensitive to the alignments of fibres. It is well known that networks with preferred fibre orientation are stiffer (in that direction) than less-oriented ones [11,12]. This results in an anisotropic mechanical response for materials subjected to loading in different directions.

As randomness in fibre distribution is concerned, its two distinct sources can be differentiated: (i) randomness in spatial alignment of fibres and (ii) randomness in their orientation. Generally, a fibre in a fibrous matrix can have any orientation (i.e. first type of randomness), and the centre of this fibre can be anywhere in the matrix (the second type) (Fig. 5.1A). Orientation distribution of fibres can be expressed in terms of the *orientation distribution function* (ODF). In some cases, fibres can be preferentially aligned in one direction with a narrow ODF (reducing to the delta function for a network of parallel fibres) (Fig. 5.1B). Such anisotropy in the orientation of fibres is the reason for direction-dependent values of stiffness and toughness [13]. The ODF was for the first time introduced by Cox [14] in order to quantify the effect of orientation distributions of fibres on elasticity and strength of fibrous mats. Gaussian and Weibull distributions can be used to model some of the random fibrous networks [15]. In the case of orientation distribution of fibres with a uniform probability density function, the fibrous network is termed *fully random* (Fig. 5.1C); otherwise, it is termed *random*. In our research, a spatial distribution of fibres is considered uniform; however, it is known that real-life systems can demonstrate a non-uniformity in spatial distribution that causes variations in localisation of fibre strain and failure [16].

5.2.2 Affinity in network deformation behaviour

In order to understand the notion of *affinity* with regard to the behaviour of random fibrous networks, one can consider a uniform stress field with a resultant displacement field $u(x)$, which is a linear function of current position vector x, at the boundaries of

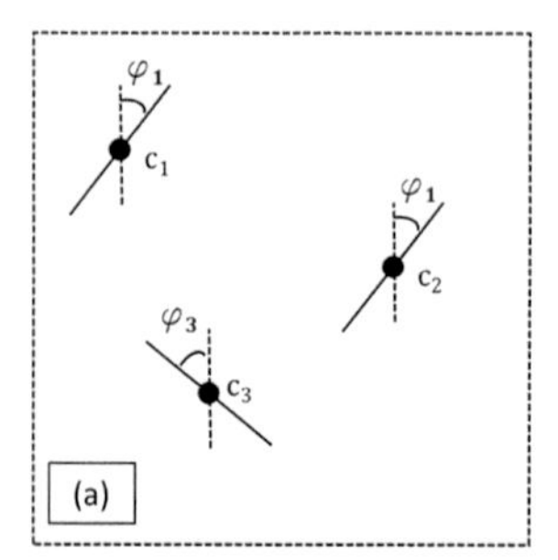

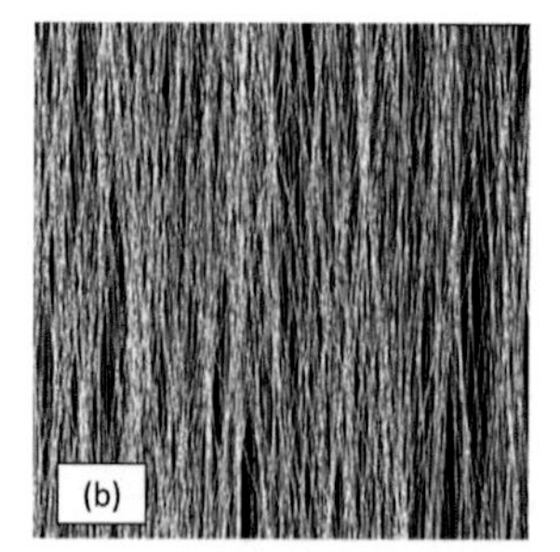

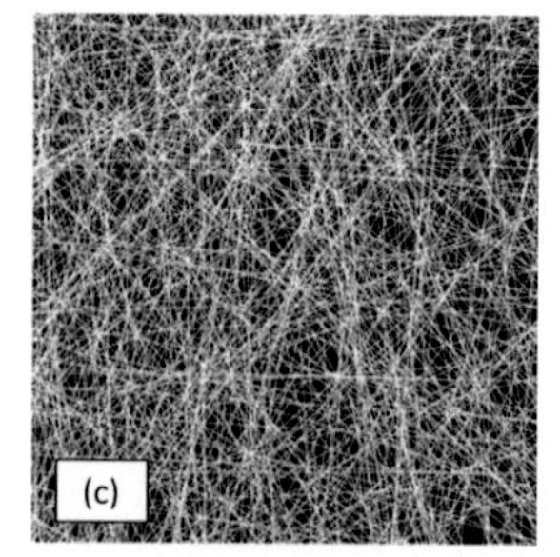

Fig. 5.1 (A) Fibres with three different configurations: reference configuration (c_1); same orientation but different centre (c_2); different orientation and centre (c_3). (B) Network with the preferential orientation of fibres (in the vertical direction); and (C) fully random fibrous network.

the network. This is known as *affine deformation*. If the network behaves differently, the type of deformation is called *non-affine*. A comparison of kinematics of affine and non-affine fibrous networks was made by Chandran and Barocas [17], and the effect of non-affine deformation behaviour on the elasticity of cross-linked fibrous networks was investigated with a scaling theory by Heussinger and Frey [18]. A degree of anisotropy in orientation distributions of fibres, i.e. more preferentially oriented fibres, causes an increase in the non-affinity of a fibrous network [19]. The scaling effect of non-affine deformations on the mechanics of random fibrous networks was investigated by Hatami-Marbini and Picu [19,20]. Elastic models with affine and non-affine deformation features were studied for polymer hydrogel fibrous networks by Wen et al. [21]. The mechanical behaviour of affine networks relied on individual filaments, whilst for the non-affine ones, the quantification of the displacements of beads placed into a flexible fibrous network was considered with confocal microscopy (Fig. 5.2) [22].

5.2.3 Non-linear behaviour and curvature of fibres

Single fibres can demonstrate a highly non-linear mechanical behaviour as a result of the non-linearity of the mechanical response of their materials, e.g. polymers. In order to assess their material properties and elucidate their mechanical behaviour, fibres can be tested with in-situ techniques employed for fibrous networks, for instance, atomic force microscopy (AFM) (see, e.g. [23]), or they can be extracted from the network material to be tested with a microtensile system [24,25]. Typical non-linear flow curves, i.e. true stress–strain curves, of polypropylene fibres extracted from a type of non-woven material are presented in Fig. 5.3A.

In addition to non-linear behaviour caused by material non-linearity, a fibre can also have a curved geometry along its longitudinal axis. This is known as *fibre*

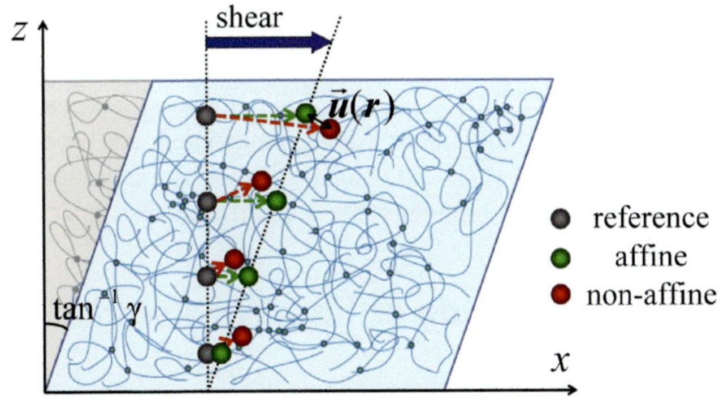

Fig. 5.2 Affine and non-affine shear deformation of polymer-gel fibrous network with entrapped tracer beads [22].
Reprinted with permission from A. Basu, Q. Wen, X. Mao, T.C. Lubensky, P.A. Janmey, A.G. Yodh, Nonaffine displacements in flexible polymer networks, Macromolecules 44 (2011) 1671–1679. Copyright (2011) American Chemical Society.

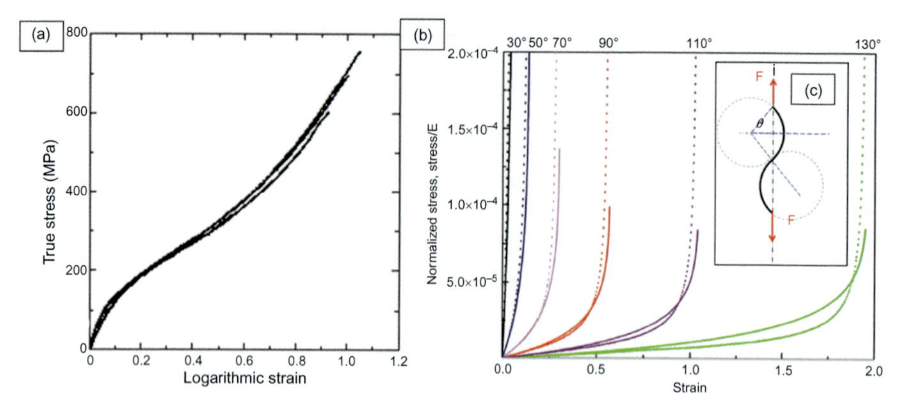

Fig. 5.3 (A) Typical non-linear true stress-strain curves of polypropylene fibres extracted from non-woven fabric [24] (B) Normalised stress-strain curves of steel wire defined by central angle θ under load F [shown in (C) radius 120 mm] for various central angles θ (*continuous lines* for experiments; *dashed lines* for analytical solution) [13].
Panel (A) Reprinted from A. Ridruejo, C. González, J. LLorca, Micromechanisms of deformation and fracture of polypropylene nonwoven fabrics, Int. J. Solids Struct. 48 (2011) 153–162. Copyright (2011), with permission from Elsevier.

curvature, or *crimp* in the case of intentional (manufactured) curvature. In electrospun fibrous networks, for instance, fibre curvature can be controlled by process parameters [26,27]; however, it is not possible in natural fibrous networks such as in bacterial cellulose hydrogels [3] and collagen fibrous networks [1]. The effect of fibre curvature on the mechanics of collagenous skin was investigated by Yang et al. [13]. Animal skin is formed by collagenous fibrous networks composed of highly curved fibres, where the degree of curvature can be defined in terms of a radius r and central angle θ under load F (see Fig. 5.3C). An analogy between curved steel fibres and collagen ones was established to derive an analytical formulation for different levels of fibre curvature based on Castigliano's theorem [13]; the tensile mechanical response of steel wires with various central angles is shown in Fig. 5.3B. Hearle and Stevenson [28] and Hearle and Newton [29] investigated the tensile properties of non-wovens with fibre curvature, with only axial forces considered and neglected bending moments. Fibres carry small-level axial loads in their curved state and, after full straightening under a load, they demonstrate a significant resistance to stretch.

5.3 Numerical modelling of fibrous networks

In order to understand and predict the mechanical behaviour of random fibrous networks, some researchers, e.g. Carlsson and Lindstrom [30] and Räisänen et al. [31], started their analysis at the fibre level. W. Batchelor [32] proposed an analytical solution for a load distribution along a half fibre in a non-woven network (Fig. 5.4A), with a force transferred to the fibre at discrete contacts with other fibres. Fibre strain was found to be highly sensitive to the configuration of contact points. In one of the earliest

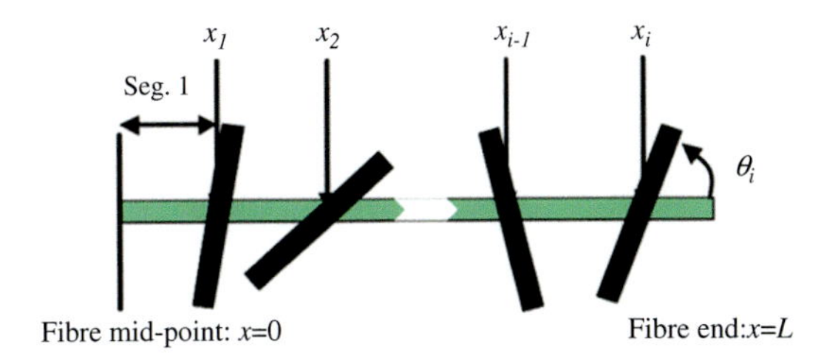

Fig. 5.4 Half-fibre of length L with i crossing fibres [32].
Reprinted from W. Batchelor, An analytical solution for the load distribution along a fibre in a nonwoven network, Mech. Mater. 40 (2008) 975–981. Copyright (2008), with permission from Elsevier.

models Cox [14] derived an analytical expression for the stiffness and strength of paper-based random fibrous networks in terms of orientation distribution of fibres and fibre strength using a shear-lag model. Räisänen et al. [31] investigated whether the shear-lag model is applicable for random fibrous networks, where the stress transfer mechanisms in analytical solution based on shear-lag and numerical investigations were compared and the research pointed out that the majority of the axial stress is transferred through fibre-to-fibre approach different from a shear-lag approach, which was first proposed by H. L. Cox.

In another study [33], 2D random fibrous networks were simulated, with the stress distribution along a fibre in good agreement with the mean-field Cox predictions. The stress-transfer mechanism was not only affected by the local fibre-segment stiffness, but also by the mechanical properties of the entire fibre. In Hearle and Newton [29] and Hearle and Stevenson [28], the tensile properties of non-woven fibrous networks were analysed in the presence of fibre curvature by considering only axial forces. Pai et al. [27] later extended the previous model with an energy-based modelling approach, incorporating the bending effects for curved fibres. Narter et al. [34] developed Cox's model further by expanding it into 3D space, with the orientation of fibres in a unit volume defined by two independent angles – θ and $\varnothing$. The main assumption was that the global and individual mechanical behaviour of constituent fibres follow the classical Hooke's law:

$$\sigma = K \cdot E, \tag{5.1}$$

where

$$K = c \int_0^{\pi} \int_0^{\pi} \omega(\theta, \varnothing) \cdot f_{i,j}(\theta, \varnothing) \sin(\theta) d\theta d\varnothing, \tag{5.2}$$

where $\omega(\theta, \varnothing)$ is the orientation distribution function of the fibre segments in the network and $f_{i,j}(\theta, \varnothing)$ is a transformation function.

Although analytical models of random fibrous networks are constrained by the need to account for complex geometrical shapes and mechanical processes, discrete or homogeneous finite-element (FE) models can provide realistic solutions for complex non-linear mechanical problems. Since random fibrous networks exhibit highly inhomogeneous fields of deformations, stresses, and strains, the FE method presents a generalised procedure for the analysis of the mechanical behaviour of such networks and obtaining these unknown fields. In a standard finite-element method, the solid volume of the structure is, first, expressed as a set of interconnected finite elements, known as the *discretisation of geometry*. Then, prescribed boundary conditions (BCs) in terms of displacements (Dirichlet BCs) and/or tractions (Neuman BCs) are applied to the respective nodes of the elements to obtain the resultant displacements, strains, and stresses. Numerical models of fibrous networks in this chapter employ mainly FE and computational fluid dynamics (CFD) simulations.

Numerical modelling approaches used for random fibrous networks can be classified into continuous, discontinuous (or discrete), and hybrid numerical modelling approaches. In order to produce continuous models, homogenisation processes are applied to random fibre networks, with no direct introduction of fibres in the numerical model. So, mechanical contribution of individual fibres, their random distribution, and their interactions are considered indirectly. Oppositely, the discontinuous modelling approach aims to elucidate randomly distributed fibrous networks by developing models directly introducing some – or all – elements of their structure. Lastly, the hybrid models combine features of both continuous and discontinuous models.

5.3.1 Continuous modelling approach

Some continuous modelling approaches for random fibrous networks can be found in the literature, ranging from volume averaging to asymptotic homogenisation and a generalised continuum theory. Throughout this chapter, the most established ones are explained with some details about modelling strategies and specific models. Many of the discussed methods, in fact, rely on building up a discrete fibre network model within a limited design space.

Some works, such as Demirci et al. [11], Isaksson et al. [35], and Martínez-Hergueta et al. [36], attempted to simulate the material response of random fibrous networks by approaching the problem by deriving the appropriate constitutive relations. Isaksson et al. [35], for instance, developed a material constitutive model incorporating isotropic elastic strain-hardening, anisotropic plasticity, and anisotropic damage (a gradient-enhanced damage formulation), in order to simulate the failure of a paper under tensile loading.

One way to simulate the fibrous networks with a continuous modelling approach is to generate discrete models of microstructure for random fibrous networks and to use numerical homogenisation to obtain resultant material properties. The unit describing the mechanical response of a fibrous structure is called a *representative-volume element* (RVE), mechanical properties of which are considered to be equal to (representative of) those of the entire fibrous network. The parameters obtained by the homogenisation of RVEs are called the *effective material properties*. Convergence studies for the RVE size or properties are carried out to ensure that the generated RVEs adequately represent mechanical responses at macro- and micro-scales.

For continuous models, after obtaining the effective material properties from an RVE-based analysis, an effective homogeneous media is inserted in place of the RVE domain. For instance, a random fibrous network can be generated using statistical data based on experiments in the FE environment. Fibres are randomly

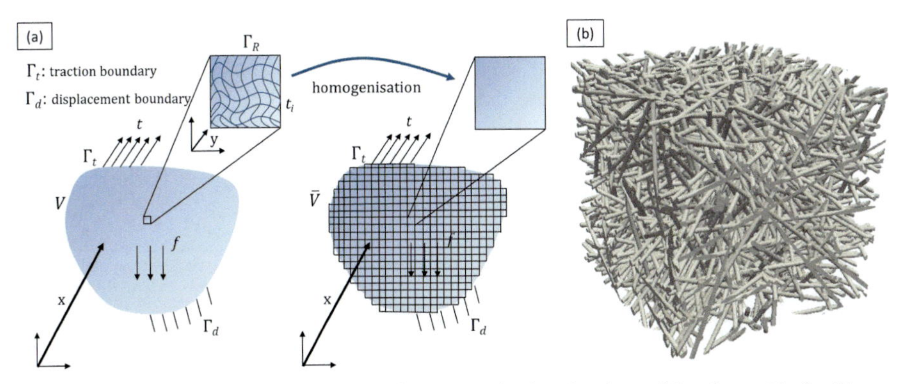

Fig. 5.5 (A) Schematic representation of homogenisation for the solid volume V of a fibrous network. (B) Fibres in a discrete fibrous network of an RVE [38].
Reprinted from J. Dirrenberger, S. Forest, D. Jeulin, Towards gigantic RVE sizes for 3D stochastic fibrous networks, Int. J. Solids Struct. 51 (2014) 359–376. Copyright (2014), with permission from Elsevier.

distributed in an RVE volume according to their orientation distribution function and material properties are assigned to individual constituent fibres; this is schematised in Fig. 5.5. This process is followed by applying a small displacement in different orthogonal directions separately in order to compute effective moduli [15]. This modelling approach accounts for the contribution of single fibres to the mechanical response of a random fibre network. For instance, linear-elastic and thermal properties of random glass–fibre networks were derived by Altendorf et al. [37] by homogenising discrete microstructures of RVEs. However, some works, such as Altendorf et al. [37], Dirrenberger et al. [38], and Gao et al. [39], indicated that the effective mechanical properties of a fibrous network strongly depended on the selected size of the RVE; so, researchers perform sensitivity analysis to define this size.

One of the homogenisation methods employed for fibrous networks is a surface-averaging approach. It relies on the application of stress and displacement fields to the surface of a chosen RVE of the studied fibrous network (shown in the zoomed-in window in Fig. 5.5A) before/after homogenisation. In surface averaging, the macroscopic average stress relation between the average stress $\overline{\sigma}$, the local stress tensor of the RVE σ, and the tractions t_i and t_j on the boundary of the RVE is expressed as

$$\overline{\sigma}=\frac{1}{V_R}\int_{V_R}\sigma dV_R=\frac{1}{V_R}\int_{\Gamma_R}\frac{1}{2}\left(t_iy_j+t_jy_i\right)d\Gamma_R \tag{5.3}$$

where y_i and y_j are the (local) coordinates at the boundary Γ_R of the RVE. The average strain tensor of the RVE is also then

$$\overline{\varepsilon}=\frac{1}{V_R}\int_{V_R}\varepsilon dV_R=\frac{1}{V_R}\int_{\Gamma_R}\frac{1}{2}\left(u_iy_j+u_jy_i\right)d\Gamma_R \tag{5.4}$$

where $\overline{\varepsilon}$ and ε are the average macroscopic strain tensor and the local strain tensor for the RVE, respectively; u_i and u_j are the displacements on the boundary of the RVE; n_i and n_j are the vectors normal to the boundary Γ_R of the RVE.

The volume-averaging method, on the other hand, assumes the total strain energy W_R of the RVE of the fibrous network is calculated from the effective medium obtained by homogenising the RVE. In this approach, the main assumption is that the strain energy of the RVE [40] is

$$W_R = \frac{1}{V_R}\int_{V_R} \sigma\varepsilon dV_R \tag{5.5}$$

A volume-averaging-based framework was used for collagen gels by Chandran and Barocas [41] and Stylianopoulos et al. [42,43] by employing finite elements at the macroscale and RVEs at the microscale (see the works for further details). Wu and Dzenis [44] suggested an effective constitutive model based on averaging the micromechanical strain energy of planar fibre networks, with fibres rigidly bonded to each other at contact points. With this model, it was possible to analyse various deformation modes of fibres (bending, stretching, and contraction).

Different from the volume-averaging approach, an asymptotic homogenisation scheme relies on relating field quantities, such as displacements, strains, and stresses, at different scales, e.g. macroscopic and microscopic [45] (Fig. 5.6). These quantities are distributed smoothly at the macroscale.

The first assumption in this scheme is for the characteristic sizes of the macroscopic and microscopic domains (H and h, respectively), with a strong separation between them ($h << H$). In the asymptotic homogenisation, a periodicity rule (i.e. assuming a periodic microstructure) should be fulfilled at the microscale in order to associate any two variables (x: macroscopic; y: microscopic) at both scales as $y = x/h$ [40].

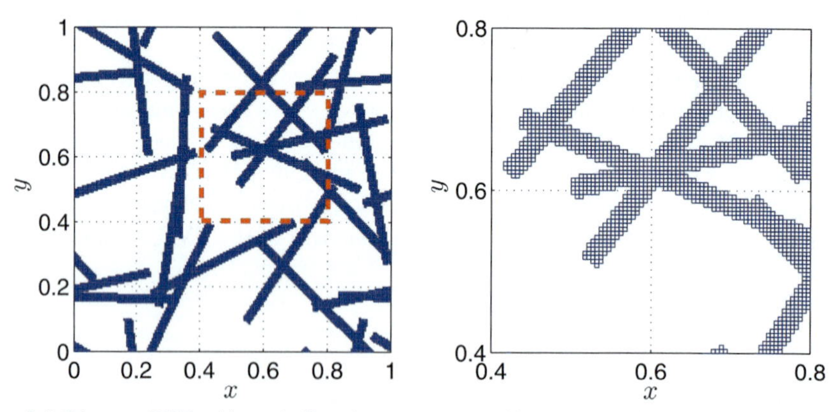

Fig. 5.6 Discrete RVE with periodic microstructure and its zoomed view [45]. Reprinted from E. Bosco, R.H.J. Peerlings, M.G.D. Geers, Asymptotic homogenisation of hygro-thermo-mechanical properties of fibrous networks, Int. J. Solids Struct. 115–116 (2017) 180–189. Copyright (2017), with permission from Elsevier.

In this approach, any field quantity, such as displacement field $u_i(x)$, can be expanded with a power series as

$$u_i(x) = u_{0i}(x, y) + h\, u_{1i}(x, y) + h^2\, u_{2i}(x, y) + \ldots \tag{5.6}$$

Bosco et al. [45] predicted the effective properties of bonded fibrous networks as well as effective hygromechanical behaviour [46] of paper sheets by using the asymptotic homogenisation. An example discrete RVE model with periodicity and a zoomed view with plane FE elements is presented in Fig. 5.6. Ghosh et al. [47] implemented a Voronoi-cell modelling scheme for an RVE of a heterogeneous material employing the same approach.

Raina and Linder [48] presented a micromechanical-based homogenisation approach for affine highly-non-linear random fibre networks at finite strains, where a preferential orientation distribution was incorporated into a microstructural model and the re-orientation of the constituent fibres was reflected into tensile deformation behaviour (see Fig. 5.7A and B). This continuous model based on the micromechanics of fibres was extended later for capturing non-affine deformation with strong discontinues for the analysis of failure of non-wovens [49]. Similarly, Silberstein et al. [50] developed a micromechanics-based constitutive model to emulate the mechanical behaviour of fibrous mats by implementing the contribution of the

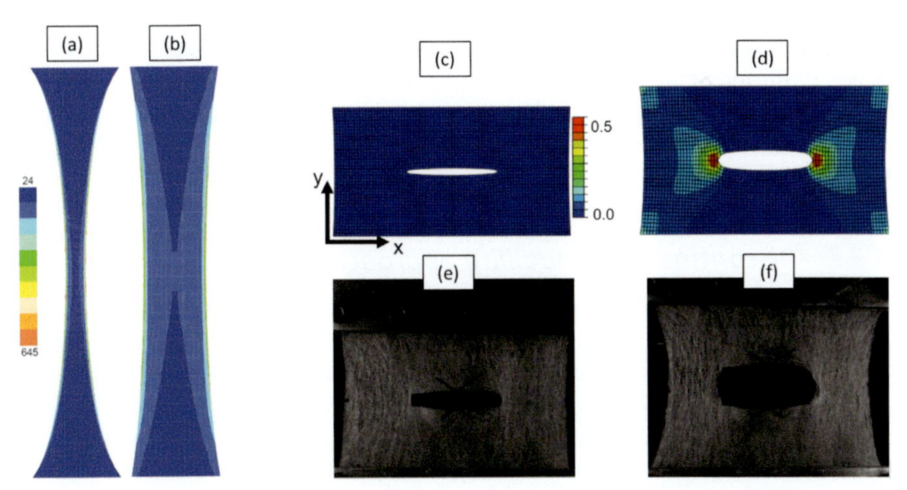

Fig. 5.7 Cauchy stress distribution in a continuous FE model of uniaxially stretched non-woven specimen with the implementation of fibre re-orientation (A) and without it (B) direction [48] and damage evolution in nonwoven with a mode-I crack at 6% and 16% extensions (C, D: simulations, E, F: experiments) [51].

(A, B) Reprinted from A. Raina, C. Linder, A homogenisation approach for nonwoven materials based on fibre undulations and reorientation, J. Mech. Phys. Solids 65 (2014) 12–34. Copyright (2014), with permission from Elsevier. (C–F) Reprinted from N. Chen, M.N. Silberstein, A micromechanics-based damage model for non-woven fibre networks, Int. J. Solids Struct. 160 (2019) 18–31. Copyright (2019), with permission from Elsevier.

elastic–plastic behaviour of individual fibres into a macroscopic membrane model, i.e. a homogenised fibre network model. This model was tested for monotonic and cyclic loading conditions.

In another study, dealing with large-strain conditions with inter-fibre bond fracture being a leading damage mechanism, the effect of bond-breaking processes between individual fibres and local damage behaviour (non-affine deformation) on macroscopic deformation was added into the constitutive model [51] (see Fig. 5.7C and D). The obtained shapes of the notched specimens were compared with respective experimental results (Fig. 5.7E and F). Karakoç [52] proposed a computational homogenisation framework considering a discrete microstructure of a network with randomly distributed fibres and fibre-to-fibre interactions. The homogenisation of mechanical behaviour of such a network was achieved by considering boundary and control nodes at the fibre scale. In another approach [36], a constitutive model was developed to simulate the mechanical behaviour of a mechanically entangled non-woven network (a specific type of a random fibre network) accounting for complex mechanics, for example, non-affine deformation, fibre straightening as well as failure mechanisms at fibre-, bundle- and macro-scales (its mechanical characterisation was presented by [53]).

Some of the suggested continuous models are listed in Table 5.1. In these models, researchers aimed to either derive their constitutive equations relying on micromechanics of fibrous networks or assess the material parameters for the constitutive models based on RVEs. Furthermore, in some cases, such as in Jin and Stanciulescu [54] and Zhang et al. [55], a random fibrous network was embedded into a matrix material. Jin and Stanciulescu [54] generated a 2D FE model of a random fibrous network for biomaterials with a random-walk algorithm and homogenised the mechanical behaviour for a discontinuous model in the matrix with a variational finite-element approach.

5.3.2 Discontinuous modelling approach

In order to understand and simulate the mechanical behaviour of random fibrous networks, characterisation of their complex microstructures is vital. The underlying reason is that the performance and adequacy of a numerical model of a discontinuous fibrous network strongly depend on the quality of representation of the microstructural geometry. As discussed, fibrous networks can have various arrangements of fibres (i.e. their orientation distributions) and different types of interactions between them (e.g. bonds, sliding contact with friction, or negligible friction).

5.3.2.1 Statistically generated fibre networks

Depending on microstructure, different schemes are, therefore, used to generate numerical models for discrete microstructures of real or artificial random fibrous networks. Picu [62] exemplified discrete fibrous networks in terms of their generation scheme: (i) honeycomb-like networks; (ii) regular networks with some random defects; (iii) random networks based on random walk algorithms; (iv) networks based

Table 5.1 Continuous models of fibrous networks.

Model	Focus	Continuous modelling approach
Rubin and Bodner [56]	Soft-tissue mechanics	Phenomenological approach
Holzapfel et al. [57]	Arterial wall	Phenomenological approach
Gasser et al. [58]	Arterial layers (collagen)	Phenomenological approach
Chandran and Barocas [41]	Collagen gel mechanics	Volume-averaging approach
Stylianopoulos et al. [42]	Electrospun fibre meshes	Volume-averaging approach
Demirci et al. [11]	Non-wovens	Phenomenological approach
Silberstein et al. [50]	Non-wovens	Constitutive model based on RVE
Raina and Linder [48,49]	Non-wovens	Constitutive model based on RVE
Liu et al. [59]	Polymeric hydrogel	Phenomenological approach
Castro et al. [60]	Collagen hydrogel	Poroelastic continuum model
Martínez-Hergueta et al. [36]	Non-wovens	Phenomenological approach
Bosco et al. [45,46]	Bonded fibrous networks and paper	Asymptotic homogenisation
Castro et al. [61]	Collagen hydrogel	Poroelastic continuum model
Chen and Silberstein [51]	Non-wovens	Constitutive modelling based on RVE
Karakoç [52]	Fibrous networks	Constitutive modelling of 3D RVE with boundary and control nodes

on random placing; and (v) networks based on triangulation algorithms (see Fig. 5.8). In honeycomb-like microstructures, the symmetry of fibre cell walls is broken and randomness is introduced by organising random displacements in the intersections of fibres. Alternatively, a random fibrous network can be implemented as a lattice-like regular fibrous network by removing some fibres randomly from the systematically oriented and placed fibres.

An example of this discrete network modelling approach was used by Wilbrink et al. [64] to model bond failure and frictional sliding of fibrous materials at the meso-scale. In the first two modelling approaches, random fibrous networks were generated based on manipulation of a symmetric lattice-like network by either deleting some fibres or moving nodes. The algorithm based on random walks can also be used for the simulation of random fibrous networks (Fig. 5.8A). It was successfully employed for 3D dense generic fibre systems [37,65], 2D curved collagen networks in the tunica adventitia [66], planar layered elastomeric fibrous biomaterials [67],

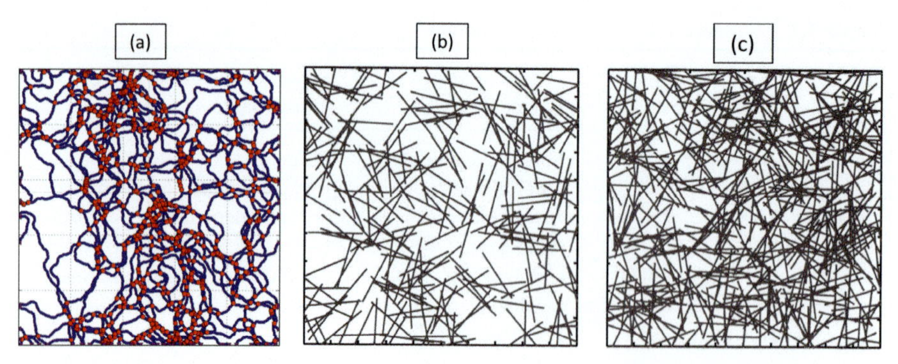

Fig. 5.8 Illustration of random fibrous networks generated with different algorithms: (A) random-walk algorithms [63]; (B) and (C) random-placing method for short fibres with different densities [44].
Reprinted from X.-F. Wu, Y.A. Dzenis, Elasticity of planar fibre networks, J. Appl. Phys. 98 (2005) 093501, with the permission of AIP Publishing.

fibrous biomaterials embedded in a matrix [54], and partially oriented microfibril stacking of Kevlar forming a single fibril [68]. In a fibrous network composed of 1D FE elements (beams or trusses) in a 3D space, for instance, a random-walk algorithm considers moving its random points along a random direction, until they cross a fibre. At that moment, the crossing point is used as a new starting point for a new random walk.

Fig. 5.9 demonstrates models of random network composed of curved fibres, generated with the random fibre deposition (models with red-coloured fibres) and the random-walk algorithm (with blue-coloured fibres). In another modelling technique, based on random placement of fibres, infinite or finite-length fibres were generated using a random selection of coordinates for their centres and they were trimmed according to a defined design domain (Fig. 5.8B and C). The users had full control over orientation of each fibre and spatial coordinates implementing the input data. Here, each fibre-to-fibre intersection of two fibres results in the formation of four branches (segments). Opposite to this, a Delaunay triangulation algorithm can result in a random fibre network with even and odd numbers of branches, connected to each intersection. So, depending on the microstructure of random fibrous networks, a suitable technique should be selected in order to reflect adequately the real microstructure.

5.3.2.2 Image-based models of fibrous networks

Another common model-generation approach is based on direct processing of material images (e.g. SEM and micro-computed tomography) in order to produce numerical models for mechanical and flow analyses. For instance, a non-woven porous material with glass fibres was simulated with an image-based micromechanical FE model to assess its compression behaviour and reveal load-transfer mechanisms, such as fibre-to-fibre contacts [70] (Fig. 5.10A). In another study, Gaiselmann et al. [71] processed micro-CT images of gas-diffusion-layer (GDL) non-wovens in

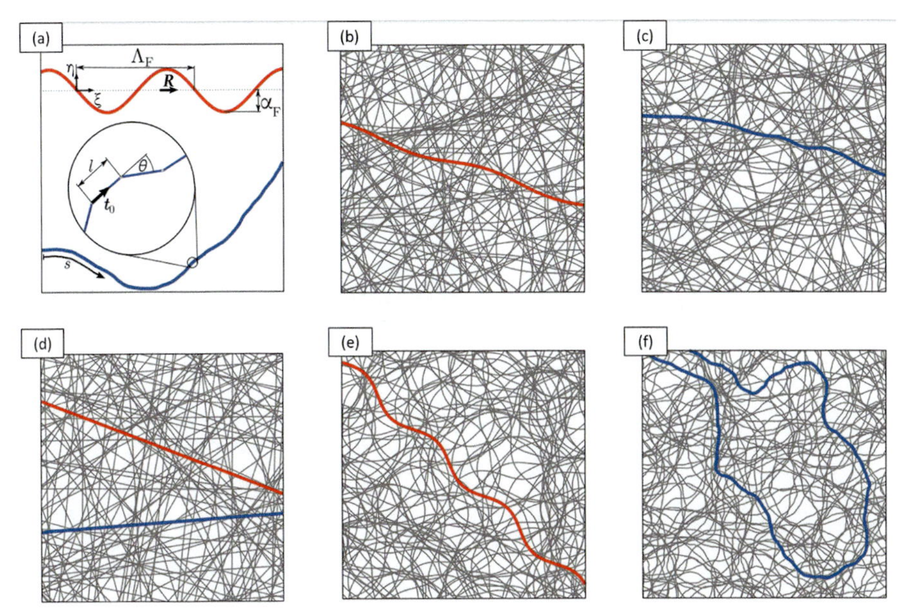

Fig. 5.9 (A) Fibres generated by using a sine function *(red)* and a stochastic approach *(blue)*; (B) two straight fibres: the *red* one generated by a sine function with an infinite wavelength and the *blue* one, stochastically with an infinite persistence length l; (B–E) *red* sinusoidal fibres with different orientation and wavelength; (C–F) stochastically generated fibres with different persistent lengths l [69].
Reprinted from S. Domaschke, M. Zündel, E. Mazza, A.E. Ehret, A 3D computational model of electrospun networks and its application to inform a reduced modelling approach, Int. J. Solids Struct. 158 (2019) 76–89. Copyright (2019), with permission from Elsevier.

uncompressed and different compression states for generation of their discrete FE models; Fig. 5.10B and C shows a CT reconstruction of a GDL and its extracted numerical model, respectively. As a relatively sparse random fibre network by Lindström et al. [72], a typical image of a fibrous collagen-I network was captured with a confocal microscope (Fig. 5.10D), and its image-based model was generated (Fig. 5.10E) together with its statistical realisations (Fig. 5.10F), by applying random processes to the acquired image data. These statistical realisations were produced with a Euclidean graph-generation algorithm and a simulated annealing method for the optimisation of microstructures. The statistical models demonstrated the same quantitative mechanical responses as the image-based model only within the elastic regime.

In another study, Alimadadi et al. [73] produced 3D FE models of wood random fibre networks by using artificial and physical realisation methods. For physical realisations (image-based models), the microstructure of the fibre network was acquired with micro-CT and then was skeletonised in order to examine the effect of microstructure on large-deformation compression behaviour and to compare the

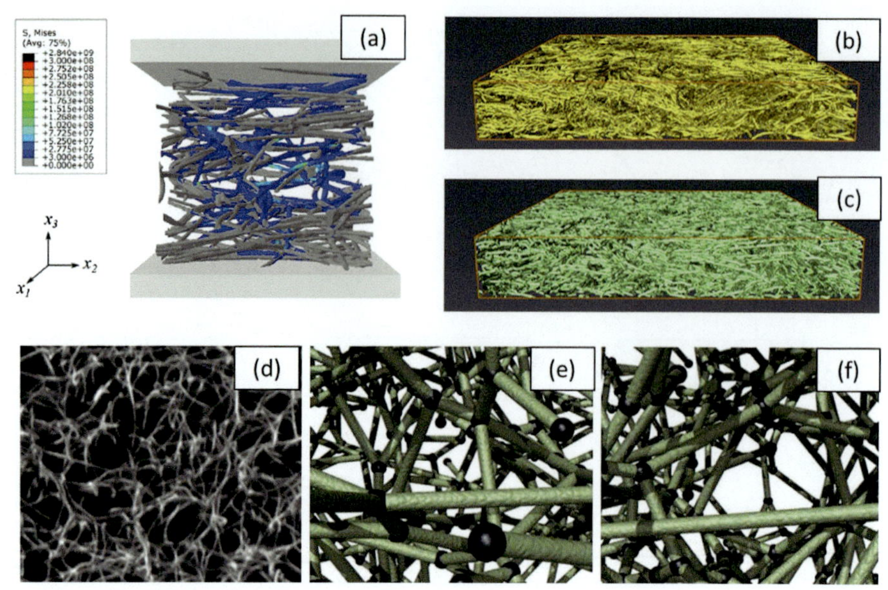

Fig. 5.10 (A) Image-based FE model of non-woven network under 10% compression with von Mises stress [70]. Micro CT volume of gas-diffusion-layer nonwoven (B) and its numerical models extracted from CT data (C) [71]. (D) Confocal-microscope image of collagen I; (E) its numerical model based on confocal microscopy data; (F) statistical realisation [72].
(A) Reprinted from Domaschke, Y. Chen, T. Siegmund, Mechanics of compaction of a porous non-woven fibre solid, Mech. Mater. 137 (2019) 103101. Copyright (2019), with permission from Elsevier. (B, C) Reprinted from G. Gaiselmann, C. Tötzke, I. Manke, W. Lehnert, V. Schmidt, 3D microstructure modelling of compressed fibre-based materials, J. Power Sourc. 257 (2014) 52–64. Copyright (2014), with permission from Elsevier. (D–F) Reprinted figure with permission from S.B. Lindström, D.A. Vader, A. Kulachenko, D.A. Weitz, Phys. Revue E 82 (2010) 051905. Copyright (2010) by the American Physical Society.

network-generation methods that caused differences even in the elastic region of deformation. Such behaviour was attributed to the non-uniformities in density and fibre orientations of the FE models generated with two different modelling approaches. Generally, microstructural features (for example, orientation and diameter distributions of fibres) can be extracted from respective images by employing automated algorithms in order to avoid the objectivity of human operators. Demirci et al. [74] suggested a Hough-transform-based algorithm for the detection of edges of fibres. In another research by D'Amore et al. [75], an algorithm extracted a microstructure of a fibrous network by accounting for unevenness in grey-intensity values. This algorithm also presented the fibre-intersection density, connectivity distribution (i.e. number of fibres at each intersection), as well as distributions of fibre diameter and orientation [76].

Another example of an image-based FE model of a fibrous network was presented by Staub et al. [77], where a fast Fourier transform solver of the Lippmann-Schwinger equations was used in order to compute the effective viscoelastic material behaviour of non-wovens. Bosbach [78] utilised nano-CT scans for reconstructing the geometry of metallic fibrous networks and converting them into discrete FE models by using medial axes of fibres. Subsequently, parameters of the deformation field were obtained from the simulation of tension, shear, and compression employing these FE models. In a similar approach, Faessel et al. [79] investigated the effect of microstructure on local thermal conductivity of random cellulosic fibrous networks with the help of CT-based FE models.

5.3.2.3 Discontinuous models of fibrous networks for biomaterials

A discrete FE modelling approach is also used to understand and model the mechanical behaviour of collagen fibre networks in various types of tissues. An adventitial collagen fibrous network, for instance, was analysed with a parametric discrete FE model, developed with an inverse random sampling with random walks [66]. This discrete model relies on morphological parameters such as orientation distribution of fibres, fibre waviness, and solid-volume fraction extracted from confocal-microscopy images. In another study, Mauri et al. [80] modelled a human amnion with a discrete random network model composed of randomly distributed finite elements representing fibre cross links and fibre segments. The same model was modified by Bircher et al. [81] for elucidating the microstructure of two biomembranes – a human amnion (cross-linked collagen fibres) and a bovine Glisson's capsule (fibre bundles) – with 2D continuum elements implemented in place of interstitial fluid. Electrospun scaffolds were modelled with a discrete FE model of a network of fibres randomly distributed in 2D to examine toughening mechanisms in fibrous networks such as formation of fibre bundles, local brittle, and ductile cracking [82].

5.3.2.4 Advanced FE models of fibrous networks

Recent advances in software and hardware underpin a trend to develop more complex finite-element models of fibrous materials. Bergström et al. [83] constructed a 3D planar fibre network in the form of a beam assembly (Fig. 5.11A), with a set of mechanical features like twisting and transverse stiffness as well as bond strength. The network was generated with an open-source discrete element method (DEM) code (ESyS-Particle) allowing the particles to interact with each other through cross links and contacts. Lindström and Uesaka [84] developed a numerical model to simulate the motion of flexible fibres in fluid flow, with fibres modelled as chains of segments interacting with each other and the surrounding fluid. The fluid motion was described by incompressible Navier-Stokes equations, solved with the immersed boundary method (IBM). The fluid-to-fibre interaction was, on the other hand, expressed in terms of viscous and drag forces, whereas the interaction between fibre segments included normal, frictional, and lubrication forces. The same IBM approach was later used by [85] and Lindström and Uesaka [86] to generate a microstructural FE model of fibrous paper. This approach can be considered as a type of fibre-deposition method within a fluid domain. It relies on consolidating randomly generated fibres by moving two mesh surfaces.

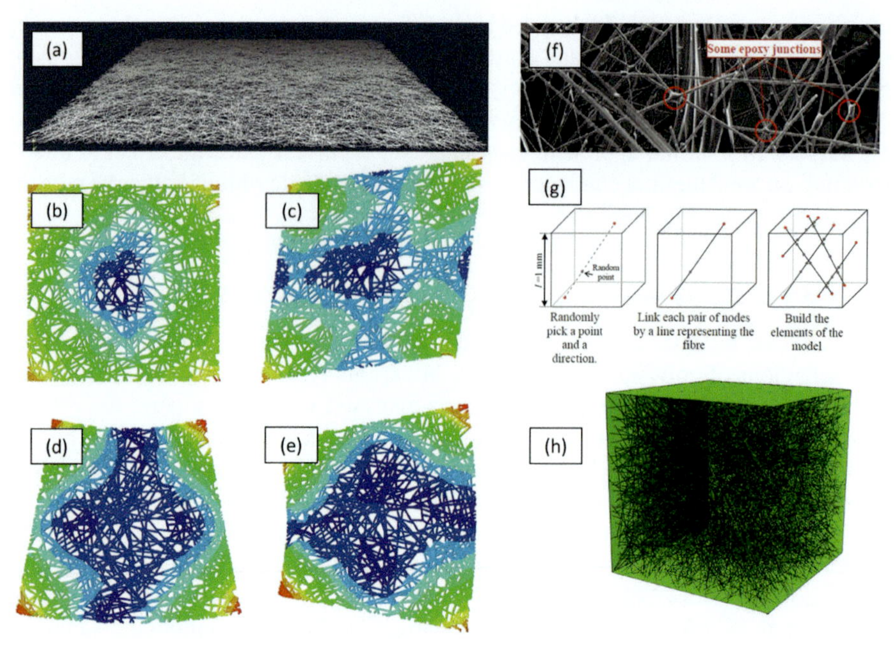

Fig. 5.11 (A) 3D model of random fibrous network generated with DEM [83]. (B–E) Deformed geometries of models after applying different displacement fields in order to identify effective constitutive coefficients [87]. SEM image of a carbon fibrous network (F), its FE model generation (G) by placing individual fibres and adding cross-links, and final discrete FE model (H) [88].

(A) Reprinted from P. Bergström, S. Hossain, T. Uesaka, Scaling behaviour of strength of 3D-, semi-flexible-, cross-linked fibre network, Int. J. Solids Struct. 166 (2019) 68–74. Copyright (2019), with permission from Elsevier. (B–E) Reprinted by permission from Springer K. Berkache, S. Deogekar, I. Goda, R.C. Picu, J.-F. Ganghoffer, Identification of equivalent couple-stress continuum models for planar random fibrous media, Contin. Mech. Thermodyn. Copyright (2019). (F–H) Reprinted from F. Chatti, D. Poquillon, C. Bouvet, G. Michon, Numerical modelling of entangled carbon fibre material under compression, Comput. Mater. Sci. 151 (2018) 14–24. Copyright (2018), with permission from Elsevier.

In another research [89], a 2D discrete random fibrous network based on the Mikado model was developed by distributing centroids of fibres through a 2D Poisson process. The fibres were rigidly cross linked by welded joints at all crossing points; Fig. 5.11B–E shows this model in its deformed configurations. The used welded joints in locations of interactions enabled transmission of both bending moments and forces between the contacting fibres.

The effects of a window size (i.e. model dimensions) and a network density on the effective mechanical moduli were investigated by Berkache et al. [87,89]. Moreover, 3D discrete RVEs of fibrous network model were developed for a new entangled cross-linked carbon-fibre material (Fig. 5.11F–H) in order to understand the effect of microstructural parameters on its macroscopic behaviour under compression [88] and shear hysteresis [90]. This material, which is basically a random fibrous network, is composed of randomly distributed fibres, which are resin cross linked at some of the fibre-to-fibre contact points. The influence of morphological parameters, such as the distance between cross links and orientation distribution of fibres, on shear stiffness was investigated in subsequent research [91].

Bircher et al. [92] combined continuous and discontinuous modelling approaches into a hybrid finite-element model as an explicit representation of a collagen fibre network of an amnion for analysis of the defect tolerance of foetal membranes. Their experimental analysis of mode-I fracture conducted on an amnion (Fig. 5.12A) was accompanied by development of the hybrid FE model (Fig. 5.12A) that allowed a study of a scatter for fibre strains in the vicinity of a notch tip (Fig. 5.12C). The same modelling approach was used by Bircher et al. [93] for assessment of a specific toughening mechanism of soft collagenous tissues as these materials demonstrate a certain level of fracture toughness and maintain their integrity even in the presence of macroscopic cracks, meaning a high level of fracture toughness or defect tolerance.

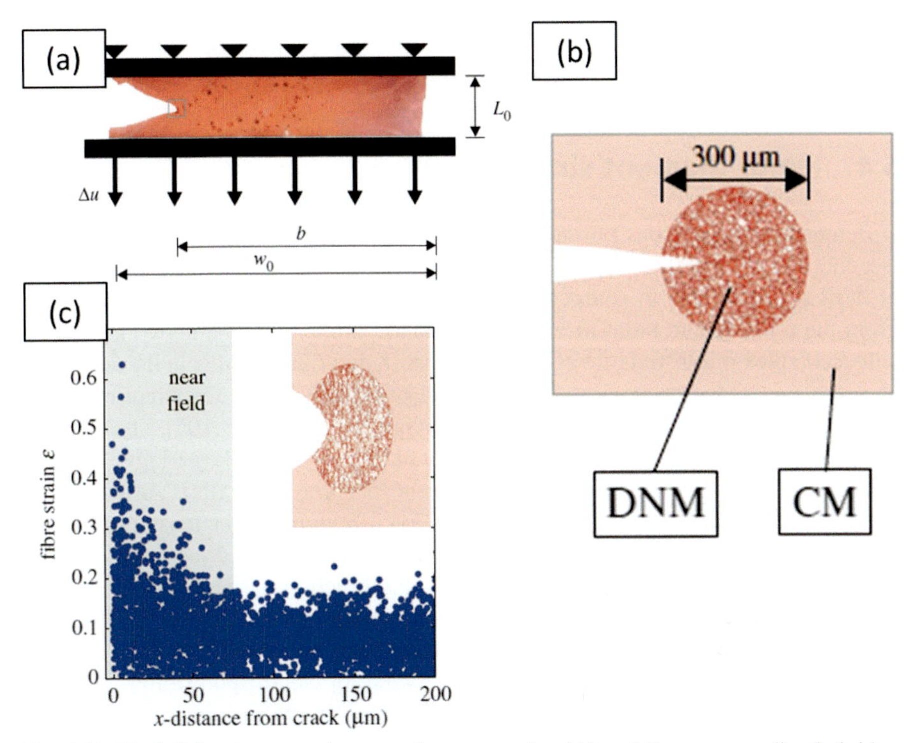

Fig. 5.12 Mode-I fracture experiment on human amnion (A) and its corresponding hybrid model (*DNM*, discrete network model; *CM*, continuum model) (B). (C) Scatter plot of fibre strain versus distance from notch tip along a line perpendicular to the loading direction obtained from simulation with a hybrid FE model [92]

Reprinted from K. Bircher, A.E. Ehret, D. Spiess, M. Ehrbar, A.P. Simões-Wüst, N. Ochsenbein-Kölble, R. Zimmermann, E. Mazza, On the defect tolerance of fetal membranes, Interf. Focus 9 (2019) 20190010. Copyright (2019), with permission from The Royal Society.

Kulachenko and Uesaka [16] used a 3D model of a disordered fibrous network to simulate processes of its deformation and failure. The model was produced by employing a fibre-deposition algorithm [94]. It was intended for paper and woven materials that are computationally expensive due to a high demand related to the fine meshing of each fibre in order to incorporate fibre-to-fibre interactions. In another study [95], a 3D particle-based fibrous network model was proposed to analyse the deformation and fracture behaviours. The fibres were presented as chains of fibre segments with spherical particles at segments' ends. When two segments intersected, i.e. in the case of a fibre-to-fibre contact, a new particle was inserted in each segment and a spring element was added to enable a force transfer between those fibres. Yang et al. [96] automated a virtual reconstruction of a microstructure of a fibreglass pack (a random fibrous network) from statistical data with a packing algorithm based on a non-uniform rational basis spline. The virtually generated network met the requirements for a fibre volume fraction and distributions of fibre diameter and orientation. Bushing-like connector elements were used in place of binder particles in the real material's microstructure connecting fibres to each other.

Table 5.2 lists some of discontinuous models of fibrous networks in terms of material, modelling approach, and fibre-to-fibre interaction.

5.4 Finite element simulations

Different from continuous FE models of fibrous networks, the simulations with their discontinuous FE models allow researchers to track the changes in the microstructure of those materials and to compute the local stress, strain, and other field quantities. From the microscopic point of view, for instance, stretching of a network complex with a complex microstructure of randomly distributed fibres results in the realisation of various deformation processes at the level of fibres such as their reorientations, straightening, and elongation until they start to fail [13,24,25,105–107]. These typical morphological changes were simulated with a 3D FE model [69] (see Fig. 5.13). In a tensile stretch, local buckling modes likely occur for some of the fibres, orthogonal to the loading direction; however, their contribution is minimal to the tensile strength of the network. This buckling is apparent when the buckled fibres are connected to the fibres, which are already stretched in the loading direction by fibre-to-fibre bonds.

In order to quantify the effect of microstructure on the mechanical response (in particular, tension and compression), two different microstructural FE models – cellular and short-fibre fibrous networks – were considered for electrospun fibrous networks by Islam and Picu [108]. Fibre-to-fibre interactions in the cellular networks were modelled as welded joints, whilst the ones in the fibrous networks were defined as freely rotating and translating cross links. A set of FE models of microstructures with different fibre volume fractions in the range from 1% to 3% experienced tensile and compressive loads was developed. Simulation results indicated that, for all the studied volume fractions, short-fibre networks with cross links demonstrated a stiffer tensile response than cellular networks. These tensile and compressive responses

Table 5.2 Discontinuous models of various fibrous networks (*fd*, fibre deposition; *cl*, cross link).

Model	Focus	Discontinuous modelling approach	Fibre interaction
Hou et al. [97]	Non-woven	2D random fd	No interaction
Kwon et al. [98]	Actin cytoskeleton	3D random fd	Beam cl
Nachtrab et al. [99]	Collagen	Poisson-Voronoi processes	Shared nodes
Altendorf and Jeulin [65] and Altendorf et al. [37]	Generic	3D random walk algorithm	No interaction
Lavrykov et al. [85]	Paper	3D random fd in fluid	Spring element
Hatami-Marbini et al. [100]	Soft tissue	2D random fd	Welded cl
Lee and Jasiuk [15]	Generic	Monte Carlo-based fd	Rigid cl
Dirrenberger et al. [38]	Generic	3D random fd	Interpenetrating fibres
Mauri et al. [80]	Human amnion	2D random fd	Pin joints
Bircher et al. [81]	Soft collagen	2D random fd	Pin joints
Gao et al. [39]	BC hydrogel	2D random fd	Shared nodes
Yang et al. [96]	Fibreglass pack	Image-based	Spring element
Goutianos et al. [101]	Cellulose nanopaper	3D random fd	Cohesive law
Deogekar and Picu [102] and Deogekar et al. [103]	Generic	Voronoi tessellation	Springs
Chatti et al. [88,91]	Entangled carbon-fibre network	3D random fd	No cl
Tyznik and Notbohm [104]	Collagen	Voronoi fd	Shared nodes
Chen and Siegmund [70]	Non-woven	Image-based	Shared nodes
Bircher et al. [93]	Soft collagen	2D random fd	Spring-like element

demonstrated various linear and non-linear regimes: (i) linear elastic; (ii) first non-linear stage; and (iii) second linear stage. Similar behaviours were also reported by Chatti et al. [88], Chen and Siegmund [70], and Yang et al. [96]. The processes of reorientation, bending, and stretching of individual fibres were responsible for the stiffening behaviour of the fibrous network under tension. In the compression case, softening response occurred in the first non-linear stage for both microstructures, but more prominent in cellular networks.

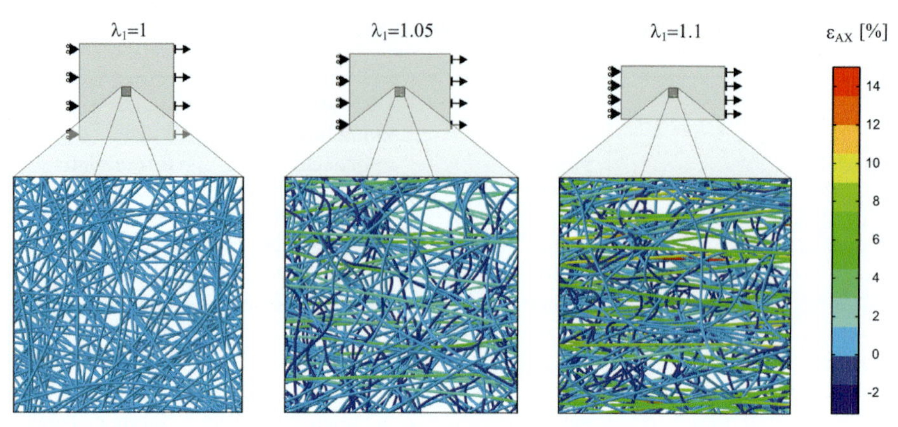

Fig. 5.13 Deformed shape of microstructure of electrospun fibrous network for various levels of stretching (from 0% to 10%); colours denote axial strain of elements [69].
Reprinted from S. Domaschke, M. Zündel, E. Mazza, A.E. Ehret, A 3D computational model of electrospun networks and its application to inform a reduced modelling approach, Int. J. Solids Struct. 158 (2019) 76–89. Copyright (2019), with permission from Elsevier.

5.4.1 Effects of window size and periodicity on mechanical properties

One of the important modelling parameters considered in FE simulations of random fibrous networks is the size of the domain of interest (i.e. window size) as it is expected to affect the calculated material properties such as effective elastic modulus. Whilst conducting numerical studies, researchers also investigated this effect on the properties of interest by modifying the size of FE models [37,38,109,110]. Here, it is worth mentioning that the required computational time to complete a converging solution for a physical problem depends on the size of the numerical model in terms of a number of finite (FEA), or volume (CFD) elements, or particles (SPH: smoothed particle hydrodynamics). A relatively large, i.e. optimum-sized, model should, therefore, be sufficient to capture the effective properties. To exemplify this, Gao et al. [39] carried out a size-effect assessment for effective mechanical properties with 2D discontinuous FE models of RVEs of a bacterial hydrogel (a random fibrous network) with various dimensions and microstructural properties of fibres. The results presented by Gao et al. [39] and another investigation by Alimadadi et al. [73] indicated that the effective modulus converges when a certain model size was reached, and a higher scatter in obtained properties was observed for the models of smaller sizes.

Tyznik and Notbohm [104] employed cylindrical RVEs for cellular fibrous networks of collagens as an alternative to traditional – rectangular (2d) or cuboid (3D) – RVEs. Cylindrical RVEs with the same microstructural parameters and various radii were generated in order to quantify the effect of the dimensionless radius of network R/l_f on its dimensionless effective modulus E_n/E_f. The computed effective modulus converged to a certain value in each model as the radius of cylindrical models increased (Fig. 5.14A).

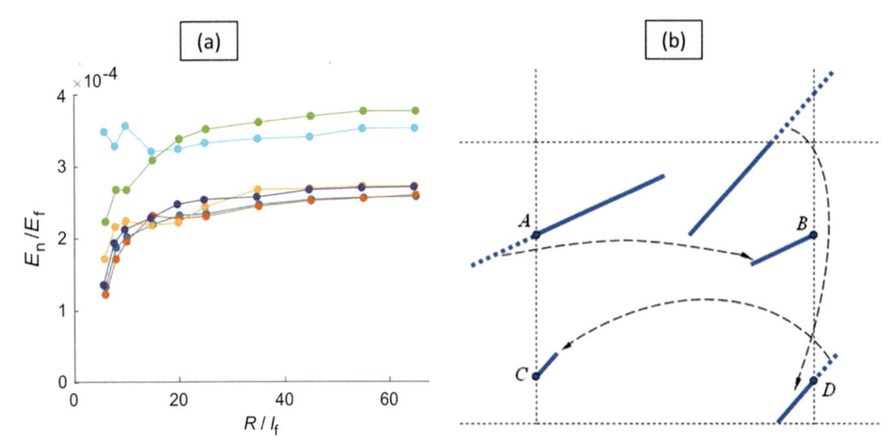

Fig. 5.14 (A) Effect of window size on dimensionless Young's modulus of random fibre networks subjected to uniaxial tension (each colour corresponds to one random network) [104]. (B) Operations with fibres to assign periodicity to an FE model of a fibrous network [111]. Panel (A): Reprinted from S. Tyznik, J. Notbohm, Length scale dependent elasticity in random three-dimensional fibre networks, Mech. Mater. 138 (2019) 103155. Copyright (2019), with permission from Elsevier. Panel (B): Reprinted from Q. Liu, Z. Lu, Z. Hu, J. Li, Finite element analysis on tensile behaviour of 3D random fibrous materials: model description and meso-level approach, Mater. Sci. Eng. A 587 (2013) 36–45. Copyright (2013), with permission from Elsevier.

As displacement boundary conditions are concerned, two types of mechanical boundary conditions can be employed in the analysis of fibrous networks: (i) periodic and (ii) non-periodic. In general, big stochastic FE models of fibrous networks are used with non-periodic boundary conditions, whereas the small-size RVEs employ mostly periodic ones. The latter aim to diminish the edge effects and computational efforts. The use of non-periodic boundary conditions in some cases, where the size of a random fibrous network is very large, minimises the edge effects. Non-periodic boundary conditions were, for instance, applied in big discontinuous models by Chatti et al. [88], Dirrenberger et al. [38], and Sozumert et al. [25]. Shahsavari and Picu [109] investigated the effect of different boundary conditions (e.g. non-periodic, half-periodic, and fully periodic) on the mechanics of a fibrous network and reported that the effect of the type of boundary conditions is more prominent in FE models of fibrous networks with smaller sizes. Further information about the implementation of periodic boundary conditions can be found, for example, in Nguyen et al. [112] and Shahsavari and Picu [109].

In order to minimise the optimum size of an FE model with converging material properties, periodicity can be applied to the microstructure whilst distributing fibres in a design domain. The resultant microstructures of fibrous networks with such distributions of fibres are still random, but periodic. Some random fibrous networks modelled with the periodic approach can be found in Ma et al. [113], Zhang et al. [114], and Zündel et al. [115]. For instance, in order to achieve an RVE network for a short-fibre random fibrous network with a periodic microstructure, such as by Zhang et al.

[114], straight fibres of equal length were first introduced with random spatial and orientation distribution by using their central points. Each fibre then was translated around the boundaries of the model, with any parts of fibres outside these boundaries trimmed (Fig. 5.14B). Further details can be found in Liu et al. [111] and Zhang et al. [114].

5.4.2 Fibre-to-fibre interactions

The interactions at fibre-to-fibre intersections are responsible for transferring normal and shear forces from one fibre to another; hence, an increase in their density leads to a higher effective elastic modulus [116]. Such interactions, therefore, play an important role in deformation, damage, and fracture mechanisms. In Table 5.2, various discrete FE models of random fibrous networks were listed with the type of interactions. It is known that both the density and type of fibre-to-fibre interactions determine the mechanical properties of the network, such as in collagens [117]. These interactions are, in some cases, defined by constitutive laws imitating contact behaviours such as adhesion in contacting fibres [118], whereas they are sometimes simplified by various cross-link elements (e.g. beams, a set of trusses or springs). For cross links, Picu [62] introduced four main groups in terms of their physical nature: (i) pin joints; (ii) actin cross links; (iii) rotating joints; and (iv) welded joints. Fibres in contact with a pin joint are only exposed to friction under deformation and, therefore, no moment is transmitted between them. Mauri et al. [80] and Bircher et al. [81] inserted pin joints into the fibres in contact points for a human amnion and soft collagen materials. An actin cross link is a joint with some actin bindings that allow fibre segments to rotate at intersections, describing the bending behaviour of fibres [98,119]. A welded joint, on the other hand, fully constrains the contacting surfaces of fibres, not allowing any relative motion.

To enable the load transfer at fibre-to-fibre contacts between 3D beams representing fibres, additional beams [113] or spring elements [85,96,102,103] were inserted to model flexible cross links in numerical simulations. Deogekar and Picu [102] modelled the bonds with multiple independent (uncoupled) springs with axial and torsional stiffnesses (Fig. 5.15A) in order to assess bond failure and its effect on mechanical properties of networks. In an image-based discrete model of a fibreglass pack [96], binder materials for fibres were modelled as spring elements (Fig. 5.15B). In a similar study [90], fibre-to-fibre bonds in entangled carbon fibres were made of deformable solidified resin, which can transfer a load between fibres. In order to replicate this deformation mechanism, springs with axial and torsional stiffnesses were introduced into the model of a fibrous network.

Mechanical properties of fibres can be obtained from experimental single-fibre tests with microtensile testers or AFM. When these testing methods become a challenge because of the size of fibres, the material properties of fibres can be computed with an inverse-parameter identification procedure on discrete FE models [120]. In a similar approach by Chen and Silberstein [121], the strength of fibre-to-fibre bonds in a non-woven fibrous network was found by optimising the parameters of an image-based FE model based on micro-tomography scans using force-displacement curves of simulations and experiments.

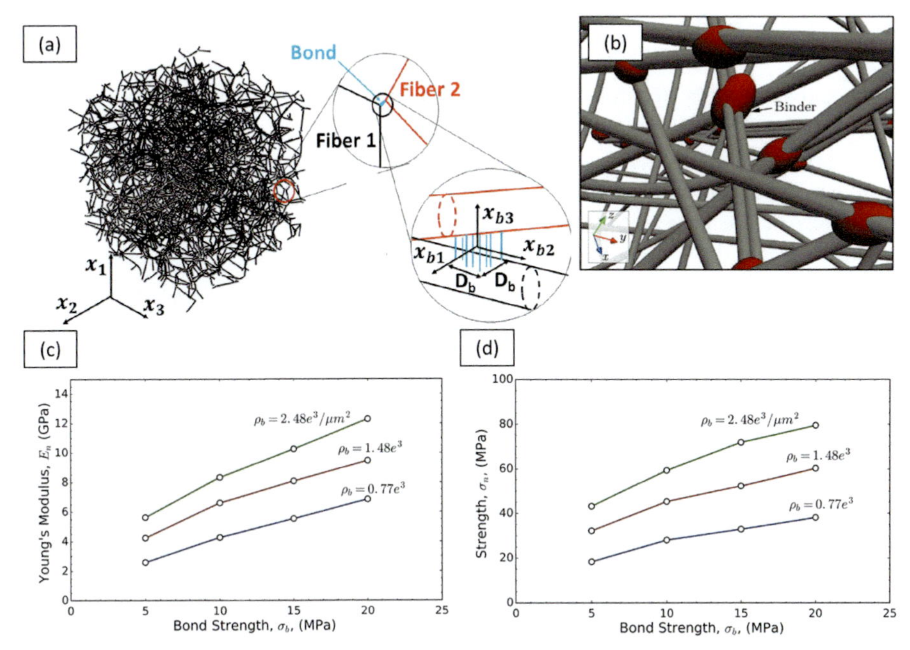

Fig. 5.15 (A) Realisation of 3D fibrous network with one of its bonds between two fibres [102]. (B) Fibre binders and bundles inside virtual fibrous network (fibreglass pack) [96]. (C, D) Effective Young's modulus and network strength with various densities and strengths of bonds [101].

Obviously, the character of fibre-to-fibre interaction affects the macroscopic (effective) mechanical properties of networks. Goutianos et al. [101] calculated the effective Young's modulus and network strength of fibrous networks with various densities and strengths of fibre-to-fibre bonds (Fig. 5.15C and D). The simulation results exhibited a strong dependency of these material properties on the bond strength. Borodulina et al. [122] produced models of 3D fibrous networks by using the 3D fibre-deposition technique of Kulachenko and Uesaka [16] in order to examine the effect of variability in properties of fibres and fibre bonds on the effective stiffness and strength of fibre networks. The numerical findings by Borodulina et al. [122] indicated that the effective strength is not noticeably affected by the strength distribution of bonds unless the distribution becomes highly asymmetric, such as in case of a Gamma L-shaped distribution.

5.5 Conclusion

In this chapter, first, the fundamental concepts (such as orientation distributions of fibres) of numerical modelling of random fibrous networks were discussed. These concepts paved a way for understanding various modelling approaches that are classified here as discontinuous, continuous, and hybrid modelling schemes. Various examples of the first two schemes were presented, though only a few models of the hybrid models were mentioned due to their limited number in the literature. Continuous models are assumed to be a pragmatic way to model mechanics of fibrous networks at the macroscale. Still, these models are limited in (or incapable of) reflecting and, hence, assessment, of the real deformation and failure mechanisms at microscale, such as failure of fibres and fibre-to-fibre bonds. The discontinuous models are, on the other hand, successful in incorporating these mechanisms by introducing both the discontinuous (and, in many cases, complex) microstructure and fibre-level features of networks. However, the constitutive relations between contacting fibres or various cross links used in place of fibre bonds that have a significant effect on properties and performance of fibrous networks in the literature could be rather simplistic, not adequately representing the respective underlying mechanisms. Thus, there is a demand for research on this topic for the development of more advanced numerical schemes for such networks.

References

[1] P. Fratzl (Ed.), Collagen: Structure and Mechanics, Springer, New York, 2008.

[2] M. Ovaska, Z. Bertalan, A. Miksic, M. Sugni, C. Di Benedetto, C. Ferrario, L. Leggio, L. Guidetti, M.J. Alava, C.A.M. La Porta, S. Zapperi, Deformation and fracture of echinoderm collagen networks, J. Mech. Behav. Biomed. Mater. 65 (2017) 42–52, https://doi.org/10.1016/j.jmbbm.2016.07.035.

[3] X. Gao, P. Kuśmierczyk, Z. Shi, C. Liu, G. Yang, I. Sevostianov, V.V. Silberschmidt, Through-thickness stress relaxation in bacterial cellulose hydrogel, J. Mech. Behav. Biomed. Mater. 59 (2016) 90–98, https://doi.org/10.1016/j.jmbbm.2015.12.021.

[4] N.S. Kalson, D.F. Holmes, A. Herchenhan, Y. Lu, T. Starborg, K.E. Kadler, Slow stretching that mimics embryonic growth rate stimulates structural and mechanical development of tendon-like tissue in vitro, Dev. Dyn. 240 (2011) 2520–2528, https://doi.org/10.1002/dvdy.22760.

[5] L.T. Choong, P. Yi, G.C. Rutledge, Three-dimensional imaging of electrospun fiber mats using confocal laser scanning microscopy and digital image analysis, J. Mater. Sci. 50 (2015) 3014–3030, https://doi.org/10.1007/s10853-015-8834-2.

[6] A. El-hadi, F. Al-Jabri, Influence of electrospinning parameters on fiber diameter and mechanical properties of poly(3-hydroxybutyrate) (PHB) and polyanilines (PANI) blends, Polymers 8 (2016) 97, https://doi.org/10.3390/polym8030097.

[7] D.H. Mueller, M. Kochmann, Numerical modeling of thermobonded nonwovens, Int. Nonwovens J. (2004), https://doi.org/10.1177/1558925004os-1300114. os-13, 1558925004os–13.

[8] J. Li, D.J. Mooney, Designing hydrogels for controlled drug delivery, Nat. Rev. Mater. 1 (2016) 16071, https://doi.org/10.1038/natrevmats.2016.71.

[9] L.G. Griffith, Tissue engineering—current challenges and expanding opportunities, Science 295 (2002) 1009–1014, https://doi.org/10.1126/science.1069210.
[10] J.P. Gong, Why are double network hydrogels so tough? Soft Matter 6 (2010) 2583, https://doi.org/10.1039/b924290b.
[11] E. Demirci, M. Acar, B. Pourdeyhimi, V.V. Silberschmidt, Finite element modelling of thermally bonded bicomponent fibre nonwovens: tensile behaviour, Comput. Mater. Sci. 50 (2011) 1286–1291, https://doi.org/10.1016/j.commatsci.2010.02.039.
[12] K. Li, R.W. Ogden, G.A. Holzapfel, Computational method for excluding fibers under compression in modeling soft fibrous solids, Eur. J. Mech. A. Solids 57 (2016) 178–193, https://doi.org/10.1016/j.euromechsol.2015.11.003.
[13] W. Yang, V.R. Sherman, B. Gludovatz, E. Schaible, P. Stewart, R.O. Ritchie, M.A. Meyers, On the tear resistance of skin, Nat. Commun. 6 (2015) 6649, https://doi.org/10.1038/ncomms7649.
[14] H.L. Cox, The elasticity and strength of paper and other fibrous materials, Br. J. Appl. Phys. 3 (1952) 72–79, https://doi.org/10.1088/0508-3443/3/3/302.
[15] Y. Lee, I. Jasiuk, Apparent elastic properties of random fiber networks, Comput. Mater. Sci. 79 (2013) 715–723, https://doi.org/10.1016/j.commatsci.2013.07.037.
[16] A. Kulachenko, T. Uesaka, Direct simulations of fiber network deformation and failure, Mech. Mater. 51 (2012) 1–14, https://doi.org/10.1016/j.mechmat.2012.03.010.
[17] P.L. Chandran, V.H. Barocas, Affine versus non-affine fibril kinematics in collagen networks: theoretical studies of network behavior, J. Biomech. Eng. 128 (2006) 259–270, https://doi.org/10.1115/1.2165699.
[18] C. Heussinger, E. Frey, Floppy modes and nonaffine deformations in random fiber networks, Phys. Rev. Lett. 97 (2006) 105501, https://doi.org/10.1103/PhysRevLett.97.105501.
[19] H. Hatami-Marbini, R.C. Picu, Effect of fiber orientation on the non-affine deformation of random fiber networks, Acta Mech. 205 (2009) 77–84, https://doi.org/10.1007/s00707-009-0170-7.
[20] H. Hatami-Marbini, R.C. Picu, Scaling of nonaffine deformation in random semiflexible fiber networks, Phys. Rev. E 77 (2008), https://doi.org/10.1103/PhysRevE.77.062103, 062103.
[21] Q. Wen, A. Basu, P.A. Janmey, A.G. Yodh, Non-affine deformations in polymer hydrogels, Soft Matter 8 (2012) 8039, https://doi.org/10.1039/c2sm25364j.
[22] A. Basu, Q. Wen, X. Mao, T.C. Lubensky, P.A. Janmey, A.G. Yodh, Nonaffine displacements in flexible polymer networks, Macromolecules 44 (2011) 1671–1679, https://doi.org/10.1021/ma1026803.
[23] Q. Cheng, S. Wang, A method for testing the elastic modulus of single cellulose fibrils via atomic force microscopy, Compos. A: Appl. Sci. Manuf. 39 (2008) 1838–1843, https://doi.org/10.1016/j.compositesa.2008.09.007.
[24] A. Ridruejo, C. González, J. LLorca, Micromechanisms of deformation and fracture of polypropylene nonwoven fabrics, Int. J. Solids Struct. 48 (2011) 153–162, https://doi.org/10.1016/j.ijsolstr.2010.09.013.
[25] Sozumert, E., Farukh, F., Sabuncuoglu, B., Demirci, E., Acar, M., Pourdeyhimi, B., Silberschmidt, V.V., 2018. Deformation and damage of random fibrous networks. Int. J. Solids Struct. S0020768318305018. doi: https://doi.org/10.1016/j.ijsolstr.2018.12.012E. Sozumert, F. Farukh, B. Sabuncuoglu, E. Demirci, M. Acar, B. Pourdeyhimi, V.V. Silberschmidt, Deformation and damage of random fibrous networks, Int. J. Solids Struct. 184 (2020) 233–247, https://doi.org/10.1016/j.ijsolstr.2018.12.012. S0020768318305018.

[26] V. Kumar, A. Rawal, Elastic moduli of electrospun mats: importance of fiber curvature and specimen dimensions, J. Mech. Behav. Biomed. Mater. 72 (2017) 6–13, https://doi.org/10.1016/j.jmbbm.2017.04.013.
[27] C.-L. Pai, M.C. Boyce, G.C. Rutledge, On the importance of fiber curvature to the elastic moduli of electrospun nonwoven fiber meshes, Polymer 52 (2011) 6126–6133, https://doi.org/10.1016/j.polymer.2011.10.055.
[28] J.W.S. Hearle, P.J. Stevenson, Studies in nonwoven fabrics: part IV: prediction of tensile properties, Text. Res. J. 34 (1964) 181–191, https://doi.org/10.1177/004051756403400301.
[29] J.W.S. Hearle, A. Newton, Nonwoven fabric studies: part XIV: derivation of generalized mechanics by the energy method, Text. Res. J. 37 (1967) 778–797, https://doi.org/10.1177/004051756703700908.
[30] L.A. Carlsson, T. Lindstrom, A shear-lag approach to the tensile strength of paper, Compos. Sci. Technol. 65 (2005) 183–189, https://doi.org/10.1016/j.compscitech.2004.06.012.
[31] V.I. Räisänen, M.J. Alava, K.J. Niskanen, R.M. Nieminen, Does the shear-lag model apply to random fiber networks? J. Mater. Res. 12 (1997) 2725–2732, https://doi.org/10.1557/JMR.1997.0363.
[32] W. Batchelor, An analytical solution for the load distribution along a fibre in a nonwoven network, Mech. Mater. 40 (2008) 975–981, https://doi.org/10.1016/j.mechmat.2008.07.003.
[33] J. Åström, S. Saarinen, K. Niskanen, J. Kurkijärvi, Microscopic mechanics of fiber networks, J. Appl. Phys. 75 (1994) 2383–2392, https://doi.org/10.1063/1.356259.
[34] M.A. Narter, S.K. Batra, D.R. Buchanan, Micromechanics of three-dimensional fibrewebs: constitutive equations, Proc. Roy. Soc. Lond. A 455 (1999) 3543–3563,- https://doi.org/10.1098/rspa.1999.0465.
[35] P. Isaksson, R. Hägglund, P. Gradin, Continuum damage mechanics applied to paper, Int. J. Solids Struct. 41 (2004) 4731–4755, https://doi.org/10.1016/j.ijsolstr.2004.02.043.
[36] F. Martínez-Hergueta, A. Ridruejo, C. González, J. LLorca, A multiscale micromechanical model of needlepunched nonwoven fabrics, Int. J. Solids Struct. 96 (2016) 81–91, https://doi.org/10.1016/j.ijsolstr.2016.06.020.
[37] H. Altendorf, D. Jeulin, F. Willot, Influence of the fiber geometry on the macroscopic elastic and thermal properties, Int. J. Solids Struct. 51 (2014) 3807–3822, https://doi.org/10.1016/j.ijsolstr.2014.05.013.
[38] J. Dirrenberger, S. Forest, D. Jeulin, Towards gigantic RVE sizes for 3D stochastic fibrous networks, Int. J. Solids Struct. 51 (2014) 359–376, https://doi.org/10.1016/j.ijsolstr.2013.10.011.
[39] X. Gao, E. Sozumert, Z. Shi, G. Yang, V.V. Silberschmidt, Assessing stiffness of nanofibres in bacterial cellulose hydrogels: numerical-experimental framework, Mater. Sci. Eng. C 77 (2017) 9–18, https://doi.org/10.1016/j.msec.2017.03.231.
[40] A.S. Phani, M.I. Hussein (Eds.), Dynamics of Lattice Materials, John Wiley & Sons, Inc, Chichester, West Sussex, United Kingdom, 2017.
[41] P.L. Chandran, V.H. Barocas, Deterministic material-based averaging theory model of collagen gel micromechanics, J. Biomech. Eng. 129 (2007) 137, https://doi.org/10.1115/1.2472369.
[42] T. Stylianopoulos, C.A. Bashur, A.S. Goldstein, S.A. Guelcher, V.H. Barocas, Computational predictions of the tensile properties of electrospun fibre meshes: effect of fibre diameter and fibre orientation, J. Mech. Behav. Biomed. Mater. 1 (2008) 326–335, https://doi.org/10.1016/j.jmbbm.2008.01.003.

[43] T. Stylianopoulos, V.H. Barocas, Volume-averaging theory for the study of the mechanics of collagen networks, Comput. Methods Appl. Mech. Eng. 196 (2007) 2981–2990, https://doi.org/10.1016/j.cma.2006.06.019.
[44] X.-F. Wu, Y.A. Dzenis, Elasticity of planar fiber networks, J. Appl. Phys. 98 (2005), https://doi.org/10.1063/1.2123369, 093501.
[45] E. Bosco, R.H.J. Peerlings, M.G.D. Geers, Asymptotic homogenization of hygro-thermo-mechanical properties of fibrous networks, Int. J. Solids Struct. 115–116 (2017) 180–189, https://doi.org/10.1016/j.ijsolstr.2017.03.015.
[46] E. Bosco, R.H.J. Peerlings, M.G.D. Geers, Hygro-mechanical properties of paper fibrous networks through asymptotic homogenization and comparison with idealized models, Mech. Mater. 108 (2017) 11–20, https://doi.org/10.1016/j.mechmat.2017.01.013.
[47] S. Ghosh, K. Lee, S. Moorthy, Two scale analysis of heterogeneous elastic-plastic materials with asymptotic homogenization and Voronoi cell finite element model, Comput. Methods Appl. Mech. Eng. 132 (1996) 63–116, https://doi.org/10.1016/0045-7825(95)00974-4.
[48] A. Raina, C. Linder, A homogenization approach for nonwoven materials based on fiber undulations and reorientation, J. Mech. Phys. Solids 65 (2014) 12–34, https://doi.org/10.1016/j.jmps.2013.12.011.
[49] A. Raina, C. Linder, A micromechanical model with strong discontinuities for failure in nonwovens at finite deformations, Int. J. Solids Struct. 75–76 (2015) 247–259, https://doi.org/10.1016/j.ijsolstr.2015.08.018.
[50] M.N. Silberstein, C.-L. Pai, G.C. Rutledge, M.C. Boyce, Elastic–plastic behavior of non-woven fibrous mats, J. Mech. Phys. Solids 60 (2012) 295–318, https://doi.org/10.1016/j.jmps.2011.10.007.
[51] N. Chen, M.N. Silberstein, A micromechanics-based damage model for non-woven fiber networks, Int. J. Solids Struct. 160 (2019) 18–31, https://doi.org/10.1016/j.ijsolstr.2018.10.009.
[52] A. Karakoç, On the computational homogenization of three-dimensional fibrous materials, Compos. Struct. 10 (2020).
[53] F. Martínez-Hergueta, A. Ridruejo, C. González, J. LLorca, Deformation and energy dissipation mechanisms of needle-punched nonwoven fabrics: a multiscale experimental analysis, Int. J. Solids Struct. 64–65 (2015) 120–131, https://doi.org/10.1016/j.ijsolstr.2015.03.018.
[54] T. Jin, I. Stanciulescu, Numerical simulation of fibrous biomaterials with randomly distributed fiber network structure, Biomech. Model. Mechanobiol. 15 (2016) 817–830, https://doi.org/10.1007/s10237-015-0725-6.
[55] L. Zhang, S.P. Lake, V.H. Barocas, M.S. Shephard, R.C. Picu, Cross-linked fiber network embedded in an elastic matrix, Soft Matter 9 (2013) 6398, https://doi.org/10.1039/c3sm50838b.
[56] M.B. Rubin, S.R. Bodner, A three-dimensional nonlinear model for dissipative response of soft tissue, Int. J. Solids Struct. 39 (2002) 5081–5099, https://doi.org/10.1016/S0020-7683(02)00237-8.
[57] G.A. Holzapfel, T.C. Gasser, R.W. Ogden, A new constitutive framework for arterial wall mechanics and a comparative study of material models, in: S.C. Cowin, J.D. Humphrey (Eds.), Cardiovascular Soft Tissue Mechanics, Kluwer Academic Publishers, Dordrecht, 2004, pp. 1–48, https://doi.org/10.1007/0-306-48389-0_1.
[58] T.C. Gasser, R.W. Ogden, G.A. Holzapfel, Hyperelastic modelling of arterial layers with distributed collagen fibre orientations, J. R. Soc. Interface 3 (2006) 15–35, https://doi.org/10.1098/rsif.2005.0073.

[59] Y. Liu, H. Zhang, J. Zhang, Y. Zheng, Transient swelling of polymeric hydrogels: a new finite element solution framework, Int. J. Solids Struct. 80 (2016) 246–260, https://doi.org/10.1016/j.ijsolstr.2015.11.010.

[60] A.P.G. Castro, P. Laity, M. Shariatzadeh, C. Wittkowske, C. Holland, D. Lacroix, Combined numerical and experimental biomechanical characterization of soft collagen hydrogel substrate, J. Mater. Sci. Mater. Med. 27 (2016) 79, https://doi.org/10.1007/s10856-016-5688-3.

[61] A.P.G. Castro, J. Yao, T. Battisti, D. Lacroix, Poroelastic modeling of highly hydrated collagen hydrogels: experimental results vs. numerical simulation with custom and commercial finite element solvers, Front. Bioeng. Biotechnol. 6 (2018) 142, https://doi.org/10.3389/fbioe.2018.00142.

[62] R.C. Picu, Mechanics of random fiber networks—a review, Soft Matter 7 (2011) 6768, https://doi.org/10.1039/c1sm05022b.

[63] M. Sacks, G. Rodin, Using simulations with realistic fibrous network geometry to determine the achievable ranges of macroscopic mechanical behaviors of elastomeric scaffolds, in: Frontiers in Bioengineering and Biotechnology. Conference Abstract: 10th World Biomaterials Congress, 2016, https://doi.org/10.3389/conf.FBIOE.2016.01.03023.

[64] D.V. Wilbrink, L.A.A. Beex, R.H.J. Peerlings, A discrete network model for bond failure and frictional sliding in fibrous materials, Int. J. Solids Struct. 50 (2013) 1354–1363, https://doi.org/10.1016/j.ijsolstr.2013.01.012.

[65] H. Altendorf, D. Jeulin, Random-walk-based stochastic modeling of three-dimensional fiber systems, Phys. Rev. E 83 (2011), https://doi.org/10.1103/PhysRevE.83.041804, 041804.

[66] V. Ayyalasomayajula, B. Pierrat, P. Badel, A computational model for understanding the micro-mechanics of collagen fiber network in the tunica adventitia, Biomech. Model. Mechanobiol. 18 (2019) 1507–1528, https://doi.org/10.1007/s10237-019-01161-1.

[67] J.B. Carleton, A. D'Amore, K.R. Feaver, G.J. Rodin, M.S. Sacks, Geometric characterization and simulation of planar layered elastomeric fibrous biomaterials, Acta Biomater. 12 (2015) 93–101, https://doi.org/10.1016/j.actbio.2014.09.049.

[68] S. Recchia, J. Zheng, A.A. Pelegri, Fiberwalk: a random walk approach to fiber representative volume element creation, Acta Mech. 225 (2014) 1301–1312, https://doi.org/10.1007/s00707-013-1069-x.

[69] S. Domaschke, M. Zündel, E. Mazza, A.E. Ehret, A 3D computational model of electrospun networks and its application to inform a reduced modelling approach, Int. J. Solids Struct. 158 (2019) 76–89, https://doi.org/10.1016/j.ijsolstr.2018.08.030.

[70] Y. Chen, T. Siegmund, Mechanics of compaction of a porous non-woven fiber solid, Mech. Mater. 137 (2019) 103101, https://doi.org/10.1016/j.mechmat.2019.103101.

[71] G. Gaiselmann, C. Tötzke, I. Manke, W. Lehnert, V. Schmidt, 3D microstructure modeling of compressed fiber-based materials, J. Power Sources 257 (2014) 52–64, https://doi.org/10.1016/j.jpowsour.2014.01.095.

[72] S.B. Lindström, D.A. Vader, A. Kulachenko, D.A. Weitz, Biopolymer network geometries: characterization, regeneration, and elastic properties, Phys. Rev. E 82 (2010), https://doi.org/10.1103/PhysRevE.82.051905, 051905.

[73] M. Alimadadi, S.B. Lindström, A. Kulachenko, Role of microstructures in the compression response of three-dimensional foam-formed wood fiber networks, Soft Matter 14 (2018) 8945–8955, https://doi.org/10.1039/C7SM02561K.

[74] E. Demirci, M. Acar, B. Pourdeyhimi, V.V. Silberschmidt, Computation of mechanical anisotropy in thermally bonded bicomponent fibre nonwovens, Comput. Mater. Sci. 52 (2012) 157–163, https://doi.org/10.1016/j.commatsci.2011.01.033.

[75] A. D'Amore, J.A. Stella, W.R. Wagner, M.S. Sacks, Characterization of the complete fiber network topology of planar fibrous tissues and scaffolds, Biomaterials 31 (2010) 5345–5354, https://doi.org/10.1016/j.biomaterials.2010.03.052.
[76] A. D'Amore, N. Amoroso, R. Gottardi, C. Hobson, C. Carruthers, S. Watkins, W.R. Wagner, M.S. Sacks, From single fiber to macro-level mechanics: a structural finite-element model for elastomeric fibrous biomaterials, J. Mech. Behav. Biomed. Mater. 39 (2014) 146–161, https://doi.org/10.1016/j.jmbbm.2014.07.016.
[77] S. Staub, H. Andrä, M. Kabel, Fast FFT based solver for rate-dependent deformations of composites and nonwovens, Int. J. Solids Struct. 154 (2018) 33–42, https://doi.org/10.1016/j.ijsolstr.2016.12.014.
[78] W.A. Bosbach, Nano-CT scans in the optimisation of purposeful experimental procedures: a study on metallic fibre networks, Med. Eng. Phys. 86 (2020) 109–121, https://doi.org/10.1016/j.medengphy.2020.10.015.
[79] M. Faessel, C. Delisée, F. Bos, P. Castéra, 3D Modelling of random cellulosic fibrous networks based on X-ray tomography and image analysis, Compos. Sci. Technol. 65 (2005) 1931–1940, https://doi.org/10.1016/j.compscitech.2004.12.038.
[80] A. Mauri, R. Hopf, A.E. Ehret, C.R. Picu, E. Mazza, A discrete network model to represent the deformation behavior of human amnion, J. Mech. Behav. Biomed. Mater. 58 (2016) 45–56, https://doi.org/10.1016/j.jmbbm.2015.11.009.
[81] K. Bircher, A.E. Ehret, E. Mazza, Microstructure based prediction of the deformation behavior of soft collagenous membranes, Soft Matter 13 (2017) 5107–5116, https://doi.org/10.1039/C7SM00101K.
[82] C.T. Koh, M.L. Oyen, Toughening in electrospun fibrous scaffolds, APL Mater. 3 (2015), https://doi.org/10.1063/1.4901450, 014908.
[83] P. Bergström, S. Hossain, T. Uesaka, Scaling behaviour of strength of 3D-, semi-flexible-, cross-linked fibre network, Int. J. Solids Struct. 166 (2019) 68–74, https://doi.org/10.1016/j.ijsolstr.2019.02.003.
[84] S.B. Lindström, T. Uesaka, Simulation of the motion of flexible fibers in viscous fluid flow, Phys. Fluids 19 (2007) 113307, https://doi.org/10.1063/1.2778937.
[85] S. Lavrykov, B.V. Ramarao, S.B. Lindström, K.M. Singh, 3D network simulations of paper structure, Nord. Pulp Pap. Res. J. 27 (2012) 256–263, https://doi.org/10.3183/npprj-2012-27-02-p256-263.
[86] S.B. Lindström, T. Uesaka, Particle-level simulation of forming of the fiber network in papermaking, Int. J. Eng. Sci. 46 (2008) 858–876, https://doi.org/10.1016/j.ijengsci.2008.03.008.
[87] K. Berkache, S. Deogekar, I. Goda, R.C. Picu, J.-F. Ganghoffer, Identification of equivalent couple-stress continuum models for planar random fibrous media, Contin. Mech. Thermodyn. 31 (2019) 1035–1050, https://doi.org/10.1007/s00161-018-0710-2.
[88] F. Chatti, D. Poquillon, C. Bouvet, G. Michon, Numerical modelling of entangled carbon fibre material under compression, Comput. Mater. Sci. 151 (2018) 14–24, https://doi.org/10.1016/j.commatsci.2018.04.045.
[89] K. Berkache, S. Deogekar, I. Goda, R.C. Picu, J.-F. Ganghoffer, Construction of second gradient continuum models for random fibrous networks and analysis of size effects, Compos. Struct. 181 (2017) 347–357, https://doi.org/10.1016/j.compstruct.2017.08.078.
[90] F. Chatti, C. Bouvet, D. Poquillon, G. Michon, Numerical modelling of shear hysteresis of entangled cross-linked carbon fibres intended for core material, Comput. Mater. Sci. 155 (2018) 350–363, https://doi.org/10.1016/j.commatsci.2018.09.005.
[91] F. Chatti, C. Bouvet, G. Michon, D. Poquillon, Numerical analysis of shear stiffness of an entangled cross-linked fibrous material, Int. J. Solids Struct. 184 (2020) 221–232, https://doi.org/10.1016/j.ijsolstr.2018.12.001.

[92] K. Bircher, A.E. Ehret, D. Spiess, M. Ehrbar, A.P. Simões-Wüst, N. Ochsenbein-Kölble, R. Zimmermann, E. Mazza, On the defect tolerance of fetal membranes, Interface Focus 9 (2019) 20190010, https://doi.org/10.1098/rsfs.2019.0010.
[93] K. Bircher, M. Zündel, M. Pensalfini, A.E. Ehret, E. Mazza, Tear resistance of soft collagenous tissues, Nat. Commun. 10 (2019) 792, https://doi.org/10.1038/s41467-019-08723-y.
[94] K.J. Niskanen, M.J. Alava, Planar random networks with flexible fibers, Phys. Rev. Lett. 73 (1994) 3475–3478, https://doi.org/10.1103/PhysRevLett.73.3475.
[95] J. Persson, P. Isaksson, A mechanical particle model for analyzing rapid deformations and fracture in 3D fiber materials with ability to handle length effects, Int. J. Solids Struct. 51 (2014) 2244–2251, https://doi.org/10.1016/j.ijsolstr.2014.02.031.
[96] M. Yang, M. Ji, E. Taghipour, S. Soghrati, Cross-linked fiberglass packs: microstructure reconstruction and finite element analysis of the micromechanical behavior, Comput. Struct. 209 (2018) 182–196, https://doi.org/10.1016/j.compstruc.2018.08.014.
[97] X. Hou, M. Acar, V.V. Silberschmidt, 2D finite element analysis of thermally bonded nonwoven materials: continuous and discontinuous models, Comput. Mater. Sci. 46 (2009) 700–707, https://doi.org/10.1016/j.commatsci.2009.07.007.
[98] R.Y. Kwon, A.J. Lew, C.R. Jacobs, A microstructurally informed model for the mechanical response of three-dimensional actin networks, Comput. Meth. Biomech. Biomed. Eng. 11 (2008) 407–418, https://doi.org/10.1080/10255840801888686.
[99] S. Nachtrab, S.C. Kapfer, C.H. Arns, M. Madadi, K. Mecke, G.E. Schröder-Turk, Morphology and linear-elastic moduli of random network solids, Adv. Mater. 23 (2011) 2633–2637, https://doi.org/10.1002/adma.201004094.
[100] H. Hatami-Marbini, A. Shahsavari, R.C. Picu, Multiscale modeling of semiflexible random fibrous structures, Comput. Aided Des. 45 (2013) 77–83, https://doi.org/10.1016/j.cad.2011.10.002.
[101] S. Goutianos, R. Mao, T. Peijs, Effect of inter-fibre bonding on the fracture of fibrous networks with strong interactions, Int. J. Solids Struct. 136–137 (2018) 271–278, https://doi.org/10.1016/j.ijsolstr.2017.12.020.
[102] S. Deogekar, R.C. Picu, On the strength of random fiber networks, J. Mech. Phys. Solids 116 (2018) 1–16, https://doi.org/10.1016/j.jmps.2018.03.026.
[103] S. Deogekar, M.R. Islam, R.C. Picu, Parameters controlling the strength of stochastic fibrous materials, Int. J. Solids Struct. 168 (2019) 194–202, https://doi.org/10.1016/j.ijsolstr.2019.03.033.
[104] S. Tyznik, J. Notbohm, Length scale dependent elasticity in random three-dimensional fiber networks, Mech. Mater. 138 (2019) 103155, https://doi.org/10.1016/j.mechmat.2019.103155.
[105] F. Farukh, E.S. Demirci, M. Acar, B. Pourdeyhimi, V.V. Silberschmidt, Characterisation and numerical modelling of complex deformation behaviour in thermally bonded nonwovens, Comput. Mater. Sci. 71 (2013) 165–171, https://doi.org/10.1016/j.commatsci.2013.01.007.
[106] X. Gao, E. Sözümert, Z. Shi, G. Yang, V.V. Silberschmidt, Mechanical modification of bacterial cellulose hydrogel under biaxial cyclic tension, Mech. Mater. 142 (2020) 103272, https://doi.org/10.1016/j.mechmat.2019.103272.
[107] B. Sabuncuoglu, M. Acar, V.V. Silberschmidt, Finite element modelling of fibrous networks: analysis of strain distribution in fibres under tensile load, Comput. Mater. Sci. 79 (2013) 143–158, https://doi.org/10.1016/j.commatsci.2013.04.063.
[108] M.R. Islam, R.C. Picu, Effect of network architecture on the mechanical behavior of random fiber networks, J. Appl. Mech. 85 (2018), https://doi.org/10.1115/1.4040245, 081011.

[109] A.S. Shahsavari, R.C. Picu, Size effect on mechanical behavior of random fiber networks, Int. J. Solids Struct. 50 (2013) 3332–3338, https://doi.org/10.1016/j.ijsolstr.2013.06.004.
[110] J. Wang, B. Yuan, R.P.S. Han, Modulus of elasticity of randomly and aligned polymeric scaffolds with fiber size dependency, J. Mech. Behav. Biomed. Mater. 77 (2018) 314–320, https://doi.org/10.1016/j.jmbbm.2017.09.016.
[111] Q. Liu, Z. Lu, Z. Hu, J. Li, Finite element analysis on tensile behaviour of 3D random fibrous materials: model description and meso-level approach, Mater. Sci. Eng. A 587 (2013) 36–45, https://doi.org/10.1016/j.msea.2013.07.087.
[112] V.-D. Nguyen, E. Béchet, C. Geuzaine, L. Noels, Imposing periodic boundary condition on arbitrary meshes by polynomial interpolation, Comput. Mater. Sci. 55 (2012) 390–406, https://doi.org/10.1016/j.commatsci.2011.10.017.
[113] Y.H. Ma, H.X. Zhu, B. Su, G.K. Hu, R. Perks, The elasto-plastic behaviour of three-dimensional stochastic fibre networks with cross-linkers, J. Mech. Phys. Solids 110 (2018) 155–172, https://doi.org/10.1016/j.jmps.2017.09.014.
[114] M. Zhang, Y. Chen, F. Chiang, P.I. Gouma, L. Wang, Modeling the large deformation and microstructure evolution of nonwoven polymer fiber networks, J. Appl. Mech. 86 (2019), https://doi.org/10.1115/1.4041677, 011010.
[115] M. Zündel, E. Mazza, A.E. Ehret, A 2.5D approach to the mechanics of electrospun fibre mats, Soft Matter 13 (2017) 6407–6421, https://doi.org/10.1039/C7SM01241A.
[116] R. Mao, S. Goutianos, W. Tu, N. Meng, S. Chen, T. Peijs, Modelling the elastic properties of cellulose nanopaper, Mater. Des. 126 (2017) 183–189, https://doi.org/10.1016/j.matdes.2017.04.050.
[117] B. Depalle, Z. Qin, S.J. Shefelbine, M.J. Buehler, Influence of cross-link structure, density and mechanical properties in the mesoscale deformation mechanisms of collagen fibrils, J. Mech. Behav. Biomed. Mater. 52 (2015) 1–13, https://doi.org/10.1016/j.jmbbm.2014.07.008.
[118] V. Negi, R.C. Picu, Mechanical behavior of cross-linked random fiber networks with inter-fiber adhesion, J. Mech. Phys. Solids 122 (2019) 418–434, https://doi.org/10.1016/j.jmps.2018.09.027.
[119] E.M. Huisman, T. van Dillen, P.R. Onck, E. Van der Giessen, Three-dimensional cross-linked F-actin networks: relation between network architecture and mechanical behavior, Phys. Rev. Lett. 99 (2007) 208103, https://doi.org/10.1103/PhysRevLett.99.208103.
[120] X. Gao, Z. Shi, C. Liu, G. Yang, I. Sevostianov, V.V. Silberschmidt, Inelastic behaviour of bacterial cellulose hydrogel: in aqua cyclic tests, Polym. Test. 44 (2015) 82–92, https://doi.org/10.1016/j.polymertesting.2015.03.021.
[121] N. Chen, M.N. Silberstein, Determination of bond strengths in non-woven fabrics: a combined experimental and computational approach, Exp. Mech. 58 (2018) 343–355, https://doi.org/10.1007/s11340-017-0346-3.
[122] S. Borodulina, H.R. Motamedian, A. Kulachenko, Effect of fiber and bond strength variations on the tensile stiffness and strength of fiber networks, Int. J. Solids Struct. 154 (2018) 19–32, https://doi.org/10.1016/j.ijsolstr.2016.12.013.

Computational homogenisation of three-dimensional fibrous materials

Alp Karakoç
Department of Communications and Networking, Aalto University, Espoo, Finland,
Department of Bioproducts and Biosystems, Aalto University, Espoo, Finland

6.1 Introduction

Fibrous materials are widely used and commercially available material systems that are used in the fields of consumer products, filtration, automotive, energy, and regenerative medicine, to name a few. They are versatile and convenient for mass production due to their thermomechanical characteristics and low weights [1–6]. They are classified based on their constituting fibre types that can be natural or man-made, the fibre bonding processes that are based on chemical, mechanical, or thermal treatments, and the fibre architecture in the form of woven or non-woven [7]. The woven architectures have repetitive patterns, which are conventionally of plain, satin, twill weaves, and their variations [8,9]. On the other hand, non-woven architecture has randomly aligned fibres; thus, repeating patterns are not observed [10].

As of 2020, owing to the preventive actions against the novel coronavirus: nCOV-Severe Acute Respiratory Syndrome Coronavirus 2 (SARS-CoV-2)- and the disease as the Coronavirus Disease 2019 (COVID-19), hygiene products and masks made out of both woven and non-woven materials of different grades – e.g. surgical masks, N95, etc. – are recently the most well-known applications of fibrous materials [11–14]. Moreover, their use in automobile panel parts, thermal insulation and filtration materials, waddings and geotextiles, building and roof coverings, sintered metallic fibres used as biomedical fabrics, fibre mats and filters used in electromagnetic shielding and fuel cell gas diffusion layers, felted or layered wood fibres used in paper and packaging products, nanocellulose fibres used in printed electronics as substrates, and bacterial nanocellulose as healing agent in regenerative medicine are prevalent examples of fibrous materials as depicted in Fig. 6.1 [15–25].

Effective characteristics of the fibrous materials at their material scales, which refers to macroscale or continuum, depend on the mechanical, morphological, and spatial parameters and interactions of the fibres at the microscale and the fibre aggregates, which is known as fibre network(s), at mesoscale [10,26–28]. Hence, modelling and coupling of the length scales of fibrous materials provide the constitutive information for deformation and damage, and further for manufacturing process simulations, which reduces the production costs and increase the material performance

Mechanics of Fibrous Networks. https://doi.org/10.1016/B978-0-12-822207-2.00015-5

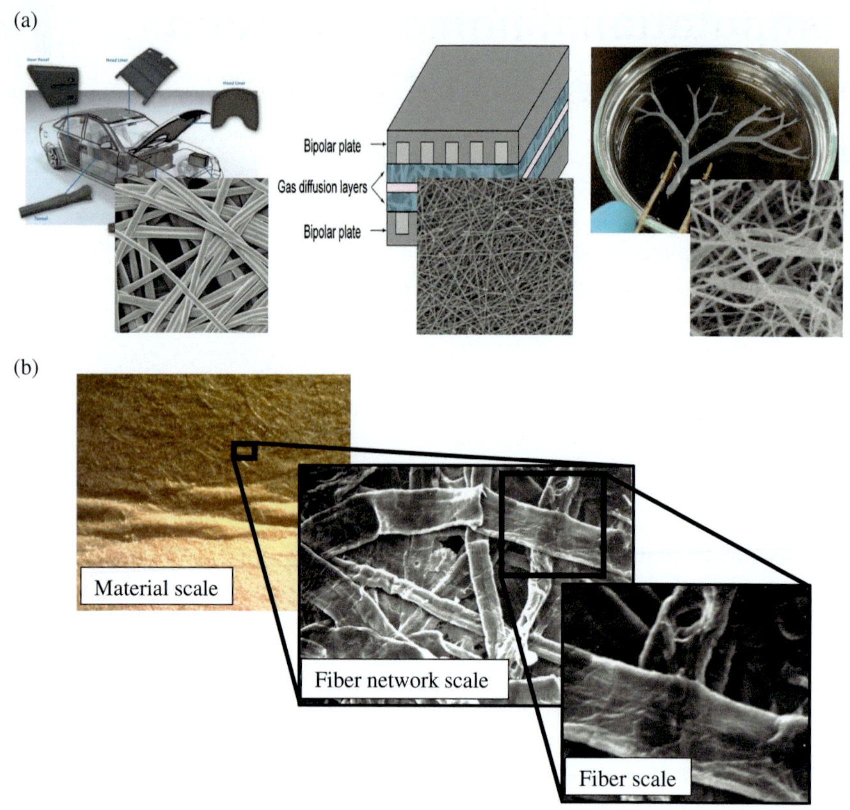

Fig. 6.1 Fibrous material applications and the length scales: (A) various engineering applications, non-woven car panel, fibre mat for gas diffusion layers and bacterial nanocellulose scaffold, and their microscope images and (B) material (macroscale), fibre network (mesoscale), and fibre (microscale) for kraft paper (Reprinted with permission).

and durability. Especially, because fibre networks represent the details and length scales of the fibrous materials well, they have been treated as the representative volume elements (RVEs) in the literature. Various fibre network models in two-dimensional (2D) and three-dimensional (3D) space have been developed by the researchers and are available in the literature [29–39]. Due to low computational costs, 2D (in-plane) models, for which the networks with thickness of order of one-tenth or less of average fibre length are often generated by random deposition techniques, have been used to determine the in-plane mechanical, thermal, and hygroscopic characteristics [40–44]. These models enabled the direct computations of the in-plane mechanical properties for the entire solution domain and the solutions to the boundary value problem (BVP) on the fibre networks at mesoscale leading to computational homogenisation investigations [45–47].

With the recent advances in the computing power and numerical optimisation routines, 3D models, for which the fibres are deposited and bend on top of each other, have been of interest to determine both in-plane and out-of-plane characteristics with better insight into fibres, fibre interactions, and fibre network scale properties [48–54]. In addition, these models have also provided better understanding of permeability characteristics owing to their integration with the flow-simulation algorithms e.g. Lattice-Boltzmann and immersed boundary methods [55–59]. In the literature, 3D fibre networks have been vastly generated in the form of geometrically repeating RVEs, on the boundaries of which linear displacements, constant tractions, and periodic boundary conditions are applied [60–64]. Under the geometric assumption of these studies, the mesh conformality of the RVE boundaries is maintained along all the boundaries. In case of mesh non-conformality or non-repeating geometries, which is often the case for the image-reconstructed fibre network domains – e.g. through micro-computed tomography (μCT) – the nodal distributions on the opposing boundaries may differ [65–67]. For such reconstructed solution domains, computational homogenisation strategies based on weak micro-periodicity of the displacement fluctuation fields can be successfully implemented to determine the effective mechanical characteristics [65,68–70].

In the context of 3D modelling investigations for fibrous materials, the following sections will elaborate the length scales and provide an overview with special focus on the computational homogenisation of non-woven fibre architecture.

6.2 Microscale: Fibres and fibre interactions

Fibres in 3D models can be described in terms of their spatial, morphological, and physical properties. As illustrated in Fig. 6.2, spatial properties refer to the fibre centroid $C(X_i, Y_i, Z_i)$ and $i \in \mathbb{Z}^+$, azimuthal (in-plane) orientation θ and polar fibre orientation φ in the XYZ-Cartesian coordinate system. The fibre orientations θ and φ are manufacturing process-dependent parameters and may differ within the material. The overall mechanical characteristics are favoured along the orientation which the most fibres align [30]. Fibre morphology at microscale can be represented with fibre curvature, length l, and cross-sectional properties including width w, height h, and

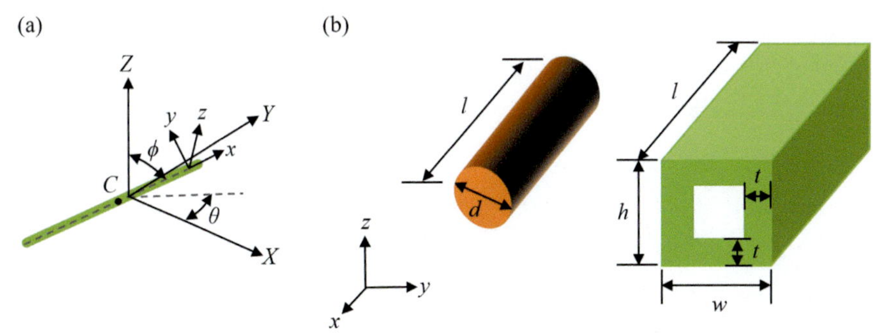

Fig. 6.2 Schematic representation of a fibre: (A) spatial properties and (B) morphological properties of a synthetic fibre with solid cross section (left) and morphological properties of a wood fibre with hollow (or collapsed) cross section (right).

wall thickness t in case of wood fibres and diameter d in case of synthetic fibres e.g. carbon or glass fibres [63,71–73]. It has been observed that l has a positive influence on the tensile strength, breaking strain, and fracture toughness of the fibrous materials [74]. On the contrary, fibre width w and fibre height h were found out to have negative effects on the elastic modulus, which can be related to the lesser number of fibres with increasing volume [75].

In addition to the spatial and morphological investigations, there has been also great interest in understanding the physical properties including mechanical, thermal, and hygroscopic characteristics. For instance, the fibres such as wood fibres exhibit anisotropic behaviour mainly due to the microfibrillar angle variations, and their deformation and damage characteristics are challenging to be accurately formulated [76–78]. These complexities have been idealised e.g. with elasto-plastic models using isotropic or kinematic hardening [79,80]. More complex models including viscoelastic, viscoplastic, and hyperelastic constitutive material models have been also developed and utilised for the sake of precise descriptions of the natural and synthetic fibre responses [81–83]. As variations of these models, hygroscopic and thermal effects have been also integrated to the constitutive equations in order to understand the effect of moisture and temperature on the mechanical and morphological properties of the fibre [84–87].

In addition to the individual fibre characteristics, fibre interactions also play a decisive role on the strength and optical, and electrical properties of the material [84]. Researchers have been systematically investigating this phenomenon by studying the effects of the network connectivity (number of fibre contacts), fibre-to-fibre bonding characteristics, and surface interactions such as van der Waals forces or molecular entanglements on the damage onset and growth [88–91]. Depending on the end-use and fibre type, fibre networks are formed through chemical, thermal, or mechanical bonding processes. For chemical bonding processes, liquid solutions or binders provide the adhesion between the fibres [27]. Thermal bonding uses controlled heating mainly for synthetic fibre networks, while compressive loads are utilised in mechanical bonding processes so as to bind and consolidate fibres [10].

Numerical models have been developed to understand the fibre interactions with focus on the bonding zones and surface interactions [31,92]. The bonding zones were characterised as point- or area-wise contacts. As illustrated in Fig. 6.3, the normal and tangential forces at these zones cause separation, slip, and reconnection of fibres,

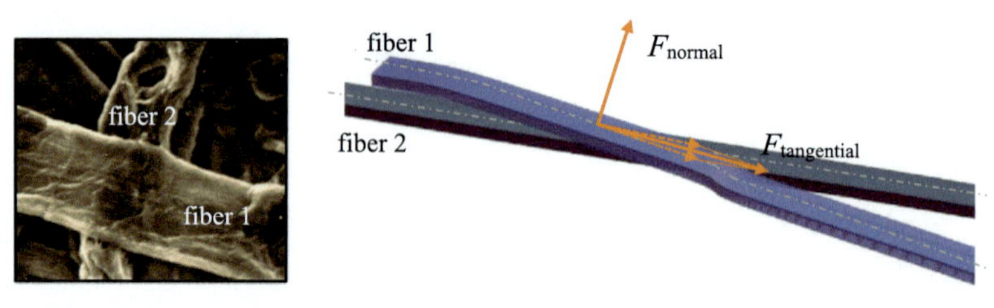

Fig. 6.3 Schematic representation of the normal and shear forces at the fibre-to-fibre bonding zone.

which are implemented in contact algorithms during analysis [1,93]. Principally, cohesive zone models (CZMs) represented with the equations coupling the applied traction with the separation have been successfully utilised for this purpose [80]. Besides, Coulomb's friction law – e.g. with general contact algorithms that are nowadays available in commercial finite element solvers including Abaqus/CAE or LS-DYNA – is applied for modelling the surface interactions [94–96].

6.3 Mesoscale: Fibre networks

Fibrous materials are formed as a result of deposition of fibres on top of each other. The deposited fibres, also termed as fibre network, inherit the spatial, morphological, and physical characteristics of fibres and their interactions. In addition, fibre volume fraction, which can be represented with the number of fibres per unit volume, is another important parameter [30]. As illustrated in the flow chart of Fig. 6.4, it can be used as the decisive parameter in fibre deposition and thus, network formation [97].

In the literature, 2D and 3D network models have been proposed, which are based on virtually generated and surface/volume reconstructed solution domains as depicted in Fig. 6.5 [15,58,64,65,75,98,99]. The network size, which is represented with network length L, width W, and thickness T, acts as the representative length scale for the fibrous materials and is one of the main concerns in these investigations and principally relies on the convergence of the material response with respect to the size. Detailed discussions on the selection of convenient representative length scales can be found elsewhere in the literature [100–104]. It is also noteworthy that the convergence may change based on the boundary conditions applied to the network e.g. linear displacements and periodic boundary conditions, etc.

To the author's knowledge, in virtual domains, the fibre deposition has been carried out by means of (I) prescribed artificial bending angles [1], (II) compression of

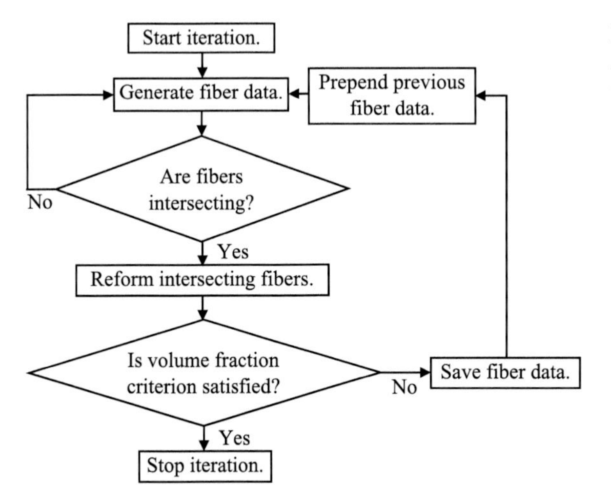

Fig. 6.4 Fibre deposition algorithm flowchart (Reprinted with permission).

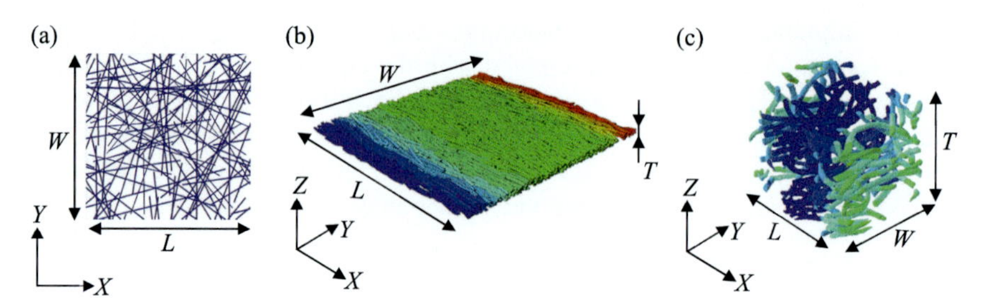

Fig. 6.5 Schematic representation of fibre network solution domain: (A) two-dimensional network, (B) three-dimensional network, and (C) volume reconstructed three-dimensional network.

straight fibres in a mould or between plates till the desired network thickness [75], and (III) immersion of multi-linked chains of rigid cylindrical parts into Newtonian fluids [58,67]. On the other hand, with the recent advances in the volume reconstruction and image processing algorithms, it has been possible to detect and label fibres (mainly synthetic fibres due to their distinguishable morphologies) [105,106]. Therefore, fibre and bonding characteristics can be integrated to the volume reconstructed domains – e.g. as depicted in Fig. 6.5C – to analyse the fibrous materials in situ.

6.4 Mesoscale to macroscale: Computational homogenisation

Multiscale modelling has been studied in depth for material systems with multiphases and structural hierarchies that are expressed in terms of length and time scales [107,108]. In this context, as also previously described, one can classify the length scales of fibrous materials as micro-, meso-, and macroscales referring to individual fibres, fibre network, and fibrous materials, respectively. These distinct length scales are uncontested by the researchers in the field and have been used for multiscale modelling purposes [109–111]. The coupling between these scales and their relation with the macroscale deformations provide optimised design and manufacturing processes, which principally reduces labour-intense mechanical testing procedures and costs. Especially, it is crucial to understand the fibrous material response at mesoscale, which includes not only fibre interaction data but also spatial, geometrical, and mechanical characteristics of individual fibres.

To the author's best knowledge, there are three approaches available in the realm of multi-scale modelling: sequential (or hierarchical), semi-concurrent, and concurrent multiscale modelling approaches [99,112,113]. However, there is no consensus on this classification and definitions vary in different studies in the literature [108]. The sequential multiscale modelling approach is based on the coupling of different length scales, where the information is passed from finer length scales to coarser scales (not vice versa) by solving microstructural BVP. By means of the averaging theorems – e.g. Hill–Mandel (or macro-homogeneity) condition, or parametric identification – simulations at the mesoscale can be used to determine the effective properties at macroscale describing the continuum behaviour [114–116]. Computational homogenisation is classified as an upscaling technique in the realm of sequential multiscale modelling, which is efficient especially to extract the effective material properties of fibrous materials [62]. In recent years, with the advances in the field, numerous studies that implement computational homogenisation methods have been conducted to incorporate the mechanical properties of fibrous materials with the fibre and fibre network characteristics. As illustrated in Fig. 6.6, common strategy in these methods is the use of geometric periodicity of the virtually generated solution domains, which enable conformal meshing or equal number of nodes matching on each opposing domain boundary. Thereafter, periodic boundary conditions are readily assigned [98,117,118].

However, geometrical periodicity definition is not likely for the volume reconstructed domains generated from tomographic images that provide near-exact microstructural data. Therefore, models using weak periodicity and (Euclidean) bipartite matching can be suitable approaches in such solution domains with non-conformal meshes [3,65,68,70,119]. In case of weak periodicity, formulation uses a mixed format handling the displacement field inside the RVE and the tractions on the RVE boundary independently [120]. Bipartite matching framework represented in Fig. 6.7 can be implemented for conformal or non-conformal meshes e.g. by defining a control node set q that can be kinematically coupled with the boundary node set p of the boundary domain $\partial\omega$ on the fibre network serving as the RVE. For the coupling pipeline, the total distance between p and q sets is minimised through Euclidean bipartite matching. The $n \times n$ distance matrix

$$\begin{bmatrix} \mathbf{d}(p_1, q_1) & \mathbf{d}(p_1, q_2) & \dots & \mathbf{d}(p_1, q_n) \\ \mathbf{d}(p_2, q_1) & \dots & \dots & \dots \\ \dots & \dots & \dots & \dots \\ \mathbf{d}(p_n, q_1) & \dots & \dots & \mathbf{d}(p_n, q_n) \end{bmatrix} \tag{6.1}$$

is generated based on the Euclidean distance $\mathbf{d}$ of each (p, q) combination with n being the set length of p (or q). Then, the optimal permutation of matched nodes is discerned based on their total Euclidean distance T through the minimisation problem [121]

$$T = \min_{\Pi} \sum \mathbf{d}(p, q) \tag{6.2}$$

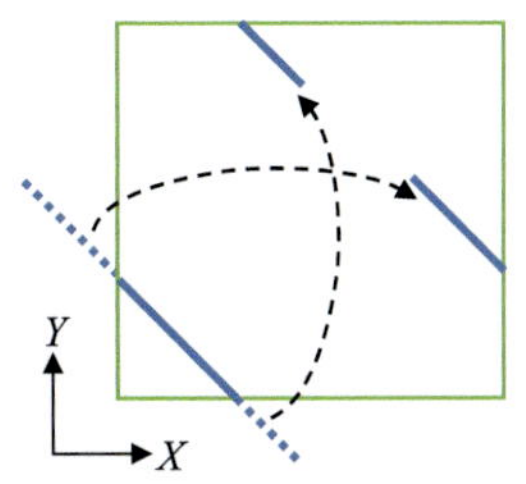

Fig. 6.6 Schematic representation of the geometrically periodic network generation where the fibre sections outside the network boundaries are trimmed and inserted back in the domain.

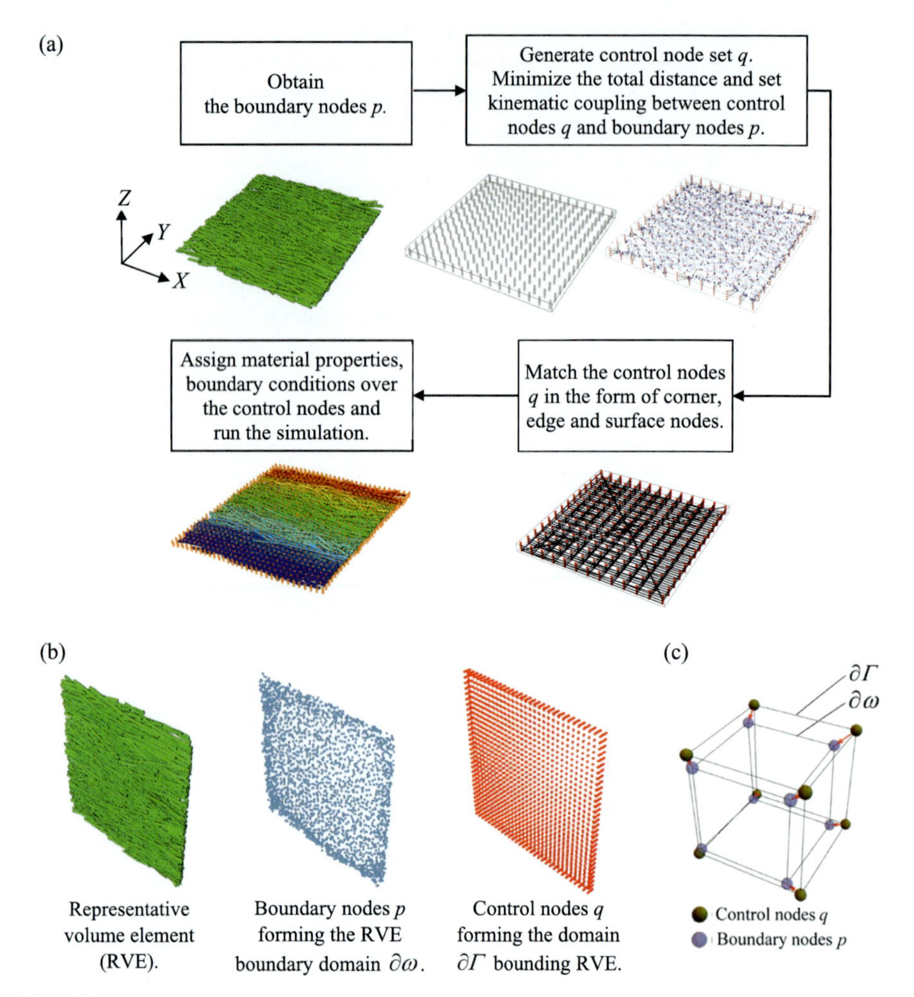

Fig. 6.7 Computational homogenisation of three-dimensional fibrous materials: (A) framework, (B) representative volume element (RVE) boundary domain represented with $\partial\omega$, domain $\partial\Gamma$ bounding RVE, boundary nodes p on $\partial\omega$ and control nodes q comprising vertices, edge, and surface nodes on $\partial\Gamma$, and (C) elaboration of linking and kinematic coupling phenomenon between control and boundary nodes (Reprinted with permission).

where $\prod$ is the permutations that abide a one-to-one correspondence [122]. The corresponding nodes of p and q are then kinematically coupled. The boundary conditions are defined on the control domain $\partial\Gamma$ comprised of q. BVP is then solved – e.g. via strain driven homogenisation – for which the macroscale strain $\mathbf{e}^{\mathrm{M}}$ is used as the driving parameter of the microscopic displacement field of the fibre network so that

$$\vec{u}^{\,\mathrm{m}} = \vec{r} \cdot \mathbf{e}^{\mathrm{M}} + \underline{\vec{u}}. \tag{6.3}$$

Here, $\vec{r}$ is the position vector between two nodes; thus, $\vec{r} \cdot \mathbf{e}^{\mathrm{M}}$ refers to the macroscopic displacement contribution and $\underline{\vec{u}}$ is the displacement fluctuations arising from heterogeneities. Continuity conditions for the displacement field are then satisfied at $\partial\Gamma$ by applying Eq. (6.3) to the control node sets q, which eliminates $\underline{\vec{u}}$. By implementing the Hill–Mandel (or macro-homogeneity) condition, coupling between the fibre network (mesoscale) and material (macroscale) can be formed. Hence, macroscale stress $\mathbf{s}^{\mathrm{M}}$ can be computed as the volume average of the microscale stress $\mathbf{s}^{\mathrm{m}}$ such that

$$\mathbf{s}^{\mathrm{M}} = \frac{1}{\Gamma}\int_{\Gamma} \mathbf{s}^{\mathrm{m}}\,\mathrm{d}\Gamma, \tag{6.4}$$

the detailed derivation of which can be found in the literature [120,123]. The given $\mathbf{e}^{\mathrm{M}}$ and the computed $\mathbf{s}^{\mathrm{M}}$, as schematised in Fig. 6.8, can be combined to determine the compliance $\mathbf{C}^{\mathrm{M}}$ e.g. by means of the least-squares minimisation of six distinct deformation modes in 3D space (three axial tension $\mathbf{e}^{\mathrm{M}}_{XX}$, $\mathbf{e}^{\mathrm{M}}_{YY}$, $\mathbf{e}^{\mathrm{M}}_{ZZ}$ and three shear $\mathbf{e}^{\mathrm{M}}_{XY}$, $\mathbf{e}^{\mathrm{M}}_{YZ}$, $\mathbf{e}^{\mathrm{M}}_{ZX}$ loading modes) as

$$\mathcal{e}(C_{11}, \ldots, C_{66}) = \sum_{i=1}^{n} \left\| \mathbf{e}^{\mathrm{M}}_{i} - \mathbf{C}^{\mathrm{M}} : \mathbf{s}^{\mathrm{M}}_{i} \right\|^{2}. \tag{6.5}$$

Here, i refers to the number of experiments. Under the assumptions of small strains and fibrous material orthotropy, $\mathbf{C}^{\mathrm{M}}$ can be expressed as

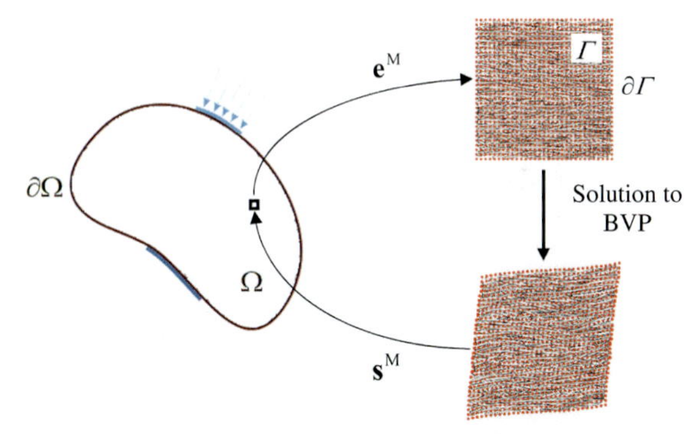

Fig. 6.8 Computational homogenisation schematics with the given macroscopic strain $\mathbf{e}^{\mathrm{M}}$ and the computed macrostress $\mathbf{s}^{\mathrm{M}}$. Here, Ω and $\partial\Omega$ represent the volume and boundary of continuum, respectively, and Γ and $\partial\Gamma$ represent the volume and boundary, respectively, of the control domain for the RVE (Reprinted with permission).

$$[\mathbf{C}^{\mathrm{M}}] = \begin{bmatrix} \frac{1}{E_X} & \frac{-\nu_{YX}}{E_Y} & \frac{-\nu_{ZX}}{E_Z} & 0 & 0 & 0 \\ \frac{-\nu_{XY}}{E_X} & \frac{1}{E_Y} & \frac{-\nu_{ZY}}{E_Z} & 0 & 0 & 0 \\ \frac{-\nu_{XZ}}{E_X} & \frac{-\nu_{YZ}}{E_Y} & \frac{1}{E_Z} & 0 & 0 & 0 \\ 0 & 0 & 0 & \frac{1}{G_{YZ}} & 0 & 0 \\ 0 & 0 & 0 & 0 & \frac{1}{G_{ZX}} & 0 \\ 0 & 0 & 0 & 0 & 0 & \frac{1}{G_{XY}} \end{bmatrix}, \tag{6.6}$$

for which E_X, E_Y, E_Z are the elastic moduli, G_{YZ}, G_{ZX}, G_{XY} are the shear moduli, and ν_{XY}, ν_{YX}, ν_{XZ}, ν_{ZX}, ν_{YZ}, ν_{ZY} are Poisson's ratios defined in the global (specimen) *XYZ*-Cartesian coordinate system as depicted in Fig. 6.7.

6.5 Case studies: Effects of fibre volume fraction and orientation variations

The above-mentioned bipartite matching approach for computational homogenisation is used to investigate the effects of fibre volume fraction V_f, as illustrated with fibre networks of Fig. 6.9, and azimuthal (in-plane) orientation variation $\Delta\theta$, as the ones represented in Fig. 6.10, on the effective elastic properties of 3D fibrous materials. As listed in Table 6.1, the effective elastic properties are normalised with respect to the elastic modulus value E_x of the fibres for appropriate comparison. The literature values for wood fibres were adapted as $E_x = 15{,}000$ MPa, $E_y = E_z = 5000$ MPa, $G_{xy} = G_{xz} = 3000$ MPa, $G_{yz} = 1080$ MPa, $\nu_{xy} = \nu_{xz} = 0.066$, and $\nu_{yz} = 0.39$, for which the subscripts x, y, and z refer to the axes of the local (fibre) *xyz*-Cartesian coordinate system as shown in Fig. 6.2 [124,125]. Here, fibres are fully bonded and deformed under maximum macroscale strain value of $\max(e_{ij}^{\mathrm{M}}) = 0.025$. For the current case studies, the constant fibre parameters are fibre length $l = 1.5$ mm, width

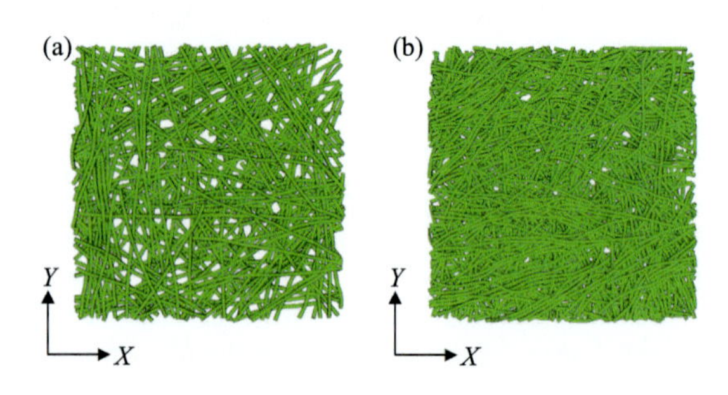

Fig. 6.9 Fibre volume fraction: (A) $V_f \approx 18\%$ and (B) $V_f \approx 30\%$, for which $\Delta\theta = \pm 90°$.

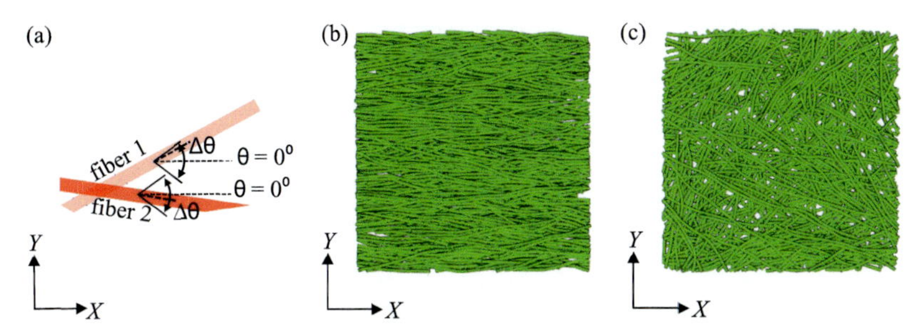

Fig. 6.10 Azimuthal (in-plane) orientation θ and its variation $\Delta\theta$: (A) schematic representation of fibre orientations, (B) fibre network with $\Delta\theta = \pm 15°$, and (C) fibre network with $\Delta\theta = \pm 90°$. Fibre volume fraction is taken as $V_f \approx 0.28$.

$w = 0.025$ mm, height $h = 0.010$ mm, wall thickness $t = 0.004$ mm, and polar orientation follows $\varphi = 0°$. The fibre network size is $L \times W \times T = 2 \times 2 \times 0.006$ mm^3.

The listed elastic properties in Table 6.1 show the inverse relationship between the material directional properties and $\Delta\theta$ e.g. highly directional material properties with $\mu(E_X/E_x) = 0.281$ and $\mu(E_Y/E_x) = 0.065$ for $\Delta\theta = \pm 15$ and (transversely) isotropic properties with $\mu(E_X/E_x) = 0.256$ and $\mu(E_Y/E_x) = 0.256$ for $\Delta\theta = \pm 90$. In addition to this trend, there is an increase in G_{YZ} and v_{XZ} with $\Delta\theta$, while G_{XZ} and v_{YZ} decrease with $\Delta\theta$. However, there is a fluctuating relationship between G_{XY}, v_{XY}, and $\Delta\theta$. This is mainly due to the combined effects of $\Delta\theta$ and V_f variations. In addition, the effect of $\Delta\theta$ on E_Z seems to be negligible showing the minimal contribution of $\Delta\theta$ on the out-of-plane deformation characteristics. In addition to the $\Delta\theta$ investigations, fibre volume fraction V_f is deduced to have positive influence on the effective in-plane elastic moduli E_X and E_Y and all the three shear moduli G_{XY}, G_{XZ}, and G_{YZ}. However, similar to the $\Delta\theta$ investigations, there is no clear influence of V_f on E_Z. This may imply that the axial tensile loading along the Z-axis is insensitive to the increase in V_f i.e. the densification of the fibre network.

6.6 Conclusions

The computational homogenisation, which is a branch of multiscale material modelling, plays a crucial role in design and development of fibrous materials towards better performance and tailorable properties. In recent years, various 3D fibre modelling methods have been developed so as to understand the effects of different length scales on the overall characteristics of fibrous materials. This chapter scrutinised these efforts, which are believed to enable better design of experiments, reduce the labour and assumptions for the empirical studies, and contribute to the development of tailorable and case-specific components with better overall performance.

Table 6.1 Normalised effective elastic properties with volume fraction V_f and azimuthal orientation variation $\Delta\theta$.

		$\frac{E_X}{E_x}$ (−) (μ, σ)	$\frac{E_Y}{E_x}$ (−) (μ, σ)	$\frac{E_Z}{E_x}$ (−) (μ, σ)	$\frac{G_{XY}}{E_x}$ (−) (μ, σ)	$\frac{G_{XZ}}{E_x}$ (−) (μ, σ)	$\frac{G_{YZ}}{E_x}$ (−) (μ, σ)	ν_{XY} (−) (μ, σ)	ν_{XZ} (−) (μ, σ)	ν_{YZ} (−) (μ, σ)
Volume fraction	$V_f \approx 0.18$	0.174	0.175	0.038	0.039	0.012	0.014	0.098	0.093	0.020
		0.001	0.001	0.001	0.001	0.001	0.001	0.002	0.001	0.001
	$V_f \approx 0.30$	0.285	0.284	0.023	0.062	0.020	0.020	0.124	0.137	0.011
		0.006	0.006	0.002	0.001	0.001	0.001	0.003	0.032	0.002
Azimuthal (in-plane) orientation variation	$\Delta\theta = \pm 15°$	0.281	0.065	0.017	0.048	0.028	0.011	0.138	0.011	0.203
		0.011	0.006	0.001	0.001	0.001	0.001	0.011	0.004	0.011
	$\Delta\theta = \pm 90°$	0.256	0.256	0.040	0.057	0.017	0.018	0.110	0.101	0.016
		0.006	0.006	0.002	0.001	0.001	0.001	0.002	0.013	0.002

$V_f \approx 0.28$ is for the $\Delta\theta$ investigations while $\Delta\theta = \pm 90°$ is for the V_f investigations. Here, μ and σ refer to the mean value and standard deviation, respectively.

References

[1] A. Kulachenko, T. Uesaka, Direct simulations of fiber network deformation and failure, Mech. Mater. (2012), https://doi.org/10.1016/j.mechmat.2012.03.010.

[2] A. Karakoç, E. Hiltunen, J. Paltakari, Geometrical and spatial effects on fiber network connectivity, Compos. Struct. 168 (2017), https://doi.org/10.1016/j.compstruct.2017.02.062.

[3] A. Karakoç, J. Paltakari, E. Taciroglu, On the computational homogenization of three-dimensional fibrous materials, Compos. Struct. 242 (2020) 112151, https://doi.org/10.1016/j.compstruct.2020.112151.

[4] W.W. Sampson, Modelling stochastic fibrous materials with, Mathematica (2013), https://doi.org/10.1017/CBO9781107415324.004.

[5] A. Ridruejo, R. Jubera, C. González, J. Llorca, Inverse notch sensitivity: cracks can make nonwoven fabrics stronger, J. Mech. Phys. Solids (2015), https://doi.org/10.1016/j.jmps.2015.01.004.

[6] Z. Cui, Y. Huang, H. Liu, Predicting the mechanical properties of brittle porous materials with various porosity and pore sizes, J. Mech. Behav. Biomed. Mater. (2017), https://doi.org/10.1016/j.jmbbm.2017.02.014.

[7] C.R. Picu, Mechanics of random fiber networks: structure–properties relation, CISM International Center for Mechanical Sciences, Courses and Lectures, 2020, https://doi.org/10.1007/978-3-030-23846-9_1.

[8] B.S. Jeon, J.H. Bae, M.W. Suh, Automatic recognition of woven fabric patterns by an artificial neural network, Text. Res. J. (2003), https://doi.org/10.1177/004051750307300714.

[9] T. Ishikawa, T.W. Chou, Stiffness and strength behaviour of woven fabric composites, J. Mater. Sci. (1982), https://doi.org/10.1007/BF01203485.

[10] F. Farukh, E. Demirci, H. Ali, M. Acar, B. Pourdeyhimi, V.V. Silberschmidt, Nonwovens modelling: a review of finite-element strategies, J. Text. Inst. (2016), https://doi.org/10.1080/00405000.2015.1022088.

[11] Organización Mundial de la Salud, Advice on the Use of Masks in the Context of COVID-19: Interim Guidance-2, Guía Interna La OMS, 2020, https://doi.org/10.1093/jiaa077.

[12] J. Donovan, S. Skotnicki-Grant, Allergic contact dermatitis from formaldehyde textile resins in surgical uniforms and nonwoven textile masks, Dermatitis (2007), https://doi.org/10.2310/6620.2007.05003.

[13] H. Laufman, W.W. Eudy, A.M. Vandernoot, C.A. Harris, D. Liu, Strike through of moist contamination by woven and nonwoven surgical materials, Ann. Surg. (1975), https://doi.org/10.1097/00000658-197506000-00018.

[14] B. Bostanci Ceran, A. Karakoç, E. Taciroğlu, Airborne pathogen projection during ophthalmic examination, Graefes Arch. Clin. Exp. Ophthalmol. (2020), https://doi.org/10.1007/s00417-020-04815-4.

[15] A. Kulachenko, T. Denoyelle, S. Galland, S.B. Lindström, Elastic properties of cellulose nanopaper, Cellulose (2012), https://doi.org/10.1007/s10570-012-9685-5.

[16] M. Özkan, M. Borghei, A. Karakoç, O.J.O.J. Rojas, J. Paltakari, Films based on crosslinked TEMPO-oxidized cellulose and predictive analysis via machine learning, Sci. Rep. 8 (2018), https://doi.org/10.1038/s41598-018-23114-x.

[17] A. Karakoc, J. Freund, Experimental studies on mechanical properties of cellular structures using Nomex (R) honeycomb cores, Compos. Struct. 94 (2012) 2017–2024.

[18] M. Özkan, K. Dimic-Misic, A. Karakoc, S.G. Hashmi, P. Lund, T. Maloney, et al., Rheological characterization of liquid electrolytes for drop-on-demand inkjet printing, Org. Electron. 38 (2016), https://doi.org/10.1016/j.orgel.2016.09.001.
[19] I.M. Hutten, Handbook of Nonwoven Filter Media, Butterworth-Heinemann, 2015. https://doi.org/10.1016/C2011-0-05753-8.
[20] J.A. Destephen, K.J. Choi, Modelling of filtration processes of fibrous filter media, Sep. Technol. (1996), https://doi.org/10.1016/0956-9618(96)00140-3.
[21] C. Veyhl, T. Fiedler, U. Jehring, O. Andersen, T. Bernthaler, I.V. Belova, et al., On the mechanical properties of sintered metallic fibre structures, Mater. Sci. Eng. A (2013), https://doi.org/10.1016/j.msea.2012.11.034.
[22] M. Alava, K. Niskanen, The physics of paper, Rep. Prog. Phys. (2006), https://doi.org/10.1088/0034-4885/69/3/R03.
[23] A. Bazylak, V. Berejnov, B. Markicevic, D. Sinton, N. Djilali, Numerical and microfluidic pore networks: towards designs for directed water transport in GDLs, Electrochim. Acta (2008), https://doi.org/10.1016/j.electacta.2008.03.078.
[24] H. Huang, A. Hagman, M. Nygårds, Quasi static analysis of creasing and folding for three paperboards, Mech. Mater. (2014), https://doi.org/10.1016/j.mechmat.2013.09.016.
[25] Y. Kiyak, B. Mazé, B. Pourdeyhimi, Microfiber nonwovens as potential membranes, Sep. Purif. Rev. (2019), https://doi.org/10.1080/15422119.2018.1479968.
[26] A. Karakoç, A fiber network model to understand the effects of fiber length and height on the deformation of fibrous materials, Res. Eng. Struct. Mater. (2016), https://doi.org/10.17515/resm2015.17ma0825.
[27] R.C. Picu, Mechanics of random fiber networks—a review, Soft Matter (2011), https://doi.org/10.1039/c1sm05022b.
[28] A. Ridruejo, C. González, J. Llorca, Damage localization and failure locus under biaxial loading in glass-fiber nonwoven felts, Int. J. Multiscale Comput. Eng. (2012), https://doi.org/10.1615/IntJMultCompEng.2012002922.
[29] J.M.R. Curto, E.L.T. Conceição, A.T.G. Portugal, R.M.S. Simões, Three dimensional modelling of fibrous materials and experimental validation, Mater. Werkst. (2011), https://doi.org/10.1002/mawe.201100790.
[30] A.I. Abd El-Rahman, C.L. Tucker, Mechanics of random discontinuous long-fiber thermoplastics—part I: generation and characterization of initial geometry, J. Appl. Mech. (2013), https://doi.org/10.1115/1.4023537.
[31] A. Torgnysdotter, A. Kulachenko, P. Gradin, L. Wågberg, Fiber/fiber crosses: finite element modeling and comparison with experiment, J. Compos. Mater. (2007), https://doi.org/10.1177/0021998306069873.
[32] M. Briane, Three models of non periodic fibrous materials obtained by homogenization, ESAIM Math. Model. Numer. Anal. (1993), https://doi.org/10.1051/m2an/1993270607591.
[33] A. Mark, E. Svenning, R. Rundqvist, F. Edelvik, E. Glatt, S. Rief, et al., Microstructure simulation of early paper forming using immersed boundary methods, Tappi J. (2011), https://doi.org/10.32964/10.32964/tj10.11.23.
[34] N. Chen, M.N. Silberstein, A micromechanics-based damage model for non-woven fiber networks, Int. J. Solids Struct. (2019), https://doi.org/10.1016/j.ijsolstr.2018.10.009.
[35] L. Wang, J. Wu, C. Chen, C. Zheng, B. Li, S.C. Joshi, et al., Progressive failure analysis of 2D woven composites at the meso-micro scale, Compos. Struct. (2017), https://doi.org/10.1016/j.compstruct.2017.07.023.

[36] M. Zako, Y. Uetsuji, T. Kurashiki, Finite element analysis of damaged woven fabric composite materials, Compos. Sci. Technol. (2003), https://doi.org/10.1016/S0266-3538(02)00211-7.
[37] E. Bosco, R.H.J. Peerlings, M.G.D. Geers, Asymptotic homogenization of hygro-thermo-mechanical properties of fibrous networks, Int. J. Solids Struct. (2017), https://doi.org/10.1016/j.ijsolstr.2017.03.015.
[38] A.S. Shahsavari, R.C. Picu, Elasticity of sparsely cross-linked random fibre networks, Philos. Mag. Lett. (2013), https://doi.org/10.1080/09500839.2013.783241.
[39] K. Berkache, S. Deogekar, I. Goda, R.C. Picu, J.F. Ganghoffer, Construction of second gradient continuum models for random fibrous networks and analysis of size effects, Compos. Struct. (2017), https://doi.org/10.1016/j.compstruct.2017.08.078.
[40] A.M. Sastry, X. Cheng, C.W. Wang, Mechanics of stochastic fibrous networks, J. Thermoplast. Compos. Mater. (1998), https://doi.org/10.1177/089270579801100307.
[41] R.M. Soszyn'ski, Simulation of two-dimensional random fiber networks, Nord. Pulp Pap. Res. J. (2007), https://doi.org/10.3183/npprj-1992-07-03-p160-162.
[42] M. Ostoja-Starzewski, D.C. Stahl, Random fiber networks and special elastic orthotropy of paper, J. Elast. (2000), https://doi.org/10.1023/A:1010844929730.
[43] E. Bosco, R.H.J. Peerlings, M.G.D. Geers, Scale effects in the hygro-thermo-mechanical response of fibrous networks, Eur. J. Mech. A. Solids (2018), https://doi.org/10.1016/j.euromechsol.2018.03.013.
[44] K.J. Niskanen, M.J. Alava, Planar random networks with flexible fibers, Phys. Rev. Lett. (1994), https://doi.org/10.1103/PhysRevLett.73.3475.
[45] C.A. Bronkhorst, Modelling paper as a two-dimensional elastic—plastic stochastic network, Int. J. Solids Struct. (2003), https://doi.org/10.1016/S0020-7683(03)00281-6.
[46] B. Sabuncuoglu, M. Acar, V.V. Silberschmidt, Finite element modelling of fibrous networks: analysis of strain distribution in fibres under tensile load, Comput. Mater. Sci. (2013), https://doi.org/10.1016/j.commatsci.2013.04.063.
[47] E. Sozumert, Y. Kiyak, E. Demirci, V.V. Silberschmidt, Effect of microstructure on porosity of random fibrous networks, J. Text. Inst. (2020), https://doi.org/10.1080/00405000.2020.1722338.
[48] H. Altendorf, F. Mathematik, 3D Morphological Analysis and Modeling of Random Fiber Networks—Applied on Glass Fiber Reinforced Composites, PhD Thesis, Fac Fachbereich Math Tech Univ Kaiserslautern, 2011.
[49] Y. Pan, Stiffness and Progressive Damage Analysis on Random Chopped Fiber Composite Using FEM (Doctoral thesis), Rutgers University, Graduate School – New Brunswick, 2010.
[50] Y. Pan, A.A. Pelegri, Progressive damage analysis of random chopped fiber composite using finite elements, J. Eng. Mater. Technol. (2011), https://doi.org/10.1115/1.4002652.
[51] Y. Pan, L. Iorga, A.A. Pelegri, Numerical generation of a random chopped fiber composite RVE and its elastic properties, Compos. Sci. Technol. (2008), https://doi.org/10.1016/j.compscitech.2008.06.007.
[52] Y. Li, S.E. Stapleton, S. Reese, J.W. Simon, Anisotropic elastic-plastic deformation of paper: out-of-plane model, Int. J. Solids Struct. (2018), https://doi.org/10.1016/j.ijsolstr.2017.10.003.
[53] G.A. Buxton, N. Clarke, Bending to stretching transition in disordered networks, Phys. Rev. Lett. (2007), https://doi.org/10.1103/PhysRevLett.98.238103.
[54] A. Rawal, V. Kumar, H. Saraswat, D. Weerasinghe, K. Wild, D. Hietel, et al., Creating three-dimensional (3D) fiber networks with out-of-plane auxetic behavior over large deformations, J. Mater. Sci. (2017), https://doi.org/10.1007/s10853-016-0547-7.

[55] A. Koponen, D. Kandhai, E. Hellén, M. Alava, A. Hoekstra, M. Kataja, et al., Permeability of three-dimensional random fiber webs, Phys. Rev. Lett. (1998), https://doi.org/10.1103/PhysRevLett.80.716.
[56] Q. Wang, B. Maze, H.V. Tafreshi, B. Pourdeyhimi, Simulating through-plane permeability of fibrous materials with different fiber lengths, Model. Simul. Mater. Sci. Eng. (2007), https://doi.org/10.1088/0965-0393/15/8/003.
[57] S. Jaganathan, H.V. Tafreshi, B. Pourdeyhimi, A case study of realistic two-scale modeling of water permeability in fibrous media, Sep. Sci. Technol. (2008), https://doi.org/10.1080/01496390802063960.
[58] S.B. Lindström, T. Uesaka, Particle-level simulation of forming of the fiber network in papermaking, Int. J. Eng. Sci. (2008), https://doi.org/10.1016/j.ijengsci.2008.03.008.
[59] J.M. Stockie, S.I. Green, Simulating the motion of flexible pulp fibres using the immersed boundary method, J. Comput. Phys. (1998), https://doi.org/10.1006/jcph.1998.6086.
[60] Q. Liu, Z. Lu, Z. Hu, J. Li, Finite element analysis on tensile behaviour of 3D random fibrous materials: model description and meso-level approach, Mater. Sci. Eng. A (2013), https://doi.org/10.1016/j.msea.2013.07.087.
[61] H.M. Inglis, P.H. Geubelle, K. Matouš, Boundary condition effects on multiscale analysis of damage localization, Philos. Mag. (2008), https://doi.org/10.1080/14786430802345645.
[62] M.G.D. Geers, V.G. Kouznetsova, W.A.M. Brekelmans, Multi-scale computational homogenization: trends and challenges, J. Comput. Appl. Math. (2010), https://doi.org/10.1016/j.cam.2009.08.077.
[63] Y. Pan, L. Iorga, A.A. Pelegri, Analysis of 3D random chopped fiber reinforced composites using FEM and random sequential adsorption, Comput. Mater. Sci. (2008), https://doi.org/10.1016/j.commatsci.2007.12.016.
[64] S. Heyden, Network Modelling for the Evaluation of Mechanical Properties of Cellulose Fluff (Doctoral thesis) Department of Mechanics and Materials, Lund University, LTH, 2000.
[65] A. Karakoç, J. Paltakari, E. Taciroglu, Data-driven computational homogenization method based on euclidean bipartite matching, J. Eng. Mech. 146 (2020) 04019132, https://doi.org/10.1061/(ASCE)EM.1943-7889.0001708.
[66] V.D. Nguyen, E. Béchet, C. Geuzaine, L. Noels, Imposing periodic boundary condition on arbitrary meshes by polynomial interpolation, Comput. Mater. Sci. (2012), https://doi.org/10.1016/j.commatsci.2011.10.017.
[67] A. Miettinen, A. Ojala, L. Wikström, R. Joffe, B. Madsen, K. Nättinen, et al., Non-destructive automatic determination of aspect ratio and cross-sectional properties of fibres, Compos. Part A Appl. Sci. Manuf. (2015), https://doi.org/10.1016/j.compositesa.2015.07.005.
[68] F. Larsson, K. Runesson, S. Saroukhani, R. Vafadari, Computational homogenization based on a weak format of micro-periodicity for RVE-problems, Comput. Methods Appl. Mech. Eng. (2011), https://doi.org/10.1016/j.cma.2010.06.023.
[69] S. Saroukhani, R. Vafadari, R. Andersson, F. Larsson, K. Runesson, On statistical strain and stress energy bounds from homogenization and virtual testing, Eur. J. Mech. A. Solids (2015), https://doi.org/10.1016/j.euromechsol.2014.11.003.
[70] C. Sandstöm, F. Larsson, K. Runesson, Weakly periodic boundary conditions for the homogenization of flow in porous media, Adv. Model. Simul. Eng. Sci. (2014), https://doi.org/10.1186/s40323-014-0012-6.
[71] R.C. Neagu, E.K. Gamstedt, Modelling of effects of ultrastructural morphology on the hygroelastic properties of wood fibres, J. Mater. Sci. (2007), https://doi.org/10.1007/s10853-006-1199-9.

[72] E. Ceretti, P.S. Ginestra, M. Ghazinejad, A. Fiorentino, M. Madou, Electrospinning and characterization of polymer–graphene powder scaffolds, CIRP Ann. Manuf. Technol. (2017), https://doi.org/10.1016/j.cirp.2017.04.122.
[73] W. Choi, S. Lee, S.H. Kim, J.H. Jang, Polydopamine inter-fiber networks: new strategy for producing rigid, sticky, 3D fluffy electrospun fibrous polycaprolactone sponges, Macromol. Biosci. (2016), https://doi.org/10.1002/mabi.201500375.
[74] Seth R.S., Chan B.K. Measuring fiber strength of papermoking pulps. Tappi J. 1999.R.S. Seth, B.K. Chan, Measuring fiber strength of papermoking pulps, Tappi J. 82 (1999) 115–120.
[75] S. Lavrykov, S.B. Lindström, K.M. Singh, B.V. Ramarao, 3D network simulations of paper structure, Nord. Pulp Pap. Res. J. (2012), https://doi.org/10.3183/NPPRJ-2012-27-02-p256-263.
[76] S.R. Djafari Petroudy, Physical and mechanical properties of natural fibers, in: Advanced High Strength Natural Fibre Composites in Construction, Woodhead Publishing, 2017, https://doi.org/10.1016/B978-0-08-100411-1.00003-0.
[77] Smook, G.A. Handbook for Pulp and Paper Technologists. 1989.G.A. Smook, Handbook for Pulp and Paper Technologists, TAPPI, 1989.
[78] H. Holik, Handbook of Paper and Board, Wiley-VCH Verlag GmbH & Co. KGaA, 2006, https://doi.org/10.1002/3527608257.
[79] S. Borodulina, A. Kulachenko, D.D. Tjahjanto, Constitutive modeling of a paper fiber in cyclic loading applications, Comput. Mater. Sci. (2015), https://doi.org/10.1016/j.commatsci.2015.08.039.
[80] Y. Li, Z. Yu, S. Reese, J.W. Simon, Evaluation of the out-of-plane response of fiber networks with a representative volume element model, Tappi J. (2017), https://doi.org/10.32964/TJ17.06.329.
[81] S. Borodulina, H.R. Motamedian, A. Kulachenko, Effect of fiber and bond strength variations on the tensile stiffness and strength of fiber networks, Int. J. Solids Struct. (2018), https://doi.org/10.1016/j.ijsolstr.2016.12.013.
[82] J. Diani, A.M. Ortega, K. Gall, S. Kasprzak, A.R. Greenberg, On the relevance of the 8-chain model and the full-network model for the deformation and failure of networks formed through photopolymerization of multifunctional monomers, J. Polym. Sci. B (2008), https://doi.org/10.1002/polb.21456.
[83] P. Boisse, Y. Aimène, A. Dogui, S. Dridi, S. Gatouillat, N. Hamila, et al., Hypoelastic, hyperelastic, discrete and semi-discrete approaches for textile composite reinforcement forming, Int. J. Mater. Form. (2010), https://doi.org/10.1007/s12289-009-0664-9.
[84] A.K. Vainio, H. Paulapuro, Interfiber bonding and fiber segment activation in paper, BioResources (2007), https://doi.org/10.15376/biores.2.3.442-458.
[85] A.M. Olsson, L. Salmén, Mechano-sorptive creep in pulp fibres and paper, Wood Sci. Technol. (2014), https://doi.org/10.1007/s00226-014-0624-5.
[86] E. Bosco, R.H.J. Peerlings, M.G.D. Geers, Explaining irreversible hygroscopic strains in paper: a multi-scale modelling study on the role of fibre activation and micro-compressions, Mech. Mater. (2015), https://doi.org/10.1016/j.mechmat.2015.07.009.
[87] Bunsell A.R. Handbook of Properties of Textile and Technical Fibres. 2018.A.R. Bunsell, Handbook of Properties of Textile and Technical Fibres, Woodhead Publishing, 2018.
[88] S. Borodulina, E.L.G. Wernersson, A. Kulachenko, C.L.L. Hendriks, Extracting fiber and network connectivity data using microtomography images of paper, Nord. Pulp Pap. Res. J. (2016), https://doi.org/10.3183/npprj-2016-31-03-p469-478.

[89] W.W. Sampson, A model for fibre contact in planar random fibre networks, J. Mater. Sci. (2004), https://doi.org/10.1023/B:JMSC.0000021453.00080.5a.
[90] R. Hägglund, P. Isaksson, On the coupling between macroscopic material degradation and interfiber bond fracture in an idealized fiber network, Int. J. Solids Struct. (2008), https://doi.org/10.1016/j.ijsolstr.2007.09.011.
[91] D. Durville, Numerical simulation of entangled materials mechanical properties, J. Mater. Sci. (2005), https://doi.org/10.1007/s10853-005-5061-2.
[92] A. Sengab, R.C. Picu, Filamentary structures that self-organize due to adhesion, Phys. Rev. E (2018), https://doi.org/10.1103/PhysRevE.97.032506.
[93] P. Litewka, P. Wriggers, Contact betweem 3D beams with rectangular cross-sections, Int. J. Numer. Methods Eng. (2002), https://doi.org/10.1002/nme.371.
[94] P. Litewka, P. Wriggers, Frictional contact between 3D beams, Comput. Mech. (2001), https://doi.org/10.1007/s004660100266.
[95] ABAQUS, Abaqus 6.14. Abaqus 614 Anal User's GuidS, 2014.
[96] Hallquist J. LS-DYNA® Theory Manual. 2006.J. Hallquist, LS-DYNA® Theory Manual, Livermore Software Technology Corporation (LSTC), Livermore, CA, USA, 2006.
[97] A. Karakoç, E. Hiltunen, J. Paltakari, Geometrical and spatial effects on fiber network connectivity, Compos. Struct. 168 (2017), https://doi.org/10.1016/j.compstruct.2017.02.062.
[98] M. Zhang, Y. Chen, F.P. Chiang, P.I. Gouma, L. Wang, Modeling the large deformation and microstructure evolution of nonwoven polymer fiber networks, J. Appl. Mech. (2019), https://doi.org/10.1115/1.4041677.
[99] H.R. Motamedian, A. Kulachenko, Simulating the hygroexpansion of paper using a 3D beam network model and concurrent multiscale approach, Int. J. Solids Struct. (2019), https://doi.org/10.1016/j.ijsolstr.2018.11.006.
[100] K. Terada, M. Hori, T. Kyoya, N. Kikuchi, Simulation of the multi-scale convergence in computational homogenization approaches, Int. J. Solids Struct. (2000), https://doi.org/10.1016/S0020-7683(98)00341-2.
[101] V. Kouznetsova, W.A.M. Brekelmans, F.P.T. Baaijens, Approach to micro-macro modeling of heterogeneous materials, Comput. Mech. (2001), https://doi.org/10.1007/s004660000212.
[102] S. Hazanov, C. Huet, Order relationships for boundary conditions effect in heterogeneous bodies smaller than the representative volume, J. Mech. Phys. Solids (1994), https://doi.org/10.1016/0022-5096(94)90022-1.
[103] S. Hazanov, M. Amieur, On overall properties of elastic heterogeneous bodies smaller than the representative volume, Int. J. Eng. Sci. (1995), https://doi.org/10.1016/0020-7225(94)00129-8.
[104] C. Huet, Application of variational concepts to size effects in elastic heterogeneous bodies, J. Mech. Phys. Solids (1990), https://doi.org/10.1016/0022-5096(90)90041-2.
[105] R. Mansour, A. Kulachenko, W. Chen, M. Olsson, Stochastic constitutive model of isotropic thin fiber networks based on stochastic volume elements, Materials (Basel) (2019), https://doi.org/10.3390/ma12030538.
[106] R. Shkarin, A. Shkarin, S. Shkarina, A. Cecilia, R.A. Surmenev, M.A. Surmeneva, et al., Quanfima: an open source Python package for automated fiber analysis of biomaterials, PLoS One (2019), https://doi.org/10.1371/journal.pone.0215137.
[107] J. Fish, Multiscale Methods: Bridging the Scales in Science and Engineering, Oxford University Press, Oxford, 2009, https://doi.org/10.1093/acprof:oso/9780199233854.001.0001.
[108] J. Fish, Practical Multiscaling, 1st ed., Wiley, 2013

[109] D.H. Mueller, M. Kochmann, Numerical modeling of thermobonded nonwovens, Int. Nonwovens J. (2004), https://doi.org/10.1177/1558925004os-1300114.
[110] S. Limem, S.B. Warner, Adhesive point-bonded spunbond fabrics, Text. Res. J. (2005), https://doi.org/10.1177/0040517505075001l2.
[111] M.N. Silberstein, C.L. Pai, G.C. Rutledge, M.C. Boyce, Elasticplastic behavior of non-woven fibrous mats, J. Mech. Phys. Solids (2012), https://doi.org/10.1016/j.jmps.2011.10.007.
[112] M. Silani, S. Ziaei-Rad, H. Talebi, T. Rabczuk, A semi-concurrent multiscale approach for modeling damage in nanocomposites, Theor. Appl. Fract. Mech. (2014), https://doi.org/10.1016/j.tafmec.2014.06.009.
[113] A. Karakoc, A. Miettinen, J. Virkajarvi, R. Joffe, Effective elastic properties of biocomposites using 3D computational homogenization and X-ray microcomputed tomography, Compos. Struct. 273 (2021) 114302.
[114] M.G.D. Geers, V.G. Kouznetsova, W.A.M. Brekelmans, Computational homogenization, CISM International Center for Mechanical Sciences, Courses and Lectures, 2010, https://doi.org/10.1007/978-3-7091-0283-1_7.
[115] M.G.D. Geers, V.G. Kouznetsova, K. Matouš, J. Yvonnet, Homogenization methods and multiscale modeling: nonlinear problems, in: Encyclopedia of Computational Mechanical, second ed., 2017, https://doi.org/10.1002/9781119176817.ecm2107.
[116] T.I. Zohdi, Homogenization methods and multiscale modeling, in: Encyclopedia Computational Mechanics, second ed., 2017, https://doi.org/10.1002/9781119176817.ecm2034.
[117] M. Danielsson, D.M. Parks, M.C. Boyce, Micromechanics, macromechanics and constitutive modeling of the elasto-viscoplastic deformation of rubber-toughened glassy polymers, J. Mech. Phys. Solids (2007), https://doi.org/10.1016/j.jmps.2006.08.006.
[118] L. Wang, M.C. Boyce, C.Y. Wen, E.L. Thomas, Plastic dissipation mechanisms in periodic microframe-structured polymers, Adv. Funct. Mater. (2009), https://doi.org/10.1002/adfm.200801483.
[119] J.M. Tyrus, M. Gosz, E. DeSantiago, A local finite element implementation for imposing periodic boundary conditions on composite micromechanical models, Int. J. Solids Struct. (2007), https://doi.org/10.1016/j.ijsolstr.2006.08.040.
[120] V.P. Nguyen, M. Stroeven, L.J. Sluys, Multiscale continuous and discontinuous modeling of heterogeneous materials: a review on recent developments, J. Multiscale Model. 03 (2011) 229–270, https://doi.org/10.1142/s1756973711000509.
[121] A. Karakoc, E. Taciroglu, Optimal automated path planning for infinitesimal and real-sized particle assemblies, AIMS Mater. Sci. (2017), https://doi.org/10.3934/matersci.2017.4.847.
[122] F. Rendl, On the Euclidean assignment problem, J. Comput. Appl. Math. (1988), https://doi.org/10.1016/0377-0427(88)90001-5.
[123] M. Petracca, L. Pelà, R. Rossi, S. Oller, G. Camata, E. Spacone, Regularization of first order computational homogenization for multiscale analysis of masonry structures, Comput. Mech. (2016), https://doi.org/10.1007/s00466-015-1230-6.
[124] R.C. Neagu, E.K. Gamstedt, M. Lindström, Influence of wood-fibre hygroexpansion on the dimensional instability of fibre mats and composites, Compos. Part A Appl. Sci. Manuf. (2005), https://doi.org/10.1016/j.compositesa.2004.10.023.
[125] A. Jäger, T. Bader, K. Hofstetter, J. Eberhardsteiner, The relation between indentation modulus, microfibril angle, and elastic properties of wood cell walls, Compos. Part A Appl. Sci. Manuf. (2011), https://doi.org/10.1016/j.compositesa.2011.02.007.

Elasto-plastic behaviour of three-dimensional stochastic fibre networks

7

*Yanhui Ma** *and Hanxing Zhu*
School of Engineering, Cardiff University, Cardiff, United Kingdom

7.1 Introduction

Fibre network materials constitute a class of highly porous materials with low density, promising for functional and structural applications. They can be made from metals, polymers, ceramics, and even composites depending on the properties required in the industry. Paper is a typical stochastic fibre network material that is, to some extent, the most familiar one to most of us. Apart from man-made engineering fibrous materials, fibre networks are abundant in the biological system. For instance, a cytoskeleton is a network of filamentous proteins consisting of filamentous actin (F-actin), microtubules, and intermediate filaments with diameters in the order of nanometres.

Structure-mechanics relationships of three-dimensional stochastic fibre networks have been little studied compared to the extensive literature on the mechanical behaviours of foams and honeycombs that are categorised as cellular materials. Connectivity can significantly affect the stiffness and strength of the porous (permeable) cellular and fibrous materials. It was found that the in-plane deformation of porous materials with high nodal connectivity is stretching dominated. In contrast, the deformation mode of low density two-dimensional hexagonal honeycomb is a combination of cell wall stretching and bending when the nodal connectivity is low [1]. Jin et al. developed a two-dimensional random micromechanics beam model to investigate the in-plane elasto-plastic behaviour of metal fibre sintered sheets [2]. In their model, all the fibres are completely overlapped with each other to form a two-dimensional stochastic fibre network, which leads to very strong bonding connections and high nodal connectivity between fibres. Xi et al. proposed that the mechanical properties of metal fibre porous materials are highly dependent on the fibre–fibre joints and the number of metallurgy nodes [3]. The mechanical properties are enhanced with increasing sintering contact points per unit volume and the bonding intensity. In addition to connectivity, relative density (also termed as solid volume fraction) is directly correlated to the mechanical properties of porous materials [4]. Traditionally, uniaxial compression and tension tests are used to extract the elasto-plastic behaviours of the fibrous or cellular materials, such as Young's modulus, Poisson's ratio, and yield stress [5].

* Current Address: Department of Ophthalmology and Visual Science, The Ohio State University, Columbus, OH, United States.

Mechanics of Fibrous Networks. https://doi.org/10.1016/B978-0-12-822207-2.00006-4

However, it is difficult to control the relative density in the manufacturing process when it comes to investigating the role of relative density in the mechanical properties of fibre network materials. It is even harder and more expensive to conduct in vivo or in vitro experiments to measure the mechanical properties of actin filament networks inside the cell. The finite element method (FEM) which was originally developed for solving solid mechanics problems offers a means to probe the mechanical properties of intricate stochastic fibrous materials by controlling the relative density and connectivity in the model. In addition, optimised design of complex porous materials can be realised by FEM.

In this chapter, we provided an overview of the elasto-plastic behaviour of three-dimensional stochastic fibre networks using micromechanics models. Specifically, a periodic representative volume element (RVE) of the three-dimensional stochastic fibre network model with periodicity was constructed using FEM. Computational simulation delineated the elasto-plastic behaviour of three-dimensional stochastic fibre networks and their relationship with relative density. Additionally, a simplified model with 3D periodicity was built to perform dimensional analysis, which in turn revealed the in-plane and out-of-plane yield strengths of the fibre network materials as closed-form functions of the relative density. Note that the sophisticated model for stochastic fibre network can better reflect the geometric feature and mechanical properties; whereas the simplified model for dimensional analysis can better capture the dominant deformation mechanisms of fibre network materials. Both models shared the same properties of solid materials.

7.2 Micromechanics models

Before applying finite element analysis, an RVE needs to be constructed. The macroscopic material is assumed to be made up of a large number of identical periodic microscopic RVEs. Thus, when the macroscopic material is deformed by a uniform state of stress or strain, the macroscopic stresses and strains can be fully determined by the microscopic stresses and strains over an RVE [6]. RVE was proposed to reveal the full-scale model by a representative ‘cell unit’ to significantly reduce the computation complexity. RVE serves as a bridge linking the macroscopic mechanical properties with their microscopic counterparts. To meet the continuity and equilibrium between any two neighbouring RVEs, the representative volume element must be periodic [7].

Choosing an appropriate boundary condition is very important in numerical simulation. Chen et al. analysed three types of boundary conditions: mixed boundary conditions, prescribed displacement boundary conditions, and periodic boundary conditions [8]. The mixed boundary condition is representative of frictionless grips. The model with mixed boundary conditions has no tangential force and the bending moment at nodes on the boundary, which have been used by Silva et al. and Triantafyllidis and Schraad [9,10]. The prescribed displacement boundary conditions impose relatively stronger restrictions, which are representative of sticking grips. The prescribed displacement boundary conditions constrain both the translation displacement and rotation of every node on the boundary. The periodic boundary conditions assume that the corresponding nodes on the opposite side of the mesh have the same expansion in the normal direction, the same displacement in the other directions and the same rotation in all the directions. Taking uniaxial loading in the x direction as an example, a tensile or compressive displacement/strain in the x direction is applied to the corresponding reference nodes on the opposite boundaries. It has been indicated that the periodic boundary conditions are more suitable than the mixed boundary conditions and prescribed displacement boundary conditions to analyse the mechanical properties of a periodic RVE [7,8].

A continuum mechanics-based, three-dimensional, periodic beam model has been constructed to describe stochastic fibre networks. Fig. 7.1 shows the RVE of stochastic fibre networks with cross linkers. A detailed description of the construction of the periodic three-dimensional periodic random beam model can be found in Ma et al. [4]. In brief, the establishment of the three-dimensional model is deduced progressively with two-dimensional coordinates being considered first. By creating $3w \times 3w$ regions simultaneously in the x direction and the y direction with a periodicity of w, the 2-D central region is selected to be periodic. To achieve periodicity in the z direction, after having built up a stochastic three-dimensional fibre network model with N complete fibres, another N fibres are continuously put up in the z direction, whose x and y coordinates are identical to those of the previous N fibres correspondingly. Ideally, the shape of the $(i + N)$th fibre is the same as that of the ith fibre when N is sufficiently large. The top layer fibre network model with N complete fibres is selected for simulation. The coordinates in 3-D space of every key point on lines can be derived by the relative positions and 2-D coordinates.

In real fibre network materials, each fibre is bent away to connect the neighbouring fibres. For simplicity, polylines are used to represent the bent fibres in the 3D periodical stochastic fibre networks. The x-y plane projections of all the fibres in undeformed fibre networks are straight lines, and their x-z and y-z plane projections are polylines. It should be noted that the smoothly curved line model could be better in describing the real structure; however, a very large number of solid elements instead of beam elements may have to be used to model the real structure, which would lead to enormous computational complexity even when the number of fibres is only 10 in the model. Alternatively, polylines could be used to represent curved lines. Ten thousand segments were examined in this polyline model [4], the mean angle of the inclined segments with the x-y plane, $\overline{\varphi}$ is around 8.5 degrees, and the relative error with 95% confidence associated with this mean angle is below 5%. The mechanical behaviour of a curved cantilever beam expressed by the trigonometric function $z = 0.075 l_c \cos(x\pi/l_c)$ was compared with that of an inclined straight cantilever beam with the same span of l_c. As shown in Table 7.1, simplification by using polylines to represent the curved fibres does not cause significant error in predicting the performance of the fibre network model, moreover, the innovative modelling method optimises the simulation with both efficiency and accuracy.

In this numerical model, a novel way to incorporate the deformation mechanism of the fibre–fibre connection was introduced to the simulation. The connection, termed as cross linker by Ma et al. [4], was implemented with an additional beam inserted at the intersection of two connected fibres in the z direction. The length of the inserted

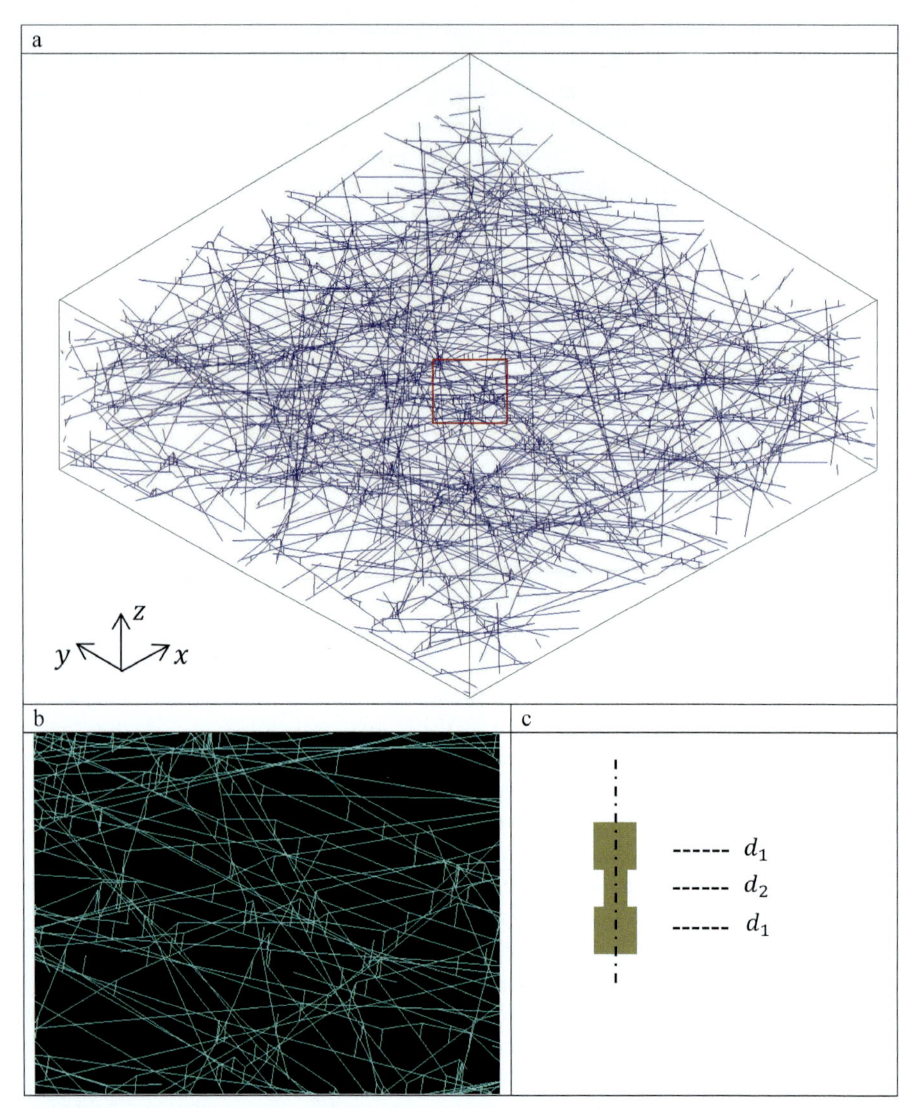

Fig. 7.1 Representative volume element of a three-dimensional stochastic fibre network model with periodicity: (A) the isometric view of the stochastic fibre network model where the fibres are represented by the polylines; (B) the connectivity in the fibre network model represented by additional beam elements inserted between the intersected fibres; (C) the cross-sectional diameters of the inserted beam elements.

Reprinted with permission from Y. Ma, H. Zhu, B. Su, G. Hu, R. Perks, The elasto-plastic behaviour of three-dimensional stochastic fibre networks with cross-linkers, J. Mech. Phys. Solids 110 (2018) 155–172.

Table 7.1 Comparison between the reaction solutions of curved cantilever beam and inclined straight cantilever beam subjected to individual displacements u_i or rotations θ_i of 0.001.

Displacement	Shape of beam	FX	FY	FZ	MX	MY	MZ
$u_x = 0.001$	Curved	0.51896	0.16312e−10	0.18808e−35	0.85954e−20	0.13294e−17	0.77844e−01
	Straight	0.57433	0.18105e−10	0.14494e−41	0.37940e−18	0.14296e−17	0.86150e−01
	Error %	10.67	–	–	–	–	10.67
$u_y = 0.001$	Curved	0.56046e−12	0.15252e−01	0.48148e−33	0.30167e−19	0.26014e−18	0.15252e−01
	Straight	0.19490e−11	0.15258e−01	0.65380e−41	0.67197e−19	0.25320e−18	0.15258e−01
	Error %	–	0.04	–	–	–	0.04
$u_z = 0.001$	Curved	0.46952e−20	0.12690e−17	0.14848e−01	0.22273e−02	0.14848e−01	0.25456e−18
	Straight	0.85055e−28	0.56703e−27	0.14924e−01	0.22386e−02	0.14924e−01	0.25751e−18
	Error %	–	–	0.51	0.51	0.51	–
$\theta_x = 0.001$	Curved	0.85944e−20	0.23228e−17	0.12145e−11	0.39764e−02	0.77691e−12	0.21869e−20
	Straight	0.34123e−18	0.22749e−17	0.79215e−12	0.39819e−02	0.45453e−12	0.12026e−19
	Error %	–	–	–	0.14	–	–
$\theta_y = 0.001$	Curved	0.40755e−22	0.11015e−19	0.64177e−12	0.39099e−12	0.50962e−02	0.87023e−19
	Straight	0.65769e−19	0.43846e−18	0.42779e−11	0.10342e−12	0.51164e−02	0.86037e−19
	Error %	–	–	–	–	0.40	–
$\theta_z = 0.001$	Curved	0.24524e−12	0.76302e−12	0.11575e−32	0.32461e−21	0.87733e−19	0.51373e−02
	Straight	0.10293e−11	0.14921e−11	0.18925e−32	0.13047e−19	0.86981e−19	0.51502e−02
	Error %	–	–	–	–	–	0.25

Note that reaction solutions (F, force; M, moment) smaller than 1e−10 were calculation noise, and their relative errors were not considered.
Adapted with permission from Y. Ma, H. Zhu, B. Su, G. Hu, R. Perks, The elasto-plastic behaviour of three-dimensional stochastic fibre networks with cross-linkers, J. Mech. Phys. Solids 110 (2018) 155–172.

beam is set as $\frac{0.95}{2}(d_{c1}+d_{c2})$, where d_{c1} and d_{c2} are the diameters of the two connected fibres. As the strength of the stochastic fibre network materials is sensitive to the cross-sectional diameter of the inserted beam, the inserted beam was divided into three sections with different cross-sectional diameters (Fig. 7.1C). The relative density (or solid volume fraction) of the fibre networks is given as

$$\rho = \frac{\sum_{i=1}^{N} L_i \times \left(\frac{1}{4}\pi d_i^2\right)}{w \times w \times t} \tag{7.1}$$

where $w \times w \times t$ is the size of the representative volume element (RVE), L_i are the fibre lengths of the individual fibres, d_i are the diameter of the circular cross-section of the fibres and N is the number of complete fibres in the RVE of fibre network model shown in Fig. 7.1. Note that the beam model (Fig. 7.1) depicts the morphology of the network with only the central axis of the beam. When considering the diameters of the beam (part of the calculation for the relative density of the fibre network), the inserted beam would be buried in the main components of the network (i.e., fibre beams). Therefore, the contribution of the cross linkers to the relative density was neglected.

7.3 Elastic behaviours

One of the significant features incorporated into our three-dimensional fibre network model is the anisotropic elasticity. Table 7.2 shows that the mean values of Young's modulus and the Poisson's ratio for 20 models are almost identical in the x and y directions, which suggests that the stochastic fibre network materials are transversely isotropic i.e., isotropic in the x-y plane with $E_x \approx E_y$ and $E_x \approx 2(1+\nu_{xy})$ G_{xy}. The relative errors with a 95% confidence interval are all below 5% for the mean results tabulated in Table 7.2.

Generalised Hooke's law, which defines the most general linear relationship between stress and strain, can be expressed as $\sigma_{ij} = C_{ijkl}\varepsilon_{kl}$, where σ_{ij} and ε_{kl} are the components of stress tensor and strain tensor, and C_{ijkl} are the components of the fourth-order stiffness tensor of the material, respectively. For orthotropic material, there are three mutually orthogonal planes of reflection symmetry. The number of independent stiffness coefficients is 9, and the stiffness tensor can be expressed as

$$C = \begin{bmatrix} C_{1111} & C_{1122} & C_{1133} & 0 & 0 & 0 \\ & C_{2222} & C_{2233} & 0 & 0 & 0 \\ & & C_{3333} & 0 & 0 & 0 \\ & & & C_{2323} & 0 & 0 \\ & symmetry & & & C_{1313} & 0 \\ & & & & & C_{1212} \end{bmatrix} \tag{7.2}$$

For transversely isotropic material, the mechanical properties are symmetric about the z axis, which is normal to the plane of isotropy (x-y plane), in addition to the three

Table 7.2 The non-dimensional Young's moduli, shear moduli, and Poisson's ratios of 20 periodic random fibre network models with $N = 200$ complete fibres, relative density $\rho = 8.8\%$, and the mean value of aspect ratio at $\overline{L}/\overline{d} = 80$.

	E_x	E_y	ν_{xy}	ν_{yx}	G_{xy}
1	8.406e−03	9.969e−03	0.230396	0.271688	3.202e−03
2	1.084e−02	1.087e−02	0.266131	0.266475	3.692e−03
3	1.128e−02	8.586e−03	0.291743	0.222034	3.463e−03
4	9.187e−03	9.106e−03	0.346416	0.342862	3.923e−03
5	1.039e−02	1.043e−02	0.309322	0.309581	4.276e−03
6	8.757e−03	1.044e−02	0.232821	0.277491	3.195e−03
7	9.418e−03	9.285e−03	0.260550	0.256878	3.407e−03
8	1.081e−02	9.897e−03	0.326866	0.299735	4.080e−03
9	9.769e−03	1.068e−02	0.290472	0.317615	3.908e−03
10	9.341e−03	9.550e−03	0.321735	0.328165	4.121e−03
11	8.873e−03	9.004e−03	0.238932	0.243409	2.922e−03
12	9.304e−03	1.187e−02	0.268266	0.342470	4.028e−03
13	9.669e−03	9.952e−03	0.305952	0.314131	4.084e−03
14	1.058e−02	7.230e−03	0.326704	0.223407	3.634e−03
15	9.579e−03	9.545e−03	0.313489	0.311736	4.022e−03
16	9.676e−03	1.089e−02	0.239218	0.268928	3.584e−03
17	9.854e−03	1.035e−02	0.263980	0.277498	3.230e−03
18	9.691e−03	9.186e−03	0.300241	0.284814	3.741e−03
19	9.674e−03	7.499e−03	0.312961	0.242252	2.988e−03
20	9.971e−03	9.459e−03	0.339308	0.320959	4.297e−03
Mean	9.753e−03	9.690e−03	0.289275	0.286106	3.690e−03
Std	7.147e−04	1.088e−03	0.035865	0.036213	4.158e−04

Note that the Young's moduli and shear moduli of the fibre network materials were normalised by Young's modulus of the solid material.

mutually orthogonal planes of reflection symmetry. The properties in the x direction are identical to those in the y direction. The number of independent stiffness coefficients is then reduced to 5, and the stiffness tensor is given by

$$C = \begin{bmatrix} C_{1111} & C_{1122} & C_{1133} & 0 & 0 & 0 \\ & C_{1111} & C_{1133} & 0 & 0 & 0 \\ & & C_{3333} & 0 & 0 & 0 \\ & & & C_{2323} & 0 & 0 \\ & \textit{symmetry} & & & C_{2323} & 0 \\ & & & & & C_{1212} \end{bmatrix} \tag{7.3}$$

where $C_{1212} = 1/2(C_{1111} - C_{1122})$.

As the fibre network materials have three orthogonal planes of symmetry, and in addition, their mechanical properties are in-plane isotropic, five independent elastic constants can be used to describe the elastic behaviours of the three-dimensional stochastic fibre networks, namely E_x, E_z, ν_{xy}, ν_{xz}, and G_{xz}. The inverse Hooke's law can be written as.

$$\begin{bmatrix} \varepsilon_{xx} \\ \varepsilon_{yy} \\ \varepsilon_{zz} \\ 2\varepsilon_{yz} \\ 2\varepsilon_{zx} \\ 2\varepsilon_{xy} \end{bmatrix} = \begin{bmatrix} \frac{1}{E_x} & -\frac{\nu_{xy}}{E_x} & -\frac{\nu_{xz}}{E_z} & 0 & 0 & 0 \\ & \frac{1}{E_x} & -\frac{\nu_{xz}}{E_z} & 0 & 0 & 0 \\ & & \frac{1}{E_z} & 0 & 0 & 0 \\ & & & \frac{1}{G_{xz}} & 0 & 0 \\ & symmetry & & & \frac{1}{G_{xz}} & 0 \\ & & & & & \frac{2(1+\nu_{xy})}{E_x} \end{bmatrix} \begin{bmatrix} \sigma_{xx} \\ \sigma_{yy} \\ \sigma_{zz} \\ \sigma_{yz} \\ \sigma_{zx} \\ \sigma_{xy} \end{bmatrix}. \quad (7.4)$$

7.4 Plastic behaviours

The yield strength is taken as the stress at the point on the stress-strain curve at which the tangent gradient is 50% of the initial gradient (i.e., Young's modulus) of uniaxial loading [4,11]. The finite element simulation results indicate that the uniaxial yield strength in the x direction is a quadratic function of the relative density of the stochastic fibre network materials when the relative density is small (i.e., $\rho \leq 0.2$). In contrast, the uniaxial yield strength in the z direction is constantly a cubic function of the relative density [4]. Besides the simulation results, dimensional analyses were performed to obtain the in-plane and out-of-plane yield strengths of the fibre network materials as closed-form functions of the relative density.

To perform dimensional analysis, one has to simplify and/or idealise the geometrical structure of the fibre networks. As the stochastic fibre network materials are transversely isotropic with the properties in the x direction the same as those in the y direction, the geometrical model for dimensional analysis is simplified, as shown in Fig. 7.2A. This model is periodic in all the x, y, and z directions. The periodicity in the z direction for the simplified model was achieved in the same way as that for the random beam model. Specifically, there are four layers in the simplified model, and the no-shown fifth layer is the same as the first layer in terms of the fibre distribution within the layer. The connection (cross linker) between fibres in the fourth and fifth layer would be isolated after removing the fifth layer, and those isolated nodes can be matched with corresponding nodes on the first layer and applied with the out-of-plane periodic boundary conditions. It is worth noting that the 3D simplified model here served to reveal the dominant deformation mechanisms for fibre network materials under different loadings, on the ground of which the yield strengths as closed-form functions of the relative density can be derived.

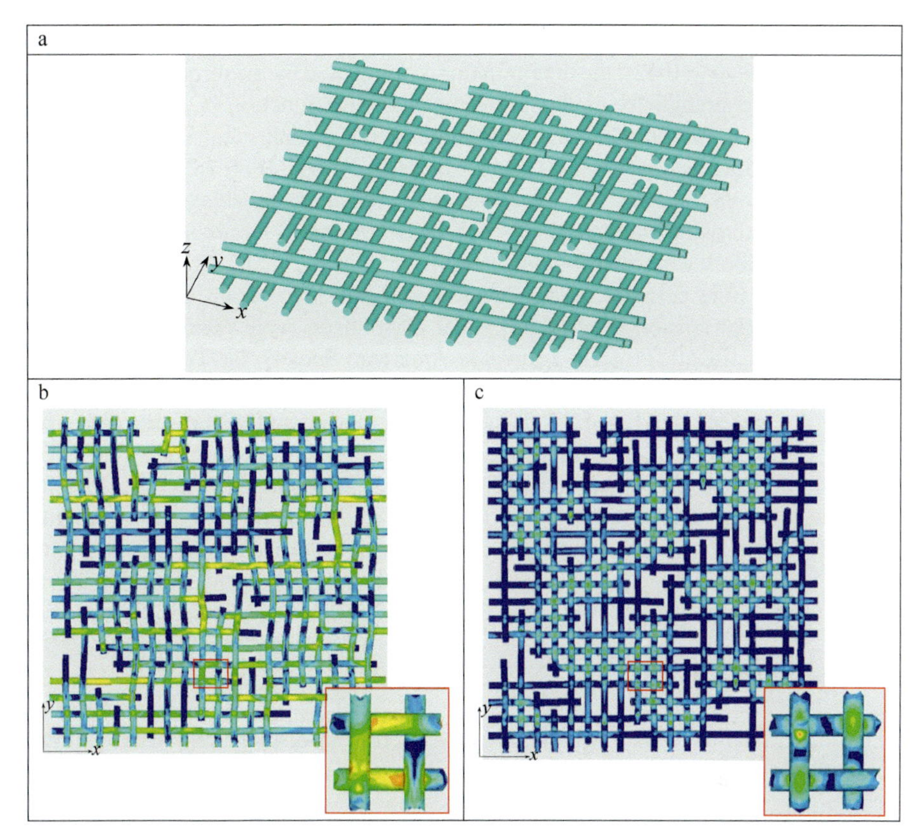

Fig. 7.2 (A) Simplified model of the fibre network materials for dimensional analysis. Contour of von Mises stress when the simplified model is subject to uniaxial tension in the x direction (B), and in the z direction (C).
Adapted with permission from Y. Ma, H. Zhu, B. Su, G. Hu, R. Perks, The elasto-plastic behaviour of three-dimensional stochastic fibre networks with cross-linkers, J. Mech. Phys. Solids 110 (2018) 155–172.

Fig. 7.2B shows the contour of von Mises stress when the low-density fibre network material undergoes uniaxial tension in the x direction. It indicates that bending and torsion are the dominant deformation mechanisms, based on which the relationship between relative density and uniaxial yield strength in the x direction was derived and takes the form [4],

$$Y_x = \frac{1.275\rho^2}{\sqrt{1 + 20.89\rho^2}}\sigma_s \tag{7.5}$$

where Y_x is the uniaxial yield strength (or tensile strength) in the x direction of the fibre network, σ_s is the yield strength of the solid material from which the fibre network is made. Eq. (7.5) suggests that when the relative density is small (i.e., $\rho \leq 0.2$), the in-plane tensile strength Y_x is a quadratic function of ρ; whilst when the relative density is large (i.e., $\rho \geq 0.22$), the denominator is dominated by the term $20.89\rho^2$ and the in-plane tensile strength Y_x gradually becomes a linear function of ρ.

The results from computational simulation using a sophisticated random model (Fig. 7.1) and dimensional analysis using a simplified model (Fig. 7.2) were compared with experimental testing. In the experiments [5], metal fibre sheets were sintered in a vacuum furnace using stainless steel fibre layers produced by air-laid web-forming technology. Fibres were distributed randomly within the metal fibre sheets, which resembles the fibre network model developed in this study. Both in-plane and out-of-plane uniaxial tension tests of the metal fibre sheets were conducted following ASTM (International, formerly known as American Society for Testing and Materials) standards. The dimensional analysis results from Eq. (7.5) for in-plane tensile strength are presented in Fig. 7.3A, in comparison to the simulation results and experimental results, showing excellent agreement.

When the simplified model is stretched by uniaxial tensile stress in the z direction, the contour of von Mises stress in the idealised fibre network materials is shown in Fig. 7.2C. As can be seen, bending is the dominant deformation mechanism. Thus, the relationship between relative density and uniaxial yield strength in the z direction (Y_z) was derived and takes the form [4]

$$Y_z = 3.58\rho^3\sigma_s \tag{7.6}$$

The dimensional analysis results from Eq. (7.6) for out-of-plane tensile strength are presented in Fig. 7.3B, in comparison to the simulation results and experimental

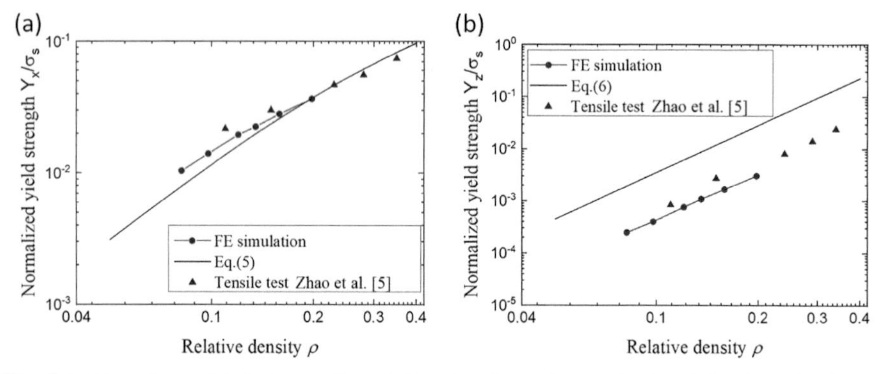

Fig. 7.3 The relationship between the relative density and the uniaxial yield strength of stochastic fibre network materials in the x direction (A) and in the z direction (B). Adapted with permission from Y. Ma, H. Zhu, B. Su, G. Hu, R. Perks, The elasto-plastic behaviour of three-dimensional stochastic fibre networks with cross-linkers, J. Mech. Phys. Solids 110 (2018) 155–172.

results [5], confirming that the difference is just a constant factor and fibre bending is indeed the dominant deformation mechanism.

The results in Fig. 7.3 demonstrate that the above dimensional analyses obtained from a simplified model have correctly captured the main deformation mechanisms when the stochastic fibre network materials undergo uniaxial in-plane or out-of-plane tension. It is noted that the relative density is assumed to be the same or uniform over the whole model of the fibre network materials in the dimensional analysis, whilst in reality, the local density could be double or half the average relative density at different locations. When the fibre network materials undergo uniaxial tension in the x direction, the fibre networks can be treated as a number of springs (with different stiffnesses) that are deformed partly in parallel and partly in series. As Y_x is proportional to ρ^2 when $\rho \leq 0.2$; and proportional to ρ when $\rho \geq 0.22$, it is less affected by the non-uniform distribution of the relative density. When the fibre network materials undergo uniaxial tension in the z direction, the fibre networks can be treated as a number of springs (with different stiffnesses) in series. As Y_z is proportional to ρ^3, the yield strength of the fibre network materials is mainly dominated by the locations where the relative density is much lower than the average. That is why the dimensional analysis results are closer to the simulation results in the x direction, but much larger than those in the z direction (Fig. 7.3).

7.5 Conclusion

A computer code has been developed to construct three-dimensional periodic random fibre network models with periodic boundary conditions. The macroscopic stress and strain of the stochastic fibre network materials have been described over a periodic representative volume element. The simulation results indicate that the stochastic fibre network materials are transversely isotropic with the properties in the x direction identical to those in the y direction. Both simulation and dimensional analysis demonstrated that the uniaxial yield strength in the x direction is a quadratic function of the relative density of the stochastic fibre network materials when the relative density is smaller and gradually becomes a linear function with increasing relative density. In contrast, the uniaxial yield strength in the z direction is constantly a cubic function of the relative density.

Adapted from this fibre network model, some other micromechanics models have been developed with an emphasis on its industrial engineering application and tissue engineering application. Specifically, the fibre network structure was integrated into composites in an attempt to enhance the stiffness and strength of the composites. The periodic representative volume element for composites reinforced by a random transversely isotropic fibre network is shown in Fig. 7.4. It has been suggested that interpenetrating composites reinforced by a

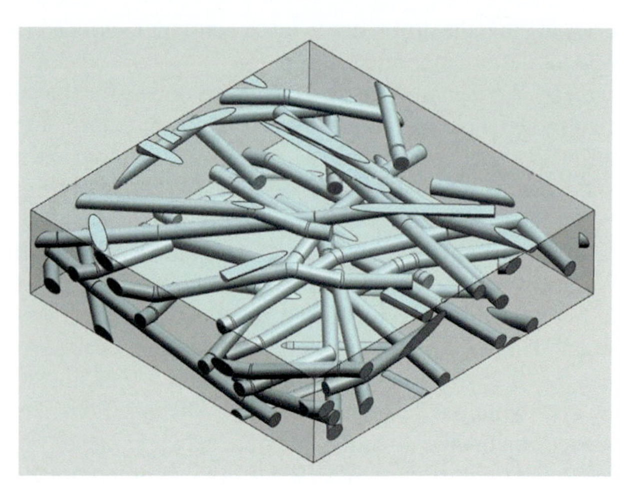

Fig. 7.4 Representative volume element of the composite reinforced by a transversely isotropic random fibre network containing 50 complete fibres.
Reprinted with permission from X. Lin, H. Zhu, X. Yuan, Z. Wang, S. Bordas, The elastic properties of composites reinforced by a transversely isotropic random fibre-network, Compos. Struct. 208 (2019) 33–44.

self-connected fibre network have superior elastic properties compared to fibre-reinforced composites without any connections amongst fibres. Additionally, fibre network composites demonstrated larger in-plane stiffness than particle composites (Glass/epoxy and Particle/matrix) [12]. A three-dimensional fibre network model has been developed to mimic the mechanical properties of actin filament (F-actin) networks cross linked by filamin A, in which the cross linkers between two distinct filaments were modelled by deformable curved elastic rods, as shown in Fig. 7.5. It was found that the cross-linker density and the actin filament volume fraction strongly affect the in-plane shear modulus of cross-linked actin filament networks (CAFNs). The simulation results suggested that the CAFNs are transversally isotropic, and the scaling relationship between the in-plane shear modulus of CAFNs and the volume fraction of actin filaments takes the form, $G_{xy} \propto \rho^{5/2}$ [13]. The investigation on the fibre network materials would facilitate the identification of conditions for the onset of inelastic behaviours and the development of models applied in tissue engineering.

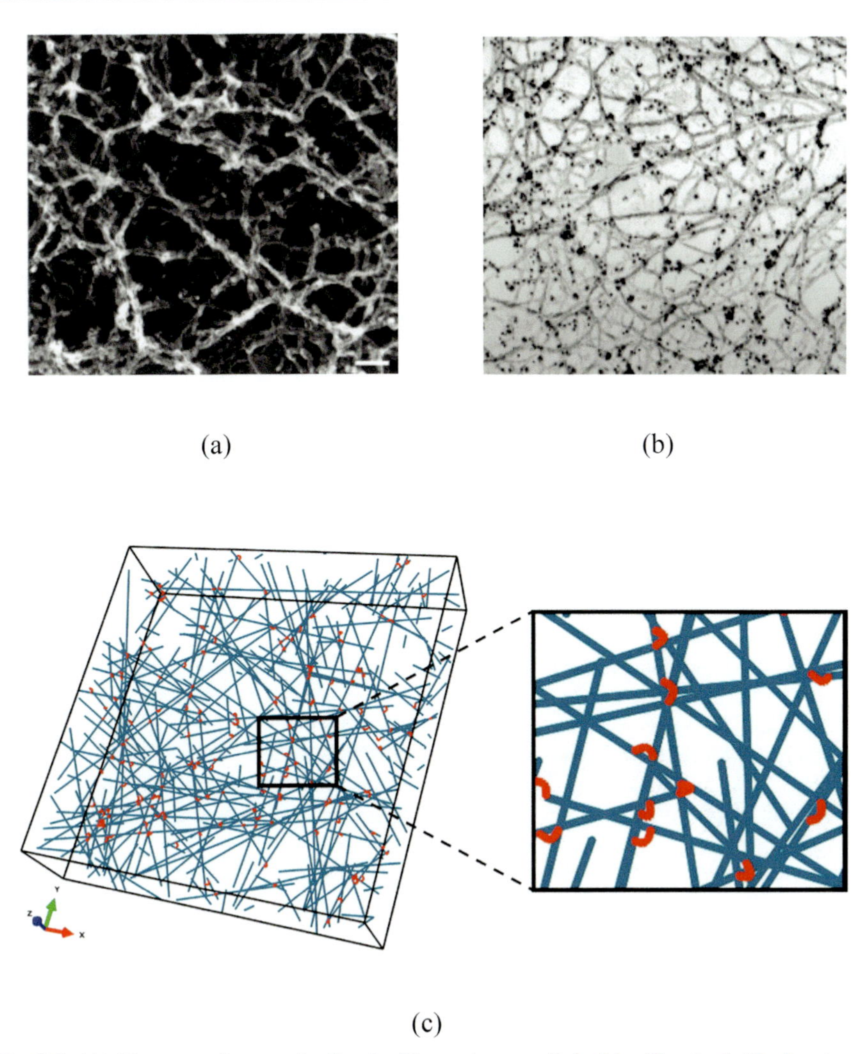

Fig. 7.5 (A) Electron micrograph of actin filaments cross linked by filamin A (the bar is 100 nm). (B) Electron micrograph of actin filament network in human blood platelet (C) Representative volume element of a 3D cross-linked actin filament network with filamin A *(red)*.
Reprinted with permission from X. Wang, H. Zhu, Y. Lu, Z. Wang, D. Kennedy, The elastic properties and deformation mechanisms of actin filament networks crosslinked by filamins, J. Mech. Behav. Biomed. Mater. 112 (2020) 104075.

Acknowledgement

Some text and figures in this chapter have appeared previously in our own work: Ma YH, Zhu HX, Su B, Hu GK, Perks R. "The elasto-plastic behaviour of three-dimensional stochastic fibre networks with cross linkers", *Journal of the Mechanics and Physics of Solids*. 2018 Jan 1;110:155–72, which has been used with permission and edited for this chapter.

References

[1] D.D. Symons, N.A. Fleck, The imperfection sensitivity of isotropic two-dimensional elastic lattices, J. Appl. Mech. 75 (2008).

[2] M.Z. Jin, C.Q. Chen, T.J. Lu, The mechanical behavior of porous metal fiber sintered sheets, J. Mech. Phys. Solids 61 (2013) 161–174.

[3] Z. Xi, J. Zhu, H. Tang, Q. Ao, H. Zhi, J. Wang, C. Li, Progress of application researches of porous fiber metals, Materials (Basel) 4 (2011) 816–824.

[4] Y. Ma, H. Zhu, B. Su, G. Hu, R. Perks, The elasto-plastic behaviour of three-dimensional stochastic fibre networks with cross-linkers, J. Mech. Phys. Solids 110 (2018) 155–172.

[5] T.F. Zhao, C.Q. Chen, Z.C. Deng, Elastoplastic properties of transversely isotropic sintered metal fiber sheets, Mater. Sci. Eng. A 662 (2016) 308–319.

[6] R. Hill, Elastic properties of reinforced solids: some theoretical principles, J. Mech. Phys. Solids 11 (1963) 357–372.

[7] H.X. Zhu, J.R. Hobdell, A.H. Windle, Effects of cell irregularity on the elastic properties of 2D Voronoi honeycombs, J. Mech. Phys. Solids 49 (2001) 857–870.

[8] C. Chen, T.J. Lu, N.A. Fleck, Effect of imperfections on the yielding of two-dimensional foams, J. Mech. Phys. Solids 47 (1999) 2235–2272.

[9] M.J. Silva, W.C. Hayes, L.J. Gibson, The effects of non-periodic microstructure on the elastic properties of two-dimensional cellular solids, Int. J. Mech. Sci. 37 (1995) 1161–1177.

[10] N. Triantafyllidis, M.W. Schraad, Onset of failure in aluminum honeycombs under general in-plane loading, J. Mech. Phys. Solids 46 (1998) 1089–1124.

[11] M.O. Withey, J. Aston, Johnson's Materials of Construction, 1918.

[12] X. Lin, H. Zhu, X. Yuan, Z. Wang, S. Bordas, The elastic properties of composites reinforced by a transversely isotropic random fibre-network, Compos. Struct. 208 (2019) 33–44.

[13] X. Wang, H. Zhu, Y. Lu, Z. Wang, D. Kennedy, The elastic properties and deformation mechanisms of actin filament networks crosslinked by filamins, J. Mech. Behav. Biomed. Mater. (2020) 104075.

Hygro-mechanics of fibrous networks: A comparison between micro-scale modelling approaches

Emanuela Bosco[a], *Ron H.J. Peerlings*[b], *Noud P.T. Schoenmakers*[b], *Nik Dave*[b], *and Marc G.D. Geers*[b]
[a]Department of the Built Environment, Eindhoven University of Technology, Eindhoven, The Netherlands, [b]Department of Mechanical Engineering, Eindhoven University of Technology, Eindhoven, The Netherlands

8.1 Introduction

Fibrous materials are widely present in several engineering applications, including paper, nonwoven, and bio-inspired systems. Among the various applications, the focus of this chapter is on paper-based fibrous networks. Paper consists of hydrophilic wood fibres that exhibit highly anisotropic deformations upon moisture content variations, whereby the transverse expansion is approximatively 20–30 times the longitudinal one [1–3]. The deformation of the individual fibres and their interactions within the fibrous network ultimately determine the effective moisture-induced response at the macroscopic, sheet level. Moisture-induced deformations typically lead to dimensional stability issues, which are critical in the areas of printing, packaging, and storage operations [4,5]. In order to accurately predict the hygro-mechanical behaviour of paper sheets as a function of the properties of the fibres and the network, a multiscale modelling approach that bridges the complex hygro-expansive phenomena at the different length scales is needed.

Several network models have been proposed in the literature that describes the effective response of fibrous materials. Among these, analytical models are typically based on the hypothesis of affine deformation of the network, whereby the local strains are assumed to be equal to the macroscopic deformation [6–9]. These models provide closed-form expressions for the effective network properties, as a function of different micro-structural features (e.g. the geometry of the fibres, and their distribution and orientation). To circumvent the fact that the assumption of affine deformation may not always be realistic, as discussed in Ref. [10], more recently, numerical or computational models of fibrous networks have been proposed [10–19]. Most of these studies represent the material as a two-dimensional (2D) domain, in which the single fibres are modelled as a series of trusses or beams, typically characterised by an isotropic constitutive response. The fibres are connected at the nodes by interfibre bonds, whose role is often not explicitly analysed.

Mechanics of Fibrous Networks. https://doi.org/10.1016/B978-0-12-822207-2.00009-X

These works mainly focus on the mechanical behaviour of the network, possibly including damage and fracture.

As for the hygroscopic behaviour of (paper) fibrous networks, most existing studies have addressed the problem through phenomenological models at the macro-scale, or via macro-scale experiments – see among the others [2,20–26]. Fewer works focus on the hygro-mechanics of paper based on a micro-structural description [27–31]. In particular, in Ref. [27], an explicit relation is given between the micro-structural parameters and the macro-scale paper hygro-expansive behaviour, where the effective hygro-expansive coefficients of paper are expressed as a function of the longitudinal and transverse hygro-expansivity of the single fibres. Specific weight factors incorporate the role of the hygro-mechanical interactions occurring at the interfibre bonds in the network. The proposed expression can capture qualitatively the trend of different experiments [22,27]; however, a quantitative estimate of the effective hygro-expansive properties for a specific geometry of the network is not provided.

This chapter precisely presents a multiscale framework that aims to investigate the effective hygro-elastic response of paper fibrous networks, by explicitly addressing the influence of individual micro-structural geometrical and material parameters. This is achieved by developing a model of the underlying fibrous network, from which the overall response is extracted by means of homogenisation. In principle, the model can be formulated considering different assumptions, which may have an influence on the results. Here, the point of departure is a highly idealised model that simplifies paper to a 2D lattice structure, represented by a periodic unit cell with two fibres in perpendicular directions [32,33]. In the proposed approach, a distinction is made between the free standing, unbonded fibre segments, and the interfibre bonds, where the fibres are considered to be perfectly bonded. Moreover, a network characterised by equal response along the two in-plane principal directions is considered, but the model may be extended to incorporate the effect of an anisotropic orientation distribution – see Ref. [32]. A hygro-elastic constitutive behaviour is adopted for the fibres. Extensions towards irreversible material behaviour can be incorporated along the lines of Ref. [34]. The model is solved analytically via homogenisation and provides the sheet-scale hygro-mechanical properties as an explicit function of the micro-structural parameters. This idealised model relies on two major assumptions, namely (i) the representation of the random network by a lattice description, and (ii) a 2D representation. The effect of these two specific assumptions on the predicted effective hygro-elastic response will be investigated by relaxing them one at a time, as explained in the following.

First, instead of the planar lattice description, a model that accounts for the *effect of in-plane randomness* of the network is introduced. This model is based on a periodic repetition of a 2D network of fibres, following the work presented in Refs. [35,36]. The network is generated by randomly positioning the fibres in the plane, according to a uniform orientation distribution. Similar models in the literature have been used in, for example, Refs. [12,37,38]. The fibres are described as 2D domains, with transversely isotropic hygro-mechanical properties. As in the lattice model, perfectly bonded fibres are considered. The effective hygro-mechanical properties of the network are obtained via asymptotic homogenisation, which is a multiscale framework that applies to heterogeneous domains characterised by an underlying periodic micro-scale structure [39–42]. This multi-scale method allows to represent the heterogeneous medium with rapidly oscillating material properties as an equivalent homogeneous solid, for which the effective material response is retrieved from the micro-scale fields through an averaging procedure that relies on rigorous mathematical principles. The method departs from writing the solution of the equilibrium problem as an asymptotic expansion, ultimately yielding explicit relations for the effective mechanical and hygro-expansive properties. These relations are finally solved numerically, typically by finite elements, for specific geometries of the underlying network domain. The hygro-elastic properties computed through this detailed network model will be compared to those obtained from the idealised lattice model.

Second, instead of the 2D representation, a *three-dimensional (3D) lattice model* is considered. Note that 3D network representations to describe the hygroscopic response of paper are a topic of current interest in the literature, see also the recent contributions [30,31]. The 3D unit cell consists of a rectangular prism composed of two families of fibres with prismatic cross section, which are orthogonally oriented. From a top view, the 3D model has a square shape, consistently with the geometry of the 2D lattice model. In the out-of-plane direction, the cell is characterised by several fibre layers stacked on top of each other. Models are considered in which the fibres are either straight, or they are partially wrapped around crossing fibres. It has been suggested in the literature that this wrap around may have a significant effect on the hygro-expansion observed at the sheet scale [22,30]. In this contribution, the degree of wrap around is thus systematically varied and its effect is studied. The 3D model is solved numerically by performing hygro-elastic simulations, from which the effective properties are computed by means of numerical homogenisation. Also in this case, a comparison will be made between the hygro-elastic properties computed through the 2D and the 3D lattice models.

This chapter is organised as follows. In Section 8.2, the reference 2D lattice model is presented. Section 8.3 describes the 2D random network model, while Section 8.4 illustrates the main features of the 3D lattice model. The influence of the effect of in-plane randomness and the 3D description on the obtained effective hygro-elastic properties is discussed in Section 8.5. Conclusions are finally given in Section 8.6.

8.2 Two-dimensional lattice model

The following concepts are based on the previous contributions from the authors [32,33]. Consider a global coordinate system (x, y, z), in which the axes are aligned with respect to the principal directions of paper, i.e. the machine direction, the cross direction, and the thickness direction. A position vector $\mathbf{x}$ is introduced, with $\mathbf{x} = x\mathbf{e}_x + y\mathbf{e}_y + z\mathbf{e}_z$, where $\mathbf{e}_i$ $(i = x, y, z)$ are the unit vectors of a Cartesian vector basis.

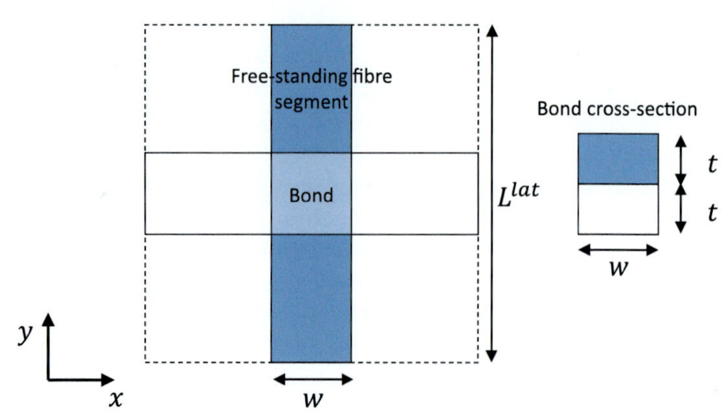

Fig. 8.1 Two-dimensional lattice model. Geometry of the reference unit cell and cross section of the bond.

8.2.1 Model geometry

The reference fibrous network description is based on the highly idealised, 2D lattice model shown in Fig. 8.1. Paper is approximated as a 2D material, whereby the fibres are mainly oriented parallel to the (x, y) plane. A 2D plane stress model is thus assumed. The model is based on the periodic repetition of a square unit cell with edge L^{lat}, composed of two fibres with orientation $\theta^{(1)} = 0$ and $\theta^{(2)} = \pi/2$, with respect to the x-axis of the reference system. The fibres have a rectangular cross section, with width w and thickness t. Here, we limit ourselves to a network characterised by equal response along the x- and y-directions, but the model may be extended to include the effect of an anisotropic orientation distribution – see Ref. [32]. In the proposed model, a distinction is made between the free standing, unbonded part of the fibres, with length $L^{\text{lat}} - w$, and the square bonding area of edge w. It is assumed that, due to moisture content or mechanical load variations, the free-standing portions of the fibres can freely deform in the lateral direction and that they interact with the bond as if they were pin-jointed. This implies that only an axial force is transferred, such that the forces acting in the free-standing fibre segments are in equilibrium with the average bond stress. On the contrary, in the bonding area, the interaction between the intersecting fibres leads to mutual hygro-mechanical constraints between the two fibres upon moisture variations. The bond is thus considered as a layered composite plate of thickness $2t$, in which the two plies are perfectly bonded and have the properties of the respective fibres.

8.2.2 Constitutive response

The constitutive response of an individual fibre is considered to be transversely isotropic [43–45]. This is described by the fourth-order compliance tensor ${}^4\mathbf{S}^f$ and by the second-order hygro-expansion tensor $\boldsymbol{\beta}^f$. The tensors ${}^4\mathbf{S}^f$ and $\boldsymbol{\beta}^f$ are expressed with respect to a local coordinate system (ℓ, t, z), defined along the principal directions of a fibre with arbitrary orientation θ, as

$$
\begin{aligned}
{}^4\mathbf{S}^f = {} & \frac{1}{E_\ell}\mathbf{e}_\ell\otimes\mathbf{e}_\ell\otimes\mathbf{e}_\ell\otimes\mathbf{e}_\ell - \frac{\nu_{\ell\mathrm{t}}}{E_\ell}\mathbf{e}_\ell\otimes\mathbf{e}_\ell\otimes\mathbf{e}_\mathrm{t}\otimes\mathbf{e}_\mathrm{t} - \frac{\nu_{\ell\mathrm{t}}}{E_\ell}\mathbf{e}_\ell\otimes\mathbf{e}_\ell\otimes\mathbf{e}_z\otimes\mathbf{e}_z \\
& - \frac{\nu_{\ell\mathrm{t}}}{E_\ell}\mathbf{e}_\mathrm{t}\otimes\mathbf{e}_\mathrm{t}\otimes\mathbf{e}_\ell\otimes\mathbf{e}_\ell + \frac{1}{E_\mathrm{t}}\mathbf{e}_\mathrm{t}\otimes\mathbf{e}_\mathrm{t}\otimes\mathbf{e}_\mathrm{t}\otimes\mathbf{e}_\mathrm{t} - \frac{\nu_{\mathrm{tz}}}{E_\mathrm{t}}\mathbf{e}_\mathrm{t}\otimes\mathbf{e}_\mathrm{t}\otimes\mathbf{e}_z\otimes\mathbf{e}_z \\
& - \frac{\nu_{\ell\mathrm{t}}}{E_\ell}\mathbf{e}_z\otimes\mathbf{e}_z\otimes\mathbf{e}_\ell\otimes\mathbf{e}_\ell\otimes - \frac{\nu_{\mathrm{tz}}}{E_\mathrm{t}}\mathbf{e}_z\otimes\mathbf{e}_z\otimes\mathbf{e}_\mathrm{t}\otimes\mathbf{e}_\mathrm{t} + \frac{1}{E_\mathrm{t}}\mathbf{e}_z\otimes\mathbf{e}_z\otimes\mathbf{e}_z\otimes\mathbf{e}_z \\
& + \frac{1}{G_{\ell\mathrm{t}}}\mathbf{e}_\ell\otimes\mathbf{e}_\mathrm{t}\otimes\mathbf{e}_\ell\otimes\mathbf{e}_\mathrm{t} + \frac{1}{G_{\ell\mathrm{t}}}\mathbf{e}_\ell\otimes\mathbf{e}_z\otimes\mathbf{e}_\ell\otimes\mathbf{e}_z + \frac{2(1+\nu_{\mathrm{tz}})}{E_\mathrm{t}}\mathbf{e}_\mathrm{t}\otimes\mathbf{e}_z\otimes\mathbf{e}_\mathrm{t}\otimes\mathbf{e}_z, \\
\boldsymbol{\beta}^f = {} & \beta_\ell\mathbf{e}_\ell\otimes\mathbf{e}_\ell + \beta_\mathrm{t}\mathbf{e}_\mathrm{t}\otimes\mathbf{e}_\mathrm{t} + \beta_\mathrm{t}\mathbf{e}_z\otimes\mathbf{e}_z\,,
\end{aligned}
\tag{8.1}
$$

whereby $\mathbf{e}_i$ $(i=\ell,\mathrm{t},z)$ are the unit vectors of the local Cartesian vector basis. The compliance tensor ${}^4\mathbf{S}^f$ relates to the stiffness tensor ${}^4\mathbf{C}^f$ via the common relation ${}^4\mathbf{S}^f = \left({}^4\mathbf{C}^f\right)^{-1}$. In Eq. (8.1), E_ℓ and E_t are the longitudinal and transverse moduli of elasticity, respectively, $G_{\ell t}$ is the shear modulus relevant for shear deformations between the longitudinal and (any) transverse direction, $\nu_{\ell t}$ is the Poisson's ratio between these directions, and ν_{tz} is the out-of-plane Poisson's ratio acting in the isotropic plane. The hygroscopic response is described through the longitudinal and transverse hygro-expansion coefficients of the fibres, β_ℓ and β_t, respectively.

8.2.3 Prediction of the hygro-elastic response via analytical homogenisation

In order to obtain the effective properties of the material, first ${}^4\mathbf{C}^f$ and $\boldsymbol{\beta}^f$ are reduced to transversely isotropic 2D tensors, according to the plane stress assumption. Moreover, the constitutive quantities (8.1) must be expressed with respect to the global reference system (x, y, z). This is done by applying a standard transformation to the constitutive tensors as a function of the orientations $\theta = [0, \pi/2]$ of the individual fibres within the lattice [33,46]. The constitutive response of the bonding region is calculated under the assumption of full kinematical compatibility between the two fibre layers (i.e. perfect bonding). In the further derivations, the corresponding elastic and hygro-expansivity tensors are denoted by the superscript "b" as ${}^4\mathbf{C}^b$ and $\boldsymbol{\beta}^b$, respectively. For the free-standing fibre segments, an additional assumption is made, whereby the free-standing fibre segments are modelled as trusses, i.e. the effect of the transverse components of the hygro-elastic deformation is neglected.

The effective hygro-mechanical properties of the reference lattice model may now be calculated analytically through homogenisation. The procedure is based on the assumption that the unit-cell strains are consistent with the applied macroscopic deformation; moreover, energy equivalence between the two scales is enforced [32,33]. The effective Young's moduli $\overline{E}^{\mathrm{lat}}_{xx}, \overline{E}^{\mathrm{lat}}_{yy}$ and hygro-expansion coefficients $\overline{\beta}^{\mathrm{lat}}_{xx}, \overline{\beta}^{\mathrm{lat}}_{yy}$ are thus obtained as

$$
\overline{E}^{\mathrm{lat}}_{xx} = \overline{E}^{\mathrm{lat}}_{yy} = \left[\frac{2(L^{\mathrm{lat}} - w)}{wE_\ell} + \frac{1}{(C^b_{xx} - \nu^b_{xy}C^b_{yx})}\right]^{-1}, \tag{8.2}
$$

$$\overline{\beta}_{xx}^{\text{lat}} = \overline{\beta}_{yy}^{\text{lat}} = \frac{1}{L^{\text{lat}}}\left[(L^{\text{lat}} - w)\beta_\ell + w\beta_{xx}^b\right]. \tag{8.3}$$

Note that the model is not isotropic, but it provides the closest estimate, which can be obtained by an orthogonal lattice model, to an isotropic response. Moreover, the considered lattice cell does not possess shear resistance. The shear stiffness can be incorporated by adding to the model two additional families of fibres, oriented at $\theta = \pm\pi/4$. The reader may refer to Ref. [32] for details.

8.3 Two-dimensional random network model

8.3.1 Model geometry

In order to investigate the role of in-plane randomness of the network on the predicted effective hygro-mechanical response, the in-plane lattice assumption characterising the reference model described in Section 8.2 is replaced by a detailed, 2D periodic network model, in which the fibres are randomly oriented. This random network model is essentially based on the work proposed in Refs. [35,36]. Despite the fact that paper micro-structures are not truly periodic, the unit cell is here assumed to be a representative volume element (RVE). An RVE is characterised by a sufficiently large size to contain enough micro-structural information, thus being representative of a stochastic network [47]. The RVE is taken to be a square domain, with edge L and area Q: $(0, L) \times (0, L)$. The network fibres present a rectangular shape, with length l and width w, and a rectangular cross section $w \times t$, where t is the thickness of the fibre. Two parameters are used to characterise each individual fibre within the network: (i) the location of the fibre's geometrical centre $\mathbf{x}_c$ in the domain Q, which is determined by a uniform random point field, and (ii) the angle θ between the x-axis and the fibre axis, which is defined according to a uniform orientation distribution. If sufficiently sampled, this thus results in an isotropic network. In order to guarantee periodicity, the portions of the fibres falling outside the square region Q along a certain edge are cut and copied into the domain at the opposite side. Moreover, since a 2D description is used, the porosity of the system cannot be explicitly taken into account. While for each material point the number of fibres through the thickness and their orientations are well defined, the specific order in which the fibres are placed in the network is not considered. *All* the fibres that overlay in a certain point of the domain are assumed to be perfectly bonded mutually. This implies that the thickness direction is collapsed into a plane, whereby its average thickness can be quantified via the areal coverage. The areal coverage $\overline{c}$ is defined as the ratio between the total in-plane area of the fibres and the area of the RVE, and is computed as

$$\overline{c} = \frac{n_f w l}{Q}, \tag{8.4}$$

with n_f the number of fibres of the network. The coverage represents the average number of fibre layers characterising the network and thus defines its average thickness to be $t\overline{c}$.

As mentioned, each fibre is modelled as a 2D, rectangular solid. The generation of a finite element mesh for such a network geometry may be challenging, especially at the interfibre bonds, where two or more fibres characterised by different orientations overlap. A simplified strategy is adopted, based on a nonconforming mesh consisting of a regular grid of square finite elements. Any fibre that contains the centroid of a finite element contributes to the stiffness of that finite element. This leads to a discretised network geometry characterised by fibres with jagged boundaries, which may affect the local stress and strain distributions. However, additional simulations not reported here have shown that, if a sufficiently fine discretisation is used, the resulting effective hygro-mechanical properties are only minimally influenced. An example of a fibrous network of coverage $\overline{c} = 1$, with cell size $L/l = 2$ and fibres of aspect ratio $l/w = 25$, is given in Fig. 8.2A; Fig. 8.2B shows a detail of the finite

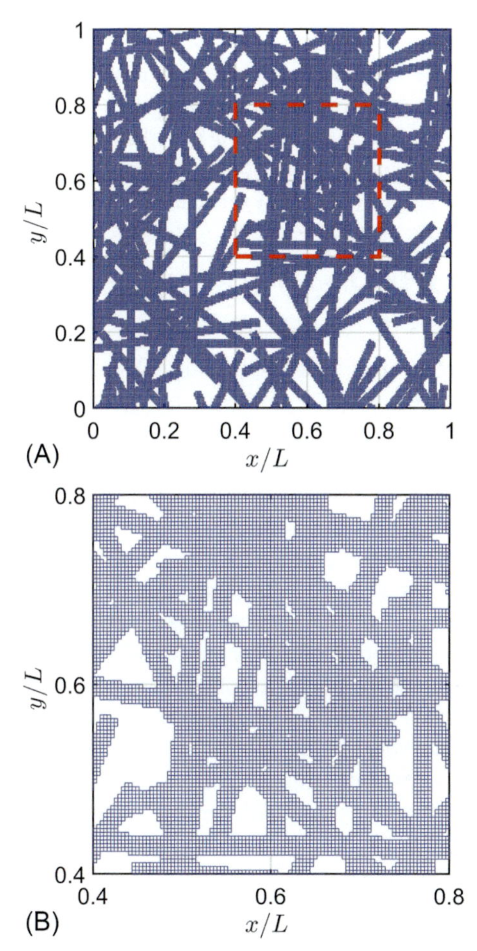

Fig. 8.2 Two-dimensional fibrous network model. (A) Example of fibrous network configuration of coverage $\overline{c} = 1$ with cell size $L/l = 2$ and fibres of aspect ratio $l/w = 25$. (B) Detail of the finite element discretisation, corresponding to the *red dashed square* in (A), for which an average of five finite elements along the fibre width has been used.

element meshing of the network, for which an average of five finite elements along the width of a fibre has been used. Note that the fine discretisation necessitated by the nonconforming mesh may result in a high computational cost. To overcome this issue, and at the same time improve the accuracy in the geometrical description, an advanced discretisation scheme can be used, in which the level-set formalism is used in combination with the extended finite element method, as has been proposed in Ref. [48].

8.3.2 Constitutive response

The constitutive response of an individual fibre is based on the transversely isotropic hygro-elastic tensors ${}^4\mathbf{C}^f$ and $\boldsymbol{\beta}^f$ given in expression (8.1), reduced to the case of a plane stress state. To derive the local properties of the network model, for each fibre oriented along a certain direction θ, the tensors ${}^4\mathbf{C}^f$ and $\boldsymbol{\beta}^f$ must be first expressed with respect to the global reference system (x, y, z). From this, for all the material points $\mathbf{x}$ within the network, the local values of the elasticity tensor ${}^4\mathbf{C}(\mathbf{x})$ and the hygro-expansion tensor $\boldsymbol{\beta}(\mathbf{x})$ can be calculated, according to the assumption of full kinematical compatibility. This is done by identifying the fibres passing through a given position $\mathbf{x}$, adding their individual contributions, and dividing the result by the average thickness of the network – see Ref. [36] for details.

8.3.3 Prediction of the hygro-elastic response via asymptotic homogenisation

The effective hygro-mechanical behaviour of the network model is obtained by using a multiscale approach based on asymptotic homogenisation [39–41]. Under the hypothesis of a strong separation between the macroscopic and the microscopic scales, asymptotic homogenisation assumes that an heterogeneous medium characterised by rapidly oscillating material properties – in this case, the hygro-elastic tensors ${}^4\mathbf{C}(\mathbf{x})$ and $\boldsymbol{\beta}(\mathbf{x})$ – can be represented as an equivalent homogeneous solid. The effective material response of this equivalent medium is retrieved from the network level fields through an averaging procedure that is based on the formulation of two sets of mathematical problems – one at the micro(meso)-scale and one at the macro-scale. The first mathematical problem focuses on the calculation of the influence functions ${}^3\mathbf{N}^1(\mathbf{x})$ and $\mathbf{b}^1(\mathbf{x})$, which are the periodic solutions of so-called cell problems. These influence functions ${}^3\mathbf{N}^1(\mathbf{x})$ and $\mathbf{b}^1(\mathbf{x})$ have the physical meaning of periodic micro-fluctuations of the displacement field experienced by the network in response to a variation of the average macroscopic strain and moisture content, respectively. These cell problems are two boundary value problems defined on the representative volume element Q, as introduced in Section 8.3.1, that depend only on the micro-structural constitutive properties ${}^4\mathbf{C}(\mathbf{x})$ and $\boldsymbol{\beta}(\mathbf{x})$:

$$\nabla \cdot ({}^4\mathbf{C}(\mathbf{x}) : (\nabla {}^3\mathbf{N}^1(\mathbf{x}) + {}^4\mathbf{I}^S)) = {}^3\mathbf{0}, \tag{8.5}$$

$$\nabla \cdot ({}^4\mathbf{C}(\mathbf{x}) : (\nabla \mathbf{b}^1(\mathbf{x}) - \boldsymbol{\beta}(\mathbf{x}))) = \mathbf{0}, \tag{8.6}$$

where $\nabla\cdot$ indicates the divergence operator and ${}^4\mathbf{I}^S$ is the fourth-order symmetric identity tensor, defined as $I^S_{ijkl} = (\delta_{il}\delta_{jk} + \delta_{ik}\delta_{jl})/2$. Eqs. (8.5), (8.6) are complemented by periodic boundary conditions and by the requirement that the functions ${}^3\mathbf{N}^1(\mathbf{x})$ and $\mathbf{b}^1(\mathbf{x})$ vanish on average on the representative volume element.

The second problem defines the effective response of the network through the properties of an equivalent homogeneous medium at the macro-scale. These macroscopic properties are the effective elastic stiffness ${}^4\overline{\mathbf{C}}$ and hygro-expansion tensor $\overline{\boldsymbol{\beta}}$, which are defined as

$$ {}^4\overline{\mathbf{C}} = \frac{1}{|Q|}\int_Q {}^4\mathbf{C}(\mathbf{x}) : (\nabla^3\mathbf{N}^1(\mathbf{x}) + {}^4\mathbf{I}^S)\mathrm{d}Q, \tag{8.7}$$

$$\overline{\boldsymbol{\beta}} = \frac{1}{|Q|}{}^4\overline{\mathbf{C}}^{-1} : \int_Q {}^4\mathbf{C}(\mathbf{x}) : (\boldsymbol{\beta}(\mathbf{x}) - \nabla \mathbf{b}^1(\mathbf{x}))\mathrm{d}Q, \tag{8.8}$$

where ∇ is the gradient operator.

In order to obtain the effective hygro-elastic constants ${}^4\overline{\mathbf{C}}$ and $\overline{\boldsymbol{\beta}}$, the cell problems (8.5), (8.6) are first solved for a given network geometry by means of a finite element simulation. The solution is next inserted into relations (8.7), (8.8). For details on the derivation of the effective properties, the solution of the cell problems, and the implementation, the reader may refer to Ref. [35].

8.4 Three-dimensional lattice model

8.4.1 Model geometry

The influence of the assumption of a 2D description on the predicted effective hygro-mechanical response is next investigated, by introducing a 3D, computational model that extends the 2D lattice model in the z-direction. In the (x, y) plane, the model consists of a square unit cell with edge L^{lat}. The unit cell is composed of two families of fibres that are orthogonally oriented, consistently with the 2D lattice model geometry shown in Fig. 8.1. The fibres have constant cross section, which is composed by a rectangle and two semicircles. Its dimensions are characterised by the thickness t and width w, where the latter is defined such that the product wt gives the cross sectional area, thus enabling a direct comparison with the 2D lattice model. In the z-direction, the cell is characterised by several fibre layers stacked on top of each other. The number of fibre layers defines the average thicknesses H of the paper sheet.

In the reference model, the fibres are assumed to be straight – see Fig. 8.3A for a sketch of the unit cell for the case of four fibre layers, i.e. $H/t = 4$. In addition, the influence of an imposed out-of-plane waviness of the fibres is explored, as illustrated, again for $H/t = 4$, in Fig. 8.3B. The rationale for the shape imposed on the fibres follows from the suggestion made in the literature that the sheet-scale hygro-expansivity

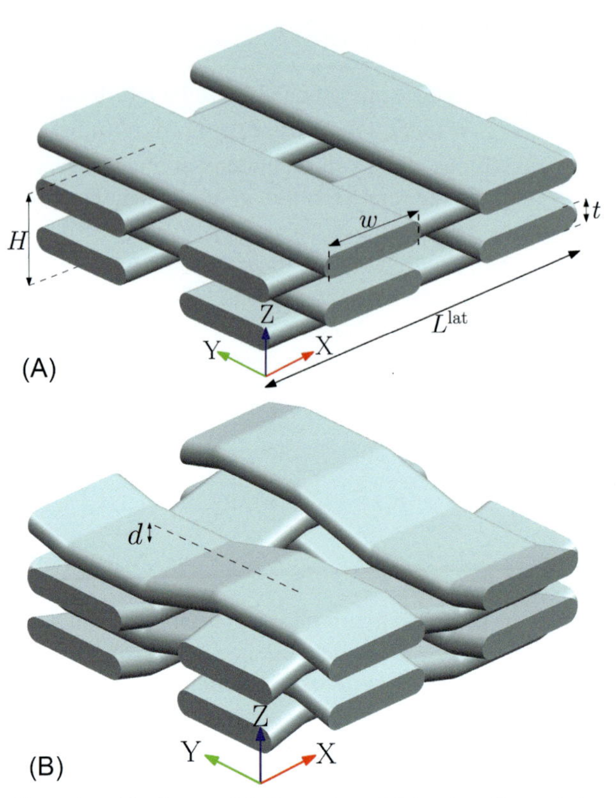

Fig. 8.3 Three-dimensional lattice model. Examples of 3D unit cell consisting of four fibre layers ($H/t = 4$) and porosity $\phi = 0.5$ with (A) no fibre wrap around, i.e. $d = 0$, and (B) full fibre wrap around, i.e. $d = t$.

may depend significantly on the degree to which the fibres are "wrapped around" each other in their bonds – and that a different degree of "wrap around" may explain the markedly different sheet-scale hygro-expansivity of freely dried and restrained dried papers [22]. The bond geometry considered in this model has been designed to exhibit such a wrap around – see Fig. 8.3B. It is worth mentioning that a recent experimental study has revealed little difference of the degree of wrap around between freely dried and restrained dried sheets [49], hence putting into question if the degree of wrap around is responsible for the different macro-scale hygro-expansivity observed for these drying conditions. Here, nevertheless, the effect of fibre wrap around is systematically explored – partially as a plausible generalisation of the (not very realistic) case of straight fibres.

The degree of fibre wrap around (or waviness) is varied in the range $0 \leq d \leq t$. For $d = 0$, the fibres are straight and no wrap around is present, as shown in Fig. 8.3A. For $d = t$, the fibres are fully wrapped around each other in the bonding region, such that the distance in the z-direction between the top faces of two adjacent bonds is equal to the fibre thickness. This extreme is illustrated in Fig. 8.3B. For given geometrical parameters L^{lat}, H, t, w, and d, the porosity ϕ of the 3D unit cell (and, hence, of the sheet) can be obtained by simple geometrical relationships not detailed here.

The proposed 3D cell is modelled using the commercial finite element simulation software Marc Mentat. Although for clarity Fig. 8.3 shows the full 3D unit cells, the simulations are performed by using a quarter of this unit cell, by exploiting the symmetry of the domain with respect to the x- and y-axes. Note that no periodicity is assumed in the thickness direction, i.e. H represents the full (average) thickness of the network (sheet). The domain is discretised by means of quadratic serendipity elements. A mesh convergence study, not reported here, has been performed to guarantee that the adopted mesh is sufficiently fine to provide homogenised hygro-elastic properties that are independent of the mesh size.

8.4.2 Constitutive model

For the 3D model, the hygro-elastic response of a single fibre is based on the (3D) transversely isotropic tensors ${}^4\mathbf{C}^f$ and $\boldsymbol{\beta}^f$ given in expression (8.1). The tensors ${}^4\mathbf{C}^f$ and $\boldsymbol{\beta}^f$ are defined directly in a local (curvilinear) coordinate frame attached to the fibre axis. In the fibre-to-fibre bonding regions, the respective fibre surfaces are assumed to be perfectly bonded (no relative displacement) wherever they touch. Unlike the 2D models, no further assumptions are made, as the hygro-mechanical interactions occurring in these regions ensue naturally from the numerical simulations.

8.4.3 Prediction of the hygro-elastic response via numerical homogenisation

The elastic and hygroscopic properties of the 3D lattice model are obtained by numerical homogenisation. First, an elastic finite element analysis is performed, for which one of the boundary planes perpendicular to the x-axis is subjected to a prescribed displacement $\bar{x}$ along the x-direction, while keeping the other boundary plane fixed in the x-direction. The average displacement in the y-direction is left unconstrained. The imposed boundary conditions require that symmetry planes initially perpendicular to the x- and y-axes remain planar upon deformation. In addition, a hygro-elastic finite element analysis is performed, in which the lattice model is subjected to a uniform moisture variation $\Delta\chi$. Boundary conditions are applied that ensure that the unit cell can expand freely in the x-, y-, and z-directions, again with the constraint that planes initially perpendicular to the x- and y-axes remain planar during deformation. From these simulations, the elastic moduli and the hygro-expansion coefficients are computed as

$$\bar{E}_{xx}^{3D} = \bar{E}_{yy}^{3D} = \frac{F_x}{H\bar{x}}, \tag{8.9}$$

$$\bar{\beta}_{xx}^{3D} = \bar{\beta}_{yy}^{3D} = \frac{\bar{x}^{\chi}}{\Delta\chi L^{\text{lat}}}, \tag{8.10}$$

where F_x is the reaction force component in the x-direction due to the prescribed displacement $\bar{x}$; $\bar{x}^{\chi}$ is the average displacement along the x-axis of the right boundary of the unit cell, upon application of a change in moisture content $\Delta\chi$.

8.5 Results

8.5.1 Geometrical and material parameters used in the simulations

In the analyses, the following material parameters are assumed. The fibre longitudinal elastic modulus is taken as $E_\ell = 35$ GPa [28]. The transverse elastic modulus E_t and the shear modulus $G_{\ell t}$ are defined with respect to the longitudinal stiffness via the ratios $E_\ell/E_t = 6$ and $E_\ell/G_{\ell t} = 10$ [50]. This leads to $E_t = 5.83$ GPa and $G_{\ell t} = 3.5$ GPa. The Poisson's ratio $\nu_{\ell t}$ is taken as $\nu_{\ell t} = 0.3$ [50]. Due to the lack of experimental data, the Poisson's ratio ν_{tz} is selected for simplicity to be equal to $\nu_{\ell t}E_t/E_\ell = 0.05$. The longitudinal hygro-expansive coefficient, expressed in terms of the fractional change in strain per unit mass variation, is taken as $\beta_\ell = 0.013$ [−] [14]. The transverse hygro-expansion is expressed as a function of the longitudinal expansivity as $\beta_t = 20\beta_\ell = 0.26$ [−] [1].

The influence of in-plane randomness of the network on the prediction of the effective hygro-elastic properties is estimated by comparing the results of the 2D idealised lattice model and the random network model. The effective properties are computed for several values of coverage, i.e. $\bar{c} = [0.25, 0.5, 1, 2, 5, 10]$. For the network model, an average fibre length $l = 3.5$ mm is assumed [51]. The width of the fibre w is defined as a function of l via the aspect ratio $l/w = 100$ [1], resulting in $w = 35$ μm. Note that it is not necessary to explicitly define the value of the fibre thickness t, since for the 2D models only the ratio between the thickness of the individual fibres and the thickness of the bonds plays a role. The unit-cell edge is selected as $L = 2l = 7$ mm, which guarantees the network model to be a representative volume element, as discussed in Ref. [35]. In order to make a comparison between the response of the lattice model with that of the network, the length L^{lat} of the edge of the lattice unit cell must be specified as a function of the coverage $\bar{c}$. Considering mass equality between the macroscopic level and the unit cell and using the coverage definition (8.4), provides the relation

$$L^{\text{lat}} = n_f^{\text{lat}} w/\bar{c}, \tag{8.11}$$

with $n_f^{\text{lat}} = 2$ the number of fibres considered in the lattice model. According to the above expression, a limit configuration is reached by the lattice model at $\bar{c} = 2$, when $L^{\text{lat}} = w$. At this coverage, the unit cell is entirely covered by the bond. The effective properties (8.2), (8.3) reach their maximum values at $\bar{c} = 2$, and remain constant for any coverage $\bar{c} \geq 2$.

The influence of including a 3D description on the effective hygro-elastic properties is further assessed by comparing the results of the 2D and the 3D lattice models. For the 3D model, a porosity $\phi = 0.5$ is adopted throughout. Moreover, several configurations characterised by a different network average thickness are considered, with $H/t = [2, 4, 10]$. Note that the value of $H/t = 10$ is in the range of many paper grades used for, for example, printing. To perform the comparison between the 2D and 3D descriptions, the same fibre dimensions in the (x, y) plane are selected for the two models. Considering the assumed porosity $\phi = 0.5$ for the 3D model, the reference 2D lattice is characterised by an equal size of the bonding region and the free-standing part of the fibre, i.e. $L^{\text{lat}} = 2w$.

8.5.2 Influence of in-plane randomness on material response

8.5.2.1 Local deformation field

The in-plane network model provides detailed insight into the local fields at the microscopic level. This cannot be achieved via the idealised lattice description. Consider a network subjected to a macroscopic variation of moisture content $\Delta\chi$, in a zero average stress state – i.e. expanding freely. Fig. 8.4 shows the local total maximum ε_{max} and minimum ε_{min} principal strains, normalised by $\beta_\ell \Delta\chi$, for a network of coverage $\bar{c} = 1$ mapped on the undeformed configuration. The strains are computed from the

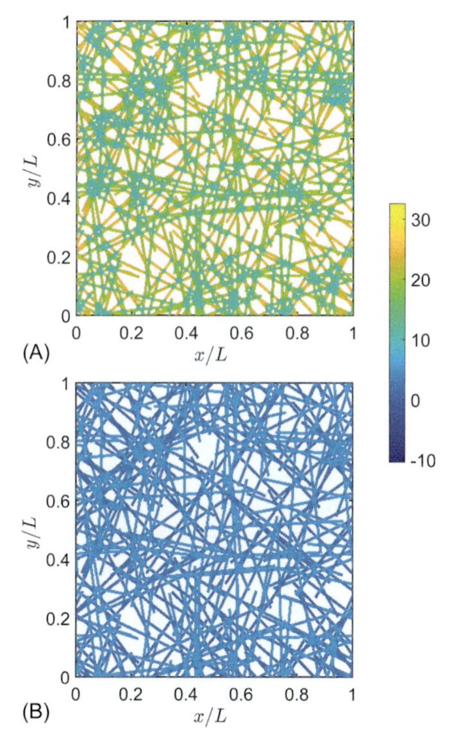

Fig. 8.4 Local deformation fields of the 2D network model. (A) Maximum principal total strain $\varepsilon_{\text{max}}/\beta_\ell\chi$ and (B) minimum principal total strain $\varepsilon_{\text{min}}/\beta_\ell\chi$ due to free hygro-expansion in a network of coverage $\bar{c} = 1$, shown on the undeformed configuration.

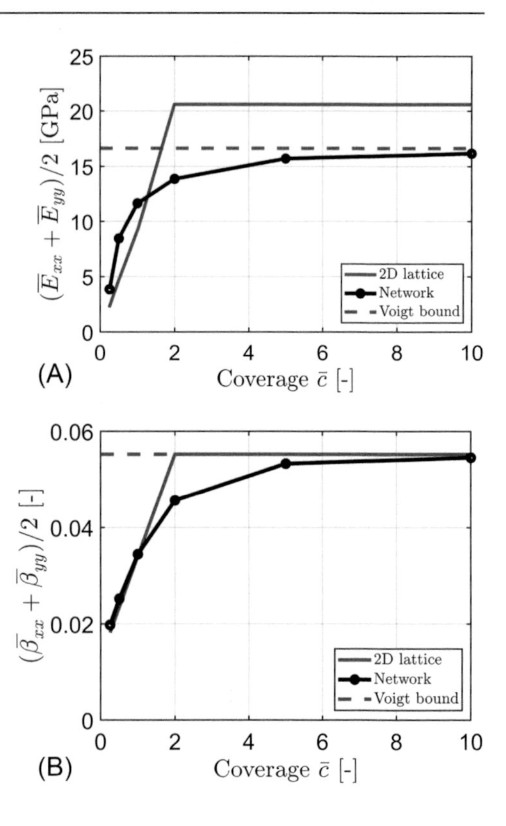

Fig. 8.5 Influence of in-plane randomness on the effective hygro-elastic properties. (A) Orientation-averaged effective elastic moduli $(\overline{E}_{xx}+\overline{E}_{yy})/2$ and (B) orientation-averaged effective hygro-expansion coefficients $(\overline{\beta}_{xx}+\overline{\beta}_{yy})/2$ as a function of the coverage $\overline{c}$, calculated with the 2D lattice model, the network model and the Voigt bound.

influence functions ${}^{3}\mathbf{N}^{1}(\mathbf{x})$ and $\mathbf{b}^{1}(\mathbf{x})$, which are the solutions of the cell problems (8.5), (8.6), following the procedure discussed in Ref. [35]. As mentioned in Section 8.3.3, tensor ${}^{3}\mathbf{N}^{1}(\mathbf{x})$ has the physical meaning of the micro-fluctuation of the displacement field due to a macroscopic deformation, while vector $\mathbf{b}^{1}(\mathbf{x})$ represents the micro-fluctuation of the displacement field corresponding to a macroscopic moisture variation in conditions of constrained expansion. From Fig. 8.4, it can be noticed that the strain distribution is relatively homogeneous in the free-standing fibre segments, as it is essentially due to the expansive contribution only. Larger fluctuations in the strain distribution can be observed in the bonding regions. These are due to internal stresses caused by the interplay between the mechanical response and the hygroscopic expansion, which is governed by the misorientation of the different fibres forming the bonds.

8.5.2.2 Effective hygro-elastic properties

Fig. 8.5 shows the effective, sheet-scale hygro-elastic constants obtained with the 2D models, as a function of the coverage $\overline{c}$. The results are presented in terms of the orientation-averaged elastic moduli $(\overline{E}_{xx}+\overline{E}_{yy})/2$ and hygro-expansive coefficients $(\overline{\beta}_{xx}+\overline{\beta}_{yy})/2$. These averages allow to minimise the effect of the randomness of the fibre orientations, which may lead to a slightly anisotropic network. Five network realisations have been considered per coverage value. The continuous black lines with dot markers represent the mean values of the properties computed via asymptotic homogenisation from these five network configurations. For the considered size of the unit cell, the maximum deviation from the effective properties obtained from an individual network realisation from the averaged value over all the realisations is approximatively 8% for coverages $\overline{c}\leq 1$; for larger values of the coverage the deviation is lower than 3%. The number of realisations used can be thus considered to be sufficient. The continuous blue lines illustrate the estimates from the reference lattice model according to the relations (8.2), (8.3). Finally, the dashed magenta lines indicate the effective hygro-elastic properties computed according to the Voigt bound. The Voigt bound is obtained by assuming an ideal composite plate with an infinite number of layers, covering all orientations through the thickness, and is calculated as

$$ {}^4\overline{\mathbf{C}}^V = \frac{1}{\pi}\int_{-\pi/2}^{\pi/2} {}^4\mathbf{C}^{f(k)}(\theta)\,\mathrm{d}\theta, \tag{8.12} $$

$$ \overline{\boldsymbol{\beta}}^V = \frac{1}{\pi}\,({}^4\overline{\mathbf{C}}^V)^{-1} : \int_{-\pi/2}^{\pi/2} {}^4\mathbf{C}^{f(k)}(\theta) : \boldsymbol{\beta}^{f(k)}(\theta)\,\mathrm{d}\theta, \tag{8.13} $$

where ${}^4\mathbf{C}^{f(k)}$ and $\boldsymbol{\beta}^{f(k)}$ are the hygro-elastic constitutive tensors referred to a fibre k oriented along the direction $\theta^{(k)}$.

Consider first Fig. 8.5A, comparing the mechanical properties. The elastic moduli computed via asymptotic homogenisation from the network simulations increase for increasing network coverage $\overline{c}$, and approach an asymptotic value. It can be observed that the network response is in general overestimated by the elastic moduli calculated via the lattice model (except at coverages $\overline{c}\leq 1$). The stiffness of the lattice model reaches its maximum value for coverages $\overline{c}\geq 2$, when the model assumes the limit configuration for which the lattice unit cell is entirely covered by the bond, corresponding to a composite plate with two fibre layers oriented at $[0, \pi/2]$. The Voigt bound captures closely the effective random network properties for high coverages, $\overline{c}\geq 5$. Note that the value of the elastic modulus calculated through the Voigt average is lower than the lattice estimate for $\overline{c}\geq 2$. This is due to the fact that the stiffness obtained via the Voigt average is isotropic, while the stiffness of a "network" with only two layers of infinitely long fibres, which are furthermore perfectly aligned in the x- and y-directions, is anisotropic. A better estimate of the network mechanical properties can be obtained when considering a lattice model that includes a third and a fourth family of fibres oriented at $\pm\pi/4$ [32]. These two additional families of fibres furthermore furnish the lattice model with a shear stiffness. Consider for instance the case of a network of coverage $\overline{c}=2$. The elastic modulus $\overline{E}_{xx}^{\mathrm{lat},4}$ predicted with a lattice model with four families of fibres is equal to

15.21 GPa, which is an intermediate value between the estimates of the reference lattice model ($\overline{E}_{xx}^{\mathrm{lat}} = 20.58$ GPa) and the random network model ($\overline{E}_{xx} = 13.87$ GPa). Moreover, at a coverage $\overline{c} = 4$, which represents the limit configuration for the lattice model of four families of fibres, the elastic modulus $\overline{E}_{xx}^{\mathrm{lat},4}$ is equal to 16.5 GPa, which practically coincides with the Voigt estimate.

The predicted effective hygro-expansive properties are shown in Fig. 8.5B. Also in this case, the network predictions show an increasing trend as a function of the coverage, and they approach an asymptotic value. The lattice model captures accurately the network hygro-expansion for low coverages ($\overline{c} \leq 1$). At higher coverages ($\overline{c} \geq 5$), the Voigt estimate closely matches the network response. In this case, the Voigt prediction coincides with that of the lattice model for $\overline{c} \geq 2$. This can be explained by the fact that, unlike the stiffness tensor, the hygro-expansion tensor is isotropic both for the case of a two-layered bond and a Voigt plate with infinite number of layers. In general, from Fig. 8.5 three regimes can be distinguished for both the mechanical and hygro-expansive properties. For sparse networks ($\overline{c} \leq 1$), there is a strong dependence of the effective properties on the coverage. Constant properties are retrieved for dense networks, with $\overline{c} \geq 5$, while in between a transition zone is observed.

8.5.3 *Influence of 3D geometry on the effective hygro-elastic properties*

8.5.3.1 *Local deformation field*

Consider for the 3D unit cell a macroscopic moisture content variation $\Delta\chi$, in free hygro-expansion conditions. Fig. 8.6 shows the total strain component ε_{xx}, normalised with respect to $\beta_\ell \Delta\chi$, mapped on the deformed configuration, for two 3D unit cells with relative thickness $H/t = 2$ and $H/t = 4$ (Fig. 8.6A and B, respectively). Note that, as discussed in Section 8.4.1, the simulation is done only on a quarter of the periodic cell – exploiting symmetry of the geometry with respect to the vertical mid-planes of the fibres. Accordingly, the computational model has half fibres (in width direction) along its lateral faces and the strain values on these faces are those in the (vertical) fibre mid-planes. The 3D description allows to capture the strain distribution through the thickness. The deformation of the free-standing fibre segments is essentially due to the longitudinal fibre expansion only, and therefore the strain is substantially homogeneous and equal to $\beta_\ell \Delta\chi$. In the bonding regions, higher strain values (up to approximately 20 times) are reached in the transverse fibre, due to the higher hygro-expansion coefficient in the transverse direction ($\beta_t = 20\beta_\ell$). At the interface, the longitudinal fibre and the transverse fibre are in competition, resulting in a bond expansion on the order of $5\beta_\ell$. Comparing Fig. 8.6A and B, it can be observed that the strain values are only mildly influenced by the relative thickness of the network; however, the model with a smaller thickness ($H/t = 2$) exhibits more bending, which has consequences on the predicted effective hygroscopic response.

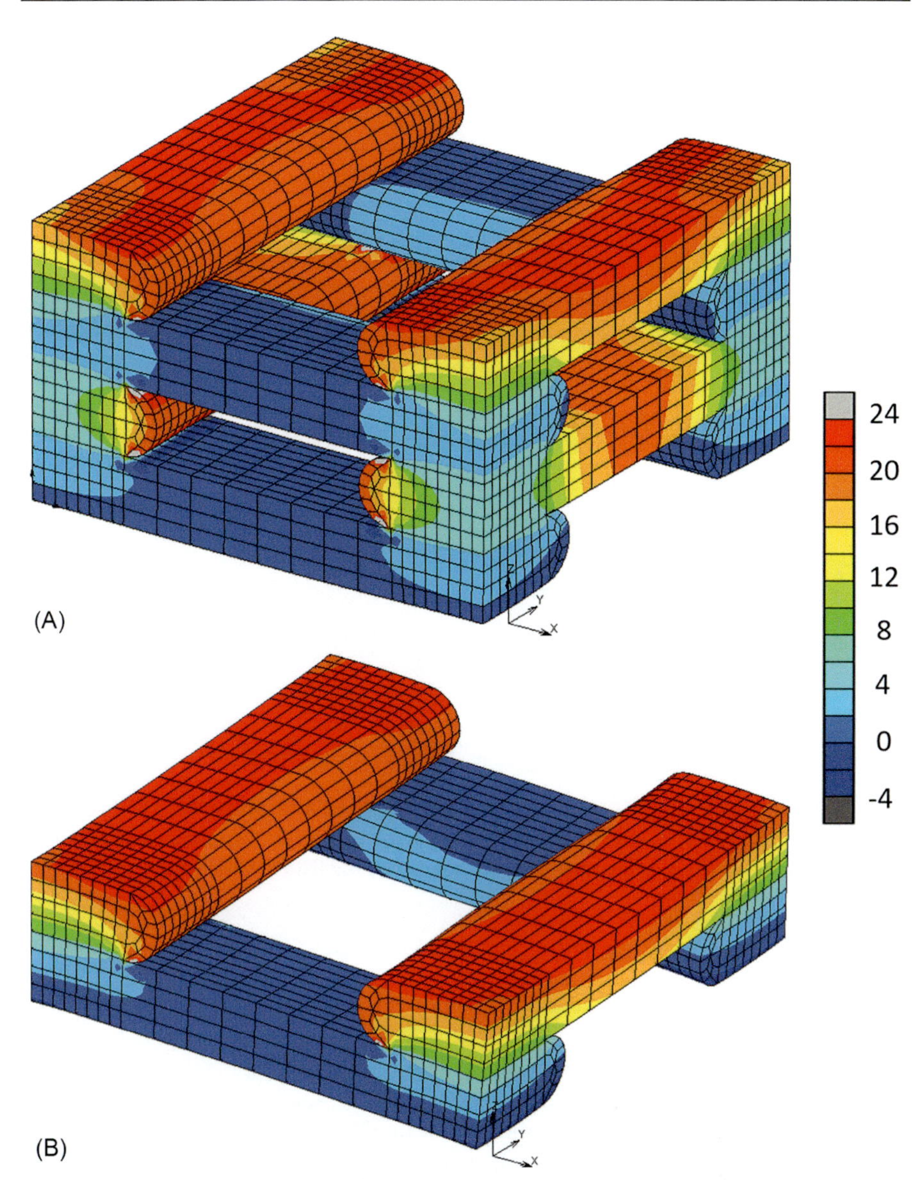

Fig. 8.6 Local deformation fields of the 3D lattice model. Total strain component $\varepsilon_{xx}/(\beta_\ell \Delta\chi)$ for unit cells with relative thickness (A) $H/t = 2$ and (B) $H/t = 4$. The simulations have been run considering a quarter of the unit cell shown in Fig. 8.3, due to the in-plane symmetry of the model.

8.5.3.2 Effective hygro-elastic properties

Fig. 8.7 shows the effective, sheet-scale hygro-elastic constants obtained with the 2D lattice model and the 3D lattice model, as a function of the degree of fibre waviness d/t. The ratio d/t ranges from 0 to 1, corresponding to the two extreme configurations

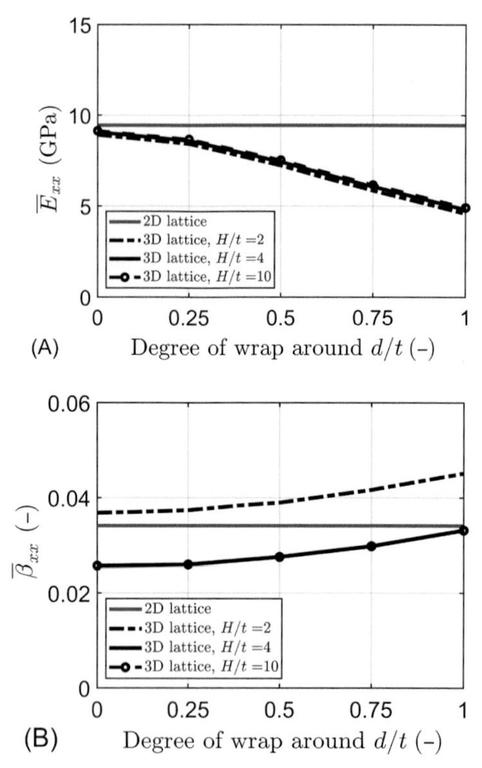

Fig. 8.7 Influence of 3D description on the effective hygro-elastic properties. (A) Effective elastic modulus $\overline{E}_{xx}$ and (B) effective hygro-expansion coefficient $\overline{\beta}_{xx}$, as a function of degree of bond wrap around d/t, calculated with the 2D lattice and 3D lattice models.

shown in Fig. 8.3. The continuous blue line represents the results of the 2D model. Black colour refers to the results of the 3D model, whereby the dot-dashed lines, the continuous lines, and the dashed lines with dot markers indicate configurations characterised by a relative average network thickness $H/t = 2$, $H/t = 4$, and $H/t = 10$, respectively.

The effective elastic moduli obtained from the 2D and the 3D lattice models are shown in Fig. 8.7A. Considering first the results of the 3D model, for a given number of fibre layers, the elastic moduli decrease for increasing degree of waviness. This is due to the fact that, with a higher degree of fibre waviness, the deformation involves more bending, partially removing the need to stretch – which involves the higher axial stiffness compared with the bending stiffness. It can be further observed that the trend is consistent for all the considered configurations ($H/t = 2, 4, 10$) and that the average network thickness plays only a minimal role in the resulting effective elastic properties. In particular, the elastic modulus slightly increases for increasing thickness of the unit cell. Finally, the 2D lattice model generally overestimates the response of the 3D configuration. This is due to the assumption of full kinematic compatibility considered in the bonding regions, which results in a stiffer response of the 2D system. With reference to the value corresponding to $H/t = 10$ and perfectly straight fibres ($d/t = 0$), the elastic modulus of the 2D model overestimates the modulus computed with the 3D model by less than 5%. For the highest degrees of fibre wrap around ($d/t = 1$), the elastic properties of the 3D lattice are overestimated by a factor of approximately 2.

The effective hygro-expansive properties are shown in Fig. 8.7B. For a fixed number of fibre layers in the out-of-plane direction, the effective hygro-expansion coefficient increases for increasing degree of fibre wrap around. The trend is independent of the number of layers. This is possibly due to the fact that a higher degree of entanglement, in combination with the high transverse hygro-expansion coefficient of the fibres, induces a larger swelling also in the longitudinal direction, in correspondence with the bonding regions. This observation is consistent with that made for a 2D model in Ref. [22]. Independently of the degree of wrap around, the model characterised by $H/t = 2$ provides a higher value of the hygro-expansion coefficients than in the case of a larger average network thickness. For $H/t \geq 4$, the response becomes independent of the thickness of the network, as evidenced by the fact that the curves for $H/t = 4$ and $H/t = 10$ coincide. The higher hygro-expansion shown by the model characterised by $H/t = 2$ is due to the fact that this system exhibits more bending – see Fig. 8.6A. Finally, the reference 2D lattice model provides a value of the effective hygro-expansion coefficient that lies in between the estimate computed with the 3D model characterised by $H/t = 2$ and with the models with higher average network thickness ($H/t \geq 4$). For $H/t \geq 4$ and fully wrapped around fibres, i.e. $d/t = 1$, the relative difference between the coefficients obtained with the 2D and the 3D models is less than 5%. For smaller values of d/t, the hygro-expansion of the 2D unit cell overestimates that of the 3D one, possibly due to the assumption of full kinematic compatibility, which leads to an overestimate of the response of the hygro-expansion of the 2D bond. At $d/t = 0$, the relative difference between the response of the two models is about 30%.

8.6 Conclusions

This chapter has proposed a multiscale approach to calculate the effective hygro-mechanical response of paper, which aims at systematically exploring how the geometrical and material characteristics of the underlying fibrous network govern the effective hygro-elastic properties. For this purpose, a highly idealised model has been initially considered, which is based on a 2D lattice unit cell consisting of two fibres oriented in orthogonal directions. The effective hygro-mechanical response of this 2D lattice model has been obtained by analytical homogenisation. The influence of the two main assumptions of this reference model, i.e. (i) the lattice idealisation and (ii) the 2D description, on the predicted effective hygro-elastic response has been explored. To this aim, an in-plane random network description has been considered, instead of the lattice structure idealisation. Asymptotic homogenisation has been used to determine the effective properties of the 2D network model. In addition, the 2D lattice model has been extended in the third direction towards a 3D unit cell, in order to assess the influence of the 3D character of the network versus the 2D idealisation of the initial reference lattice model. In particular, the effect of sheet thickness and the out-of-plane waviness of the fibres have been explored. In this case, the effective properties have been obtained by numerical homogenisation.

The results of all the proposed models have been analysed, revealing that the assumption of a 2D lattice model has a stronger influence on the effective mechanical response rather than on the effective hygroscopic behaviour. Focusing first on the comparison between the 2D lattice and network models, the in-plane randomness significantly affects the predicted mechanical properties. For practically relevant values of the network thickness ($\overline{c} \geq 5$), the 2D idealised lattice model overestimates the mechanical response compared to the random network description of about 30%. Note, however, that a lattice model that includes a third and a fourth family of fibres (oriented at $\pm\, \pi/4$) leads to a much closer estimate of the elastic properties of the random network. The in-plane randomness has in general a limited influence on the hygroscopic properties. The idealised lattice and the network models present a similar trend in the predicted effective hygro-expansive coefficients, whereby at both low coverages ($\overline{c} \leq 1$) and high coverages ($\overline{c} \geq 5$) they also show quantitative agreement. At intermediate coverage values, the in plane randomness has a somewhat larger influence, and the lattice model over-predicts the response of the network. Considering finally the influence of the 3D description, the elastic response of the 3D unit-cell model is generally strongly overestimated by the 2D lattice – depending on the degree of wrap around, of less than 5% (for the configuration with straight fibres), up to a factor of approximately 2 (for fully wrapped around fibre). As for the hygroscopic properties, the 2D lattice model approximates the effective hygro-expansion of 3D unit cells within a range of 30% for realistic average network thicknesses. For the particular case of fibres with full degree of wrap around ($d/t = 1$), the estimates from the two models are very close. Certainly, the assumption of fibres that are perfectly aligned in the 3D model may have an effect on the results. Breaking this regularity may have more effect than the waviness considered here. Finally, the ultimate assessment of the effect of micro-structural features on the predicted hygro-elastic response must be done by considering a 3D random network model. This may be the subject of the future work.

References

[1] K. Niskanen, Paper Physics, Fapet Oy Helsinki, Finland, 1998.

[2] P.A. Larsson, L. Wagberg, Influence of fibre-fibre joint properties on the dimensional stability of paper, Cellulose 15 (4) (2008) 515–525.

[3] N.H. Vonk, M.G.D. Geers, J.P.M. Hoefnagels, Full-field hygro-expansion characterization of single softwood and hardwood pulp fibers, Nordic Pulp Paper Res. J. 36 (1) (2021) 61–74.

[4] S. Douezan, M. Wyart, F. Brochard-Wyart, D. Cuvelier, Curling instability induced by swelling, Soft Matter 7 (4) (2011) 1506–1511.

[5] E. Bosco, R.H.J. Peerlings, M.G.D. Geers, Local network effects on hygroscopic expansion in digital ink-jet printing, Nordic Pulp Paper Res. J. 31 (4) (2016) 684–691.
[6] H.L. Cox, The elasticity and strength of paper and other fibrous materials, Br. J. Appl. Phys. 3 (3) (1952) 72.
[7] J. Astrom, S. Saarinen, K. Niskanen, J. Kurkijarvi, Microscopic mechanics of fiber networks, J. Appl. Phys. 75 (5) (1994) 2383–2392.
[8] X.F. Wu, Y.A. Dzenis, Elasticity of planar fiber networks, J. Appl. Phys. 98 (9) (2005) 093501.
[9] D. Tsarouchas, A.E. Markaki, Extraction of fibre network architecture by X-ray tomography and prediction of elastic properties using an affine analytical model, Acta Mater. 59 (18) (2011) 6989–7002.
[10] H. Hatami-Marbini, R.C. Picu, Heterogeneous long-range correlated deformation of semi-flexible random fiber networks, Phys. Rev. E 80 (4) (2009) 046703.
[11] J. Wilhelm, E. Frey, Elasticity of stiff polymer networks, Phys. Rev. Lett. 91 (10) (2003) 108103.
[12] C.A. Bronkhorst, Modelling paper as a two-dimensional elastic-plastic stochastic network, Int. J. Solids Struct. 40 (20) (2003) 5441–5454.
[13] R. Hagglund, P. Isaksson, On the coupling between macroscopic material degradation and interfiber bond fracture in an idealized fiber network, Int. J. Solids Struct. 45 (3–4) (2008) 868–878.
[14] J. Strömbro, P. Gudmundson, Mechano-sorptive creep under compressive loading—a micromechanical model, Int. J. Solids Struct. 45 (9) (2008) 2420–2450.
[15] A. Kulachenko, T. Uesaka, Direct simulations of fiber network deformation and failure, Mech. Mater. 51 (2012) 1–14.
[16] A.S. Shahsavari, R.C. Picu, Size effect on mechanical behavior of random fiber networks, Int. J. Solids Struct. 50 (20–21) (2013) 3332–3338.
[17] Z. Lu, M. Zhu, Q. Liu, Size-dependent mechanical properties of 2D random nanofibre networks, J. Phys. D Appl. Phys. 47 (6) (2014) 065310.
[18] J. Dirrenberger, S. Forest, D. Jeulin, Towards gigantic RVE sizes for 3D stochastic fibrous networks, Int. J. Solids Struct. 51 (2) (2014) 359–376.
[19] H. Reda, K. Berkache, J.F. Ganghoffer, H. Lakiss, Dynamical properties of random fibrous networks based on generalized continuum mechanics, Waves Random Complex Media 30 (1) (2020) 27–53.
[20] T. Uesaka, C. Moss, Y. Nanri, The characterization of hygroexpansivity of paper, J. Pulp Paper Sci. 18 (1) (1992) J11–J16.
[21] Y. Nanri, T. Uesaka, Dimensional stability of mechanical pulps—drying shrinkage and hygroexpansivity, Tappi J. 76 (6) (1993) 62–66.
[22] T. Uesaka, D. Qi, Hygroexpansivity of paper—effects of fibre to fibre bonding, J. Pulp Paper Sci. 20 (6) (1994) J175–J179.
[23] K.J. Niskanen, S.J. Kuskowski, C.A. Bronkhorst, Dynamic hygroexpansion of paperboards, Nordic Pulp Paper Res. J. 12 (19974) 103–110.
[24] A. Torgnysdotter, L. Wagberg, Tailoring of fibre/fibre joints in order to avoid the negative impacts of drying on paper properties, Nordic Pulp Paper Res. J. 21 (2006) 411–418.
[25] A.-L. Erkkilä, T. Leppänen, M. Ora, T. Tuovinen, A. Puurtinen, Hygroexpansivity of anisotropic sheets, Nordic Pulp Paper Res. J. 30 (2) (2015) 326–335.
[26] C.G.V. der Sman, E. Bosco, R.H.J. Peerlings, A model for moisture-induced dimensional instability in printing paper, Nordic Pulp Paper Res. J. 31 (4) (2016) 676–683.
[27] T. Uesaka, General formula for hygroexpansion of paper, J. Mater. Sci. 29 (9) (1994) 2373–2377.

[28] C. Sellén, P. Isaksson, A mechanical model for dimensional instability in moisture-sensitive fiber networks, J. Compos. Mater. 48 (3) (2014) 277–289.
[29] E. Bosco, R.H.J. Peerlings, M.G.D. Geers, Scale effects in the hygro-thermo-mechanical response of fibrous networks, Eur. J. Mech. A/Solids 71 (2018) 113–121.
[30] A. Brandberg, H.R. Motamedian, A. Kulachenko, U. Hirn, The role of the fiber and the bond in the hygroexpansion and curl of thin freely dried paper sheets, Int. J. Solids Struct. 193–194 (2020) 302–313.
[31] H.R. Motamedian, A. Kulachenko, Simulating the hygroexpansion of paper using a 3D beam network model and concurrent multiscale approach, Int. J. Solids Struct. 161 (2019) 23–41.
[32] E. Bosco, R.H.J. Peerlings, M.G.D. Geers, Predicting hygro-elastic properties of paper sheets based on an idealized model of the underlying fibrous network, Int. J. Solids Struct. 56–57 (2015) 43–52.
[33] E. Bosco, M.V. Bastawrous, R.H.J. Peerlings, J.P.H. Hoefnagels, M.G.D. Geers, Bridging network properties to the effective hygro-expansive behaviour of paper: experiments and modelling, Philos. Mag. 95 (28–30) (2015) 3385–3401.
[34] E. Bosco, R.H.J. Peerlings, M.G.D. Geers, Explaining irreversible hygroscopic strains in paper: a multi-scale modelling study on the role of fibre activation and micro-compressions, Mech. Mater. 91 (Part 1) (2015) 76–94.
[35] E. Bosco, R.H.J. Peerlings, M.G.D. Geers, Asymptotic homogenization of hygro-thermo-mechanical properties of fibrous networks, Int. J. Solids Struct. 115–116 (2017) 180–189.
[36] E. Bosco, R.H.J. Peerlings, M.G.D. Geers, Hygro-mechanical properties of paper fibrous networks through asymptotic homogenization and comparison with idealized models, Mech. Mater. 108 (2017) 11–20.
[37] C.T.J. Dodson, Spatial variability and the theory of sampling in random fibrous networks, J. R. Stat. Soc. B (Methodol.) 33 (1) (1971) 88–94.
[38] W.W. Sampson, Modelling Stochastic Fibrous Materials With Mathematica, Springer, London, 2009.
[39] E. Sanchez Palencia, Nonhomogeneous Media and Vibration Theory, Springer, Berlin, 1980.
[40] N.S. Bakhvalov, G. Panasenko, Homogenisation: Averaging Processes in Periodic Media, Kluwer Academic Publishers, 1989.
[41] J.M. Guedes, N. Kikuchi, Preprocessing and postprocessing for materials based on the homogenization method with adaptive finite element methods, Comput. Methods Appl. Mech. Eng. 83 (2) (1990) 143–198.
[42] R.H.J. Peerlings, N.A. Fleck, Computational evaluation of strain gradient elasticity constants, Int. J. Multiscale Comput. Eng. 2 (2004) 599–619.
[43] A. Bergander, L. Salmén, Cell wall properties and their effects on the mechanical properties of fibers, J. Mater. Sci. 37 (1) (2002) 151–156.
[44] R.C. Neagu, E.K. Gamstedt, Modelling of effects of ultrastructural morphology on the hygroelastic properties of wood fibres, J. Mater. Sci. 42 (24) (2007) 10254–10274.
[45] S. Borodulina, A. Kulachenko, D. Tjahjanto, Constitutive modeling of a paper fiber in cyclic loading applications, Comput. Mater. Sci. 110 (2015) 227–240.
[46] D. Roylance, Mechanics of Materials, Wiley & Sons, 1996.
[47] W.J. Drugan, J.R. Willis, A micromechanics-based nonlocal constitutive equation and estimates of representative volume element size for elastic composites, J. Mech. Phys. Solids 44 (4) (1996) 497–524.
[48] P. Samantray, R.H.J. Peerlings, E. Bosco, M.G.D. Geers, T.J. Massart, O. Rokoš, Level set-based extended finite element modeling of the response of fibrous networks under

hygroscopic swelling, J. Appl. Mech. Trans. ASME 87 (10) (2020), https://doi.org/10.1115/1.4047573.
[49] G. Urstöger, A. Kulachenko, R. Schennach, U. Hirn, Microstructure and mechanical properties of free and restrained dried paper: a comprehensive investigation, Cellulose 27 (2020) 8567–8583.
[50] K. Schulgasser, D.H. Page, The influence of transverse fibre properties on the in-plane elastic behaviour of paper, Compos. Sci. Technol. 32 (4) (1988) 279–292.
[51] M. Jajcinovic, W.J. Fischer, U. Hirn, W. Bauer, Strength of individual hardwood fibres and fibre to fibre joints, Cellulose 23 (3) (2016) 2049–2060.

Deformation and damage of random fibrous networks

Emrah Sozumert[a,b], Vincenzo Cucumazzo[a], and Vadim V. Silberschmidt[a]
[a]Wolfson School of Mechanical, Electrical and Manufacturing Engineering, Loughborough University, Leicestershire, United Kingdom, [b]College of Engineering, Swansea University, Swansea, United Kingdom

9.1 Introduction

9.1.1 Background

Understanding and characterising of deformation and damage behaviour of random fibrous networks is important since they can be found in multiple natural and engineering systems, for instance, in an amniotic sac as a protective environment for developing embryos in the human body [1] or electrospun networks used as hosts for cells for the treatment of skin injuries [2]. In previous investigations, researchers focused on the notch behaviour of fibrous networks such as bacterial hydrogel [3], non-wovens [4,5], electrospun networks [6], human amniotic membranes [7], and soft collagenous tissues [8].

In contrast to traditional continuous materials, the presence of a notch in a random fibrous network can even improve the material stiffness [3,4]. In cases when the size of a crack (i.e. notch) was less than a critical length, a notch insensitivity for the random fibrous network was observed [9].

Apparently, the microstructure of fibrous networks influences their deformation and damage mechanisms. The character of orientation distribution of fibres in a network is one of the key anisotropy parameters affecting material behaviour [10,11]; it can be expressed in terms of an orientation distribution function (ODF) [12]. It is an important input parameter for the generation of microstructural numerical models similar to the actual microstructure of random fibrous networks. For instance, the impact of the ODF on effective elastic material properties was computationally investigated with various ODFs, considering isotropic- or preferentially oriented random fibres [13]. The ODF and nonaffine characteristics of fibrous networks might also control the orientation of failure planes (e.g. the path of crack growth) [14]. Simulations of the fracture behaviour of multilayer fibrous networks [15] revealed that the stress distribution and energy dissipation differed depending on the microstructure of each layer. Similarly, deformation, damage, and fracture of bacterial hydrogels were experimentally investigated with the presence of artificial notches under tensile loading [3] and biaxial loading [16]. As a collagenous fibrous network, rabbit skins were tested under tensile loading to analyse the main stages of deformation and damage [17]:

Mechanics of Fibrous Networks. https://doi.org/10.1016/B978-0-12-822207-2.00002-7

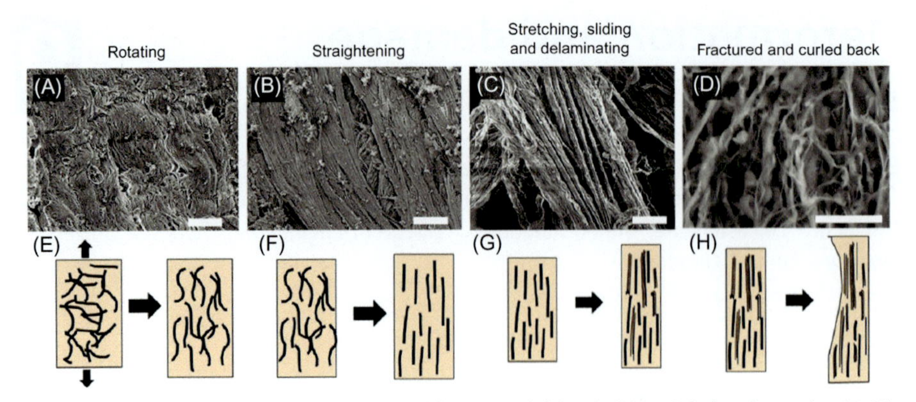

Fig. 9.1 Stages of deformation, damage, and fracture of skin (A–D) and their schematics (E–H).

(i) rotation of fibres; (ii) straightening of fibres; (iii) stretching, sliding, and delamination; and (iv) fractured and curled back of fibres (Fig. 9.1).

In electrospun fibrous scaffolds (networks) with a mode-I crack, fibre bundles formed in front of the crack tip, in which fibres had higher stress concentration [18]. Electrospun nanofibre networks demonstrated higher fracture toughness (resistance to fracture) as crack length increased [19].

To study the deformation and damage of random fibrous networks at multiple scales, a thermally bonded non-woven was chosen as an example material for experimental and numerical studies in this research. The specimens without artificial cracks or notches (termed *virgin specimens*) and with mode-I crack were tested in tension. Macroscopic and microscopic observations of their response to stretching are reported together with statistical analysis for constituent fibres of the chosen type of fibrous network.

9.1.2 Aim and objectives

The aim of this study is to investigate the deformation and damage of random fibrous networks at multiple scales using experimental and numerical methods. In experiments and numerical simulations, the focus is on fundamental damage related to mode-I cracks introduced in thermally bonded non-woven specimens perpendicular to their loading direction. In order to quantify the effect of a notch on deformation and damage behaviours, specimens without any artificial cracks were also analysed numerically and experimentally. This allows us to measure stress and strain localisations around notches at the microscale (fibre scale). Individual fibres were extracted from non-woven samples and tested in a testing machine to acquire time-dependent elasto-plastic properties. Subsequently, tensile tests of rectangular specimens of non-wovens were performed to obtain data for deformation, damage initiation, and growth at the fabric scale. Material properties derived from single-fibre tests were employed as input parameters for finite-element (FE) models of random fibrous networks. The models also accounted for microstructural features of fibrous networks, such as fibre diameters and orientation distributions of fibres, obtained from SEM images by direct measurement and image-processing, respectively. They were incorporated into the FE models using an automated parametric modelling approach without a notch and with it, mimicking the actual microstructure of the chosen fibrous network material. These discrete FE models were used to simulate the stretching along a specific direction [machine direction (MD) of non-woven] to quantify the deformation and fracture behaviours at multiple scales (global and local).

9.2 Experimentation

9.2.1 Material

A spun-bonded hot-calendered 25 g/m^2 non-woven composed of bi-component core-sheath fibres made of 70% polypropylene and 30% polyethylene (70/30 PP/PE) is used throughout this study as an example of a random fibrous network. Spun-bonding is a direct one-step manufacturing process that converts a raw polymer (or a mixture of polymers), in the form of chips or granules, into the final product, with produced polymer fibres randomly laid onto a conveyor belt. Subsequently, they are bonded employing an embossed roll (calender) by applying heat and pressures to loose fibres. The obtained material's microstructure is a heterogeneous medium that presents compact and stiff regions named *bond points* (with a pattern transferred from the calender), connected by a fibrous matrix (Fig. 9.2A). These bond-point regions can have various shapes (elliptical ones are shown in Fig. 9.2A) and they might be distributed according to a specific spatial pattern. As is well known, the mechanical properties, such as failure strength, of thermally bonded non-wovens are strongly controlled by web-bonding parameters, e.g. bonding temperature.

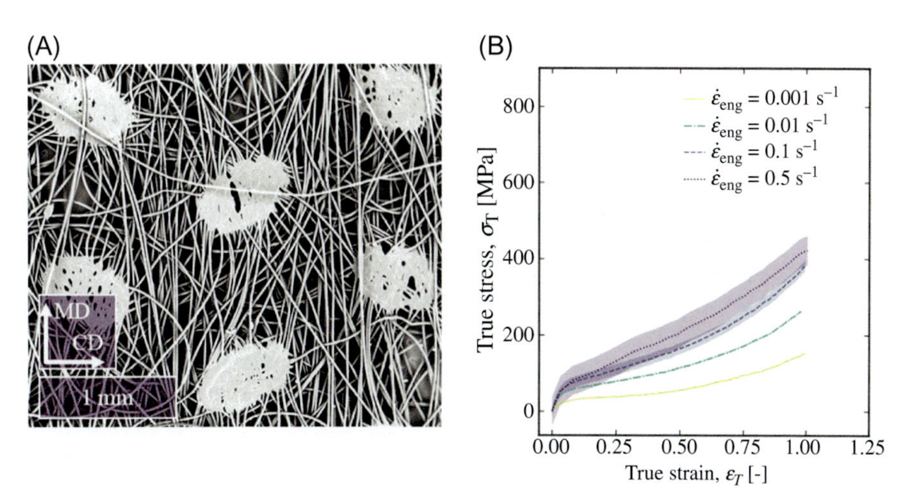

Fig. 9.2 (A) SEM image of 25 gsm thermally bonded non-woven. (B) True stress–strain curves of individual fibres at 0.001, 0.01, 0.1, 0.5, and 1 s^{-1} strain rate (clouds demonstrate the scatter).

9.2.2 Experimental procedure

Having a heterogeneous microstructure due to a manufacturing process, the random fibrous network requires a multiscale characterisation [experimental tests at micro- and macroscales (fibre and fabric scales, respectively)] to investigate deformation and damage behaviours. The mechanical behaviour of a fibrous network at the micro-level is governed by the mechanical properties of fibres; therefore, material properties of individual fibres should be assessed and quantified, for instance, with micro-tensile tests. Morphological (such as fibre diameter) and material properties (such as elastic–plastic parameters) of single fibres are used in the generation of FE models of the fibrous networks and implemented into constitutive models imitating mechanical behaviour at the fibre scale. At the macro-level, non-woven fabrics as a fibrous network were tested under uniaxial tension for the subsequent validation of the FE models and investigation of their damage behaviour with virgin and notched samples. The evolution of deformation and damage of networks were examined under large strains by conducting detailed experimental and numerical investigations.

9.2.3 Single-fibre tests: Microscale

Material properties of individual fibres can be obtained with multiple experimental testing methods; however, the key problem is to select the most suitable method according to, for instance, the fibre diameter. The first way to assess these properties is to test the fibres with a micro-tensile tester. As the material properties of polymeric fibres depend on pressure, temperature, and humidity, they are altered during the process of the thermal web bonding of fibres. Mueller and Kochmann [20] found that the failure strength of fibres changed significantly during bonding, and a similar loss in strength was also reported by Farukh et al. [21]. Therefore, only the material properties of processed fibres (i.e. fibres after the bonding process) can be used in numerical simulations to represent the correct failure trend at the fibre scale. In order to assess the material properties of fibres, individual fibres were carefully extracted from the 25 gsm non-woven fabric using a tweezer, preventing the fibres from any early damage. The processed fibres were attached to double-side sticky labels at their ends to improve the friction between them and the clamps in the tensile testing machine as well as to avoid the damage in fibres by the clamps. The extracted fibres were tested with displacement control in a universal testing system Instron Micro-tester 5944 equipped with a high-precision load cell with a load capacity of ± 5 N. The tests were performed at four different engineering strain rates $\dot{\varepsilon}_{eng}$ – 0.001, 0.01, 0.1, and 0.5 s^{-1} – in order to quantify the elastic–plastic material properties. Eight fibre tests were performed for each strain rate. The true stress–strain curves (Fig. 9.2B) were calculated based on the force-displacement readings of the testing system and the assumption of a perfectly circular fibre cross section with fibre diameter of 17 μm, which was measured from SEM images. In some cases, it was not easy to perform single-fibre tests in a standard tensile-testing machine due to the small diameter or length of fibres; therefore, experimental tests based on atomic force microscopy were used to extract the material properties [22].

In numerical investigations, the Young's modulus at strain rate 0.01 s^{-1} (1390 MPa) and the Poisson's ratio of 0.42, obtained from our single-fibre tests, were used together with the elastic–plastic curve measured at the same strain rate. The effect of time dependency on mechanical behaviour was outside the scope of this research.

9.2.4 Fabric tests: Macroscale

At the macroscale, rectangular specimens of 25 gsm non-woven fabric were investigated via tensile testing to understand the deformation and damage behaviour of random fibrous networks and to validate the numerical models. The virgin specimens (i.e. without a notch) and those with slit-notch geometry (mode I) were subjected to tension in MD. The specimens were in the form of rectangular strips with dimensions of the gauge area 25 mm × 40 mm (i.e. excluding the parts fixed in clamps). With the selected specimen size, all the features involved in deformation, damage, failure, and notch-induced behaviours were captured. Enhanced specimen-to-grip adhesion was guaranteed by attaching a fine layer of sandpaper onto the serrated rubber clamps. The five specimens for each case of virgin geometry and notches were tested in a universal tensile testing system (Instron 5944) with a 2 kN load cell at a strain rate of 0.1 s^{-1}. As is well known, the manufacturing process of non-woven materials results in anisotropy in three main orthogonal directions, defined with respect to the direction of conveyor belt during the manufacture: machine direction (MD), cross direction (CD), and thickness direction (TD). The abbreviations of these orthogonal directions are used throughout this research.

9.3 Numerical investigations

In this research, an implicit finite-element method was used to simulate the deformation and damage response of the selected random fibrous network. Generally, such networks can be modelled with discontinuous, continuous, or hybrid modelling approaches. Continuous models rely on the homogenisation of fibrous microstructure and, therefore, can simulate mostly macroscopic behaviour. In discontinuous modelling, random fibres are directly implemented into the volume of interest to elucidate the actual microstructure and microstructural mechanical behaviour. The main advantage of the discontinuous modelling approach is that it allows us to capture the local processes of fibre stretch, rotation, and failure modes in microstructure and reflect the effect of these modes on global (macroscopic) mechanical behaviour.

9.3.1 Finite-element modelling of random fibrous networks

A finite-element model representing a discrete microstructure of the selected random fibrous network was developed in this study by following a modelling strategy similar to that in Sozumert et al. [5]. Geometric features – fibre diameter, dimensions and orientations of bond point, and their spatial pattern – were determined from processing of SEM images of non-wovens. The complex microstructure with randomly distributed fibres is one of the most characteristic features of a fibrous network, affecting its mechanical properties significantly [23]. As orientation distribution of fibres is one of the dominant factors affecting the stiffness depending on the stretch direction (i.e. anisotropic mechanical response), it was carefully computed using an in-house code based on the Hough-transform-based algorithm [10] and SEM images; the basic steps involved in this algorithm are demonstrated in Fig. 9.3. By detecting the fibre edges in SEM (or X-ray computed tomography) images of the thermally bonded non-woven, the overall orientation distribution of the area of interest was automatically obtained in Matlab. A typical result of the ODF of the studied fibrous network, computed with this algorithm, is presented in Fig. 9.4. Using a different algorithm

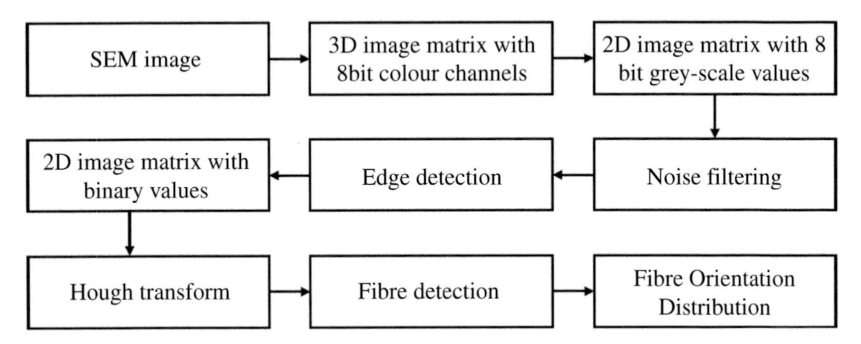

Fig. 9.3 Image-processing algorithm to compute the orientation distribution of fibres from SEM or X-ray images.
Adapted from E. Demirci, M. Acar, B. Pourdeyhimi, V.V. Silberschmidt, Computation of mechanical anisotropy in thermally bonded bicomponent fibre nonwovens, Comput. Mater. Sci. 52 (2012) 157–163. https://doi.org/10.1016/j.commatsci.2011.01.033.

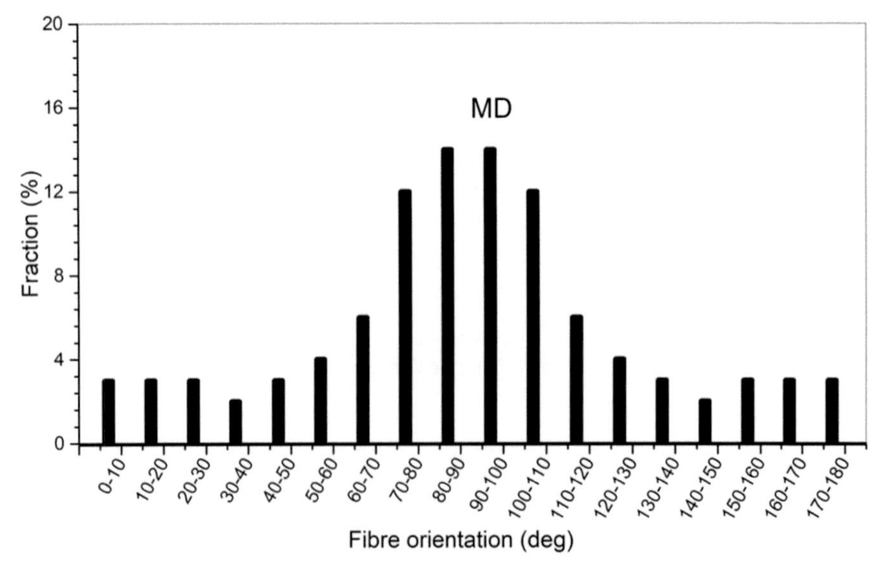

Fig. 9.4 Computed orientation distribution function for fibres of 25 gsm thermally bonded non-woven (*MD*, machine direction).

written in Python in a finite-element environment (specifically, in MSC Marc), constituent fibres were randomly distributed spatially and their orientations were assigned to match the overall orientation distribution of fibres and considering the total weight of the fibre mass in the specimen. The main steps of this algorithm are explained in Fig. 9.5. After the distribution of fibres, bond-point geometries were introduced following the assigned bond pattern and their geometries (elliptical bond points were used in this study). Fibre geometries were trimmed at the boundaries of bond points, with fibre parts within the bond points removed: it was assumed that fibres were completely melted in bond points due to high temperature and pressure during the

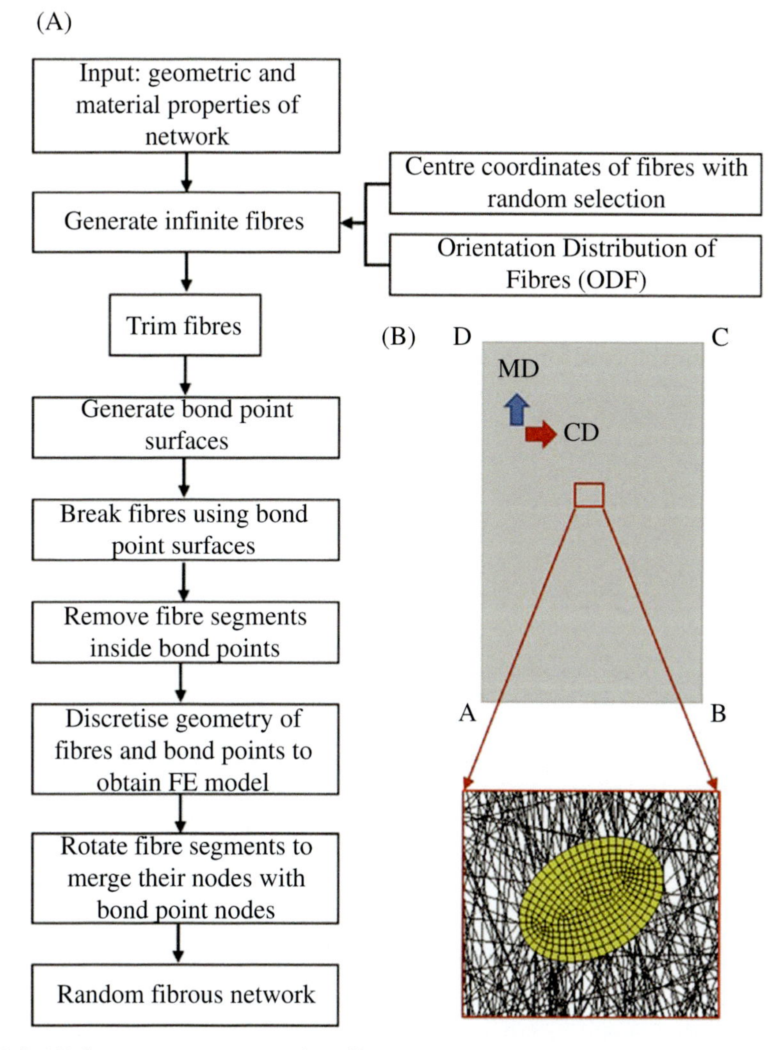

Fig. 9.5 (A) Steps to generate a random fibrous network. (B) FE model with zoomed-in view of bond point and fibres.

manufacturing process and, thus, could be modelled as continuum domains. The heterogenous microstructural model was discretised with two-node beam and thin-shell elements for the fibre segments and bond points, respectively, due to the relatively low thickness of the latter compared to their length and width. Both fibres and bond points of the studied non-woven were represented one to one in the developed FE models.

9.3.2 Finite-element formulations

In the employed FE software, Marc Mentat fibres and bond points were modelled with two-node beam and thin-shell elements, respectively, with the help of the Python script and a parametric modelling approach. In FEM, the weak formulation of a momentum equation is solved in place of the strong form one. In the weak formulation for a momentum equation for a beam-shell system, the total internal virtual work, δW_{int}, which is the summation of the contribution of beam and shell elements, is equal to total external virtual work δW_{ext}:

$$\begin{aligned}\delta W_{int} &= \delta W_{ext}, \\ \delta W_{int} &= \sum_{i=1}^{n} \delta W_{int}^{beam} + \sum_{i=1}^{m} \delta W_{int}^{shell}.\end{aligned} \tag{9.1}$$

The internal virtual work for a beam – in the absence of transverse shear terms – i.e. Euler beam, and for a shell element based on a discrete Kirchhoff theory in the absence of shear deformation in x and in y (in-plane) is given by

$$\begin{aligned}\delta W_{int}^{beam} &= \int_0^L \left(N_z \delta\varepsilon_z + M_x \delta\kappa_x + M_y \delta\kappa_y + T_z \delta\kappa_z\right) dz, \\ \delta W_{int}^{shell} &= \int_0^L \left(M_x \delta\kappa_x + 2M_{xy} \delta\kappa_{xy} + M_y \delta\kappa_y\right) dA.\end{aligned} \tag{9.2}$$

The virtual work of an Euler beam is defined by the work done by the axial force N_z, the bending moments M_x and M_y, the torsion T_z, where, in a local coordinate system of a beam, the z-axis coincides with the longitudinal axis. Additionally, the virtual work of a shell is the sum of the virtual works done by the bending moments M_x and M_y, and the twisting moment M_{xy}, where κ_x, κ_y, and κ_{xy} are curvatures. The material behaviour of single fibres was incorporated in the incremental form of elastic–plastic stress–strain relations [24].

9.3.3 Assumptions, boundary conditions, and solver

Fibre curvature and fibre-to-fibre interactions were neglected in this research. It was reported that the curvature of fibres can increase the toughness of the overall fibrous network [5]. However, incorporating this nonlinearity into the FE model can cause convergence problems of finite-element simulations.

Let us denote the corners of the representation of the FE model with letters A to D (see. Fig. 9.5B). In FE simulations, whilst stretching the model (i.e. the domain ABCD in Fig. 9.5B), the nodes at the edge AB were fixed in orthogonal directions, though they were free to rotate. However, a linearly increasing displacement field (i.e. constant velocity) was applied at the edge CD as a prescribed boundary condition and, similar to AB, the nodes at the edge CD rotated freely. Employment of these boundary conditions reflected the conditions of the rubber grips in real-life experiments. Thanks to a low effective strain rate over the specimens, FE simulations were carried out with an implicit solver (Full Newton Solver) under quasi-static loading conditions by using a large-deformation formulation.

9.4 Results and discussions

9.4.1 Macroscopic response of random fibrous networks: Experiments

In order to assess the effect of anisotropy due to the random orientation of fibres, the previous literature about deformation and damage mechanisms of a low-density non-woven [21,25] can be used. Images of the low-density thermally point-bonded non-woven in the undeformed state and under 60% extension in MD and CD are demonstrated in Fig. 9.6A and B, respectively. The major difference in the deformed overall shapes results from the differences in the alignment of constituent fibres in both orthogonal directions, resulting in different effective Poisson's ratios. Additionally, random characters of distribution of fibres and their contacts with bond points caused the loss of symmetry in the bond-point pattern, observed in the unstretched configuration (Fig. 9.6C, left), due to unequal rotation of fibres towards the loading direction (along MD).

Some studies [4] reported that the presence of a crack can make random fibrous networks stronger, or the network might be insensitive to the crack [9]. However, there is a question of applicability of this statement to all the cases of such networks. So, a notch-sensitivity analysis for the selected fibrous network was experimentally carried out in this section. Artificial cracks (notches) with lengths of 2, 4, and 6 mm were inserted at the non-woven specimens' mid-height using a thin blade. The size of the notched specimens was the same as those of virgin specimens (i.e. 25 mm $\times$ 40 mm). The levels of force and extension of the specimens were recorded during the tests; the former were normalised by the width of the unnotched part of specimens. The resultant (mean) normalised force vs engineering strain curves of the virgin and notched specimens are illustrated in Fig. 9.7 for three different orientations θ with regard to MD: along MD ($\theta = 0°$), CD ($\theta = 90°$), and at $\theta = 45°$. The strengthening effect was observed in the presence of a notch, supporting the previous observations [4]. A plausible explanation of this phenomenon could be that, at some point, the crack propagating towards the centre of the fabric encountered a bundle of aligned fibres (due to a marked necking effect) that acted as a crack arrestor, thus, enhancing the load-bearing capacity of the material. An increase in the notch length

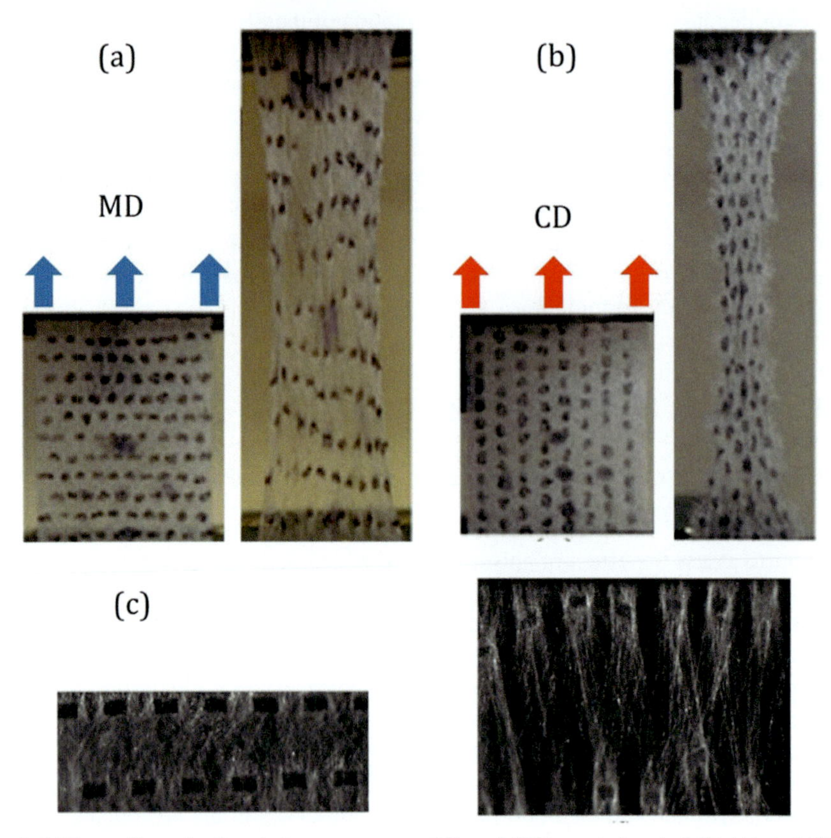

Fig. 9.6 Thermally point-bonded non-woven at 0% and 60% extension in MD (A) and CD (B) (bonded points were manually marked) [25]. (C) Fibres and bond points before and after a uniaxial stretch in MD [21].

triggered a weakening mechanism in the material as the specimens were stretched in MD; this effect was not apparent in CD.

For a better understanding of mode-I behaviour, a notch analysis was performed for notched and virgin samples. In order to make a comparison and quantify it, the notion of a notch strength ratio R was introduced as

$$R = \frac{\sigma_u^{not}}{\sigma_u^{unot}} \tag{9.3}$$

where σ_u^{not} and σ_u^{unot} are the ultimate strength (fracture stress) of the notched and virgin specimens. The computed notch strength ratios are illustrated in Fig. 9.8. Obviously, in the case $R = 1$, the random fibrous network can be considered notch-insensitive, whilst the case $R > 1$ corresponds to notch-strengthening behaviour. The experiments (Fig. 9.8) did not produce a universal type of response to notches, demonstrating that notch sensitivity depends on notch size as well as the orientation of specimens.

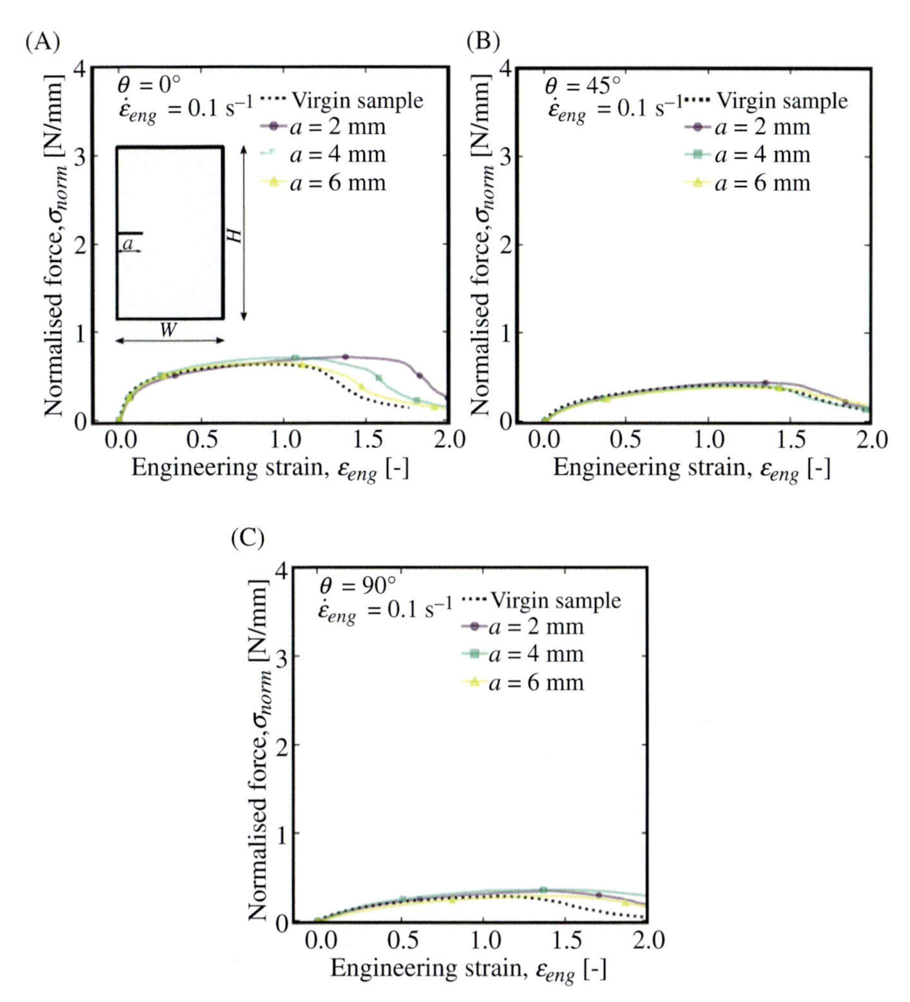

Fig. 9.7 Normalised force vs engineering strain for virgin and notched samples of non-woven fibrous networks (notch length a:2, 4, or 6 mm) for specimens orientated along MD (A), at 45° to MD (B) and along CD (C).

9.4.2 Deformation and damage evolution: FE simulations

Tensile behaviour of non-woven specimens without initial notches and with mode-I cracks (notches) of length ranging from 2 to 6 mm were successfully simulated with our developed FE approach. It included a damage criterion for single fibres, expressed as

$$\text{If } D = \frac{\varepsilon_a}{\varepsilon_f} = 1.0, \text{ fibre fails.} \tag{9.4}$$

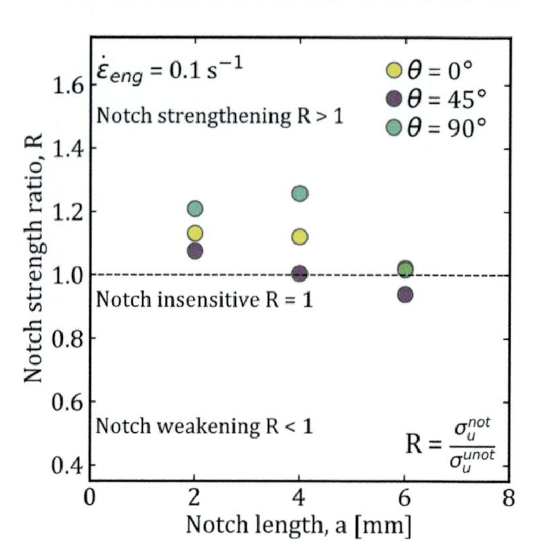

Fig. 9.8 Notch-sensitivity analysis for thermally bonded non-woven with different notch length a:2, 4, or 6 mm for three different orientations θ with regard to MD: along MD ($\theta = 0°$), CD ($\theta = 90°$), and at $\theta = 45$ ° .

In simulations, the constituent fibres were assumed to have a fully isotropic material behaviour, and their failure stress ε_f was taken 1.0 from the single-fibre tests. In the FE models, each Euler beam element represented a fibre segment and, as the strain ε_a along the beam's longitudinal axis exceeded the failure stress ε_f, it was assumed that the fibre (or beam element) failed and it was deleted. Deformation and damage patterns of virgin and notched specimens (with initial notch lengths of 2, 4, and 6 mm) for 0% and 60% extensions are presented in Fig. 9.9. Obviously, the FE models were capable of capturing the main deformation and damage features: (i) damage growth with stretching; (ii) mode-I crack opening; and (iii) global necking behaviour of specimens.

As observed in both experiments and simulations of the non-woven fabric, at the early stage of tensile stretch, the notches grew nearly similarly in MD and CD. But after the realignment of fibres by rotation towards the loading direction was completed, the rate of damage growth in MD samples was much faster than that in CD ones. This was related to the difference in the realisation of damage mechanisms for the two orthogonal directions. The damage growth in CD was controlled more by the fibre failure, whilst the one in CD was a result of the crack opening.

9.4.3 Microscopic analysis: FE simulations

As a random fibrous network is stretched, various physical processes can occur in the microstructure [21,26]: (i) debonding of fibres under low level of strain; (ii) reorientations of fibres towards the loading direction; (iii) straightening of fibres; (iv) stretching; and (v) failure of individual fibres. Damage starts at the local level with failure of fibres and, as a fibrous network is stretched further, it spreads to the global level. Ensuring the damage zone in a certain area makes the analyses easier; this can be

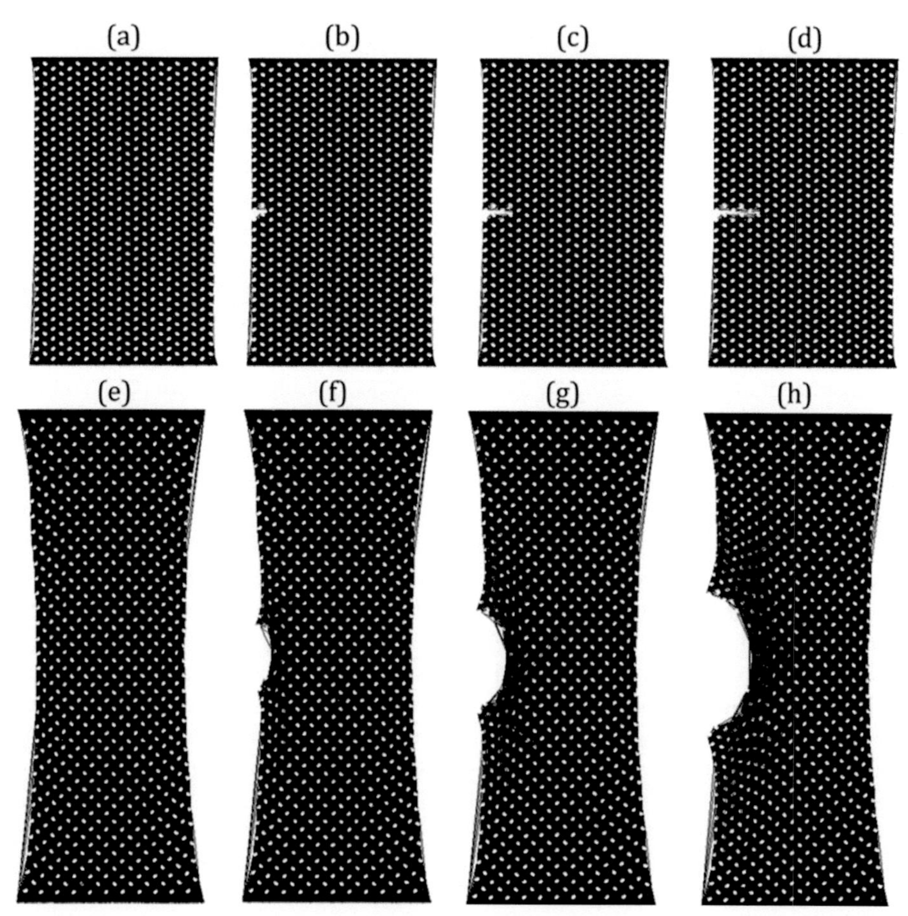

Fig. 9.9 Results of FE simulations for deformation and damage patterns of virgin and notched specimens with 2, 4, and 6 mm crack (notch) lengths for 0% (A–D) and 60% (E–H) extensions.

achieved by inserting an artificial macroscopic defect (mode-I crack in our case). As opposed to the studied case, the previous analysis by Sozumert et al. [5] investigated the effect of notch shape on damage and deformation. Notches of different shapes – slit, circular, and square – were inserted into a similar thermally bonded non-woven material with rectangular bond points.

The notch shape (such as circular or rectangular) is known to control the deformation and damage mechanisms [4,27]. Its orientation with respect to the loading and main (manufacturing-induced) directions affects the tensile strength of the fibrous network. This is related to the mutual orientation of fibres and notches; for instance, deformation and damage behaviours are effectively insensitive to a slit notch if its direction is parallel to the preferential orientation of fibres. A negligibly small contribution to the specimen's load-carrying capacity was observed for fibrous networks, stretched in MD, with a slit notch parallel to the loading direction [5].

As discussed, the mechanical behaviour of a thermally bonded fibrous network is mainly controlled by the properties of fibres and bonding regions. The effect of notch on the deformation and damage of fibrous networks with brittle fibres was a focus of some studies such as [4,6,28] whilst for fibrous networks with ductile fibres it was analysed by Ridruejo et al. [26] and Sozumert et al. [5]. Ductile fibres in random fibrous networks have a propensity for dispersing strain energy in larger areas than brittle fibres, reducing stress concentrations around notches and blunting the notch tips. In contrast to ductile fibres, brittle fibres concentrate the stresses close to the notch tips.

As soon as a crack develops in a stretched material, it keeps growing as the material continues to deform. The shape and orientation of the crack in fibrous networks change continuously during the deformation process, making any measurement rather cumbersome. Also, such measurements should be made in situ to avoid the material's spring-back effect as soon as it is unclamped. The apparent advantage of the simulation approach is detailed information about all the parameters during the entire deformation process.

In the developed FE models of random networks, fibres along the designated paths were tracked using an in-house Python code that read the FE result file, in order to assess the localised strain distributions; sample tracked paths in this analysis are illustrated in Fig. 9.10. They were chosen in the centre of the specimens. For the virgin specimen, the centre line was divided into 24 equal-length segments, from L_v to R (Fig. 9.10A), and the fibres crossing this line were separately tracked for assessment of their strains. In the mode-I specimens, the fibres were tracked for the line starting from the crack tip (i.e. 2, 4, or 6 mm from the left-hand side, such as L_a in Fig. 9.10B), and the total horizontal path was then split into 24 equal-length segments. The distributions of fibre strains (mean and normalised by the effective strain of the entire specimen) calculated with FE simulations for virgin samples and those with mode-I cracks at 60% extension differed significantly (Fig. 9.10C). Fibre strains in the former only demonstrated some fluctuation in the distribution along the horizontal path as a result of random orientations of fibres. The introduction of a mode-I crack into the specimens resulted in apparent notch sensitivity. As the notch length increased from 2 to 6 mm, the mean strain in the vicinity of the crack tip increased significantly. It was observed that the length of the path with excessive fibre strains (as compared to those in the other parts of the specimen) was comparable to the initial notch length.

9.5 Conclusions

Experimental and numerical methods for characterising deformation and damage behaviour of a random fibrous network were presented in this chapter using the sample material – thermally bonded non-woven fabric. Analysis of the literature demonstrated that specific features of transition of deformation and damage from microscale (fibre level) to macroscale (fabric level) depend on the microstructure of fibrous networks. Hence, an adequate numerical approach should directly incorporate such microstructures, presupposing the use of discrete (discontinuous) modelling

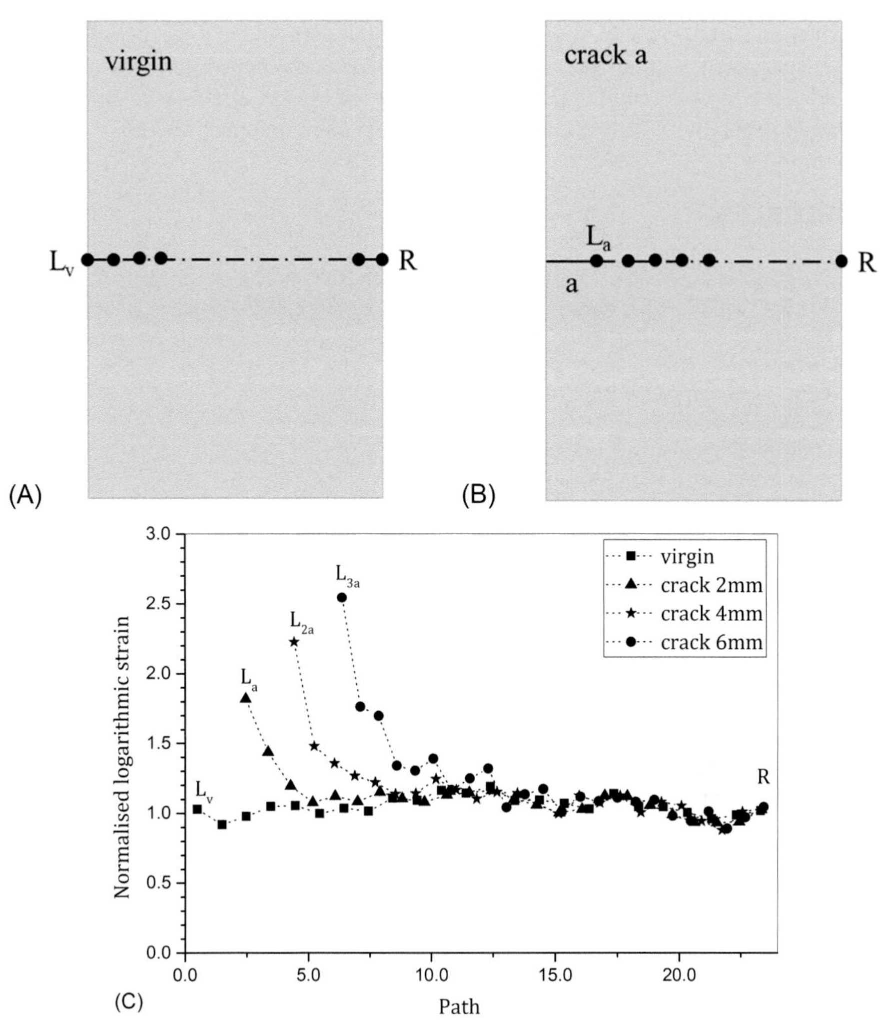

Fig. 9.10 Paths selected for strain analysis in the virgin specimen (A) and one with mode-I 2-mm-long notch (B). (C) FE-based calculated distributions of fibre strains at 60% extension for virgin specimen and specimens with various lengths of the initial mode-I notch.

approaches, in contrast to traditional (continuous) ones. Three main factors should be considered in these models: (i) single-fibre mechanical properties; (ii) orientation distribution of fibres; and (iii) character and spatial distribution of fibre-to-fibre interaction (bonding or contact). To analyse the damage in the studied fibrous materials, polymer fibres extracted from it were tested for assessing their elastoplastic properties. A realistic FE model of the fibrous network reflecting the true anisotropic nature of its microstructure was generated with an in-house computer algorithm written in Python. Numerical simulations successfully reproduced the evolution of deformation and damage at micro- and macroscales, captured in experiments on nonwoven specimens without initial defects as well as with mode-I notches of various initial lengths. It was found that both the distribution of fibre strains and the character of strength sensitivity of the studied material depended on crack lengths.

References

[1] W. Buerzle, E. Mazza, On the deformation behavior of human amnion, J. Biomech. 46 (2013) 1777–1783, https://doi.org/10.1016/j.jbiomech.2013.05.018.

[2] M. Norouzi, S.M. Boroujeni, N. Omidvarkordshouli, M. Soleimani, Advances in skin regeneration: application of electrospun scaffolds, Adv. Healthc. Mater. 4 (2015) 1114–1133, https://doi.org/10.1002/adhm.201500001.

[3] X. Gao, Z. Shi, C. Liu, G. Yang, V.V. Silberschmidt, Fracture behaviour of bacterial cellulose hydrogel: microstructural effect, Procedia Struct. Integr. 2 (2016) 1237–1243, https://doi.org/10.1016/j.prostr.2016.06.158.

[4] A. Ridruejo, R. Jubera, C. González, J. LLorca, Inverse notch sensitivity: cracks can make nonwoven fabrics stronger, J. Mech. Phys. Solids 77 (2015) 61–69, https://doi.org/10.1016/j.jmps.2015.01.004.

[5] E. Sozumert, F. Farukh, B. Sabuncuoglu, E. Demirci, M. Acar, B. Pourdeyhimi, V.V. Silberschmidt, Deformation and damage of random fibrous networks, Int. J. Solids Struct. (2018), https://doi.org/10.1016/j.ijsolstr.2018.12.012. S0020768318305018.

[6] C.T. Koh, M.L. Oyen, Branching toughens fibrous networks, J. Mech. Behav. Biomed. Mater. 12 (2012) 74–82, https://doi.org/10.1016/j.jmbbm.2012.03.011.

[7] C.T. Koh, K. Tonsomboon, M.L. Oyen, Fracture toughness of human amniotic membranes, Interface Focus 9 (2019) 20190012, https://doi.org/10.1098/rsfs.2019.0012.

[8] K. Bircher, M. Zündel, M. Pensalfini, A.E. Ehret, E. Mazza, Tear resistance of soft collagenous tissues, Nat. Commun. 10 (2019) 792, https://doi.org/10.1038/s41467-019-08723-y.

[9] Y. Zhang, Z. Lu, Z. Yang, D. Zhang, Fracture behavior of fibrous network materials: crack insensitivity and toughening mechanism, Int. J. Mech. Sci. 188 (2020), https://doi.org/10.1016/j.ijmecsci.2020.105910, 105910.

[10] E. Demirci, M. Acar, B. Pourdeyhimi, V.V. Silberschmidt, Computation of mechanical anisotropy in thermally bonded bicomponent fibre nonwovens, Comput. Mater. Sci. 52 (2012) 157–163, https://doi.org/10.1016/j.commatsci.2011.01.033.

[11] E. Demirci, M. Acar, B. Pourdeyhimi, V.V. Silberschmidt, Finite element modelling of thermally bonded bicomponent fibre nonwovens: tensile behaviour, Comput. Mater. Sci. 50 (2011) 1286–1291, https://doi.org/10.1016/j.commatsci.2010.02.039.

[12] H.L. Cox, The elasticity and strength of paper and other fibrous materials, Br. J. Appl. Phys. 3 (1952) 72–79, https://doi.org/10.1088/0508-3443/3/3/302.

[13] Y. Lee, I. Jasiuk, Apparent elastic properties of random fiber networks, Comput. Mater. Sci. 79 (2013) 715–723, https://doi.org/10.1016/j.commatsci.2013.07.037.

[14] Y. Chen, A. Ridruejo, C. González, J. Llorca, T. Siegmund, Notch effect in failure of fiberglass nonwoven materials, Int. J. Solids Struct. 96 (2016) 254–264, https://doi.org/10.1016/j.ijsolstr.2016.06.004.

[15] W. Khoo, S. Chung, S.C. Lim, C.Y. Low, J.M. Shapiro, C.T. Koh, Fracture behavior of multilayer fibrous scaffolds featuring microstructural gradients, Mater. Des. 184 (2019), https://doi.org/10.1016/j.matdes.2019.108184, 108184.

[16] X. Gao, E. Sözümert, Z. Shi, G. Yang, V.V. Silberschmidt, Mechanical modification of bacterial cellulose hydrogel under biaxial cyclic tension, Mech. Mater. 142 (2020), https://doi.org/10.1016/j.mechmat.2019.103272, 103272.
[17] W. Yang, V.R. Sherman, B. Gludovatz, E. Schaible, P. Stewart, R.O. Ritchie, M.A. Meyers, On the tear resistance of skin, Nat. Commun. 6 (2015) 6649, https://doi.org/10.1038/ncomms7649.
[18] C.T. Koh, M.L. Oyen, Toughening in electrospun fibrous scaffolds, APL Mater. 3 (2015), https://doi.org/10.1063/1.4901450, 014908.
[19] U. Stachewicz, I. Peker, W. Tu, A.H. Barber, Stress delocalisation in crack tolerant electrospun nanofiber networks, ACS Appl. Mater. Interfaces 3 (2011) 1991–1996, https://doi.org/10.1021/am2002444.
[20] D.H. Mueller, M. Kochmann, Numerical modeling of thermobonded nonwovens, Int. Nonwovens J. 13 (2004) 56–64, https://doi.org/10.1177/1558925004os-1300114.
[21] F. Farukh, E. Demirci, M. Acar, B. Pourdeyhimi, V.V. Silberschmidt, Meso-scale deformation and damage in thermally bonded nonwovens, J. Mater. Sci. 48 (2013) 2334–2345, https://doi.org/10.1007/s10853-012-7013-y.
[22] Q. Cheng, S. Wang, A method for testing the elastic modulus of single cellulose fibrils via atomic force microscopy, Compos. A: Appl. Sci. Manuf. 39 (2008) 1838–1843, https://doi.org/10.1016/j.compositesa.2008.09.007.
[23] X. Hou, M. Acar, V.V. Silberschmidt, Finite element simulation of low-density thermally bonded nonwoven materials: effects of orientation distribution function and arrangement of bond points, Comput. Mater. Sci. 50 (2011) 1292–1298, https://doi.org/10.1016/j.commatsci.2010.03.009.
[24] T. Belytschko, W.K. Liu, B. Moran, K. Elkhodary, Nonlinear Finite Elements for Continua and Structures, second ed., Wiley, 2014.
[25] X. Hou, M. Acar, V.V. Silberschmidt, Non-uniformity of deformation in low-density thermally point bonded nonwoven material: effect of microstructure, J. Mater. Sci. 46 (2011) 307–315, https://doi.org/10.1007/s10853-010-4800-1.
[26] A. Ridruejo, C. González, J. LLorca, Micromechanisms of deformation and fracture of polypropylene nonwoven fabrics, Int. J. Solids Struct. 48 (2011) 153–162, https://doi.org/10.1016/j.ijsolstr.2010.09.013.
[27] A. Rawal, S.K. Patel, V. Kumar, H. Saraswat, M.A. Sayeed, Damage analysis and notch sensitivity of hybrid needlepunched nonwoven materials, Text. Res. J. 83 (2013) 1103–1112, https://doi.org/10.1177/0040517512467063.
[28] P. Isaksson, R. Hägglund, Structural effects on deformation and fracture of random fiber networks and consequences on continuum models, Int. J. Solids Struct. 46 (2009) 2320–2329, https://doi.org/10.1016/j.ijsolstr.2009.01.027.

Time-dependent statistical failure of fibre networks: Distributions, size scaling, and effects of disorders

Tetsu Uesaka
Department of Chemical Engineering, Mid Sweden University, Sundsvall, Sweden

10.1 Introduction

Time-dependent failure is probably the most common mode of failures of many materials in end-use. It typically refers to creep and fatigue failures, in which cases a relatively low level of load is applied to the structural member for a prolonged period. However, other loading conditions, such as the monotonically increasing tensile/compression loading, also exhibit time-dependency of failure in a different timescale. Therefore, the investigation of time-dependent failure involves a material response to general loading histories (creep, fatigue, monotonic, random, impact loading, etc.). This further means that we need to consider failure as a 'process': The process starts with the initiation of a single damage, somewhere in the material, followed by the creation of more damage sites, the damage growth, the formation of damage clusters, and their coalescence, whilst interacting with complex material structures [1]. This process ends with an abrupt growth of one of the (largest) damage clusters (avalanche-like failure). Inspecting each sub-process, we can easily recognise that this process is highly stochastic: First, the underlying molecular process (thermal motions) is stochastic, and second, the structure is also disordered in different length scales.

The manifestation of such stochastic nature of time-dependent failure may be seen, for example, in creep failures. Under a constant load, the material fails at a certain time, which is called lifetime. It is known that the lifetime exhibits enormous variability, e.g. for industrial fibrous materials, a coefficient of variation (standard deviation/mean) of 100% or even more [2]. The same is true under cyclic loading conditions (fatigue). This contrasts with normal strength (short-term strength), which typically shows the variation of only 4%–10% [3,4] (see also Fig. 10.1.)

Because of the highly skewed distribution and its extremely long tail in the longer lifetime range, the mean of the lifetime loses its usual practical significance. Although the stochastic aspect of creep lifetime and fatigue lifetime has been recognised for many years in the field, the majority of works still focus on the mean strength and the mean lifetime, and enormous uncertainties associated with the lifetime have been largely ignored. A basic question is what causes such large uncertainty of the lifetime

Mechanics of Fibrous Networks. https://doi.org/10.1016/B978-0-12-822207-2.00004-0

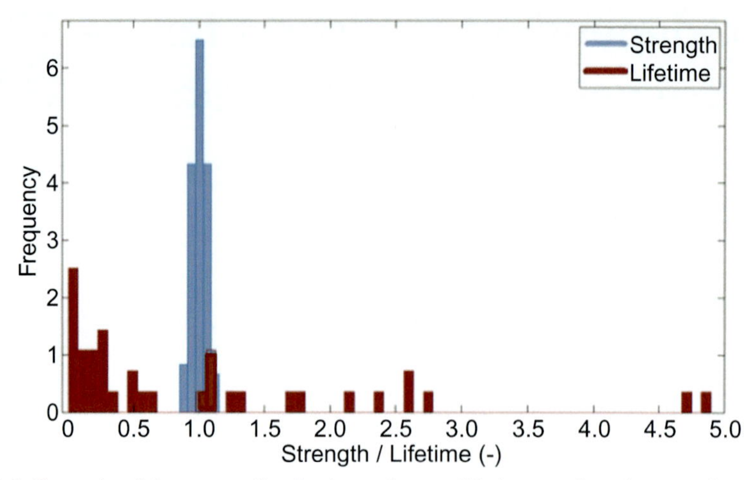

Fig. 10.1 Example of frequency distributions of creep lifetimes and static strengths of paperboard. The x-axis is normalised by the means of lifetime and strength, respectively.

and how to reduce the uncertainty in both materials and structures since the uncertainty is a major stumbling block in the selection and design of materials for structural members.

In this chapter, we discuss the basic mechanism of time-dependent statistical failures (TDSF) of a fibre network and the origin of the uncertainty of a lifetime from a statistical mechanics viewpoint. The fibre network is a ubiquitous structure widely seen in industrial materials (paper, nonwoven, and fibre-based composites), biological tissues (collagen and cytoskeleton), and polymers (rubber and hydrogels). Therefore, we consider here the fibre network as a paradigm of those different classes of materials, and try to extract general principles underlying time-dependent failure phenomena.

For this purpose, we begin with discussing a general formulation of TDSF of a 'single fibre', based on the seminal work by Coleman [5]. This formulation provides important insights into understanding the behaviour of a fibre network. We then consider, theoretically, some of the relations which transcend over the hierarchical structure (from a single fibre to a network), and define three material parameters that characterise TDSF. Based on this theoretical consideration, we use Monte-Carlo simulations to investigate the damage evolution leading to avalanche failures, the size-scaling of the lifetime distribution, particularly the validity of the weakest-link scaling (WLS), and the asymptotic lifetime distribution. The numerical study is then applied to establish the relationship between fibre properties and network properties; specifically, we discuss how the brittleness of a single fibre influences the short-term strength, the failure mode, and the lifetime distribution. We also discuss how different types of disorders affect the lifetime distribution. In the last section, we discuss the experimental characterisation, including a method to determine the material parameters and a comparison of the material parameters for various fibre-based materials.

10.2 Formulation of time-dependent statistical failures of a single fibre

We consider a single fibre subjected to a general loading history $f(t)$, such as shown in Fig. 10.2.

The problem is how to obtain the distribution of the time to failure (lifetime) t_B with the information on $f(t)$. Coleman is probably the first one who has constructed a phenomenological but axiomatic theory for this problem [5,6]. Later, his approach has been utilised and extended by various researchers, e.g. Christensen [7], Curtin et al. [8], and Phoenix and Beyerlein [9]. His formulation is based on three postulates: (1) the WLS (weakest-link scaling), (2) a breakdown rule (the damage evolution), and (3) a probabilistic failure criterion.

First, consider a single fibre of length l, consisting of M imaginary fibre elements whose length is l_0 ($M = l/l_0$). The WLS is expressed by the following equation:

$$1 - F_l(t) = [1 - F_0(t)]^M, \tag{10.1}$$

where $F_l(t)$ is the probability that the single fibre fails at less than time t, i.e. the cumulative distribution function of lifetimes of the single fibre, and $F_0(t)$ is the corresponding probability for the imaginary fibre elements. This equation is simply an expression of the fact that the failure of the fibre is controlled by the 'weakest element' of the fibre so that the probability that the entire fibre fails at less than time t is a combined probability event that all M elements fail at less than time t. An underlying assumption is that fibre element failures are a statistically independent event controlled by the same distribution function $F_0(t)$. This implies that the size of the 'element' must be sufficiently large to contain a typical damage cluster that triggers the entire failure. Usually, this is not a problem for a one-dimensional (1D) element (i.e. fibre), but, for a 2D- or 3D-elements, there must be a lower bound to the size of the element. The latter is the question for a fibre network, as will be discussed later.

The second postulate, a breakdown rule, states that the growth rate of the damage $\Omega(t)$ in the element is a function of a force $f(t)$. One of the frequently used forms is a power-law form:

$$\frac{d\Omega(t)}{dt} = \kappa(f(t)) = c \cdot f(t)^{\rho} \tag{10.2}$$

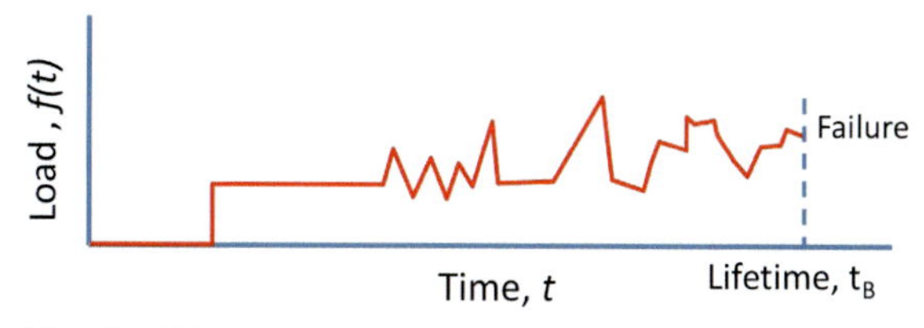

Fig. 10.2 General loading history.

That is, the damage growth rate at time t increases in proportion to the ρ-th power of the force $f(t)$. Another form often used in the polymer science community is the exponential form of the damage evolution [5]. It was sometimes claimed that, as the exponential form is based on the absolute reaction rate theory, it is a nonempirical and a better form than the power-law form. However, Phoenix and Tierney [10] counter-argued that the power-law form is, in fact, a better approximation of typical thermal-activation-energy functions. They have provided a molecular-interpretation of the exponent ρ that appears in Eq. (10.2):

$$\rho = \frac{U_0}{kT}, \tag{10.3}$$

where U_0 is a fitting parameter of the potential barrier function, k is the Boltzmann constant, and T is the absolute temperature. As seen in Eqs (10.2) and (10.3), the parameter ρ represents important characteristics of material failures. First, it is the sensitivity of the damage growth to the force [Eq. (10.2)]. Second, in Eq. (10.3), ρ is the ratio of the potential energy contribution to the kinetic energy contribution when a damage grows. If the kinetic energy component decreases (e.g. due to the temperature decrease), ρ goes up, meaning that the material fails in a more brittle way (less kinetic energy dissipation). Such material is, at the same time, more durable, like ceramics, as compared with thermoplastic polymers. It should be noted that Coleman's original breaking rule involves not only the dependence on the force $f(t)$, but also on the damage state $\Omega(t)$. The most notable of such form is the damage evolution rule used by Kachanov [11]. It was shown [2] that the dependence on the damage state is the second-order effect, and for small damage states $\Omega(t)$, i.e. for brittle and quasi-brittle failures, the form of Eq. (10.2) is recovered.

The third postulate is the probabilistic failure criterion. We can generally assume that the probability that the element fails at a lifetime less than t depends on how much damage is accumulated by that time, i.e. $\Omega(t)$, and it increases with $\Omega(t)$.

$$F_0(t) = \Psi(\Omega(t)) \tag{10.4}$$

According to Eq. (10.1), $F_l(t)$ depends on only small values of $F_0(t)$ for a large M. It is, therefore, sufficient to approximate $F_0(t)$ for a small value of Ω. By assuming the smoothness of the function $F_0(t)$ up to $(\beta - 1)$-th order at $\Omega = 0$, we have

$$\lim_{\Omega \to 0} \Psi(\Omega(t)) = C \cdot \Omega^{\beta} + o(\Omega^{\beta}) \tag{10.5}$$

Using these three postulates and taking the limiting form for $M \to \infty$, we find the cumulative distribution function of lifetimes of the single fibre:

$$\lim_{M \to \infty} F_l(t) = 1 - \exp\{-MC\Omega^{\beta}\} = 1 - \exp\left\{-A\left[\int_{s=0}^{t} f(s)^{\rho} ds\right]^{\beta}\right\} \tag{10.6}$$

where A absorbs all constants ($A = MCc^{\beta}$). For later use, we recast Eq. (10.6) as

$$F_f(t) = 1 - \exp\left\{-A\left[\int_{s=0}^{t} f(s)^{\rho_f} ds\right]^{\beta_f}\right\}$$
$$= 1 - \exp\left\{-\left[\int_{u=0}^{\frac{t}{t_0}} \left(\frac{f(s)}{T_c}\right)^{\rho_f} du\right]^{\beta_f}\right\}, \tag{10.7}$$

where $A = t_0^{-\beta} T_c^{-\rho\beta}$, t_0 is an appropriately chosen unit time, and T_c, ρ_f, and β_f are the material constants that characterise TDSF of the single fibre. With the knowledge of these three material constants, and by specifying a loading history $f(s)$, we can completely determine the distribution function of lifetimes through Eq. (10.7). We discuss these material constants to a greater extent in later sections.

In the case of creep, $f(s) = f_c$, Eq. (10.7) becomes

$$F_f(t) = 1 - \exp\left\{-\left(\frac{f_c}{T_c}\right)^{\rho_f} t^{\beta_f}\right\}. \tag{10.8}$$

The creep lifetime distribution for a single fibre is, in fact, the Weibull distribution. (Note that 't' is used here for denoting the non-dimensional time.) In the case of a constant-loading-rate test, Eq. (10.7) predicts that the strength also follows Weibull distribution. (However, for a general loading history, the distribution is, of course, not necessarily Weibull.)

10.3 Formulation of time-dependent statistical failure of fibre network – Theoretical consideration

We now consider a network of fibres whose cumulative distribution function of lifetimes is given by Eq. (10.6). This means that we consider a central-force fibre network. Rewriting the equation for the i-th fibre in the network, we have

$$F_i(t) = 1 - \exp\left\{-A_i\left[\int_{s=0}^{t} f_i(s)^{\rho_f} ds\right]^{\beta_f}\right\}, \tag{10.9}$$

where A_i depends on the specific fibre i, as its length (the node-to-node distance) varies from one fibre to another for a random fibre network. The force f_i, of course, also varies depending on the position and the orientation of the i-th fibre in the network. Here, the material parameters ρ_f and β_f are assumed to be the same for all fibres.

Taking the non-dimensional time $\hat{t} = s/t$, Eq. (10.9) is rewritten as

$$F_i(t) = 1 - \exp\left\{-A_i t^{\beta_f}\left[\int_{\hat{t}=0}^{1} f_i(\hat{t})^{\rho_f} d\hat{t}\right]^{\beta_f}\right\}. \tag{10.10}$$

In the case of a linear elastic system, the local force f_i developed in the i-th fibre is a linear transformation of a global force (the force applied to the network). Therefore, by expressing $f_i(t) = K_i(t)S(T)$, where $S(t)$ the force applied to the network and $K_i(t)$ a force concentration function, we can further rewrite the above equation as

$$F_i(t) = 1 - \exp\left\{-A_i t^{\beta_f}\left[\int_{\hat{t}=0}^{1} S(\hat{t})^{\rho_f} K_i(\hat{t})d\hat{t}\right]^{\beta_f}\right\}. \quad (10.11)$$

Here we find two variables, t^{β_f} and S^{ρ_f}, that are not dependent on the specific fibre 'i'. These are called global variables, and these variables should appear in the expression of the distribution function of the fibre network. This provides an important constraint to the functional form of the distribution function of the fibre network.

There are many possible forms, but one of the forms which is frequently used for the distribution function of the network is the same form as the one used for a single fibre [i.e. Eq. (10.6)]. Denoting the quantities for the network with the subscript 's' and rewriting the equation in the same way as Eq. (10.7), we have

$$F_s(t) = 1 - \exp\left\{-A_s t^{\beta_s}\left[\int_{\hat{t}=0}^{1} S(\hat{t})^{\rho_s} d\hat{t}\right]^{\beta_s}\right\} \quad (10.12)$$

Comparing Eqs (10.11) and (10.12) and applying the constraint, we find the following relations between the single fibre and the network for the parameters β and ρ:

$$\beta_s = a\beta_f, \quad (10.13)$$

and

$$\rho_s = b\rho_f, \quad (10.14)$$

where a and b are functions of network structures, and potentially other fibre properties except its own (i.e. a is not a function of β_f, and b is not a function of ρ_f). These equations provide an important link between the fibre properties (β_f and ρ_f) and network properties (β_s and ρ_s). In the next section, we will examine whether this is the case or not.

For the later use, we, again, express Eq. (10.12) in the same way as the case of the single fibre [Eq. (10.7)]:

$$F_s(t) = 1 - \exp\left\{-\left[\int_{s=0}^{t}\left(\frac{S(s)}{S_c}\right)^{\rho_s} ds\right]^{\beta_s}\right\}, \quad (10.15)$$

where S_c is now called 'characteristic strength' of the fibre network, which is the force that causes a 63% (=1 − 1/e) of the samples to fail within the unit time. The 't' is understood as a non-dimensional time.

The applicability of Eq. (10.15) may be examined by performing tests under a specific loading history. The most common test is the creep test, where $S(t) = S_0U(t)$ with S_0 the applied load, and $U(T)$ the Heaviside step function. Expressing Eq. (10.15) in the Weibull format, we obtain

$$\ln\left(\ln(1 - F_s(t))^{-1} \right) = \rho_s\beta_s \ln(S_c) + \beta_s \ln(t) + \rho_s\beta_s \ln(S_0) \tag{10.16}$$

Plotting the left-hand side against $\ln(t)$ [or against $\ln(S_0)$] should give a linear relation, and thus this procedure is used for experimental verifications of Eq. (10.15). Eq. (10.16) is also useful for determining the parameters S_c, β_s, and ρ_s. In fact, the relation Eq. (10.16) has been widely observed in experimental studies of fibre bundles and uniaxial fibre-reinforced composites [12,13], and of paperboards (fibre networks) [14]. We discuss more details of the applicability of Eq. (10.15) in the subsequent section.

10.4 Monte-Carlo simulations of creep failures of fibre network

10.4.1 Model description

Determining, experimentally, an exact form of distribution functions is difficult, and examining size-scaling laws is even more difficult. This is because the number of tests performed in a usual experiment is limited to, at most, 100, and size-scaling experiments require changing not only test configurations but also test machines. In this sense, numerical experiments have some advantages, though a range of difficulties still exists. For example, determining a low tail of the distribution function (e.g. up to 10^{-3} or even 10^{-6}) requires thousands or millions of statistical representations of failure events. At the same time, it requires monitoring the damage evolution process for each failure. These tasks easily amount to a very unrealistic computational cost, even if it is feasible.

Here, we introduce a simple central-force, triangular lattice network, shown in Fig. 10.3A. This model is much simplified from the full fibre network models, e.g., Bergström et al. [15] and Kulachenko and Uesaka [16], but it still retains the essential network mechanics (long-range correlation) and rich statistical mechanics.

A single fibre in the network is assumed to fail according to Coleman's relation, Eq. (10.8), where we assume a special case $\beta_f = 1$ (i.e. a memory-less or Markov process) [17]. To simulate a uniaxial creep test, a constant tensile force is applied at the top and bottom boundaries, and traction-free boundaries are assumed for the sides. The size of the network can be changed either by the number of stacks m (length) or the number of fibres in the horizontal direction n (width). A periodic boundary condition was not used to avoid the introduction of any artificial length scale.

The simulation proceeds as follows: We first apply a dead load at the top and bottom boundaries, and calculate the forces of individual fibres. Then, we generate random numbers (F_f) between 0 and 1 for each fibre, and determine the lifetime values of

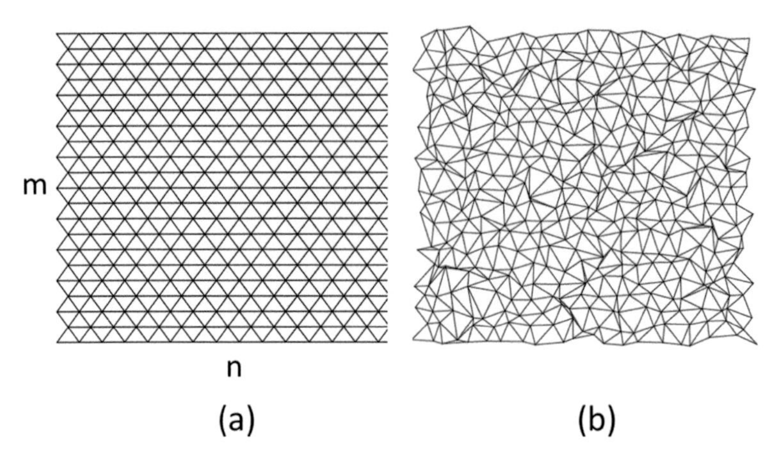

Fig. 10.3 Triangular lattice networks. (A) Regular network and (B) disordered network. The size parameters m and n denote the number of stacks in the vertical direction and the number of fibres in the horizontal direction, respectively.

the individual fibres by solving Eq. (10.8) for t. We choose the fibre with the shortest lifetime, and set the modulus of the chosen fibre to zero (or a small number), and recalculate the new state of mechanical equilibrium of the system. Since new loading histories are now given to the surviving fibres, the lifetime values are updated according to Eq. (10.8) [17]. From this new set of lifetime values of the surviving fibres, we again choose the fibre with the shortest lifetime as the second fibre to break. This process is repeated until an avalanche failure takes place.

With this model, fibre properties can be changed, both locally and globally, for example, elastic stiffness of a single fibre, its threshold strength, and the parameters, ρ_f. In addition, we can create a disordered fibre network, such as shown in Fig. 10.3B. This network is still a triangular lattice, but the position of the node is displaced according to a specified statistical distribution.

10.4.2 Damage evolution

The evolution of damage in a single fibre is controlled by Eq. (10.2), which can be rewritten in terms of the newly introduced parameters as

$$\frac{d\Omega(t)}{dt} = \left(\frac{f(t)}{T_c}\right)^{\rho_f} \tag{10.17}$$

where $T_c = (ct_0)^{-1/\rho}$, the characteristic strength of the fibre, and t here is again understood as non-dimensional time. As soon as the local force $f(t)$ exceeds T_c, the damage grows rapidly, especially for a high value of ρ_f. Therefore, the parameter ρ_f indicates how quickly the damage grows around the threshold strength T_c. Fig. 10.4 shows an example of the state of damage just before the avalanche failure of the fibre network [2].

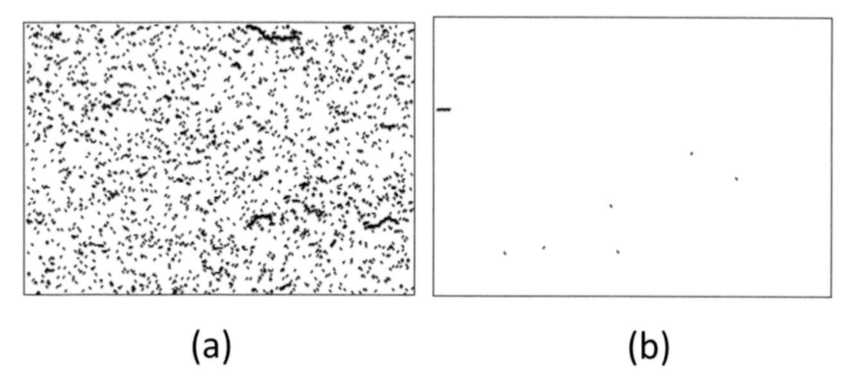

Fig. 10.4 Damage evolution in the fibre networks made from different brittleness fibres: (A) $\rho_f = 5$ (less brittle fibres), and (B) $\rho_f = 20$ (more brittle fibres).

In the case of $\rho_f = 5$, a large number of damage sites are created prior to the avalanche failure (i.e. extensive energy dissipation before the final failure). The system failure exhibits a more ductile feature. However, in the case of $\rho_f = 20$, only several damage sites are created prior to the avalanche failure. The failure started, in a very brittle manner, from the largest damage cluster located at the left upper edge. This example shows that the parameter ρ_f represents brittleness of the fibre: the higher the value of ρ_f, the more the fibre is brittle. This conforms to the molecular interpretation of ρ [see Eq. (10.3)] by Phoenix and Tierney [10]. Note that higher brittleness is not necessarily an unfavourable property. For example, Eq. (10.8) can be solved for median lifetime t_m [i.e. $F_f(t_m) = 1/2$], yielding

$$t_m = \left(\ln 2 \cdot \frac{T_c}{f_c} \right)^{\rho_f/\beta_f} \tag{10.18}$$

At a given threshold strength T_c, creep load f_c, and the parameter β_f, materials with higher brittleness ρ_f have longer median lifetimes t_m, i.e. more 'durable' [14,18]. (Note that, in creep tests, the 'median' lifetime is more easily and accurately obtained than the 'mean' lifetime, because of the typical skewed distribution of lifetimes and the presence of extremely long tail of the lifetime.)

10.4.3 Size scaling and asymptotic distribution

We first examine how the creep lifetime distribution function depends on the size of the fibre network. For this purpose, we change the size N (the total number of fibres) both in the length and width directions. The brittleness parameter ρ_f is set to brittle ($\rho_f = 50$) and quasi-brittle ($\rho_f = 10$) cases. In Fig. 10.5, the lifetime distribution functions with different sizes are plotted in a special format, the so-called Weibull format. In this format, if the size dependence follows the weakest-link scaling [Eq. (10.1)], all curves should form a single master curve. Besides, if the distribution is Weibull distribution, then each plot should be linear.

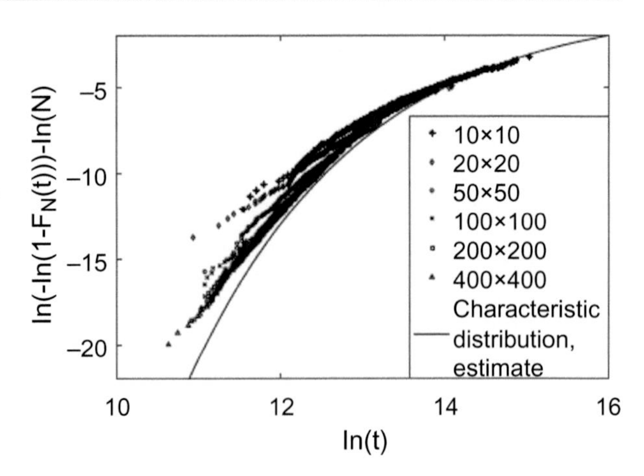

Fig. 10.5 Size dependence of the distribution function of creep lifetimes. The distribution functions are plotted in a Weibull format so that if the size dependence follows the weakest-link scaling, all curves form a single master curve. The *solid curve* represents the characteristic distribution function estimated, as described in the main text.

First, the curves corresponding to different sizes of the networks tend to overlap with increasing the size, but a single master curve has not been obtained yet with current system sizes. Each distribution curve was approximately linear, at least within the tested probability range. Generally, as the size increases, each curve tends to increase its slope and tends to shift downwards. To quantify the size dependence, we fitted each distribution curve with Eq. (10.16), and determined the material parameters, S_c, ρ_s, and β_s. Fig. 10.6 shows the results.

As expected, the characteristic strength of the network S_c decreases with size, the brittleness parameter ρ_s remains constant, and the parameter β_s increases with size. Each parameter has a unique functional relation with the size N. For example,

$$\ln\left(\frac{S_c}{T_c}\right) = a_s \ln[\ln(N)] + b_s, \tag{10.19}$$

$$\rho_s = 9.9942 \approx \rho_f = 10, \text{and} \tag{10.20}$$

$$\beta_s = a_b \ln(N) + b_b, \tag{10.21}$$

where $a_s = -0.2116$, $b_s = 0.5261$, $a_b = 0.4536$, and $b_b = -0.1205$.

An interesting question is whether the [WLS, Eq. (10.1)] emerges or not as the system size further increases, and if so, what kind of function is asymptotically approached. Denote the lifetime distribution function for the system size N as $F_N(t)$; then the WLS is stated:

$$\lim_{N\to\infty} [1 - F_N(t)]^{1/N} = 1 - W(t), \tag{10.22}$$

where $W(t)$ is the function that is asymptotically approached, called the characteristic distribution function. This is equivalent to the distribution function of an imaginary

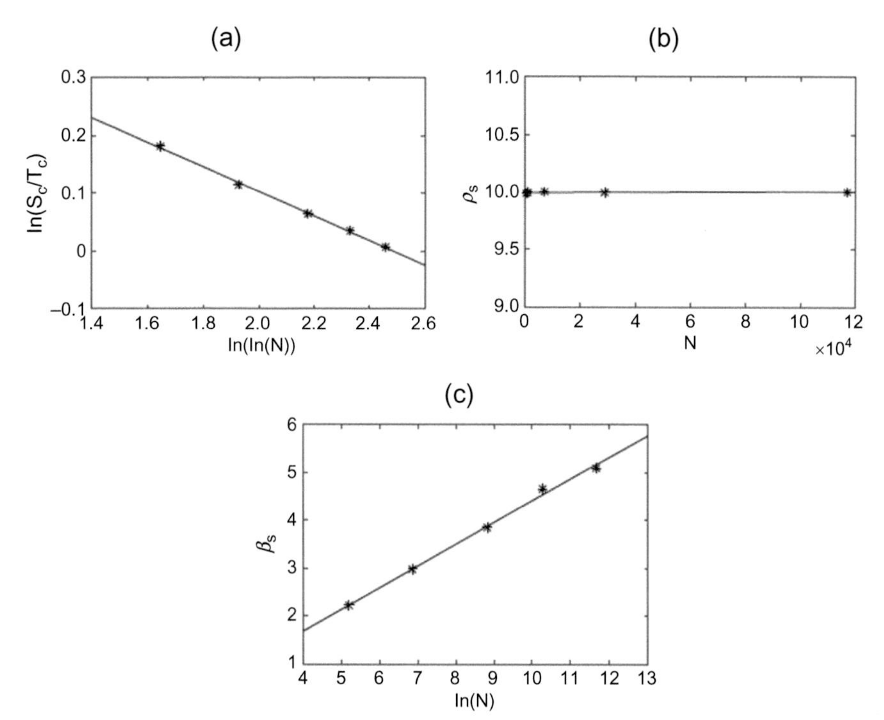

Fig. 10.6 The dependence of the material parameters, S_c, ρ_s, and β_s on the system size N the total number of fibres.

element in Eq. (10.1). Since the left-hand side of the above equation approaches $1 - W(t)$ very slowly, we use the following condition to express the WLS:

$$[1-F_N(t)]^{1/N} = [1-F_{N+1}(t)]^{1/(N+1)}, \tag{10.23}$$

as $N \to \infty$. Using Eqs (10.19) through (10.21) and Eq. (10.15), we can express $F_N(t)$ as a function of N. Then, solving Eq. (10.23) for N and inserting it into the expression of $F_N(t)$, we find $W(t)$, as

$$W(t) \approx \exp\left\{-pt^{-\frac{1}{q}} + d\right\}, \tag{10.24}$$

where $p = 3.72 \times 10^3$, $q = 2.12$, and $d = -0.0099$. Therefore, the distribution function for a large system when the WLS emerges is given by Mattsson and Uesaka [17] as

$$F_N(t) = 1 - \exp\left\{-rN\exp\left\{-pt^{-\frac{1}{q}} + d\right\}\right\}, \tag{10.25}$$

where $r = \exp\{d\} \cong 1$. As also seen in Fig. 10.5, this distribution is clearly not Weibull distribution, though the distribution functions for smaller size systems look, approximately, Weibull distribution within the limited ranges. The distribution of the form Eq. (10.25) was first found by Duxbury, Leath, and Bearle for static strength by using simple random fuse models [19], later for fibre bundle models [20,21]. This is called DLB-type distribution. Now we have just seen the same distribution also for creep lifetimes of the fibre network. In the earlier studies, Duxbury and coworkers assumed various pre-existing disorders, such as percolation disorders, threshold strength distributions, and crack length distributions, for the purpose of generating strength distributions. However, Eq. (10.25) was obtained by introducing only the underlying stochastic process of fibre (or bond) failures, without any pre-defined disorders in the network.

An interesting question may arise: If the DLB-type distribution is the consequence of the weakest-link theorem for a large system, why should it not be one of the extreme-value distributions, i.e. Weibull, Gumbel, or Fréchet distributions, which have been discussed for decades? The recent literature showed that the DLB distribution does converge to the Gumbel form, instead of the Weibull form [22]. However, the rate at which the distribution converges to the limiting distribution is extremely slow. This extreme slowness makes the resulted extreme distribution much less practical. The reason for the extreme slowness may be found in one of the postulates of the weakest-link theorem: The system is divided into the elements with "independent and identically distributed random variables." Such a system may not universally exist, particularly in a disordered medium with long-range elastic interactions [22]. The ubiquitous nature of Weibull distribution, therefore, may not be due to the natural consequence of an extreme-value distribution, but rather a result of a specific selection of the defect size distribution (an algebraic form), as discussed by Curtin and Scher [23]. As seen in Fig. 10.5, Weibull distribution is not accurate in both upper and lower tails [24]. However, from an engineering viewpoint, it is considered as a conservative estimate of the low-tail distribution.

The above discussions concern brittle or quasi-brittle cases when ρ_f is large (e.g. $\rho_f \geq 5$). How about cases where fibres are more ductile (or subjected to a higher temperature or higher humidity)? As we will see in the next section, with decreasing the parameter ρ_f, the parameter β_s of the fibre network consistently increases. Note that the parameter β_s is a Weibull shaped parameter, or often called the Weibull modulus for creep failure. In other words, the higher the β_s, the more symmetric and more uniform the distribution is. This means that the lifetime distribution of the fibre network made from more ductile fibres becomes more symmetric and narrower, approaching to a Gaussian distribution. Indeed, analytical solutions for a fibre bundle model showed the Gaussian distribution [25]. For the fibre bundle with 'equal load sharing', i.e. the load is shared equally by neighbouring fibres next to a broken fibre and with $\rho_f = 1$, the distribution asymptotically follows a Gaussian distribution with $N \to \infty$, and its mean $Ex\{t\}$ and variance $Var\{t\}$ are

$$Ex\{t\} \cong \frac{1}{L}, \text{and} \tag{10.26}$$

$$Var\{t\} \cong \frac{1}{NL^2}, \tag{10.27}$$

where N is the number of fibres in the bundle, and L is the (nondimensional) load per fibre [25]. It is interesting to note that, in the ductile limit, the mean lifetime does not depend on the size of the system, and the distribution becomes narrower and narrower with increasing the system size N. That is failure events become more deterministic as the system size grows. Qualitatively, these features are seen also in the DLB distribution for the brittle- and quasi-brittle cases. More importantly, these results are distinct from what Weibull statistics predict.

10.4.4 Effects of the brittleness parameter of a single fibre ρ_f on the network properties

As already seen in the previous section, the brittleness parameter ρ_f of a single fibre influences many of the statistical failure properties of a fibre network. Here, we look at impacts on the three network properties, the characteristic strength S_c, the brittleness ρ_s, and the Weibull modulus β_s of the network, as presented in Fig. 10.7.

First, the characteristic strength increases quickly with increasing fibre brittleness and then plateaus in the high brittleness range. This trend may be understood with Eq. (10.17) and Fig. 10.4. For ductile or less brittle fibres, damages start appearing at forces lower than T_c, and by the time the network reaches the avalanche failure stage, a large number of failure sites are accumulated (Fig. 10.4A). This means that

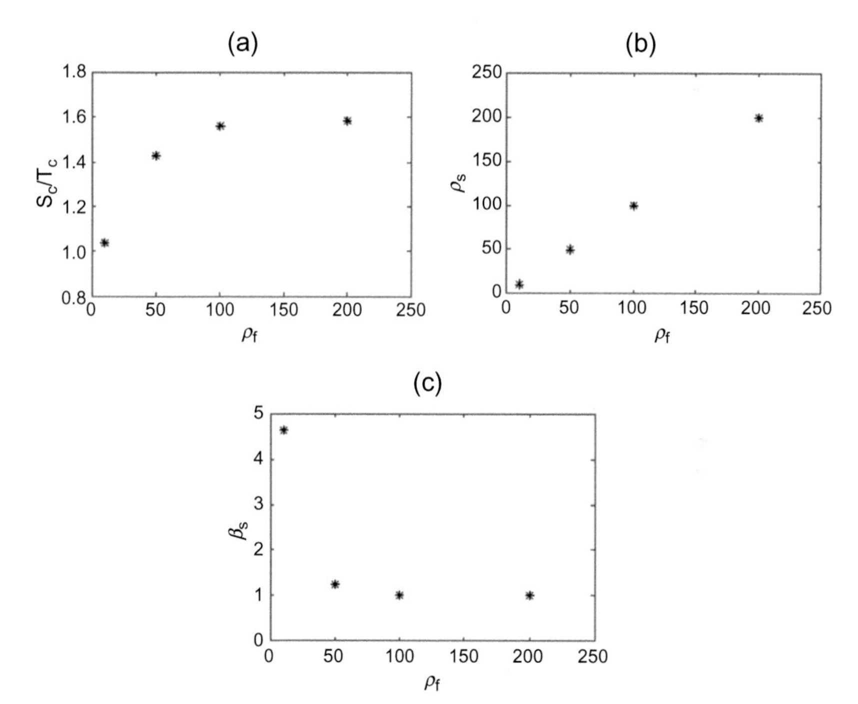

Fig. 10.7 Effects of the brittleness parameter of a single fibre ρ_f on the material parameters of the network.

the load-carrying capacity is significantly reduced at the time of the network failure. This results in a lower threshold strength S_c. With increasing fibre brittleness, fewer and fewer failure sites are created before the avalanche failure (Fig. 10.4B), and thus the load-carrying capacity is better maintained to the end. Therefore, with increasing fibre brittleness, the characteristic strength of the network approaches the strength of the intact network.

Second, the brittleness parameter of the network ρ_s is entirely controlled by fibre brittleness ρ_f, and $\rho_s = \rho_f$ [see Fig. 10.7B.] In other words, in Eq. (10.14), $b = 1$. As shown in Fig. 10.6B and also in the later section, the brittleness of the network is determined by the fibre brittleness, and not by its structures. In the case of fibre-reinforced polymer composites, however, the brittleness of the matrix component indeed affects the brittleness of the composites [26].

Third, the Weibull modulus of the network β_s decreases sharply with increasing fibre brittleness ρ_f, and then approaches unity, which is the assumed value of Weibull modulus of the single fibre (Fig. 10.7). This is because, for brittle failures, only a few failure sites are created before the system failure, and the lifetime distribution of the network is effectively controlled by the distribution of the lifetimes of the first fibre failure. This is because the second, third and subsequent failures occur almost immediately. The trend, "the higher the brittleness, the lower the Weibull modulus (the lower the reliability)," is widely seen in many fibrous materials. We see examples in the last section of this chapter.

10.4.5 Impacts of structure disorders on the network properties

We have seen so far responses of a regular lattice network (triangular lattice) without any disorders. We now investigate the impacts of structural disorders, such as shown in Fig. 10.3B, on the three network properties. In this example, the position of each node is displaced by an amount controlled by a Gaussian random variable of zero mean and a specified standard deviation (SD). Each time such a network is created (realised), creep failure simulations are performed to determine the network properties, S_c, ρ_s, and β_s (Fig. 10.8).

With increasing the degree of the disorder (SD), the characteristic strength S_c decreased, as may be expected. Interestingly, the network brittleness ρ_s remains constant with $\rho_s = \rho_f$, that is, the network brittleness is independent of the structural disorder. The Weibull modulus of the network β_s decreases with increasing the SD, and, most importantly, β_s becomes even less than unity, which is the value of the single fibre Weibull modulus β_f. For fibre bundle models and regular lattice networks, it was shown that $\beta_s \geq \beta_f$, i.e. $a \geq 1$ in Eq. (10.13). In other words, the load-sharing network structures make the network more reliable (a higher Weibull modulus) in creep failure performance than the single fibre counterpart. However, the introduction of the structural disorder reverses the effect. It should be noted that, experimentally, the values of β_s less than unity are ubiquitously seen for fibres and fibre-reinforced polymer composites. (We will discuss it in the next section.)

Fig. 10.9 compares the effects of different types of disorders on the Weibull modulus of a creep lifetime distribution. The stiffness disorders are created by randomly

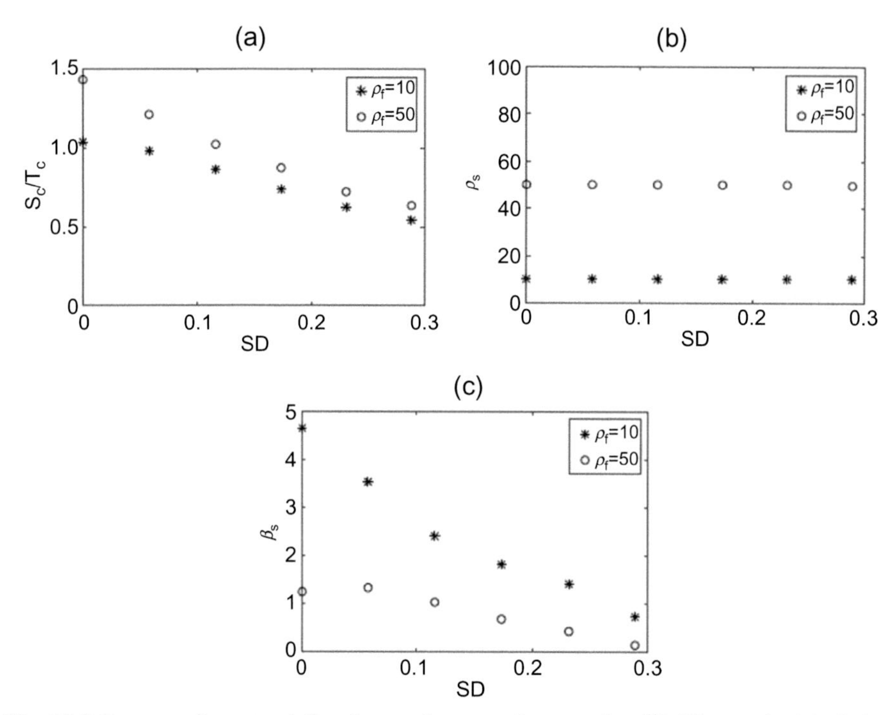

Fig. 10.8 Impacts of structural disorders on the network properties. SD: The standard deviation of the node displacement divided by the original fibre length.

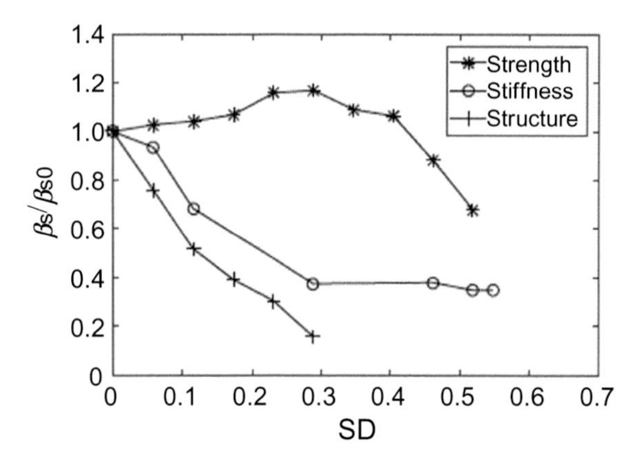

Fig. 10.9 Effects of different types of disorder on the Weibull modulus of a creep lifetime distribution. SD: The standard deviation of the corresponding disorder. β_{s0}: The Weibull modulus when there are no disorders in the structure, fibre stiffness, and fibre characteristic strength.

varying the cross-sectional area of the truss (fibre) element, and the strength disorders are created by varying T_c.

Interestingly, the small strength disorders initially have an enhancing effect of reliability (a higher Weibull modulus), whereas both stiffness and structural disorders quickly deteriorate the reliability. Since both stiffness and structural disorders cause stress disorders, the result indicates that the system's reliability β_s may be closely related to the stress uniformity within the structure.

10.5 Experimental determination of the material parameters

As shown in Eq. (10.15), to characterise the TDSF of fibrous materials subjected to various loading histories, three material-specific parameters are required: S_c the characteristic strength, which is numerically very close to the usual mean strength of the network, ρ_s the brittleness parameter, and β_s the Weibull modulus for a creep lifetime distribution (a measure of reliability). It is important to re-emphasise that once these three parameters are determined, one can find the distribution function of lifetimes for any loading histories.

Experimentally, these parameters can be determined by two methods: One uses a series of creep failure tests under different loads, and the data are non-linear fitted with Eq. (10.16). (This method was used also in numerical creep tests, as discussed in the previous section.) The other method uses the tensile/compression/shear test at a constant-loading rate to determine a strength distribution. The loading rate is then varied to allow the determination of the parameters by the following equation.

$$\log(-\log(1-G(f_B))) = \log(f_B)\beta_s(\rho_s+1) - \log(S_c)\beta_s\rho_s - \log(\rho_s+1)\beta_s - \log(\alpha)\beta_s, \tag{10.28}$$

where f_B is the strength and α is the loading rate. Typical data of a creep lifetime distribution and a strength distribution are shown in Figs 10.10 and 10.11, respectively [27]. The samples are commercially made containerboards (flutes, basis weight 140 g/m^2), the specimens were cut into a size of 25 mm $\times$ 105 mm, and testing was done in a compression mode for both cases. Although these tests are equivalent in terms of numerical data treatments, practical aspects of testing are different. First, creep tests inherently take time, because of the long-tails of the lifetime distribution (particularly, the longer end of the lifetime distribution), as seen in Fig. 10.10. This means that the tests require strict control and maintenance of environmental conditions and testing electronics continuously for a prolonged period (e.g. a few months or more). This also means that there are greater risks of disruptions and interruptions of data logging. It further affects the number of samples for the determination of the distribution function. For constant-rate loading tests, however, as long as the testing machine provides accurate load control in a wide range of loading rate, which most of the modern machines do, the total testing time is reasonable (a few days to several days), and there are much fewer disturbances and uncertainties.

Typical errors associated with the determination of the material parameters are, 0.03% for the characteristic strength, and 2%–6% for ρ_s and β_s. However, it was shown that the method is sensitive to any slightly biassed sampling of test specimens so that a strict random sampling procedure must be designed and executed.

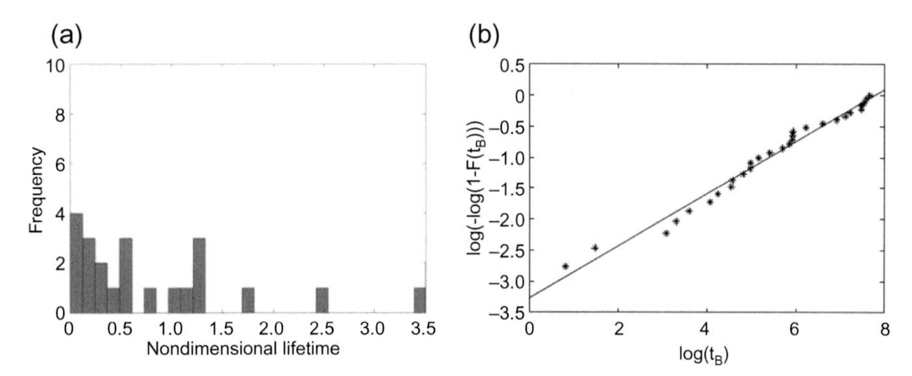

Fig. 10.10 Frequency diagram of creep lifetimes and the corresponding Weibull plot of the lifetime distribution.

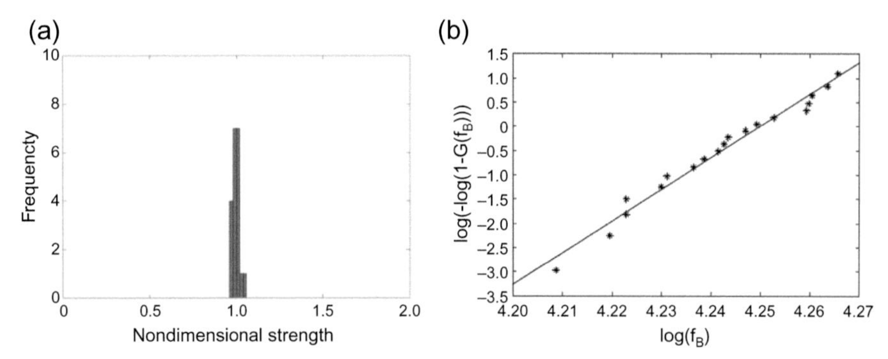

Fig. 10.11 Frequency diagram of strengths under constant loading rate, and the corresponding Weibull plot of the strength distribution.

As found in Figs 10.10 and 10.11, as well as in Fig. 10.1, there are dramatic differences between creep lifetime distributions and strength distributions: Creep lifetime data normally scatter far more than strength data. Such a difference is predicted by Eq. (10.15), and it is closely related to the brittleness of the system (fibre). The relationship between the Weibull modulus for a creep lifetime distribution, β_s and the Weibull modulus for a strength distribution, m_s, is given by

$$m_s = (\rho_s + 1)\beta_s. \tag{10.29}$$

Since most of the brittle or quasi-brittle materials show ρ_s in the range of 20–60 (or more), the strength distribution is much narrower (a higher Weibull modulus) than that for the lifetime.

Fig. 10.12 shows comparisons of the Weibull modulus and the brittleness parameter for carbon fibre, glass fibre, fibre-reinforced composites and fibre networks (paperboards). First, there is a general relationship between the Weibull shape

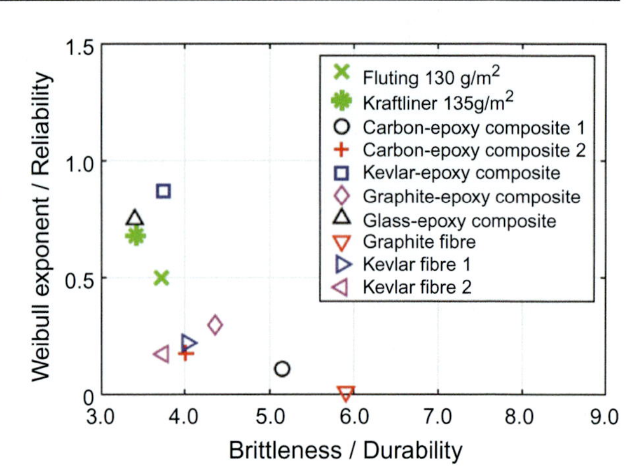

Fig. 10.12 Comparisons of the Weibull modulus and brittleness parameter for different fibres and composite materials. The Weibull modulus (alternatively called Weibull exponent) is a measure of reliability of materials and the brittleness parameter is a measure of durability. Flutes and kraft liners are types of the paperboards (fibre networks) used for container boxes.

parameter and the brittleness: the higher the brittleness, the lower the Weibull modulus, as also seen in Fig. 10.7C. Typically, carbon fibre and its composites tend to be more brittle (and more durable) but less reliable (a lower Weibull modulus). However, polymer-based fibres and their composites tend to be less brittle (and less durable), but more reliable (a higher Weibull modulus). Comparing single fibres and composites (fibre bundles and network), one finds that the load-sharing structures give more reliability. This suggests that, for high-stiffness, brittle fibres (typically carbon fibres), it is better to form bundle, network, or woven structures to compensate for their inherent low reliability.

10.6 Concluding remarks

Strength is probably one of the most researched subjects since the days of Leonardo Da Vinci [28]. Despite this long history, one still has to pose the following basic questions, 'What is a strong material?': something strong in a lab test which normally takes seconds or minutes ('strong' materials); something that withstands for a long time (durable materials); or something that consistently performs all the time (reliable materials)? Time-dependent, statistical failures that we have discussed in this chapter touches upon all of these questions.

Fibre network is a good paradigm for investigating the subject. This is because, first, the basic element of the network is a fibre which can be, approximately, treated as a 1D-element. Second, the network structure is configured in a variety of ways to model different types of materials. It also represents broad classes of industrial materials and biological materials.

According to Coleman's formulation, TDSFs are characterised by three material parameters: the characteristic strength, which represents a short-term strength, the brittleness parameter, which represents both brittleness and durability, and Weibull modulus, which indicates reliability. At a given loading history, with these three parameters, one can determine the distribution functions of lifetimes of the material.

The distribution function of creep lifetimes (and 'strength' measured under various conditions) is inherently size-dependent. Unlike a 1D material (single fibre), the WLS emerges extremely slowly in the case of a network (2D and 3D). As a result, an extreme-value distribution (EVD) is approached even more slowly, and this extreme slowness precludes the use of WLS and EVD for engineering applications for failure problems. The recent studies from a statistical mechanics viewpoint conclude that the ubiquitous nature of Weibull distribution for many materials is due to its specific (or fortunate) choice of the element distribution function (algebraic form), rather than the nature of the EVD. In practice, Weibull distribution allows us to make a conservative estimate of failure probabilities in the lower tail of lifetime (or strength) distributions.

The relationship between fibre properties and network properties may be compared in terms of the three parameters. The characteristic strength of the network is size dependent, and with increasing fibre brittleness, it increases and then plateaus. The brittleness of the network is very much controlled by fibre brittleness. The Weibull modulus (reliability) of the network is affected by the size and also the network structures. Load-sharing structures, such as fibre bundles and networks, increase the Weibull modulus and thus enhance reliability, whereas stress distributions due to disordered structures or fibre stiffness variations decrease the parameter and degrade the reliability.

Acknowledgements

Much of the content in this chapter is based on the work done by Dr. Amanda Mattsson (currently with California Institute of Technology, on leave from Mid Sweden University) when she was a PhD student. The author acknowledges her dedication and hard work shown during the period, and is honoured to have been her supervisor as well as her co-worker.

References

[1] W.A. Curtin, Stochastic damage evolution and failure in fibre-reinforced composites, Adv. Appl. Mech. 36 (1999) 163–253.

[2] A. Mattsson, T. Uesaka, Time-dependent statistical failure of fiber networks, Phys. Rev. E 92 (2015) 42158.

[3] R.M. Christensen, Exploration of ductile, brittle failure characteristics through a two-parameter yield/failure criterion, Mater. Sci. Eng. A 394 (2005) 417–424.

[4] D.T. Hristopulos, T. Uesaka, Factors that control the tensile strength distribution in paper, in: 2003 International Paper Physics Conference. PAPTAC, Victoria, BC, Canada, 2003.

[5] B.D. Coleman, Time dependence of mechanical breakdown phenomena, J. Appl. Phys. 27 (1956) 862–866.

[6] B.D. Coleman, On the strength of classical fibres and fibre bundles, J. Mech. Phys. Solids 7 (1958) 60–70.

[7] R.M. Christensen, A probabilistic treatment of creep rupture behavior for polymers and other materials, Mech. Time-Depend. Mater. 8 (2004) 1–15.

[8] W.A. Curtin, M. Pamel, H. Scher, Time-dependent damage evolution and failure in materials. II. Simulations, Phys. Rev. B 55 (1997) 12051.

[9] S.L. Phoenix, I.J. Beyerlein, Distributions and size scalings for strength in a one-dimensional random lattice with load redistribution to nearest and next-nearest neighbors, Phys. Rev. E 62 (2000) 1622.
[10] S.L. Phoenix, L.J. Tierney, A statistical model for the time dependent failure of unidirectional composite materials under local elastic load-sharing among fibers, Eng. Fract. Mech. 18 (1983) 193–215.
[11] L.M. Kachanov, Introduction to Continuum Damage Mechanics, Mechanics, Springer, Dordrecht, Lanchester, UK, 1986, https://doi.org/10.1007/978-94-017-1957-5.
[12] H. Otani, S.L. Phoenix, P. Petrina, Matrix effects on lifetime statistics for carbon fibre-epoxy microcomposites in creep rupture, J. Mater. Sci. 26 (1991) 1955–1970.
[13] S.L. Phoenix, P. Schwartz, H.H. Robinson, Statistics for the strength and lifetime in creep-rupture of model carbon/epoxy composites, Compos. Sci. Technol. 32 (1988) 81–120, https://doi.org/10.1016/0266-3538(88)90001-2.
[14] A. Mattsson, T. Uesaka, Time-dependent, statistical failure of paperboard in compression, in: S.J. I'Anson (Ed.), Advances in Pulp and Paper Research, Transactions of the 15th Fundamental Research Symposium, The Pulp and Paper Fundamental Research Society, Cambridge, UK, 2013, pp. 711–734.
[15] P. Bergström, S. Hossain, T. Uesaka, Scaling behaviour of strength of 3D-, semi-flexible-, cross-linked fibre network, Int. J. Solids Struct. (2019) 68–74.
[16] A. Kulachenko, T. Uesaka, Direct simulations of fibre network deformation and failure, J. Mech. Mater. 51 (2012) 1–14.
[17] A. Mattsson, T. Uesaka, Time-dependent breakdown of fiber networks: uncertainty of lifetime, Phys. Rev. E 95 (2017) 53005, https://doi.org/10.1103/PhysRevE.95.053005.
[18] B.D. Coleman, Statistics and time dependence of mechanical breakdown in fibres, J. Appl. Phys. 29 (1958) 968–983.
[19] P.M. Duxbury, P.L. Leath, P.D. Beale, Breakdown properties of quenched random systems: the random-fuse network, Phys. Rev. B 36 (1987) 367.
[20] P.M. Duxbury, S.G. Kim, P.L. Leath, Size effects and statistics of fracture in random materials, Mater. Sci. Eng. A176 (1994) 25–31.
[21] P.M. Duxbury, P.L. Leath, Exactly solvable models of material breakdown, Phys. Rev. B 49 (1994) 12676–12687.
[22] A. Shekhawat, Fracture in Disordered Brittle Media, Cornell University, 2013.
[23] W.A. Curtin, H. Scher, Algebraic scaling of material strength, Phys. Rev. B 45 (1992) 2620.
[24] Z.P. Bažnt, Scaling theory for quasibrittle structural failure, PNAS 101 (2004) 13400–13407.
[25] W.I. Newman, S.L. Phoenix, Time-dependent fibre bundles with local load sharing, Phys. Rev. E 63 (2001) 21507–21520.
[26] S. Mahesh, S.L. Phoenix, Lifetime distributions for unidirectional fibrous composites under creep-rupture loading, Int. J. Fract. 127 (2004) 303–360, https://doi.org/10.1023/B:FRAC.0000037675.72446.7c.
[27] A. Mattsson, T. Uesaka, Characterisation of time-dependent, statistical failure of cellulose fibre networks, Cellulose (2018), https://doi.org/10.1007/s10570-018-1776-5.
[28] J.R. Lund, J.P. Byrne, Leonardo Da Vinci's tensile strength tests: implications for the discovery of engineering mechanics, Civ. Eng. Syst. 18 (2001) 243–250.

Ballistic response of needlepunched nonwovens

11

Francisca Martínez-Hergueta[a], Alvaro Ridruejo[b], Carlos González[b,c], and Javier Llorca[b,c]
[a]School of Engineering, Institute for Infrastructure and Environment, The University of Edinburgh, William Rankine Building, Edinburgh, United Kingdom, [b]Department of Materials Science, Technical University of Madrid, School of Civil Engineering, Madrid, Spain, [c]IMDEA Materials Institute, Getafe, Madrid, Spain

11.1 Introduction

Dry fabrics are a lightweight solution for ballistic protection used in the defence sector for a large variety of applications including soft vehicle and body armour [1, 2]. Depending on their architecture, the fabrics can be classified into two categories: wovens and nonwovens, see Fig. 11.1. In woven fabrics fibres are bundled in yarns weaved into a regular pattern, meanwhile nonwovens present a random fibre network connected through local bonds consolidated by thermal fusion, chemical binding, or mechanical entanglement [3]. These bonds determine the interaction between fibres and have a primary role in the deformation and ductility of the nonwoven. In particular, the needlepunched mechanical consolidation process results in a lower stiffness and strength (as well as processing cost), but much higher strain to failure than their woven counterparts, resulting in outstanding ductility and energy absorption capacity [4–6], with an excellent ballistic performance against shrapnel and small calibres [7, 8], making nonwovens a perfect cushion layer for soft-body armour [9–13].

Although the impact response of dry woven fabrics based on Kevlar and Dyneema fibres is detailed reported in the literature and their use is common in the defence sector [14–16], the technology readiness level of nonwoven fabrics is still immature and their applications are limited. It is possible to find in the literature a handful of examples focused on damage reduction of rear components [17, 18] or sandwich cores [19, 20], nevertheless, the lack of knowledge regarding the ballistic performance of needlepunched nonwovens hinders the development of predictive design tools and delays their wider implementation in the defence and transport sector. The mechanical response of needlepunched nonwovens subjected to quasistatic tensile loads has been studied previously [21, 22]. The stiffness of the material is directly proportional to the percentage of fibres oriented with the loading direction and the bond density. The manufacturing process can also induce an anisotropic mechanical response by introducing a heterogeneous connectivity between the fibres. Upon uniaxial deformation, the random fibre network evolves exhibiting progressive fibre uncurling and rotation. This microstructural evolution leads to a nonlinear response of the material, with

Mechanics of Fibrous Networks. https://doi.org/10.1016/B978-0-12-822207-2.00011-8

Fig. 11.1 Ultra-high molecular weight polyethylene fabrics. (A) Three harness satin dry woven fabric composed of Dyneema SK65 fibre yarns and (B) needlepunched nonwoven fabric composed of Dyneema SK75 fibres.

increasing tangent stiffness with the deformation. The ductility at large strains is mainly controlled by frictional deformation micromechanisms such as fibre sliding, slippage, and pull-out from the entanglement points. These findings are the basis to develop sound, physically based, constitutive models to predict the mechanical response of needlepunched nonwovens [23]; however, further efforts are needed to characterise the wave propagation phenomenon and determine the influence of high-strain rates in the mechanical response of the material.

This chapter aims at reviewing the response of needlepunched nonwovens subjected to ballistic impact. The dynamic characterisation of the material is accomplished through split-Hopkinson bar and impact experiments. These procure the wave propagation phenomenon, the dominant deformation and failure micromechanisms at high-strain rates, and the residual velocity curves, including the ballistic limit. In addition, a finite element digital twin is developed in the software Abaqus/Explicit to explore potential applications in the defence and transport sectors. The numerical model is validated against experimental results and it is employed to analyse the potential of the material to improve the ballistic response of conventional targets such as dry woven fabrics and metal sheets. In both cases, the addition of the nonwoven layer increases substantially the specific energy absorption capacity of the targets with a negligible increment of the total areal weight of the shield.

11.2 Experimental characterisation

11.2.1 Material

The two-dimensional (2D) needlepunched nonwoven is a DSM product commercialised under the trademark Fraglight NW201. It is composed of Dyneema SK75 ultra-high molecular weight polyethylene fibres (UHMWPE) with an approximate length of ≈ 60 mm. The stochastic nature of the material results in a variable thickness of ≈ 1.5–2 mm and an areal weight of $\rho \approx$ 190–220 g/m^2. The batt is manufactured on a moving bed by continuous fibre deposition and mechanically entangled by oscillatory barbed needles [3]. This manufacturing process induces two principal material directions along the bed or machine direction (MD) and the transverse direction (TD), see Fig. 11.2. The difference in mechanical properties between perpendicular directions was pronounced, with higher stiffness and strength along the TD [22].

11.2.2 Experimental techniques

11.2.2.1 In-plane dynamic tests

The dynamic testing of the nonwoven fabric was conducted on a split-Hopkinson tensile bar (SHTB) device specially designed for low-impedance fabrics with large representative volume elements. The input and output bars comprised of aluminium 7075-T6 alloy hollow tubes of 50.8 mm outer diameter, 1.651 mm wall thickness, and 2.7 m in length. The pulse duration was $T \approx 1$ ms. The fabric was gripped by conical clamps allowing a maximum specimen width and gauge length of 35 × 35 mm^2. The samples were stretched at the orientation transverse to the roll direction (TD), the stiffest direction of the material, see Fig. 11.3B. This experimental set-up imposed strain rates of $\dot{\varepsilon} \approx 400$ s^{-1}, four orders of magnitude higher than the respective quasistatic rate [22]. Further details of this SHB equipment are available in Refs. [24, 25].

The bars were instrumented with three strain gauges, as shown in Fig. 11.3A. Amplifiers and high-frequency oscilloscopes were used to record the signal on the order of millivolts. The main outcomes were forces and velocities of the bars. The specimens were speckled with a random pattern and full-field displacement measurements were carried out via high-speed photography, employing an ultra-high-speed Kirana camera operated at 50,000 fps. Two-dimensional digital image correlation analysis was performed using the commercial software Vic2D.

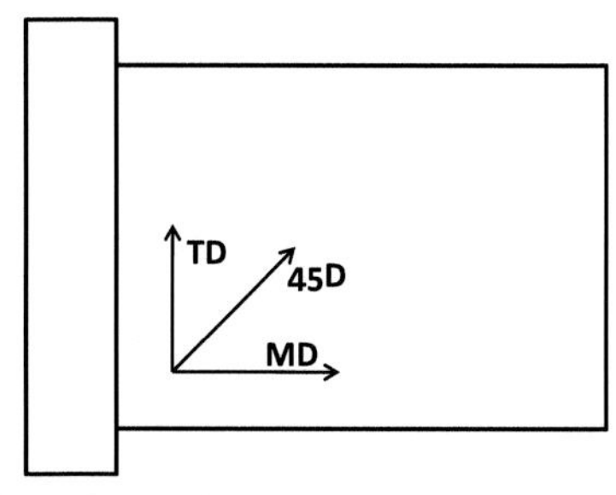

Fig. 11.2 Main material orientations of the fabric. *MD*, machine direction; *TD*, transverse direction.

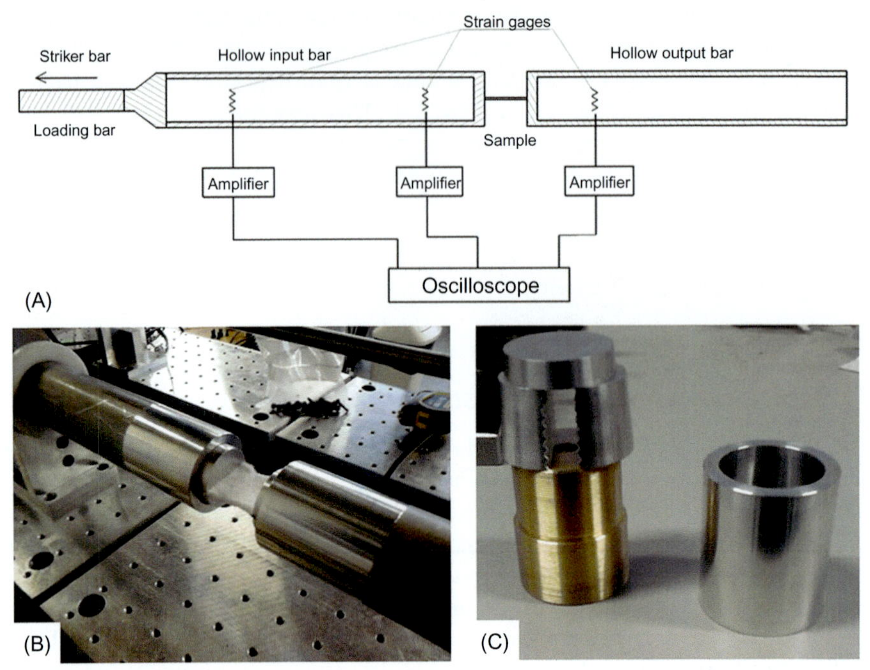

Fig. 11.3 Split-Hopkinson tensile bar experimental set-up. (A) Schematic of the different components of the SHTB, (B) gripped specimen between bar ends, and (C) aluminium and brass components of the grips [25].

11.2.2.2 Ballistic tests

The ballistic experiments were carried out on specimens of $500 \times 500\ \mathrm{mm}^2$ fully clamped along their four edges on a metallic frame with a free surface of $350 \times 350\ \mathrm{mm}^2$. The projectiles were steel spheres of 5.5 mm in diameter (calibre 0.22) and 0.706 g of mass. They were propelled by a SABRE A1 + gas gun at velocities between 270 and 400 m/s and impact energies between 25 and 55 J. The initial and residual velocities were monitored by a high-speed Phantom V12 camera with resolutions of 512×256 and 512×512 pixels and rate acquisition between 20,000 and 40,000 fps depending on the test duration. Further details of the experimental set-up are available in Ref. [26].

11.2.3 Experimental results

11.2.3.1 Dynamic tensile response

The tensile testing at high-strain rates was conducted on the SHTB apparatus presented in Section 11.2.2.1. The forces were registered by the strain gauges, while the deformation was obtained by digital image correlation. The left-hand side of Fig. 11.4 shows the evolution of the longitudinal deformation at different stages of

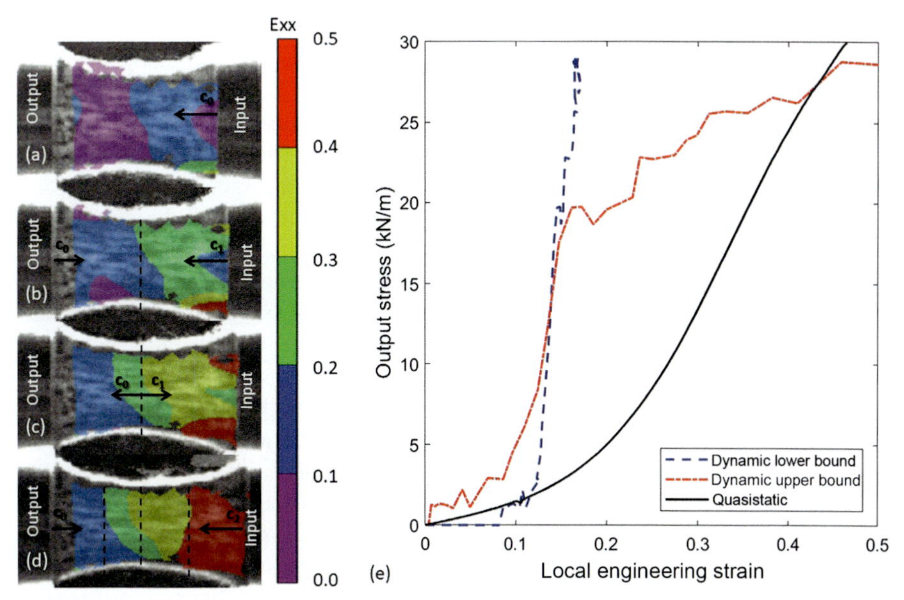

Fig. 11.4 Evolution of the longitudinal strain ε_x and wave rebound phenomenon. (A) $t = 0.2$ms, (B) $t = 0.4$ ms, (C) $t = 0.6$ ms, and (D) $t = 0.9$ ms. (E) Output stress and local engineering strain at the output interface, the lower bound (*blue dashed line*) and input interface, the upper bound (*red dashed line*). Quasistatic stress–strain curve has been included for comparison purposes [25].

the experiment, exhibiting large strain gradients across the specimen, in contrast to the characteristic homogeneous strain distributions obtained at quasistatic loading regimes before the onset of damage in this material [22]. At the initial stages of the loading process ($t = 0.4$ ms), the material exhibited an oscillating deformation around the 20% as a result of the wave reflection process. After a certain instant in time ($t = 0.6$ ms), the evolution of the strain nearby the output bar froze at a maximum value of 18% leading to a heterogeneous strain field across the specimen with an steep increment of deformation and damage localisation nearby the input bar, see Fig. 11.4D. This strain gradient indicated a lack of dynamic force equilibrium between the bars during the whole duration of the experiment, usually observed during the first stages of the dynamic experiment before the input and output forces overlap [27].

The dynamic deformation induced progressive fibre straightening, rotation, and sliding with the loading direction at microstructural level, as previously registered for quasistatic uniaxial deformation [22]. This evolution resulted in a nonlinear pseudo-plastic mechanical response of the material with an inherent increment of the tangent stiffness. The fibre orientation distribution function was proportional to the wave speed of the material, that also increased with the applied deformation. The initial fibre curvature and the random fibre orientation decreased the effective longitudinal propagation velocity, so the progressive fibre alignment increased the apparent wave speed. In addition, the magnitude of the tensile waves decreased with the distance from the input interface due to two different sources of mechanical dissipation: (i) the frictional nature of the deformation micromechanisms and (ii) the partial wave reflection at the entanglement points.

As a result of the variable stiffness and wave propagation velocity with the applied deformation, heterogeneous strain gradients appeared on the specimen during dynamic testing. The tensile pulse first propagated into the specimen with velocity c_0, see left-hand side of Fig. 11.4. After a certain amount of time, the tensile pulse reached the output interface, meanwhile faster waves with a higher stress magnitude appeared at the input interface due to the nonlinear increment of stiffness with fibre realignment ($c_1 > c_0$). Once the transmitted and reflected waves arrived at the same material point, they created a macromechanical interface with an impedance mismatch at both sides due to the differences in microstructural evolution, preventing the propagation of larger strain waves into the left side of the specimen. The reflections were repeated with waves of higher stress magnitude and velocity over the test duration ($c_2 > c_1$), generating additional mismatch fronts.

The output forces were analysed to determine the influence of the high-strain rates in the mechanical response of the fabric. Fig. 11.4E plots the output stress monitored by the strain gauge versus the local longitudinal strains registered by DIC at the input and output interfaces. The lack of a constant strain over the specimen hindered the acquisition of a conventional stress–strain curve. These curves are compared against the quasistatic stress–strain constitutive relationship. A sudden increase in stiffness was registered, reaching a stress value of 30 kN/m within the range 15%–17% deformation, which indicated a significant strain rate sensitivity of the frictional deformation micromechanisms. Further information is available in Ref. [25].

11.2.3.2 Ballistic performance

The ballistic experimental campaign was conducted to evaluate the ballistic response of the material, obtain the residual velocity curve, and identify the ballistic limit, v_{50}. Fig. 11.5 shows an example of the deformation of the fabric subjected to an impact velocity around the ballistic limit ($v = 338$ m/s). The differences in wave propagation velocities across perpendicular directions (MD and TD) resulted in an elliptical cross section and the transverse wave that followed the deflection of the layer exhibited a cone profile. The needlepunched nonwoven absorbed the kinetic energy of the projectile by in-plane deformation and momentum transfer. The load was transferred through the mechanical entanglements inducing fibre uncurling, rotation, and sliding towards the impact point, as previously appreciated during quasistatic and dynamic uniaxial tensile testing. For impact velocities above the ballistic limit, large fibre pull-out and final tearing of the layer was registered, see Fig. 11.5C. In all the cases, the damage was localised at the impact point and large strain gradients were exhibited, similar to the ones observed during the split-Hopkinson bar experiments. The thermal softening of the directly impacted fibres was also registered and validated through differential scanning calorimetry. This resulted in a lower energy absorption capacity at velocities above the v_{50}. Further information is available in Ref. [26].

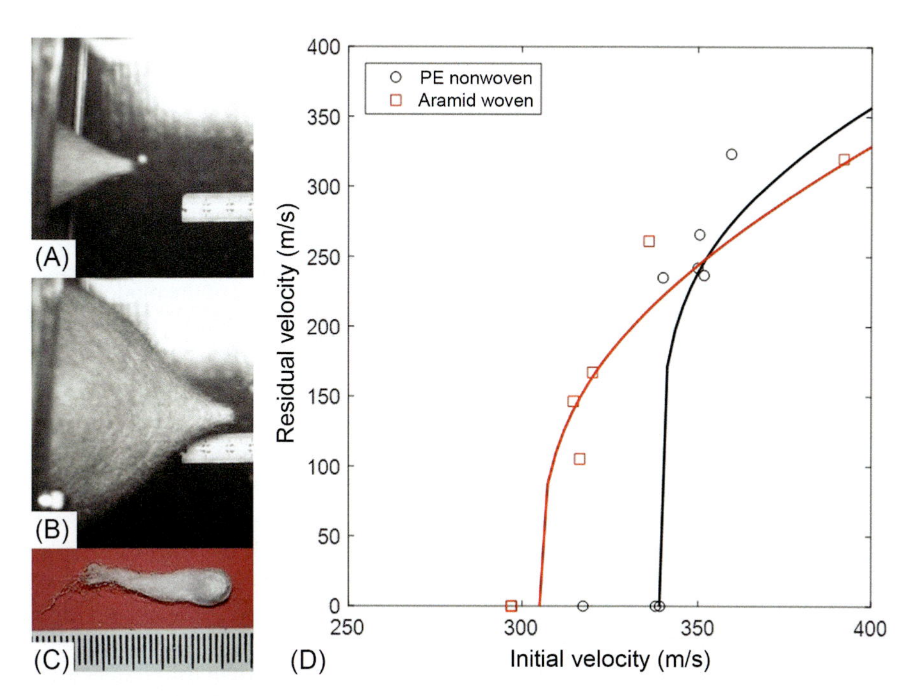

Fig. 11.5 Deformation of the polyethylene nonwoven fabric during impact at the ballistic limit $v = 338$ m/s. (A) $t = 175$μs. (B) $t = 1025$μs. (C) Extracted fibres around the projectile after impact. (D) Residual curves of the nonwoven fabric (one layer with equivalent areal weight 200 g/m^2) and comparison with a conventional woven aramid laminate (four layers with equivalent areal weight equivalent areal weight 922 g/m^2). The symbols stand for experimental results and the lines for Eq. (11.1) [26].

The residual velocity curve of the material is shown in Fig. 11.5D. The experimental results were postprocessed to obtain the least squares fitting with the Lambert equation,

$$V_{res} = \left(V_{ini}^n - V_{50}^n\right)^{1/n} \tag{11.1}$$

where V_{res} and V_{ini} are the residual and initial impact velocities, respectively. The ballistic response of the material was compared against the response of a conventional dry woven aramid target composed of four layers of Kevlar KM2 fibres with an equivalent areal weight of 920 g/m^2. The nonwoven exhibited a ballistic limit of 339 m/s and an energy absorption capacity of 40 J, 10 J higher than the maximum capacity of the aramid protection, even though the nonwoven protection was four times lighter. This experimental campaign probes the advantage of nonwovens over wovens against ballistic impact for small calibres and shrapnel.

11.3 Numerical simulation

11.3.1 Numerical implementation

A finite element digital twin of the needlepunched nonwoven was implemented in the software Abaqus/Explicit to predict the ballistic response of the material. The multiscale constitutive model developed by Martinez-Hergueta et al. [23] was implemented as a VUMAT subroutine within the framework of large deformations taking as reference the unstressed state of the material. The detailed description of the model is available in Refs. [23, 28]. The implementation for dynamic analysis incorporated one modification at Gauss point level, such that each mesodomain of the fibre network was described by 65 sets of fibres with different orientation instead of 33 to reduce the numerical instabilities. The strain rate dependency previously observed during the dynamic characterisation was accounted by two different material parameters; (i) the fibre pull-out length, L_{po}, that increased up to the total length of the fibre of 60 mm, in agreement with the full fibre pull-out registered experimentally, see Fig. 11.5C and (ii) the pull-out strength, σ_{po}, randomly defined within the range [0.3–1.7] GPa, to fit the ballistic limit of the fabric. An element deletion criteria (average damage variable $D > 0.99$) was defined to replicate the penetration of the target by fibre disentanglement.

The free area of the fabric of 350×350 mm^2 was discretised with reduced integration M3D4R membrane elements, with enhanced hourglass control, and second-order accuracy, see Fig. 11.6. The target was fully clamped along the four edges. The impact point was discretised with a fine mesh composed of elements of 1 mm^2, and a coarser mesh was implemented with the distance from the impact zone to decrease the number of degrees of freedom. The projectile was modelled as a solid rigid steel sphere of 5.5 mm in diameter and 7.85 g/cm^3 density. The tangential friction coefficient between the sphere and the fabric was set as 0.1.

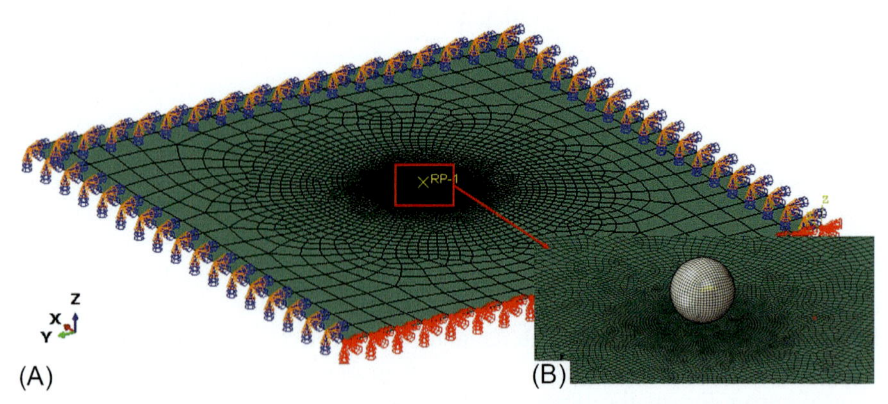

Fig. 11.6 (A) Mesh strategy and boundary conditions and (B) mesh refinement at the impact point.

11.3.2 Validation

The performance of the digital twin was analysed to determine if it captured accurately the energy absorption mechanisms of the nonwoven material. The left-hand side of Fig. 11.7 shows the numerical comparison of the experimental and numerical deflection of the layer during an impact above the ballistic limit, showing very good agreement in terms of damage prediction. An in-plane tensile wave stretched the layer towards the impact point, while the transverse wave captured the deflection with elliptical cross section previously observed during the experimental campaign. The region of the material within the bounds of the in-plane longitudinal tensile wave dissipated the impact energy through the deformation mechanisms aforementioned; fibre uncurling, rotation, and sliding. The model also predicted the strain gradients localised at the impact point and the fibre disentanglement resulting in tearing and penetration of the ply.

The digital twin also captured the energy absorption capacity of the material and the ballistic limit. Fig. 11.7G shows the residual velocity curves and compares the experimental and numerical residual velocities, V_{res}, as a function of the initial impact velocities, V_{ini}. The experimental scattering was reproduced by the stochastic definition of the pull-out strength as mentioned in Section 11.3.1. Although the model was able to capture the ballistic limit, it overestimated the energy absorption capacity above that threshold, as it did not considered the thermal softening of the Dyneema fibres registered during ballistic impact. Further information regarding the validation of the digital twin might be consulted in Ref. [28].

11.4 Case study 1: Ballistic response of hybrid nonwoven/woven targets

Dry woven and nonwoven fabrics can be combined to create soft-body armour protections with improved ballistic performance to arrest a large range of calibre sizes. The nonwoven is usually placed in the frontal face to act as cushion layer and enhance the load transfer into the woven fabric [17, 18, 29]. In this section, we aim to investigate the synergistic contribution between the nonwoven and woven layers against impact, from an experimental and numerical point of view. A ballistic experimental campaign has been conducted to determine the ballistic limit and the energy absorption capacity of the hybrid target. The role of each layer at different stages of the impact process was ascertained using a digital twin that provided separately information on kinetic and strain energy absorbed as function of time.

11.4.1 Materials

The Dyneema hybrid target was composed of a nonwoven (NW) layer and four rear harness satin woven fabrics (W) with stacking sequence ($NW/0_W/90_W/0_W/90_W$). The areal density of the shield was 920 g/m^2, where each individual layer of the woven Dyneema fabric had an areal density of $\approx$ 180 g/m^2 and thickness of $\approx$ 0.5 mm.

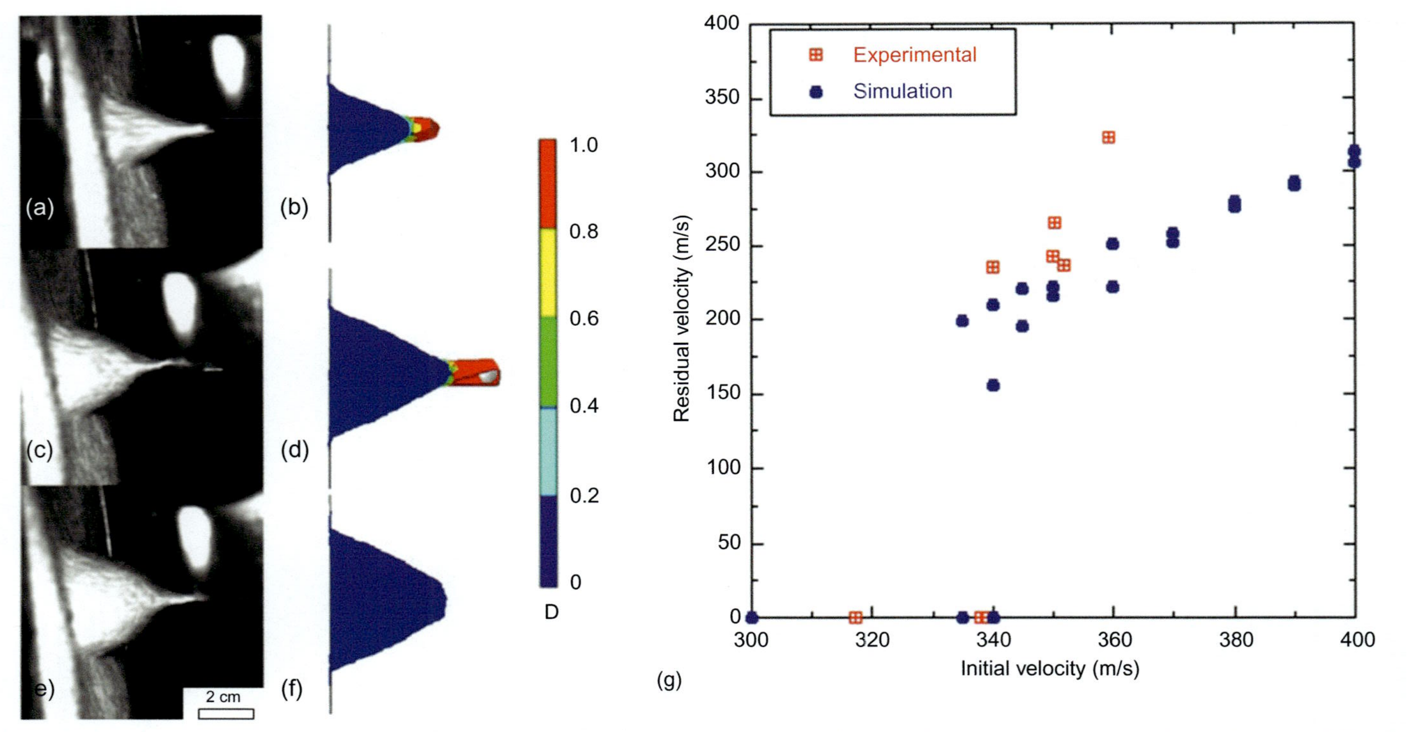

Fig. 11.7 Correlation between the experimental deflection and the damage contour plot of the needlepunched nonwoven during an impact at 360 m/s. (A, B) $t = 100$ μs. (C, D) $t = 175$ μs and (E, F) red colour stands for the disentangled fabric. (G) Experimental and numerical residual numerical residual velocity curves.

Reproduced with permission from F. Martínez-Hergueta, A. Ridruejo, C. González, J. Llorca, Numerical simulation of the ballistic response of needle-punched nonwoven fabrics, Int. J. Solids Struct. 106 (2017) 56–67.

The full description of the experimental set-up which comprised a gas gun and high-speed camera is available in Section 11.2.2.2. The size of the projectile (5.5 mm diameter) was approximately four times the width of the woven Dyneema yarns, easily promoting yarn sliding during ballistic impact due to the low friction coefficient of the Dyneema fibres. Further details of this experiment are available in Ref. [30].

11.4.2 Experimental results

The impact response of the hybrid target is shown in Fig. 11.8 during an impact at 370 m/s. Initially, all layers deflected together until the yarns of the rear woven layers slid and the nonwoven got confined into the resulting gap. The confinement led to a high level of fibre alignment with the loading direction followed by a large percentage of extracted fibres. This resulted in a ductile response of the nonwoven layer with large deformation before final disentanglement, inducing a very high energy dissipation. The hybrid configuration presented superior energy absorption capacity when compared against the single nonwoven fabric. The residual velocity curves are shown in Fig. 11.8D. The additional four layers of woven Dyneema increased the ballistic limit of the nonwoven by ≈ 30 m/s. It should be noted that the specific energy absorption capacity was higher for the single nonwoven considering the reduced weight; however, the areal weight of the hybrid shield was comparable to the previous dry woven aramid shield, see Fig. 11.5, and still exhibited a higher energy absorption

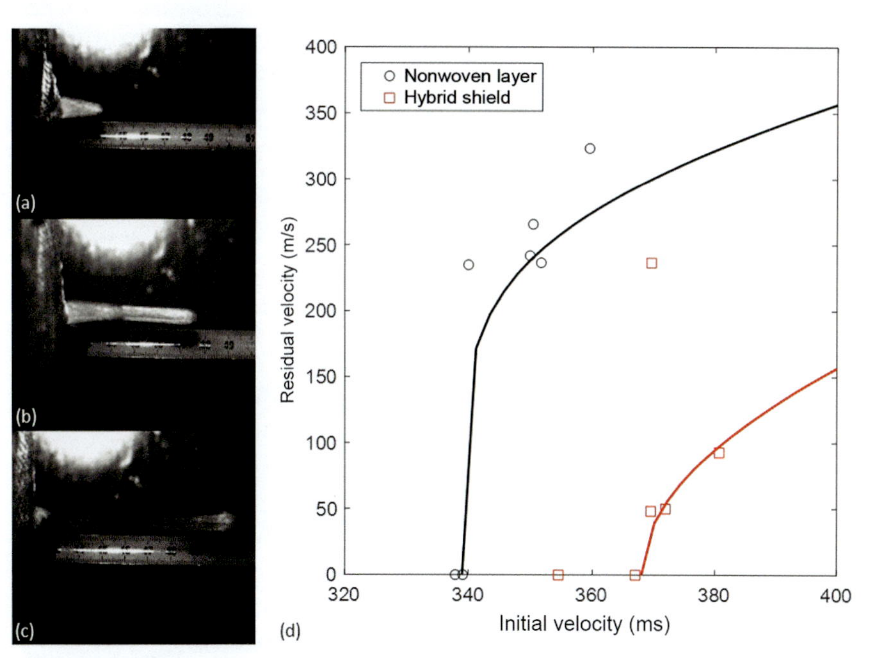

Fig. 11.8 Deflection of the hybrid shield for an impact velocity of 370 m/s. (A) $t = 150\mu s$. (B) $t = 600\mu s$. (C) $t = 1350\mu s$. (D) Residual velocity curves of the hybrid shield (920 g/m^2) and comparison with the nonwoven layer (200 g/m^2) [30].

capacity and ballistic limit than the previous [1, 16]. Furthermore, the addition of the nonwoven layer to the woven shield permitted the load transfer into the woven yarns and avoided the slippage failure mode previously observed during the ballistic impact of the single woven layers [30].

11.4.3 Numerical implementation

The analysis of the previous ballistic response was conducted by means of a digital twin based on a finite element model. The nonwoven ply was implemented following the methodology exposed in Section 11.3.1, meanwhile, the woven fabric layers were modelled at mesoscale level, discretising the weaved yarns with C3D8 solid elements [31, 32], see Fig. 11.9. The elastic constants of the transversely isotropic yarn model are available in Table 11.1.

The nonwoven and the woven layers were combined to replicate the shield with stacking sequence ($NW/0_W/90_W/0_W/90_W$). The dimensions of the nonwoven were set to $350 \times 350\,mm^2$, meanwhile, the woven fabrics were limited to $100 \times 100\,mm^2$ to reduce the computational cost. All fabrics were spaced by a gap of

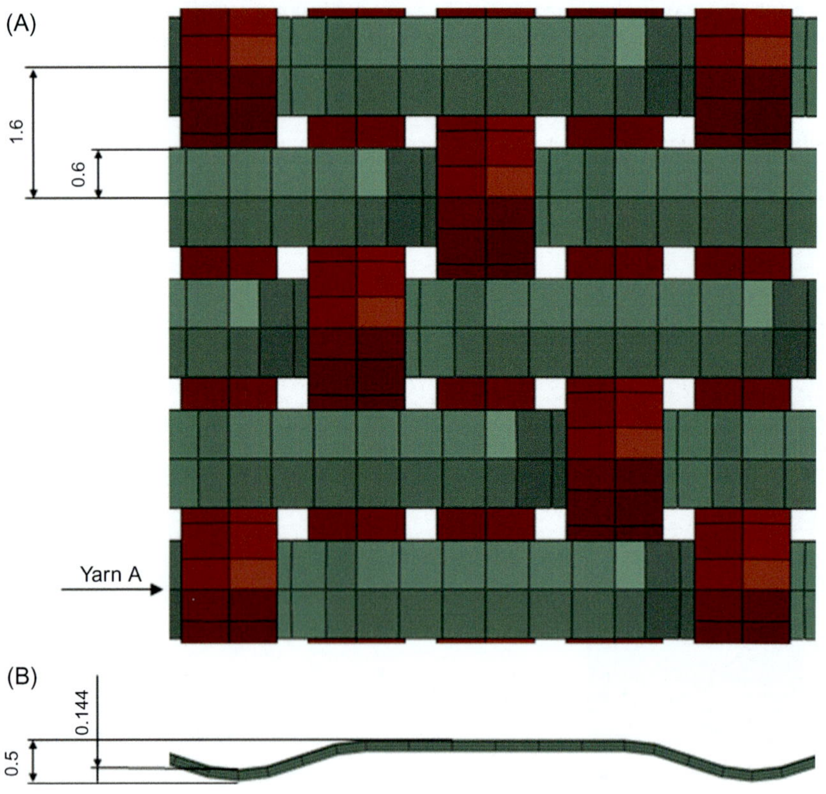

Fig. 11.9 Three harness satin pattern and mesoscale implementation of the woven layers. (A) Plan view. (B) Lateral view of yarn A. Measurements in mm [30].

Table 11.1 Elastic constants of the polyethylene SK65 yarns [32].

Density (ρ_f)	970 kg/m^3
Longitudinal elastic modulus (E_1)	95 GPa
Transverse elastic modulus (E_2)	9.5 GPa
Poisson's ratio (ν_{12})	0
Shear modulus (G_{12})	0.95 GPa

0.025 mm to avoid initial contact between layers and were constraint along the edges. The contact interaction between all the elements (the projectile, the yarns, and the nonwoven) was modelled by a Coulomb friction algorithm with a friction coefficient of 0.0075. Further details are available in Ref. [30].

11.4.4 Simulation results

The correlation between experimental and numerical results is shown in Fig. 11.10. A good agreement was found in terms of residual velocities and ballistic limit, such that the experimental results were obtained within the numerical scattering. Furthermore, the digital twin reproduced the improved behaviour of the hybrid shield with respect to the nonwoven and woven layers individually, and the complex interaction between the nonwoven and the weaved yarns. The left-hand side of Fig. 11.10 shows a comparison between the predicted and the experimental response at an impact velocity of 380 m/s. At the early stages of the impact ($t < 20$ μs), all the plies deflected together. The main energy absorption mechanisms at that stage were the kinetic and elastic energies transmitted to the rear woven layers. After yarn sliding and during the confinement stage,

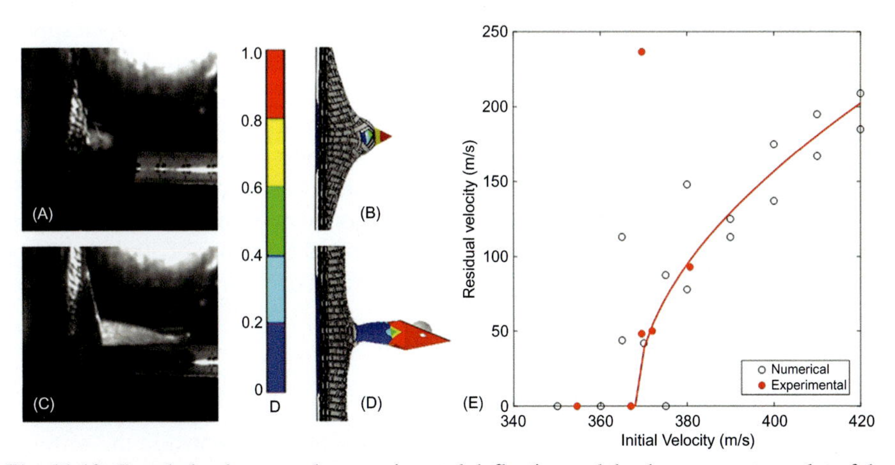

Fig. 11.10 Correlation between the experimental deflection and the damage contour plot of the hybrid shield for an impact at 380 m/s. (A, B) $t = 25$ μs and (C, D) $t = 150$ μs. The *red colour* stands for the disentangled fibre network. (E) Comparison between experimental and predicted residual velocities as function of the impact velocity [30].

elastic deformation and kinetic energy transmitted to the nonwoven layer became the predominant dissipation mechanism. Final penetration of the shield occurred by extensive fibre disentanglement. The major contribution to the energy dissipation happened at the very early stages of the impact, where a reduction of the 60% of the kinetic energy of the projectile was registered, three times higher than the energy absorbed by the baseline woven target. After the woven yarns slipped, an additional 20% of kinetic energy was dissipated due to the localised fibre pull-out. The hybrid configuration outperformed the capacity of each individual ply due to the synergistic interaction between layers. The nonwoven layer was an optimal material to transfer the load to the adjacent woven layers, drastically rising the energy absorption capacity of the woven target for these small calibres.

11.5 Case study 2: Ballistic response of multilayered metal/nonwoven shields

Dry fabrics can be also included in conventional hollow metallic components to improve the ballistic performance with a negligible increase of structural weight. The concept has been proven valid for turbine barriers, where the addition of Kevlar woven fabrics improved the ballistic performance against projectiles [33, 34]. This section aims to investigate the performance of a multilayered shield based on steel sheets, nonwovens, and air gaps.

11.5.1 Materials

The nonwoven fabric was combined with a commercial bake hardening steel 260BH manufactured by ArcelorMittal [35]. The alloy is conventionally used in the automotive industry to manufacture components such as vehicle doors. It is designed specifically for structural applications and exhibits a high ductility, with high dent and impact resistance. The vehicle door was modelled as a hollow target with a total width of 80 mm comprised of two thin steel plates with 0.7 mm thickness with and equivalent areal weight of 11 kg/m^2. This configuration allows the incorporation of additional internal nonwoven fabrics in between the steel plates to increase the energy absorption capacity of the baseline structure. A maximum of three nonwoven layers was evaluated, regularly spaced with gaps of 10 mm and a distance of 30 mm with the steel plates, see Fig. 11.11. This resulted in an increment of 5.5% of the areal weight of the component, up to 11.6 kg/m^2.

11.5.2 Numerical implementation

A digital twin of the previous multilayered shield was implemented in the software Abaqus/Explicit to study the ballistic response of the target. The mechanical response of the nonwoven layers was implemented as described in Section 11.3.1. The constitutive behaviour of the bake hardening steel was modelled as a standard isotropic elasto-plastic material available in the Abaqus/Explicit library based on the ductile

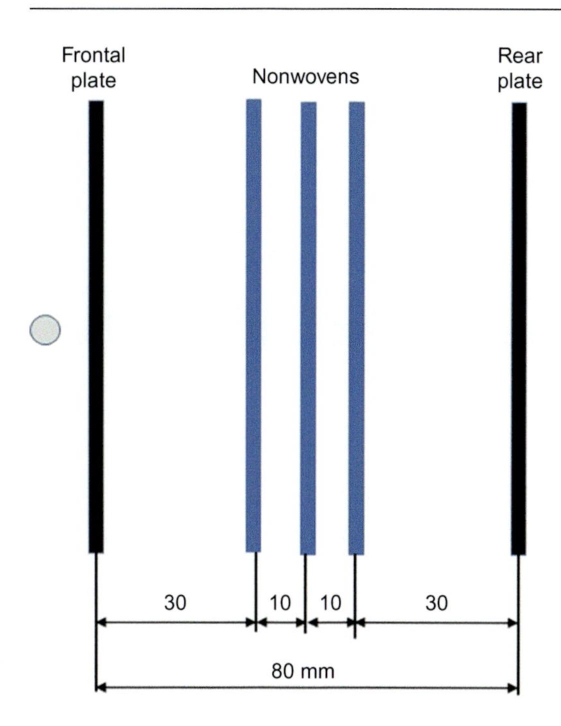

Fig. 11.11 Multilayered configuration composed of two steel plates and three nonwoven plies. Measurements in millimetres.

failure criterion proposed in Refs. [36–38]. The model incorporated a linear hardening with a Von Mises yield surface which evolved as function of the applied plastic deformation, ε^{pl}, following the expression:

$$\sigma_y = \sigma_y^0 + H\varepsilon^{pl} \tag{11.2}$$

where σ_y^0 stood for the initial value of the yield strength and H for the hardening modulus. The onset of damage was determined by the ultimate strength of the material, σ_0. This value was combined with the hardening modulus to obtain the equivalent plastic strain to failure. Once the failure enveloped was overtaken, the material degradation was implemented by a phenomenological softening of the undamaged stress tensor, $\overline{\sigma}$, such as

$$\boldsymbol{\sigma} = (1 - d)\overline{\boldsymbol{\sigma}} \tag{11.3}$$

where the damage variable, d, followed a conventional Lemaitre continuum damage model to ensure the dissipated energy was equal to the fracture toughness of the material, Γ, and avoid any size mesh dependency [39, 40]. Further information about the material model is available in Ref. [38] and material properties are summarised in Table 11.2. More details of the implementation are available in Ref. [41].

Table 11.2 Material parameters for Steel 260BH [41].

Parameter	Value
Density, ρ	7850 kg/m^3
Young's modulus, E	200 GPa
Poisson's ratio, ν	0.27
Hardening modulus, H	2.5 GPa
Ultimate strength, σ_0	400 MPa
Yield strength, σ_y^0	280 MPa
Strain to failure, ε_D^{pl}	0.3
Strain rate, $\dot{\varepsilon}^{pl}$	10^6
Triaxiality, ξ	0.8
Fracture toughness Γ	0.072 J/mm^2

The door vehicle with the internal nonwoven layers was simplified as a 350 × 350 mm^2 target to reduce the computational cost. A higher mesh density was also defined around the impact point, with finite elements of 1 mm^2 area. All the layers were fully clamped on the edges during the impact simulation. The nonwoven layers were modelled using the approach in Section 11.3.1; meanwhile, the thin steel plates were modelled with S4R shell elements, reduced integration, hourglass control, and finite membrane strain. A rigid steel sphere of 5.5 mm in diameter as in the previous ballistic studies was implemented to simulate the projectile. The penetration of the layers was reproduced by the deletion of the fully damaged elements, once the damage variable achieved the threshold $D > 0.99$. The contact between steel plates and nonwoven layers was implemented by a softened tangential contact behaviour with an sticking friction slope of $\kappa = 0.001$ and a friction coefficient $\mu = 0.1$.

11.5.3 Simulation results

The digital twin was able to reproduce the ballistic response of the multilayered shield, and the interaction between the layers. The right-hand side of Fig. 11.12 shows the penetration sequence. Initially, the projectile pierced the frontal steel plate causing a minimum plastic deformation located at the impact point and hit the rest of the layers and rear metal plate, rising the strain and kinetic energy absorbed by the system and decreasing the kinetic energy of the projectile. The progressive deformation induced damage on the nonwoven layers, nevertheless, final disentanglement was inhibited by the rear metal plate, that contributed structurally delaying the penetration of the layers and increasing the energy absorbed by the target, see Fig. 11.12A. For impact velocities just below the ballistic limit, the penetration of the rear steel plate was predicted before the full nonwoven disentanglement took place, resulting in a similar confinement to the one appreciated in the hybrid woven/nonwoven shield, see Section 11.4.2. This confinement resulted in a high volume of fibre alignment, increasing the energy absorbed locally due to fibre pull-out.

The ballistic performance of the multilayered component was compared in terms of energy absorption capacity against the predictions for its individual components; the steel plates and the nonwoven layers. Fig. 11.12D compares the residual velocity

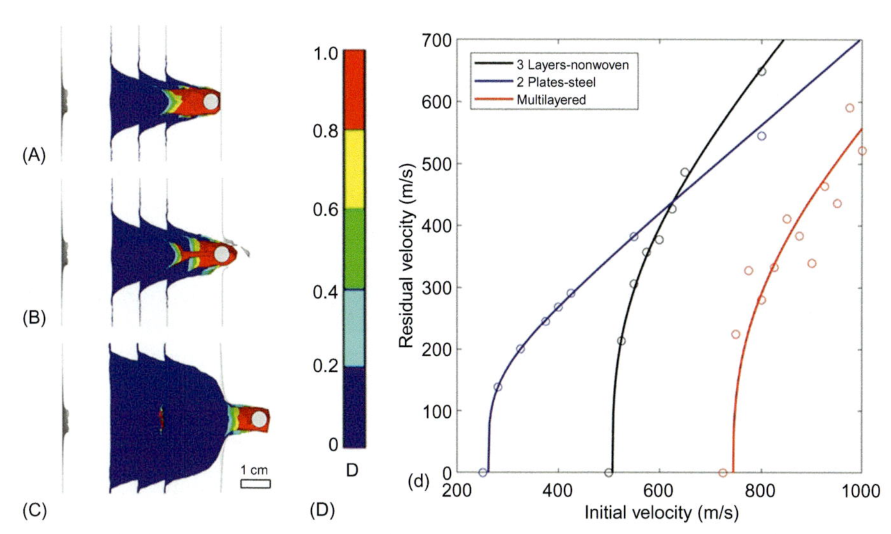

Fig. 11.12 Ballistic response of the multilayered shield impacted at 725 m/s. (A) $t = 90\mu s$, (B) $t = 135\mu s$, and (C) $t = 225\mu s$. Contour plots of the damage variable. (D) Residual velocity curves for the multilayered shield and comparison against the baseline configuration (two plates of steel) and the three layers of nonwoven reinforcement [41].

curves. The steel plates with an equivalent weight of 11 kg/m^2 presented the lowest ballistic limit with a maximum energy absorption capacity of 19.5 J, characteristic of thin metallic plates. On the other hand, the three nonwoven layers had an equivalent energy absorption capacity of 94.5 J. The multilayered shield surpassed the ballistic performance of their individual components, with a maximum energy absorption capacity of 195.9 J, almost doubling the capacity of the sum of the previous. This represented a massive improvement of specific energy absorption capacity over a factor of 8 when compared against the conventional door vehicle with a negligible increment of total areal weight of 5.5%. The large deformation of the nonwoven layers and their synergistic interaction with the rear metal sheet led to this outstanding improvement of ballistic performance.

11.6 Conclusions

The ballistic response of a Dyneema needlepunched nonwoven fabric composed of a random network of long fibres entangled through a mechanical process has been reviewed, with special focus on the main deformation and failure mechanisms, and their contribution to the overall energy absorption capacity. The ballistic performance of the material was analysed in terms of residual velocity curves and ballistic limit. In addition, a digital twin by means of a high fidelity multiscale finite element model was developed to provide further insight into physical events at microscale level that could not be revealed by high-speed imaging. The numerical model was able to capture the main deformation micromechanisms such as fibre reorientation and localisation of damage and was validated and calibrated against experimental data to ensure the robustness of the approach.

The in-plane dynamic response under high-strain rates of the material was determined using a split-Hopkinson tensile bar, specially designed for this purpose. The apparatus included a high-sensitivity output bar and had a long pulse duration to register the response of the low-impedance fibre network under large deformations. A high-speed camera registered the deformation micromechanisms under high-strain rates, which comprised fibre straightening, rotation, and sliding with the loading direction. These mechanisms were previously observed during the in-plane tensile quasistatic characterisation of the material. The main difference between the dynamic and the quasistatic loading regimes was the heterogeneous strain gradient developed at high-strain rates across the gauge length, not registered previously during the quasistatic analysis before the onset of damage. This heterogeneous strain field was a direct consequence of the dissipative nature of the frictional deformation processes and the internal impedance fronts originated in the material due to the differences in microstructural evolution. As a result, the wave propagation phenomenon prevented the propagation of larger strain waves away from the loading point, resulting in large strain gradients located nearby the input edge. The frictional mechanisms between the entangled fibres presented a strong strain rate dependency with a significant increment of stiffness for low applied strains.

The ballistic response of the nonwoven layer was characterised by a combination of experimental and numerical analyses. The nonwoven layer was subjected to ballistic impact by a small steel sphere (5.5 mm in diameter) and presented outstanding energy absorption capacity compared to a conventional aramid woven target four times heavier than the novel nonwoven solution. During impact, the energy of the projectile was accommodated by the formation of a cone of deformed material with an elliptical cross section due to the different wave propagation speed along material directions. The deformation induced the same micromechanisms observed during dynamic in-plane deformation, with a pronounced fibre rotation and realignment towards the impact point, and a sharp strain gradient with localised damage. The high ductility of the material resulted in its outstanding energy absorption capacity. The energy was dissipated by the tensile deformation of the fabric within the elliptical region defined by the wave speed of the material. The penetration of the ply occurred due to fibre disentanglement, with a large volume of fibres extracted from the layer at very large strain values.

The potential of the material to improve the ballistic response of conventional barriers was analysed with two additional case studies. In the first one, the nonwoven layer was used to improve the ballistic performance of a conventional Dyneema woven target, with a small ratio between projectile diameter and yarn width easily promoting yarn sliding and a poor ballistic performance. The nonwoven was positioned as a frontal layer to redistribute the load over the adjacent woven plies. As a result, a drastic increment of the ballistic limit and the energy absorption with respect to the original woven configuration was observed. During the first stages of the impact, the energy was transmitted to the woven yarns through the nonwoven fabric, rising by a factor of 3 the kinetic and strain energy absorbed by the woven shield. At a second stage, the projectile slipped through the yarns of the woven fabrics; however, the large ductility of the nonwoven resulted in a large fibre confinement with an additional energy absorption due to the large volume of extracted fibres.

The second case study focused on the improvement of the mechanical performance of conventional metal components in the transport sector such as a vehicle door. Three additional nonwoven layers were incorporated, spaced by gaps of 10 mm. The hybrid system outperformed the baseline steel plates and nonwoven layers, resulting in an outstanding energy absorption capacity, about twice the sum of the energies dissipated by its individual components. Furthermore, the hybrid shield increased the energy absorption capacity of the baseline steel plates over a factor of 8, with a negligible increment of areal weight of a 5.5%.

The digital twin provided high-fidelity predictions of the different configurations in terms of ballistic limit and deformation mechanisms, with valuable insight regarding the evolution of energy absorption for each component of the shields. It has been proven as a useful numerical tool that could be used in future design and optimisation tasks; however, the current approach to simulate the ballistic response of needlepunched nonwoven fabrics presents certain limitations and challenges that still need to be addressed. In particular, (i) the implementation of the strain rate dependency of the material is phenomenological and has been fitted against experimental results due to the lack of a high strain rate stress–strain relationship; (ii) the wave propagation phenomenon of the fibre network could not be accurately represented by the continuous mesodomain, so the simulations could not replicate the scattering and dissipation observed during the split-Hopkinson bar experiments; and (iii) a thermo-mechanical constitutive model needs to be implemented to capture the thermal softening of the fibres during ballistic impact.

This chapter demonstrates that the Dyneema needlepunched nonwoven fabric presents a lightweight solution to arrest small fragments, providing a low-cost alternative to improve the ballistic performance of conventional shields. It demonstrates it is possible to improve the impact response of civil automotive and aerospace components without penalising the fuel consumption to protect passengers against shrapnel. The nonwoven layer can be also incorporated into soft-body armour protections without reducing the users comfort, at a minimal increment of areal weight.

Acknowledgements

This research was supported by the Royal Society through the grant (RGS/R2/180091) and the start-up funding for recently appointed lectures of the University of Edinburgh. The collaboration of Mr. J. Vila-Ortega, Mr. S. Carter, Dr. F. Gálvez, Dr. A. Pellegrino, and Prof. Nik Petrinic is gratefully acknowledged.

References

[1] M.A. Abtew, F. Boussu, P. Bruniaux, C. Loghin, I. Cristian, Ballistic impact mechanisms—a review on textiles and fibre-reinforced composites impact responses, Compos. Struct. 223 (2019) 110966, https://doi.org/10.1016/j.compstruct.2019.110966.

[2] U. Mawkhlieng, A. Majumdar, A. Laha, A review of fibrous materials for soft body armour applications, RSC Adv. 10 (2) (2020) 1066–1086.
[3] S.J. Russell, Handbook of Nonwovens, The Textile Institute, Woodhead Publishing, 2007.
[4] R.C. Laible, M.C. Henry, A review of the development of ballistic needle-punched felts, Clothing and Personal Life Support Equipment Laboratory. U.S. Army Natick Laboratories, 1969. Technical Report No. C/PSEL-TS-167.
[5] S.J. Russell, A. Pourmohammadi, I. Ezra, M. Jacobs, Formation and properties of fluid jet entangled HMPE impact resistant fabrics, Compos. Sci. Technol. 65 (6) (2005) 899–907, https://doi.org/10.1016/j.compscitech.2004.10.015.
[6] G.A. Thomas, Non-woven fabrics for military applications, in: Military Textiles, The Textile Institute, Woodhead Publishing, 2008, pp. 44–47.
[7] T.W. Ipson, E.P. Wittrock, Response of non-woven synthetic fiber textiles to ballistic impact, Denver Research Institute, 1966. Technical Report No. TR-67-8-CM.
[8] H.L. Thomas, A. Bhatnagar, L.L. Wagner, Needle-punched non-woven for high fragment protection, in: 14th International Conference of Composite Materials, San Diego, CA, 2003.
[9] S.H. Lee, T.J. Kang, Mechanical and impact properties of needle punched nonwoven composites, J. Compos. Mater. 34 (10) (2000) 816–840, https://doi.org/10.1177/002199830003401001.
[10] J.-H. Lin, C.-H. Hsu, H.-H. Meng, Process of preparing a nonwoven/filament/woven-fabric sandwich structure with cushioning effect of ballistic resistance, Fibres Text. East. Eur. 13 (2005) 43–47.
[11] P.R.S. Reddy, T.S. Reddy, I. Srikanth, J. Kushwaha, V. Madhu, Development of cost effective personnel armour through structural hybridization, Defence Technol. 16 (6) (2020) 1089–1097, https://doi.org/10.1016/j.dt.2019.12.004.
[12] R. Yan, Q. Zhang, B. Shi, Z. Qin, S. Wei, L. Jia, Investigating the integral-structure of HRBP/CHP/CF consisting of non-woven flexible inter/intra-ply hybrid composites: compression, puncture-resistance, electromagnetic interference shielding effectiveness, Compos. Struct. 248 (2020) 112501.
[13] G.A. Thomas, Non-woven fabrics for military applications, in: Military Textiles, Woodhead Publishing, 2008, pp. 17–48.
[14] D. Naik, S. Sankaran, B. Mobasher, S. Rajan, J. Pereira, Development of reliable modeling methodologies for fan blade out containment analysis. Part I: experimental studies, Int. J. Impact Eng. 36 (1) (2009) 1–11.
[15] P.M. Cunniff, An analysis of the system effects in woven fabrics under ballistic impact, Text. Res. J. 62 (9) (1992) 495–509.
[16] A. Tabiei, G. Nilakantan, Ballistic impact of dry woven fabric composites: a review, Appl. Mech. Rev. 61 (1) (2008). 010801-1-13.
[17] C.C. Lin, C.C. Huang, Y.L. Chen, C.W. Lou, C.M. Lin, C.H. Hsu, J.H. Lin, Ballistic-resistant stainless steel mesh compound nonwoven fabric, Fibers Polym. 9 (6) (2009) 761–767.
[18] C.C. Lin, C.M. Lin, C.C. Huang, C.W. Lou, H.H. Meng, C.H. Hsu, J.H. Lin, Elucidating the design and impact properties of composite nonwoven fabrics with various filaments in bulletproof vest cushion layer, Text. Res. J. 79 (3) (2009) 268–274.
[19] J.H. Lin, C.H. Hsu, H.H. Meng, Process of preparing a nonwoven/filament/woven-fabric sandwich structure with cushioning effect of ballistic resistance, Fibres Text. 13 (2005) 43–47.
[20] A. Shahdin, L. Mezeix, C. Bouvet, J. Morlier, Y. Gourinat, Fabrication and mechanical testing of glass fiber entangled sandwich beams: a comparison with honeycomb and foam sandwich beams, Compos. Struct. 90 (4) (2009) 404–412.

[21] S. Chocron, A. Pintor, D. Cendón, C. Roselló, V. Sanchez-Galvez, Characterization of Fraglight non-woven felt and simulation of FSP's impact in it, Technical University of Madrid, 2002. Technical Report.
[22] F. Martínez-Hergueta, A. Ridruejo, C. González, J. Llorca, Deformation and energy dissipation mechanisms of needle-punched nonwoven fabrics: a multiscale experimental analysis, Int. J. Solids Struct. 64 (2015) 120–131.
[23] F. Martínez-Hergueta, A. Ridruejo, C. González, J. Llorca, A multiscale micromechanical model of needlepunched nonwoven fabrics, Int. J. Solids Struct. 96 (2016) 81–91.
[24] R. Gerlach, C. Kettenbeil, N. Petrinic, A new split Hopkinson tensile bar design, Int. J. Impact Eng. 50 (2012) 63–67.
[25] F. Martínez-Hergueta, A. Pellegrino, A. Ridruejo, N. Petrinic, C. Gonzalez, J. Llorca, Dynamic tensile testing of needlepunched nonwoven fabrics, Appl. Sci. 10 (15) (2020) 5081.
[26] F. Martínez-Hergueta, A. Ridruejo, F. Gálvez, C. González, J. Llorca, Influence of fiber orientation on the ballistic performance of needlepunched nonwoven fabrics, Mech. Mater. 94 (2016) 106–116.
[27] F. Martínez-Hergueta, D. Ares, A. Ridruejo, J. Wiegand, N. Petrinic, Modelling the in-plane strain rate dependent behaviour of woven composites with special emphasis on the non-linear shear response, Compos. Struct. 210 (2019) 840–857.
[28] F. Martínez-Hergueta, A. Ridruejo, C. González, J. Llorca, Numerical simulation of the ballistic response of needle-punched nonwoven fabrics, Int. J. Solids Struct. 106 (2017) 56–67.
[29] H.B. Kocer, Laminated and Hybrid Soft Armor Systems for Ballistic Applications (Ph.D. thesis), Auburn University, 2007.
[30] F. Martínez-Hergueta, A. Ridruejo, C. González, J. Llorca, Ballistic performance of hybrid nonwoven/woven polyethylene fabric shields, Int. J. Impact Eng. 111 (2018) 55–65.
[31] S. Chocron, E. Figueroa, N. King, T. Kirchdoerfer, A.E. Nicholls, E. Sagebiel, C. Weiss, C.J. Freitas, Modeling and validation of full fabric targets under ballistic impact, Compos. Sci. Technol. 70 (13) (2010) 2012–2022, https://doi.org/10.1016/j.compscitech.2010.07.025.
[32] S. Chocron, T. Kirchdoerfer, N. King, C.J. Freitas, Modeling of fabric impact with high speed imaging and nickel-chromium wires validation, J. Appl. Mech. 78 (5) (2011) 051007.
[33] D.A. Shockey, D.C. Erlich, J.W. Simons, Lightweight fragment barriers for commercial aircraft, in: 18th International Symposium on Ballistics, 1999, pp. 15–19.
[34] Z. Stahlecker, B. Mobasher, S.D. Rajan, J.M. Pereira, Development of reliable modeling methodologies for engine fan blade out containment analysis. Part II: finite element analysis, Int. J. Impact Eng. 36 (3) (2009) 447–459.
[35] ArcelorMittal, Bake hardening steels, 2018. *http://corporate.arcelormittal.com/*.
[36] H. Hooputra, H. Gese, H. Dell, H. Werner, A comprehensive failure model for crashworthiness simulation of aluminium extrusions, Int. J. Crashworthiness 9 (5) (2004) 449–464.
[37] R. Kiran, K. Khandelwal, A triaxiality and Lode parameter dependent ductile fracture criterion, Eng. Fract. Mech. 128 (2014) 121–138.
[38] Dassault Systemes, Abaqus 6.12 User's Manual, 2012.
[39] J. Lemaitre, A Course on Damage Mechanics, Springer Science & Business Media, 2012.
[40] Z.P. Bazant, J. Planas, Fracture and Size Effect in Concrete and Other Quasibrittle Materials, vol. 16, CRC Press, 1997.
[41] J. Vila-Ortega, A. Ridruejo, F. Martínez-Hergueta, Multiscale numerical optimisation of hybrid metal/nonwoven shields for ballistic protection, Int. J. Impact Eng. 138 (2020) 103478.

Numerical analysis of the mechanical behaviour of an entangled cross-linked fibrous network

12

Fadhel Chatti[a], *Christophe Bouvet*[a], *and Dominique Poquillon*[b]
[a]Université de Toulouse, Institut Clément Ader, ISAE-SUPAERO—UPS—IMT Mines Albi—INSA, Toulouse, France, [b]CIRIMAT, Université de Toulouse, INP-ENSIACET, Toulouse, France

12.1 Introduction

Fibrous materials are largely used in applications such as the core materials of sandwich structures, insulation structures, or lightweight structures. This family of materials allows for a wide variety of architectures. One of the difficulties of this type of material, or more precisely of this type of structure because a fibrous material is both a material and a structure, is to link the properties of the basic components with the properties of the global fibrous material. This link is very important because one of the main purposes of this type of material is to be a 'material by design'. Indeed, fibrous materials are manufactured by mixing components, mainly fibres. But some other parts, such as resins, spherical particles, or powders, can be added. It is relatively easy to add these to improve the targeted properties, e.g. the insulation properties or the mechanical ones.

To link the properties of the basic components of a fibrous material to the global ones, some models are needed. In this chapter, a numerical model of the mechanical property – more particularly of the compressive stiffness of a fibrous material – is proposed. Unlike many analytical models available in the literature, the numerical model is based on physical data. It can therefore easily be used to illustrate the effects of different parameters. For example, in the present study, this model is used to simulate the compression of an entangled cross-linked material manufactured with carbon fibre and a spray of epoxy resin to link the fibres together. It is relatively easy and useful to change some parameters such as the fibre characteristics, the volume fraction, or the distribution of the fibre orientations to study their effects on the global mechanical behaviour of the fibrous material. The ultimate goal of the simulation is to optimise the material for a given application. For example, one can imagine adding metallic fibres to a fibrous material made of carbon fibres to increase the electrical conductivity. It is also possible to add Kevlar fibres in the core of the fibrous material of a sandwich structure to increase the damage resistance.

Mechanics of Fibrous Networks. https://doi.org/10.1016/B978-0-12-822207-2.00013-1

This chapter is divided into four parts. The first part is a literature review focusing on some analytical or numerical models devoted to the mechanical behaviour of fibrous materials. The effect of first-order parameters such as the volume fraction of fibres, the average length between contacts, fibre diameter, and Young's modulus of fibres will be presented. The second part presents the main characteristics of the developed numerical model. In particular, the choice of the representative volume element (RVE) and the identification of the epoxy cross-links will be discussed. Then, the third part deals with using the model to highlight the effect of some morphological and mechanical property characteristics. For example, the effects of the distance between cross-links, their stiffness, the fibre diameter, the Young's modulus, or the volume fraction will be explained and compared to the previous models available in the literature. Finally, this chapter will be concluded and some prospects will be proposed.

12.2 Modelling of the entangled fibrous networks

12.2.1 Analytical models

The first models, which were developed to understand the mechanical behaviour of fibrous materials, are analytical models based on the flexibility of entanglement due to the bending of fibres. The other mechanisms such as compression, torsion, or traction in the fibres are neglected. Analytical approaches have been developed from the analysis of the deformation of an elementary cell. The latter is composed of a segment of fibre held by two contacts and stressed in bending. The discrete element method [1], as well as the finite element method, have also been used to model the mechanical behaviour of entangled materials (see next section).

Van Wyk is the first author to develop a uniaxial compression model for a three-dimensional (3D) random fibre network [2] His model only takes into account one mode of deformation: bending. Neither the friction between the fibres nor the torsion of the fibres is modelled. The author has proposed the following expression, which presents the relationship between the pressure P and the volume fraction f:

$$P = kE_f\left(f^3 - f_0^3\right) \tag{12.1}$$

where k is an empirical constant, E_f is the elastic modulus of the fibre, and f and f_0 are, respectively, the compressed and initial volume fractions. To obtain this equation, Van Wyk began his study by focusing on the relationship between the increase in pressure dP and the deflection $d\delta$ of the assembly:

$$dP = \frac{bN_fL}{V\bar{l}}\,d\delta \tag{12.2}$$

where $\bar{l}$ is the length between two consecutive contacts, b is the height of the elementary cell, and N_f is the number of fibres of length L in a volume V. Therefore, $\frac{N_fL}{V}$ represents the fibre length per unit volume.

Van Wyk's model is valid only for fibre networks with a moderate volume fraction of less than 10%.

Komori and Makishima have proposed an approach, which is used to predict the number of contacts between fibres for different distributions of orientations [3]. They assumed that in the case of sufficiently long fibres ($L/D \gg 1$), the interactions due to the ends of the fibres can be neglected. The number of contacts $\overline{n}$ in a fibre network is evaluated as follows:

$$\overline{n} = \frac{V}{2DNL_t I} \tag{12.3}$$

where D is the fibre diameter, NL_t is the total length of fibre per unit volume, and I is the average value of $\sin\theta$, which characterises the distribution of fibre orientations. The application of this approach in the case of a random fibre network provides results consistent with those found by Van Wyk [2]. In the case of a sheet structure having all the fibres roughly parallel to the plane of the sheet, the approach has shown good agreement with the results obtained by Kallmes and Corte [4]. Ning Pan has shown that the number of contacts determined from Komori and Makishima's approach [3] is overestimated [5]. In particular, this author questions the expression of the probability P that two fibres will touch when a test fibre is placed in a volume V and another fibre is inserted randomly:

$$P = \frac{2DL_t^2}{V} I \tag{12.4}$$

He determined that the assumption about the probability of establishing contact with the added fibre could not be independent and equal for all the other $N-1$ fibres. This assumption is indeed not correct. So, Ning Pan showed that only the first of the $N-1$ fibres could follow this expression of the probability of coming into contact with an added fibre. Ning Pan then proposed a modified approach to predict the morphological characteristics of the fibre networks [5] such as the number of contacts:

$$\overline{n} = \frac{8sfI}{\pi + 4f\psi} \tag{12.5}$$

where $s = \frac{L_t}{D}$ is the shape ratio of the fibre, which is equal to $\frac{\pi^2}{4}$ in the case of a random 3D fibre network, f is the volume fraction and ψ is a function depending of the density of fibre orientation.

Lee and Carnaby evaluated the energetic approach [6] developed to analyse the compression of a 3D random fibre arrangement [7]. In this theory, the total energy E_{tot} is equal to the sum of the energies due to bending E_B, tensioning E_T, and sliding between the fibres E_T. This energy is calculated as follows:

$$E_{tot} = \sum_{i=1}^{N_B} E_{Bi} + \sum_{i=1}^{N_T} E_{Ti} + \sum_{i=1}^{N_S} E_{Si} \tag{12.6}$$

where the energies E_B and E_T are calculated from the classical energy equation developed in a curved beam. In the case of sliding segments, the evolution of the bending energy is calculated before the tensioning of the segments. Even if a slip can occur after the tensioning, no energy is added since the work due to friction is not taken into account by the authors. To evaluate the proposed model, the authors carried out various studies by varying only one parameter each time amongst those controlling the compression energy. They concluded that the interpretation of these results in terms of experimental behaviour was subtle: in real experiments, it is effectively not possible to modify the fibre whilst the other geometric parameters remain unchanged [7].

Komori and Itoh have proposed an expression describing the compression behaviour of a set of isotropic fibres [8,9]:

$$\sigma = \frac{4}{9} E_f \left(f^3 - f_0^3 \right) \tag{12.7}$$

This model is based on the imposed deformation. The bending energy of the system is deducted from each new distribution of the fibre orientations obtained after each imposed compression increment. For reasons of simplification, the model does not take into account the effect of sliding. Komori and colleagues improved this first model 1 year later by introducing the effect of the initial tortuosity, generally defined as the average of the length between the ends of the fibres over their total length [9].

Toll developed a micromechanical theory based on a statistical study of the distribution of contacts to understand the compression behaviour of a random fibre network in 2D or 3D [10]. He adopted practically the same hypotheses as those of Van Wyk. An equation allowing for the determination of the number of contacts $\overline{n}$ has been proposed to describe the microstructure:

$$\overline{n} = \frac{\pi}{8} f r g \tag{12.8}$$

where f is the volume fraction of the fibre, g is a constant dependent on the distribution of the fibre orientation, and r is the shape ratio of the fibre.

This approach was then used to evaluate the pressure P depending on the volume fraction:

$$P = E_f \left(f^n - f_0^n \right) \tag{12.9}$$

where n depends on the dimension of the fibre network. In 3D, this is equal to 3 whilst in 2D it is equal to approximately 5.

This approach remains limited because the author assumes that there is no statistical correlation between the length of the fibre and the distribution of the incremental forces. In fact, this assumption cannot be valid when the individual fibres differ greatly in stiffness or size.

In 1997, Gibson and Ashby proposed a model studying the behaviour of a cellular material (Fig. 12.1) whose cells can be considered as fibres linked together by

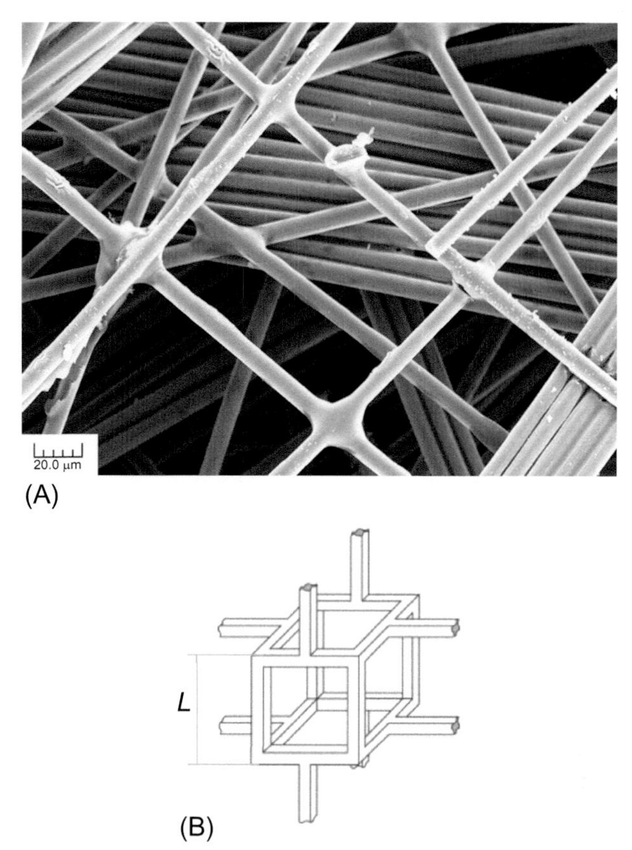

Fig. 12.1 (A) SEM observation of a cross-linked entangled material [12] and (B) cell of a cellular material with open porosity [11].

junctions (Fig. 12.1A). This model can then be applied to materials entangled with permanent links. This model provides an expression for the elastic modulus, written as follows [11]:

$$E = \frac{3\pi E_f}{4\left(\frac{L}{D}\right)^4} \tag{12.10}$$

For low-density materials, the modelling of this cellular architecture only takes into account the bending of the edges. The contribution of compression is added to bending, but it remains negligible as long as the density remains low.

Markaki and Clyne proposed another approach, which consists in determining the overall behaviour of the fibrous material from the behaviour of a single fibre whilst using a homogenisation method [13]. They postulated that all fibres are in series and subjected to the same force W. This force, applied to each fibre of the assembly,

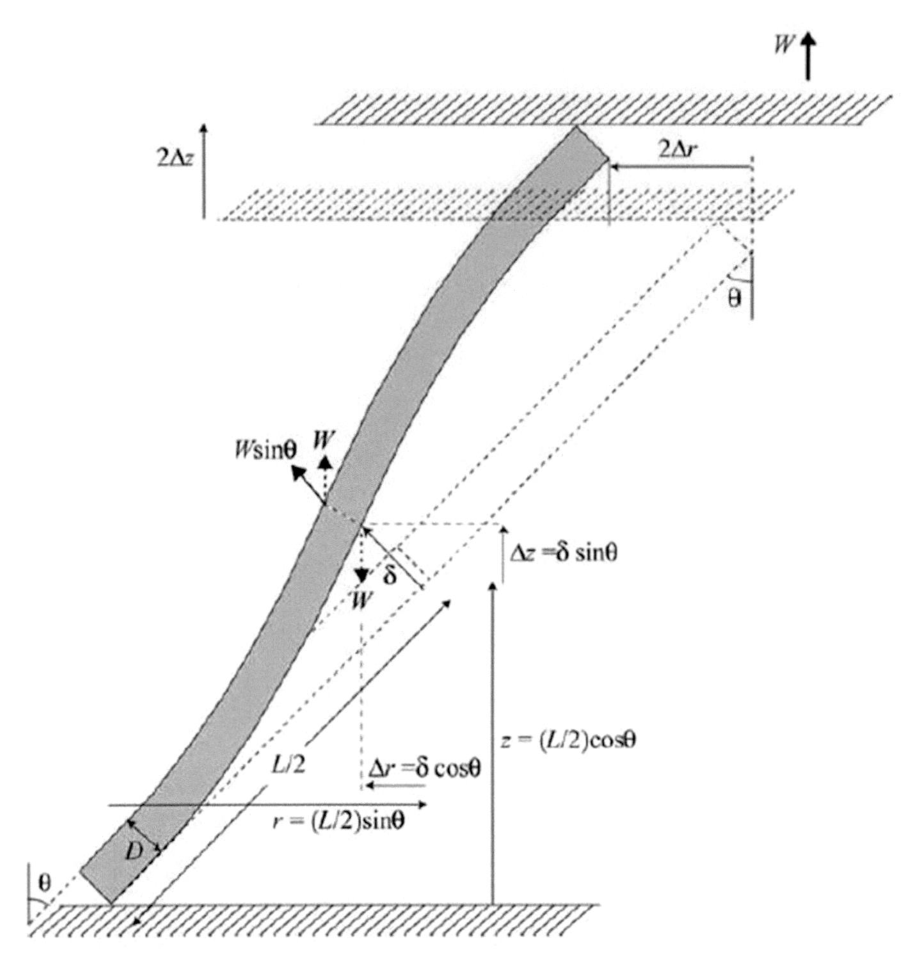

Fig. 12.2 Principle of the modelling of a single fibre [13].

induces only its flexural deformation (Fig. 12.2). This loading is separated into two contributions: a compression force and a bending force. Only the part due to bending is taken into account by the authors. The bending force induces a deflection which is characterised by a displacement Δz. The boundary conditions imposed by the authors suppose that the beams are embedded in such a way that the stiffness of the junction is supposed infinite.

The stress σ is evaluated from the force W applied to the set of fibres crossing a given section:

$$\sigma = NW \tag{12.11}$$

where N represents the number of fibres crossing a given section.

Compression strain is calculated in turn from displacement Δz:

$$\varepsilon = \frac{<\Delta z>}{<Z>} \tag{12.12}$$

where $<Z>$ is the average height of the fibres and $<\Delta z>$ is the sum of the displacements of the fibres, which takes into account their probability of presence.

Δz is determined depending both on the force W, the fibre orientation θ, and its materials resistance. The fibre is modelled as a beam of circular section (diameter D), that is:

$$\Delta z = \frac{16WL^3(\sin\theta)^2}{3E_f\pi D^4} \tag{12.13}$$

where L is the distance between junctions assumed to be constant for all fibres and E_f is the Young's modulus of the fibre.

The expression of Young's modulus deduced from this approach and proposed by Markaki and Clyne is

$$E_{MC} = \frac{\sigma}{\varepsilon} = \frac{9fE_f}{32\left(\frac{L}{D}\right)^2} \tag{12.14}$$

This model is applied only in the cases of small deformations. Indeed, it does not take into account the creation of new contacts during loading. In addition, the blockages at the junctions are infinitely rigid.

12.2.2 Numerical models

In 2002, Beil and Roberts developed a numerical model allowing for the prediction of the compression behaviour of a network of fibres from its morphological properties [14–16]. This model takes into account the static friction μ_s and the sliding friction μ_k between fibres. Two forces have been defined to model the contact interactions: F_r corresponds to the repulsion force, and the friction force F_r is calculated as follows:

$$F_r = Cd_c \tag{12.15}$$

where C is a positive constant and d_c corresponds to the distance between points in contact. In addition, the friction force F_f is equal to:

$$F_f = \mu_k F_r \tag{12.16}$$

when applied to a sliding contact.

Beil and Roberts have discretised the fibres into segments described by two vectors defining their sections and their curvilinear abscissa [14]. To assimilate the fictitious extension of the fibres outside the cubic volume modelled, they introduced two types of springs at the ends of each fibre. Those of low stiffness were applied to limit the movements in the plane of each surface, whilst those of high stiffness make it possible to limit the movements perpendicular to the faces of the cube.

The results of the numerical simulations showed a nonlinear behaviour of the fibre network with a number of contacts, which is higher during the deformation compared to what was predicted by Van Wyk. Obtaining these results required a very high calculation cost even for a small fibre volume fraction, not exceeding 1%.

Beil and Roberts studied the hysteresis phenomenon in compression through the analysis of the energy of the fibre network [15]. Fig. 12.3 shows the curve obtained for the sixth loading/unloading cycle. Their mathematical model is able to predict the energy loss due to friction. Energy dissipation decreases as the cycle progresses. However, this decrease is attenuated after a few cycles. This result is not in agreement with the experimental results. The authors believe that this may be due to the omission of viscoelastic effects.

Durville proposed a general approach for simulating the mechanical behaviour of an entangled fibre network [17]. This approach is suitable for large deformation, and it takes into account contact-friction interactions between fibres. To do this, he developed an in-house finite element code optimising contact management. To determine the contact elements, which couple two particles of material, an intermediate geometry is constructed in each region where two pieces of fibre are close. A contact between them can then be established. Each particle is defined, as in Beil and Roberts' study [14,15], by its curvilinear abscissa and its two coordinates in the crossing section.

The proximity zones are defined as parts of mean lines of beams, which are close to each other. For each pair of fibres in the network, the author calculates the distances between the points regularly distributed on one of the average lines and the closest points corresponding on the other line to determine the zones of proximity. The curvilinear abscissa intervals delimited by successive pairs of neighbouring points define a pair of parts of proximity zones (Fig. 12.4). The contact friction is introduced through a modified Coulomb's law, which allows small reversible displacements

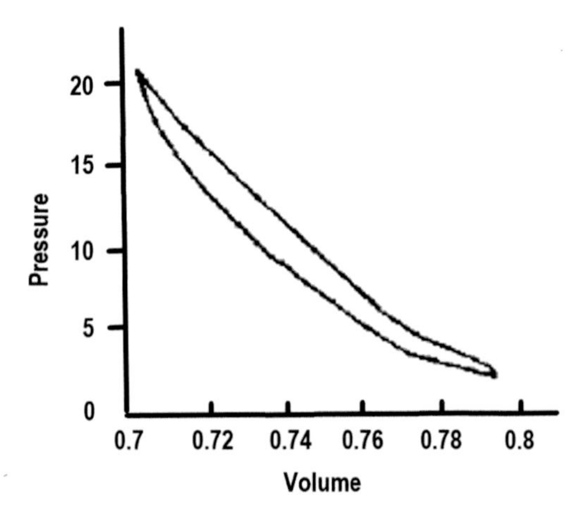

Fig. 12.3 Hysteresis of the pressure observed during the sixth loading/unloading cycle as a function of the volume [15].

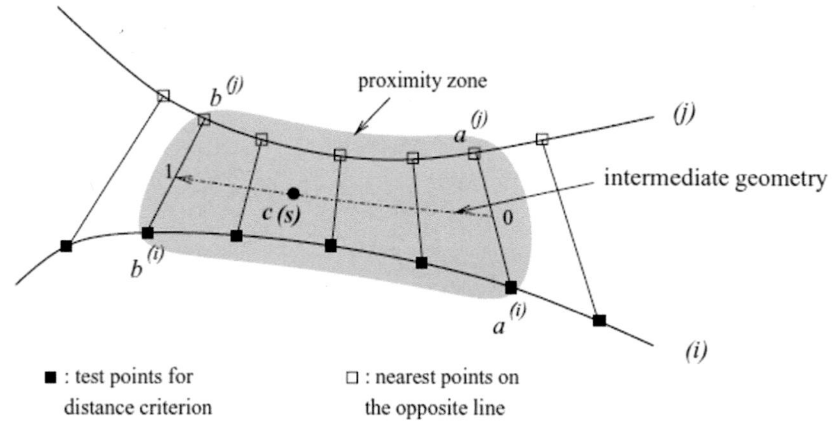

Fig. 12.4 Diagram of nodal discretisation of fibres and of contact between fibres [17].

before sliding. If the tangential component of displacement ΔU_T is less than the adjustable parameter corresponding to the reversible tangential displacement $u_{T,\ rev}$, then:

$$R_T = \frac{\mu \|R_N\|}{u_{T,rev}} \Delta U_T \tag{12.17}$$

Else:

$$R_T = \frac{\mu \|R_N\|}{\|\Delta U_T\|} \Delta U_T \tag{12.18}$$

where R_T is the tangential component in contact due to friction, R_N is the normal component of the reaction in contact, and μ is the friction coefficient.

For small deformations, the compressive force is equal to the relative variation of the volume fraction at the power 3/2.

In 2005, Rodney and his colleagues also presented a study based on the techniques used for modelling the mechanical behaviour of long polymer chains [18]. Their model is based on molecular dynamics, which consists in discretising fibres into a series of spherical nodes (pearl necklace) (Fig. 12.5). Each spherical node has three degrees of freedom corresponding to displacements in the three directions of space.

The tensile and bending behaviour of the fibres, as well as the non-penetration between fibres, are taken into account in this model. The potential energy of the system is then calculated as follows:

$$\begin{aligned} E = & \sum_{(i,i+1)consecutive} \frac{K_s}{2}\left(1 - \frac{r_{i,i+1}}{D}\right)^2 + \sum_{(i-1,i,i+1)consecutive} \frac{K_B}{2}(\theta_i - \pi)^2 \\ & + \sum_{(i,j)nonconsecutive} \frac{K_I}{2} H(D - r_{i,j})\left(1 - \frac{r_{i,j}}{D}\right)^{5/2} \end{aligned} \tag{12.19}$$

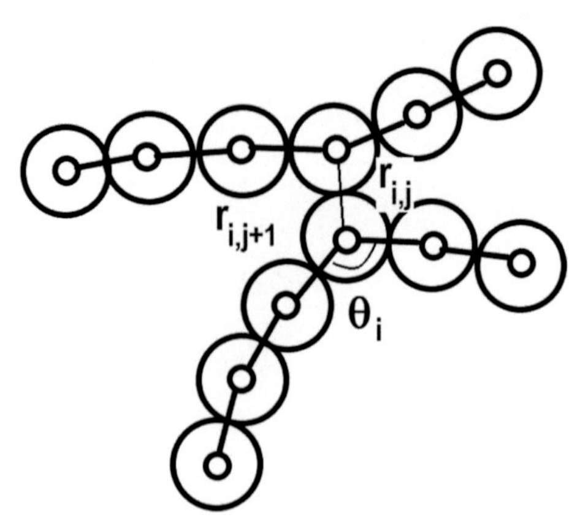

Fig. 12.5 Fibre nodal discretisation scheme [18].

where K_s is related to the tensile stiffness of the fibres, $r_{i,\ i+1}$ is the distance between consecutive nodes, D is the fibre diameter, K_B is related to the bending stiffness of fibre, θ_i is the angle formed by three consecutive nodes, and $H(D - r_{i,\ j})$ is the Heaviside function.

The first sum in Eq. (12.19) corresponds to the tensile stiffness of the fibres, which is modelled by springs between consecutive nodes. The second sum corresponds to the bending stiffness of the fibres, which is introduced through angular springs between the pairs of consecutive elements. The third sum corresponds to the modelling of the interactions between fibres by pairs of repulsive potentials.

In the case of a 3D random fibre network, the model has an exponent of 3 between the volume fraction and the compression pressure, which is in line with Van Wyk's theoretical approach. The number of contacts per fibre increases linearly with the aspect ratio of fibre L/D.

This model is limited due to its very high calculation cost. Indeed, the increase in the aspect ratio of the fibres is followed by a linear increase in the number of elements, and therefore the calculation time increases sharply. To reduce the calculation cost of this model, Barbier and his colleagues proposed a method which involves replacing spherical nodes by cylindrical elements of the same diameter as shown on Fig. 12.6 [19].

The compression of the assembly is done in stages, and the system is relaxed after each loading step to achieve balance. The number of relaxation steps depends on the length of the segments; it varies between 75,000 and 300,000. The frictional force is

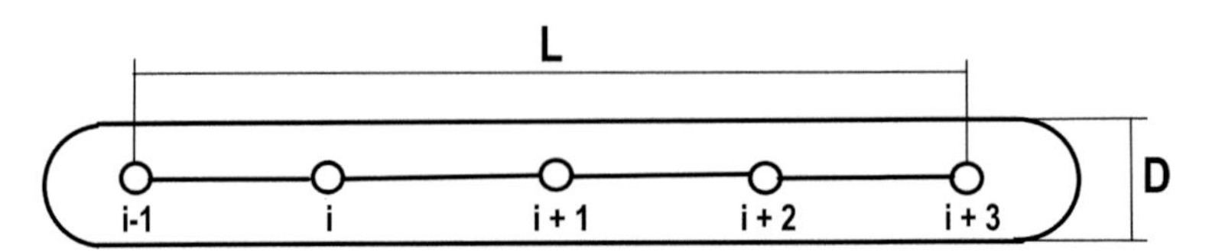

Fig. 12.6 Diagram of a fibre of diameter D composed of four segments of length l [19].

modelled by a spring generated each time a contact is detected between two points of contact of two fibre surfaces. Coulomb-type friction was introduced at the contact points.

This work made it possible to determine the evolution of the pressure P as a function of the relative density of the fibres f during loading. This evolution is well described by a power law:

$$P = (f - f_c)^m \tag{12.20}$$

where f_c represents the volume fraction of fibres at mechanical percolation, with an exponent m close to 3 in agreement with Van Wyk's theory in the case of isostatic compression and an exponent between 2 and 3 in the case of shear.

Fig. 12.7 shows the effect of static friction on the loading/unloading compression cycle. When the friction coefficient increases, the densification appears earlier and the hysteresis also increases. The authors found that regardless of the coefficient of friction value used, the bending energy is always greater than the tensile energy [19]. This result is consistent with previous studies which assume that the main mode of deformation of all the fibres is their bending around the contact points.

Abd El-Rahman and Tucker [20,21] proposed a finite element model to study the mechanical behaviour in compression of a network of fibres. Their model can be used for fibres with an aspect ratio equal to 100 and a fibre volume fraction of up to 25%. A special algorithm, using the sequential random adsorption process, has made it possible to create an initial structure of non-crossed, straight, and random fibres [20]. From this structure, a unit cell with periodic boundary conditions is built. Two types of periodic boundary conditions have been studied: partial and whole. The authors

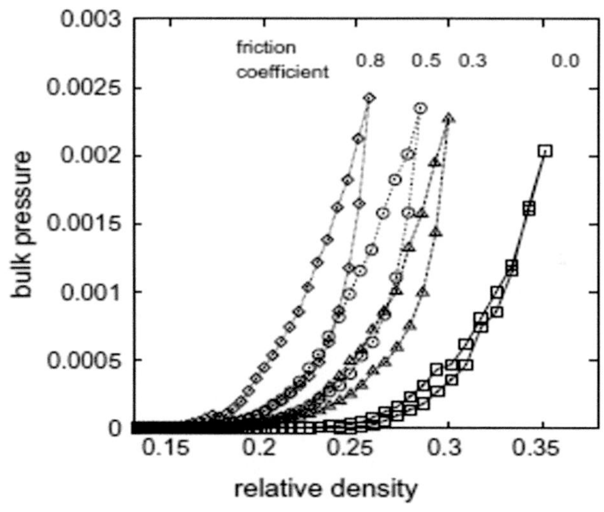

Fig. 12.7 Influence of the friction coefficient on the pressure curve as a function of the relative density [19].

showed that there was a significant difference in the compressive stress between the two boundary conditions used. This difference increases when the aspect ratio decreases. Numerical results are also affected by the fibre output from the simulation volume. This output of fibres can have an impact on the compression stress, the distribution of the fibres, and the number of contacts.

Fig. 12.8A shows that the stress curve as a function of the volume fraction correlates well with that of Van Wyk in the case where the friction coefficient is very large

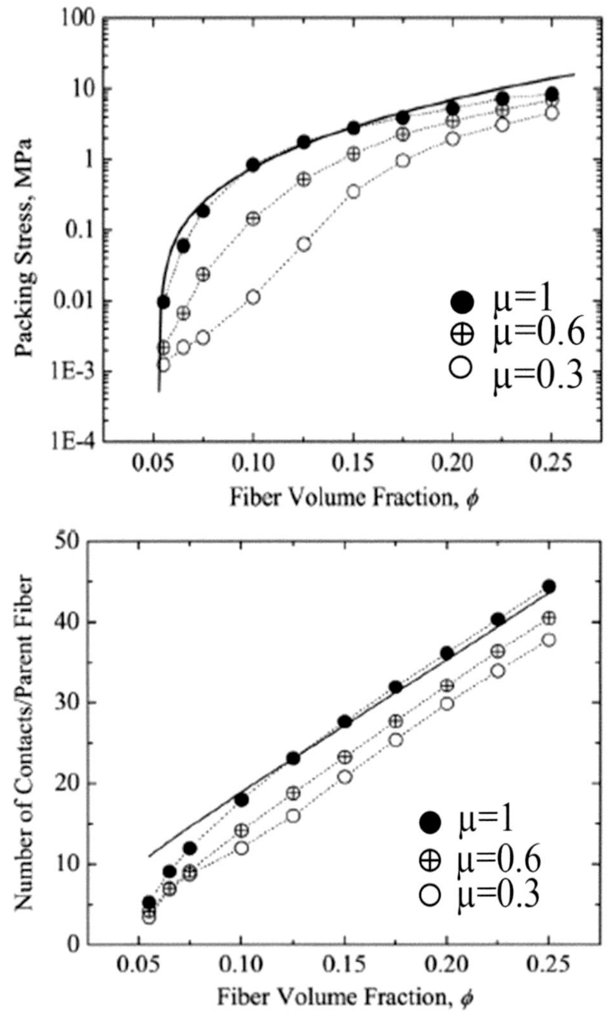

Fig. 12.8 Effect of the friction coefficient μ on (A) the mechanical behaviour of a fibre network and (B) the number of contacts [21].

(close to 1). The developed model validates Van Wyk's theory for moderate volume fractions of less than 15%. The model also has good agreement with Toll's theory in the calculation of the number of contacts when the coefficient of friction introduced is equal to 1 (Fig. 12.8B).

Recently, Zhang and his colleagues presented a finite element model with periodicity to predict the mechanical properties of CBCF 'carbon-bonded carbon fibre composites' (Fig. 12.9) [22].

In this work, a uniform random distribution was adopted to describe the distribution of the directions of the fibres in the plane, and a Gaussian distribution was adopted for the out-of-plane distribution. Each fibre is considered as an aggregation of points:

$$C_i = \left\{ K_i^r \middle| \overrightarrow{OK_i^r} = F\left(\lambda, \vec{r}, \vec{L_i}\right), \lambda \in \left[-\frac{L}{2}, \frac{L}{2}\right] \text{ and } \left|\vec{r}\right| \in \left[-\frac{D}{2}, \frac{D}{2}\right] \right\} \tag{12.21}$$

where K_t is the point i of the fibre, λ is the curvilinear coordinate of the fibre of length L, and $\vec{r}$ is the position of an element of the fibre of diameter D.

To avoid overlapping fibres, a criterion has been introduced to distinguish the relative position between two neighbouring fibres. The mathematical implementation of non-overlap between fibres is expressed as follows:

$$C_i \cap C_j = \phi \tag{12.22}$$

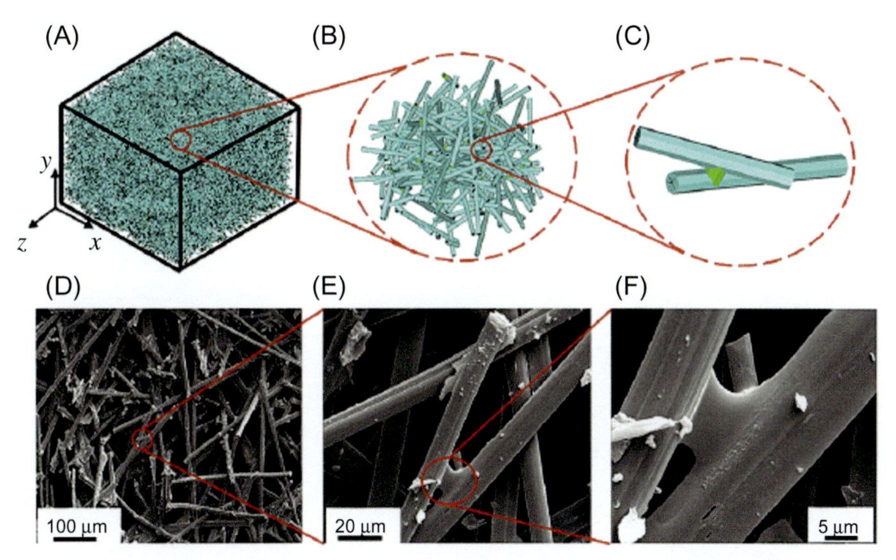

Fig. 12.9 Geometric model of random fibre network: (A) model of the representative elementary volume, (B) close-up view of the network, (C) enlarged image of two bonded fibres, with fibres coloured in *cyan* and the bonding material in *green*, and (D)–(F) SEM images of the microstructure in CBCFs at different scales [22].

This means that there is no common element between the two aggregations corresponding to two fibres. The two fibres are connected by a Timoshenko-type beam each time the distance between them satisfies the following criterion:

$$0 < d_{min}(C_i, C_j) \leq 0.5D \tag{12.23}$$

Fig. 12.10A shows that the elastic modulus of CBCF does not increase linearly with density but rather follows a power law dependence with an exponent of 3. Increasing the fibre length improves the mechanical properties of CBCF. A study of the effect of fibre orientations on compressive stiffness was also carried out by the authors. The standard deviation defining the distribution of fibre orientations relative to the out-of-plane direction was used to study this effect. Fig. 12.10B shows that the increase in the standard deviation generates a reduction of the in-plane elastic modulus and increases the out-of-plane one. Logically, the larger the proportion of fibres with directions in the vicinity of the compression direction, the more the elastic modulus of the CBCF increases.

In 2017, Ma and his colleagues also developed a finite element model to describe the mechanical behaviour of 'porous metal fibre sintered sheets' (MFSS) [23]. The digital geometry (Fig. 12.11B) was constructed from results of the analysis of a sample by X-ray tomography (Fig. 12.11A). This provides access to the architecture of the fibre network.

The building of the geometry of the digital model is gradually deduced thanks to the two-dimensional coordinates previously known. To build the model of the periodic stochastic fibre network in a two-dimensional square domain (the x–y plane) with each side equal to 1 mm, the diameter d, the x and y coordinates of the centre of the fibre, the orientation θ, and the length of the fibre L are all specified by independent random numbers that are generated digitally. The two x and y coordinates are controlled so that they are between 0 and 1. The fibre diameter varies between 8.6 and 13.6 μm (the average diameter is $d = 11.1$ μm) whilst the fibre length varies between

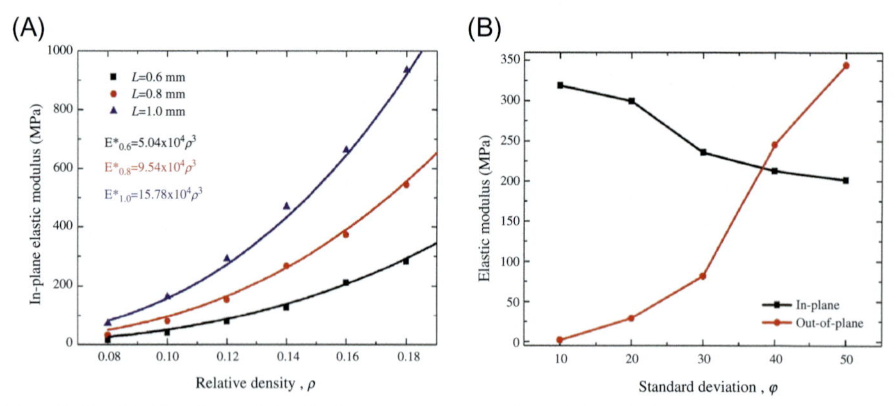

Fig. 12.10 (A) Elastic modulus in the plane as a function of the relative density with different lengths of fibres and (B) influence of the standard deviation on the elastic modulus [22].

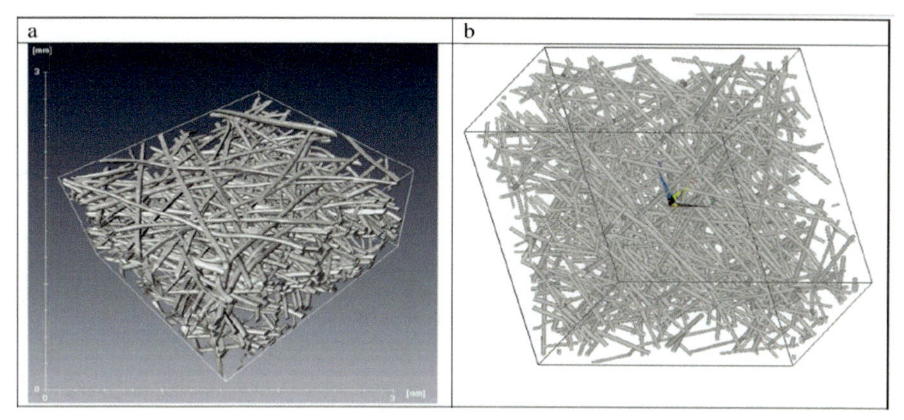

Fig. 12.11 (A) Rebuilding of geometry by tomography and (B) 3D digital model geometry [23].

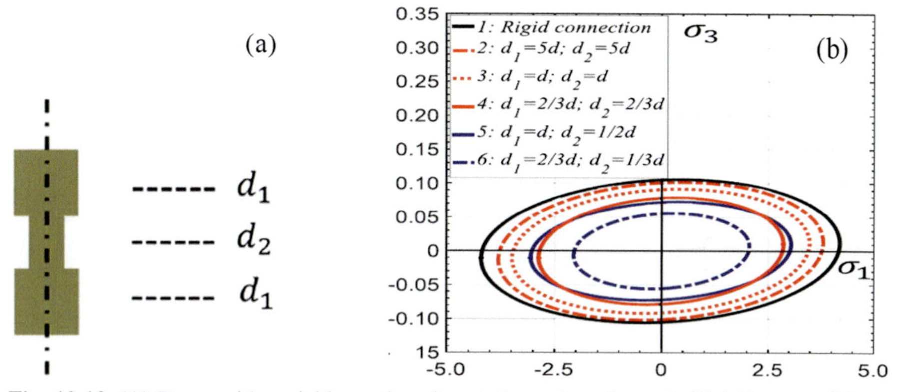

Fig. 12.12 (A) Beam with variable sections inserted as a junction and (B) influence of the different types of the junctions on the elastic limit of the MFSSs [23].

0.8 and 1.2 mm (the average length is $L = 1$ mm). The 3D model, which is based on the 2D model, requires a more complex generation in terms of intersections between fibres. The 2D model provides access to all the data on the possible intersections between the fibres. However, the management of the positions of the fibres takes into account the growth of the model in the out-of-plane direction. The fibres are stacked one by one until N fibres are generated. This process induces an anisotropic geometry with an out-of-plane behaviour different from that in the plane. In this work, the junctions between fibres are modelled by beams, which account for the different deformation mechanisms. The beams can have variable sections, as illustrated in Fig. 12.12A, to vary the stiffness of the joint. Fig. 12.12B shows the effect of the junctions on the elastic limit in the plane of anisotropy of the MFSSs according to the diameters of variable sections d_1 and d_2 (Fig. 12.12A).

12.2.3 Conclusion

The study of fibrous materials began several decades ago, but despite the large number of publications on the subject, the mechanical behaviour of sets of fibres during loading is still poorly known and understood. Analytical models are unable to describe in detail the evolution of the morphology during loading (number of contacts, distribution of fibre orientations, etc.) as well as the effect of this evolution on the overall behaviour of the fibrous material studied. Thanks to the evolution of computer resources, detailed descriptions of the instantaneous deformation of fibre networks, with or without permanent links, during loading are increasingly used in the literature. Due to limitations in the computational capabilities, it is not yet possible to study whole samples of several cubic centimetres containing tens of thousands of fibres. Thus, the mechanical behaviours of the samples are determined using an RVE. The numerical resolution methods primarily used in the literature are the finite element method and the discrete element method. In this work, detailed in the next paragraph, we used Abaqus commercial code, which is a finite element code, to determine the RVE and model its behaviour in different configurations.

Table 12.1 presents the characteristics of certain RVEs used in finite element models existing in the literature.

Numerical studies dealing with the compression of entangled fibre network without cross-linking do not describe the evolution of microscopic parameters for a large number of fibres or for volume fractions greater than 10% [14,15,17]. Abd El-Rahman and Tucker have recently succeeded in developing a numerical model capable of going up to more than 20% by volume fraction to describe the evolution of the structure of entangled fibre networks [21]. Unfortunately, during this study, the authors did not propose experimental validation of numerical simulations, e.g. in terms of the stress/strain curve. This point seems important, and in this work we will endeavour to compare the data from numerical simulations with experimental data.

12.3 Numerical generation of a fibre network

The numerical model of the compression behaviour of the entangled cross-linked fibrous material was based on the experimental identifications. First of all, the fibre network used at the beginning of fabrication is characterised by an isotropic

Table 12.1 Characteristics of some RVEs using the finite element method in the literature.

Model	RVE size (mm^3)	Fibre diameter (μm)	Initial fibre volume fraction (%)
Abd El-Rahman and Tucker [20,21]	$2 \times 2 \times 2$ and $3 \times 3 \times 3$	18	5
Zhang et al. [22]	$1.5 \times 1 \times 1.5$	9	14
Ma et al. [23]	$1 \times 1 \times 1$	11.1 (average)	8.6

distribution of the fibre orientation and an equivalence of all directions [24]. Then, the orientation of the fibres is modified where the network is compressed manually in a closed mould. On the other hand, the contacts are cross-linked by the epoxy resin after the introduction of fibres in the mould. At the end of manufacturing process, a transversely isotropic fibrous material is obtained. Two numerical simulations of compression were therefore done. In the first one, all the contacts were free to slide (no friction was taken into account), and the initial distribution of the fibre orientation was isotropic. As in the process, the volume fraction of the fibre network increased from 1.7% to 8.5%. In the second simulation, the network was initially anisotropic and some of contacts were blocked. It was then submitted to numerical compression tests.

A sample of carbon fibrous material with 8.5% of volume fraction can contain about 3.35 million 7-μm-diameter fibres. Due to computational limitations of supercomputers, a representative volume element (RVE) was used to simulate the compression stiffness of an entangled cross-linked fibrous material.

12.3.1 Generation of fibres in the RVE

The fibre network is generated in the RVE with an in-house pre-processing code written in FORTRAN. The RVE is cubic box with a size a, and it is studied in the next section (Fig. 12.13). The generation of the cross-linked fibre assembly is as follows. First, a point, which is represented in red in Figs 12.13–12.15, was randomly picked in the box. Then, a couple of angles (θ, φ) were chosen to create the fibre distribution. The polar angle θ defines the fibre direction relative to the compression z-axis, and the azimuthal angle φ represents the orientation of the fibre against the x-axis. The fibre direction is presented by the dashed green line in Fig. 12.14B. The intersection of the fibre direction and the cubic box induces the creation of two nodes, which are the extremities of the fibre. These nodes fix the length of fibres and are presented as purple

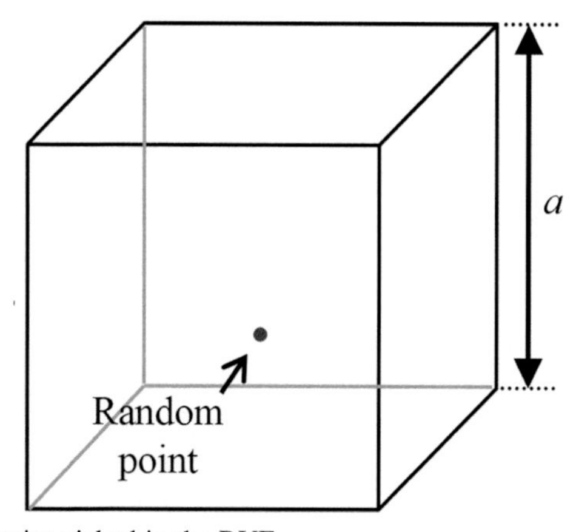

Fig. 12.13 Random point picked in the RVE.

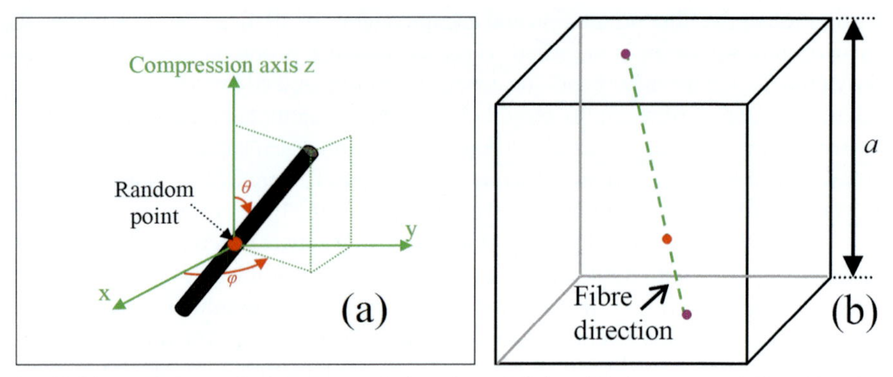

Fig. 12.14 (A) Polar angle θ and azimuthal angle φ defining (B) the generated fibre direction.

Fig. 12.15 Two nodes representing the extremities of numerical fibre.

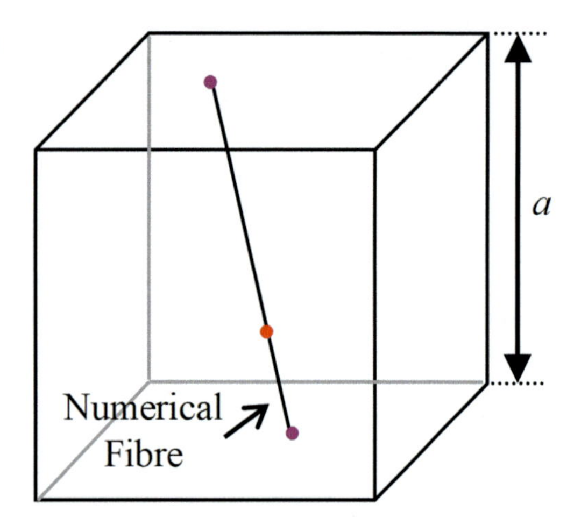

points in Figs 12.14B and 12.15. The fibres were generated in the RVE box with respect to the initial distribution of the fibre orientation that was chosen. When the distance between two fibres is smaller than the fibre diameter, this means that the two fibres intersect. Then two nodes are created, corresponding to the two closest points on the two fibres (cf. cyan points of Fig. 12.16). These nodes correspond to the extremities of 3D Timoshenko beam elements which model the fibres, and they are presented by cyan points in Fig. 12.16. The number of elements in each fibre is variable; it depends on both the orientation of fibres and the number of contacts. The last step of RVE generation consists of inserting springs that model the epoxy cross-links and bond some nodes of contact (cyan nodes). In Fig. 12.16, the springs are represented in brown.

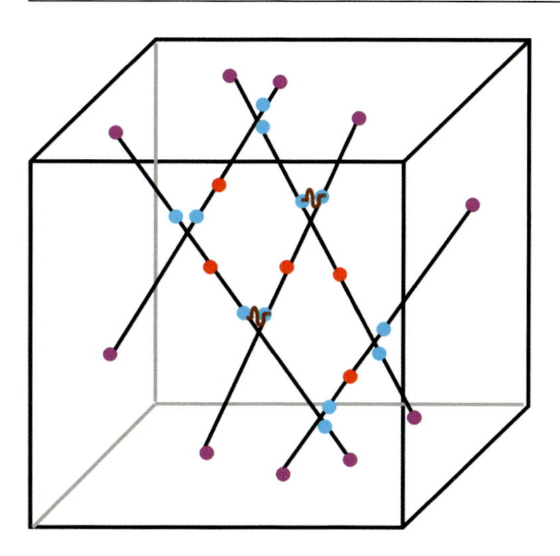

Fig. 12.16 Beam elements of the numerical model and insertion of springs modelling epoxy junctions.

12.3.1.1 Choice of the initial parameters of RVE

Size of RVE

The choice of the size of the RVE is very important to guarantee morphological isotropy for each generation of the fibre network. The determination of the appropriate size of the RVE is based on two important morphological parameters:

- The average distance between the contacts: Many studies have confirmed that this parameter has a very important influence on the global behaviour of the entangled fibrous material [25–27]. In a previous study [25], six box sizes were tested: 0.25, 0.5, 0.75, 1, 1.25, and 1.5 mm. For each cube dimension, 10 statistical realisations were carried out to test the representative of the generated architecture. In the previous simulations, it was proven that the morphological parameter is more stable and closer to that of the real material where the size of the simulation box increases. We can conclude from the former study that the side length of 1 mm is appropriate for the RVE with 6.5% of volume fraction. In the preliminary simulation, this same length guaranteed the morphological isotropy in terms of the average distance between contacts even if the volume fraction of the RVE was equal to 8.5%.
- The distribution of the fibre directions: Ten random statistical realisations of RVE were carried out to verify the morphological isotropy in terms of this parameter. The tested RVE is characterised by 1 mm size, 8.5% of volume fraction, and an isotropic direction distribution. For each statistical realisation, the orientations of the fibres by the polar angle θ were grouped into 36 intervals of 10 degrees (Fig. 12.17). The comparison between the numerical curves and the theoretical one (coloured in black) shows a good agreement between the theory and the numerical generation of the fibre network (Fig. 12.18).

The generation of the geometry in a cubic box of 1 mm was deemed available because it can guarantee an acceptable compromise between the representativeness of the fibres generated for the simulation and the computation cost. This study of orientation distribution is not the only criterion to prove the representativity of the RVE, but it can present a good assumption about it. The size of 1 mm was therefore used for the rest of the study.

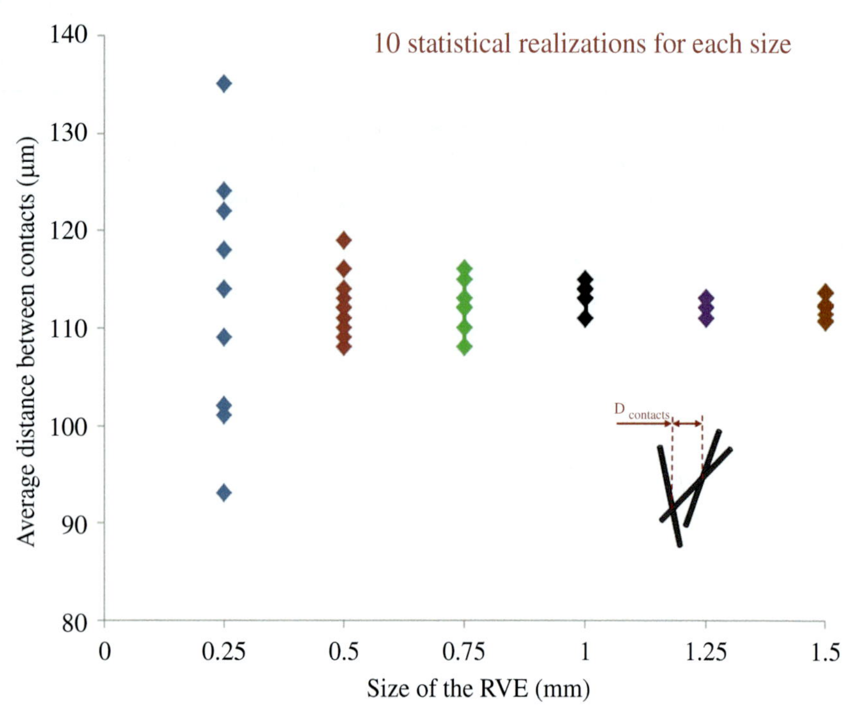

Fig. 12.17 Investigation of the average distance between contacts for a material with a volume fraction of 6% [25].

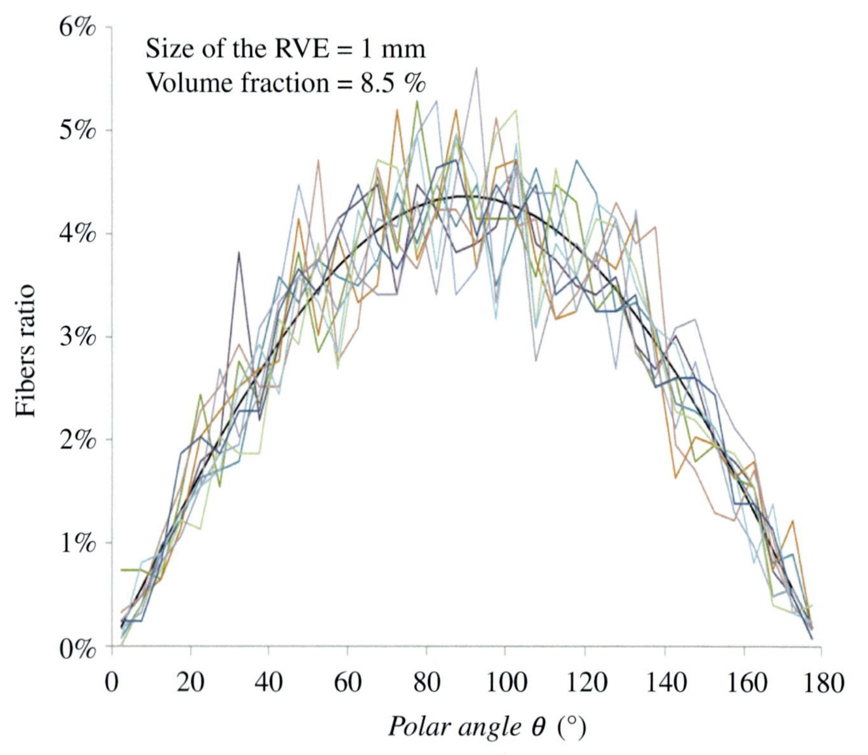

Fig. 12.18 Distribution of the fibre directions for the 10 random geometries generated for a cube with a 1-mm side length and a volume fraction of 8.5%.

Proportion of cross-linked contact

At the end of manufacturing process, the resin has polymerised and some of contacts between fibres can be blocked, as shown in Fig. 12.19A. The generation of cross-links is based on the distance between the resin junctions observed in the real sample. Mezeix et al. [12,24,28], who used the SEM analysis to determine the average distance between epoxy junctions for the same entangled cross-linked carbon fibres, found an average value equal to 120^{+140}_{-70} µm. The heterogeneity of the material can explain the uncertainty of this value. The heterogeneity is due to the occasional imperfect separation of the yarn and the random spray of epoxy during the manufacturing. Then, two out of three contacts were blocked with springs to numerically reach this average distance between cross-links. Fig. 12.19B shows that the insertion of a spring was managed as follows: two successive contacts were blocked, the next was left free, and so on until the end of the fibre; then, the positioning of springs continued with the next fibre to be generated.

12.3.2 Identification of epoxy joint stiffness

In the actual modelling, springs were added to model the polymerised epoxy junctions that blocked some of the contacts between fibres.

A new separate numerical model is used to determine the stiffness that is used in the springs of the principal model of the entangled cross-linked fibres (Fig. 12.20B). It is composed of two fibres bonded by an epoxy cross-link. This model is based on the approach of Mezeix et al. [28,29]. However, it is more developed, and a more advanced identification was carried out. Two types of stiffness (tension and torsion) related to the behaviour of epoxy junction were identified. In reality, there are other types of stiffness that characterise the behaviour of the epoxy junction: one torsion stiffness and two

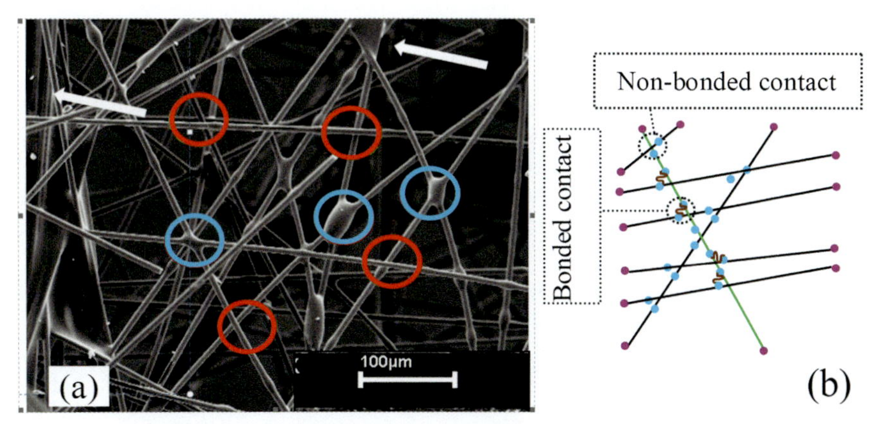

Fig. 12.19 (A) SEM observation of entangled cross-linked carbon fibres with a volume fraction of 8.5%. The *red circles* indicate non-bonded contacts, the *cyan circles* indicate cross-linked fibres by the epoxy resin, and the *white arrows* indicate that some of the yarn segments are not perfectly separated; (B) the method used to generate springs linking the contacts of one fibre (*green* one) with the other fibres (two out of three contacts are blocked).

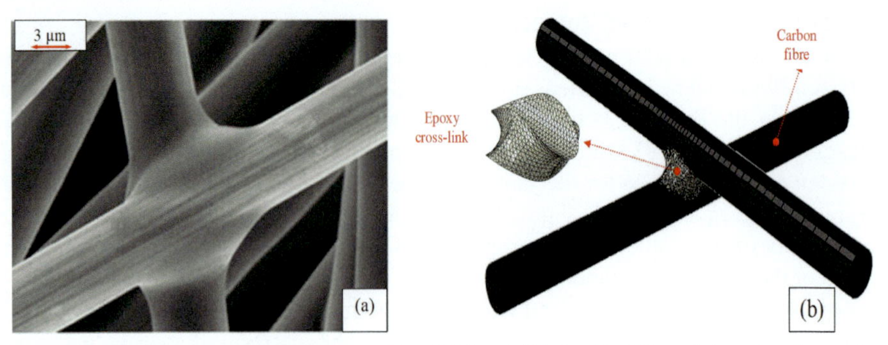

Fig. 12.20 (A) Epoxy cross-link observed by SEM and (B) the 3D FE model used to determine the spring stiffness.

bending stiffnesses. However, to obtain an isotropic stiffness matrix, just two stiffnesses are taken into account in this model: one tension stiffness and one torsion stiffness. This hypothesis makes it possible to simplify the model. The new separate model is developed on ABAQUS/Standard and based on the SEM observations. The real sample of entangled cross-linked material presents different geometries of epoxy junctions, and the angles between two cross-linked fibres are highly dispersed (Fig. 12.18A). In this work, it was assumed that the angle between the two cross-linked fibres was a right angle to simplify the problem. CATIA V5 was used to construct the geometry of the resin cross-link instead of ABAQUS because it is more appropriate for such complex geometry. Fig. 12.20B shows good resemblance between the numerical geometry of the epoxy cross-link (Fig. 12.20A) and the real one. The two carbon fibres and the epoxy joint were modelled by 3D solid elements. The epoxy junction was discretised in fine mesh because it presents a small complex geometry.

12.3.2.1 Tension spring stiffness

The global displacement for the extreme nodes of the fibre u due to tension load F was composed of one resin junction, $u_{junction}$, and one displacement, u_{fibre} (Fig. 12.21), as illustrated in the following equation:

$$u = 2 \times u_{fibre} + u_{junction} \tag{12.24}$$

The relation between the tension load F and the junction displacement $u_{junction}$ is shown in the following equation:

$$F = K_{ten} \times u_{junction} \tag{12.25}$$

where the load F is defined as follows:

$$\begin{cases} F = K_{w1} \times u \\ F = \dfrac{24EI}{L^3} \times u_{fibre} \end{cases} \tag{12.26}$$

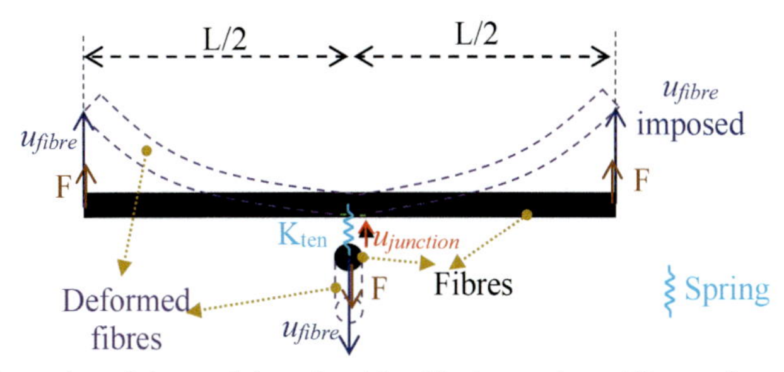

Fig. 12.21 Illustration of the model used to identify the tension stiffness of a spring (it can be noticed that the force that acts downwards is $2F$ because F is acts at each end of the bottom fibre).

where K_{w1} represents the tension stiffness of the whole model identified through the numerical simulation, L is the length of the fibre, I represents the moment of inertia of the fibre, and E is the Young's modulus of the fibre.

Then, an analytical model based on classical beam theory is used to determine the tension spring stiffness K_{ten}:

$$K_{ten} = \frac{1}{\dfrac{1}{K_{w1}} - \dfrac{L^3}{12EI}} \tag{12.27}$$

The value of tension stiffness, K_{ten}, found by this approach is 0.4 N/mm.

12.3.2.2 Torsion spring stiffness

The global angular displacement for the extreme nodes of the fibre $\Delta\alpha$ due to torsion load M was determined from the resin junction $\Delta\theta$ and one of fibres $\Delta\varphi$ (Fig. 12.22), as illustrated in the following equation:

$$\Delta\alpha = 2 \times \Delta\varphi + \Delta\theta \tag{12.28}$$

The relation between the global moment M and $\Delta\theta$ is shown by the following equation:

$$M = K_{tor} \times \Delta\theta \tag{12.29}$$

The moment M is identified as following:

$$\begin{cases} M = 2 \times K_{w2} \times \Delta\alpha \\ M = \dfrac{6EI}{L} \times \Delta\varphi \end{cases} \tag{12.30}$$

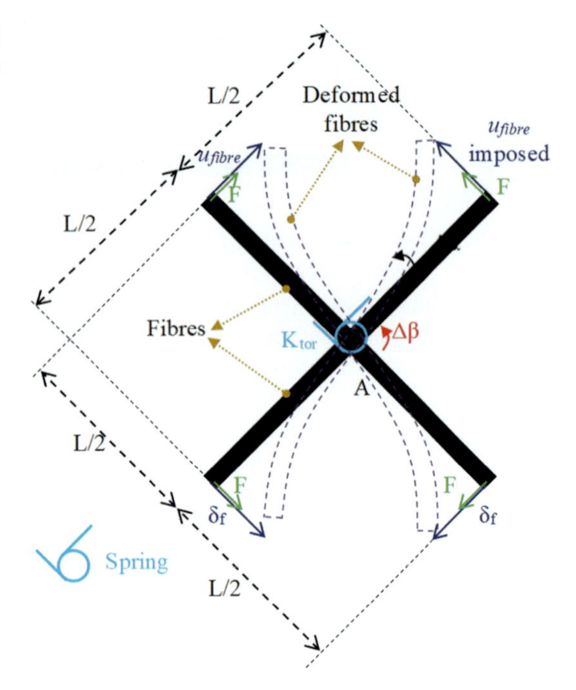

Fig. 12.22 Illustration of the model used to identify the torsion stiffness of a spring.

where K_{w2} is the torsion stiffness of whole model that is determined through the numerical simulation. An analytical model based on classical beam theory was then used to determine the torsion spring stiffness K_{tor}:

$$K_{tor} = \frac{1}{\frac{1}{2 \times K_{w2}} - \frac{L}{3EI}} \tag{12.31}$$

The value of torsion spring stiffness, K_{tor}, identified with this idealised geometry is 3×10^{-4} N mm.

12.4 Numerical analysis of compressive stiffness of a cross-linked fibrous material

12.4.1 Numerical modelling in compression

A parametric study of the Young's modulus in compression of an entangled and cross-linked material was carried out in this part. To be consistent with the actual conditions of the tests presented in Fig. 12.23, two rigid surfaces representing the two plates of the experimental compression device were used to apply the load. The RVE is placed between these two rigid surfaces. The lower rigid surface is fixed whilst the upper

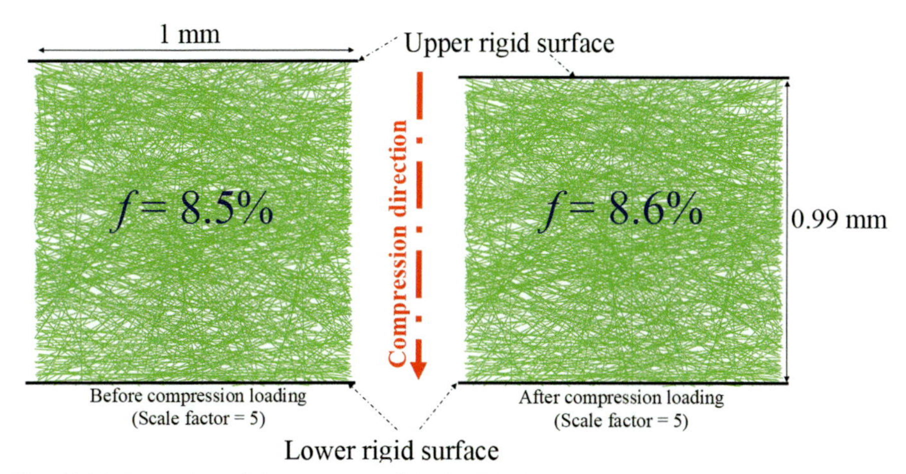

Fig. 12.23 Front view of the geometry of the RVE before and after compression for a strain of 1% with an initial volume fraction of fibres of 8.5%.

rigid surface is in movement in the direction of compression. Contact, without friction, is also managed between the rigid surfaces and the extreme nodes of the fibres whose directions cross the rigid faces. Fig. 12.23 shows the RVE before and after loading. A total of 2430 carbon fibres with a diameter of 7 μm are generated in the RVE of size $1 \times 1 \times 1$ mm^3 in order to obtain a volume fraction equal to 8.5%. The distribution of fibre orientations is taken isotropically. A total of 26,987 spring elements modelling the epoxy junctions were generated. These spring elements make it possible to block all the contacts existing between the fibres. The contacts are designated whenever a distance between the two axes of two fibres is observed that is less than the diameter. Blocking all the contacts generates an average distance between junctions equal to approximately 80 μm. The behaviour law of these springs was described in the previous section. Two stiffnesses are introduced into the springs: the stiffness in tension is equal to 0.4 N/mm whilst that in torsion is equal to 3×10^{-4} N m. The total duration of the loading step is equal to 1 s, and the initial time increment chosen is 10^{-3} s. The time increment authorised by the code is between 10^{-8} s and 10^{-3} s. A compression of 1% is therefore carried out in at least 1000 steps.

The result of the numerical simulation gives a compression modulus equal to 63 MPa. This value can be compared (Fig. 12.24) to those from two analytical models introduced in Section 12.2. These two models are based on the bending of fibres:

$$\text{Gibson and Ashby (1997)}: E_{GA} = \frac{3\pi E_f}{4\left(\frac{L}{D}\right)^4} \quad (12.32)$$

$$\text{Markaki and Clyne (2003)}: E_{MC} = \frac{9fE_f}{32\left(\frac{L}{D}\right)^2} \quad (12.33)$$

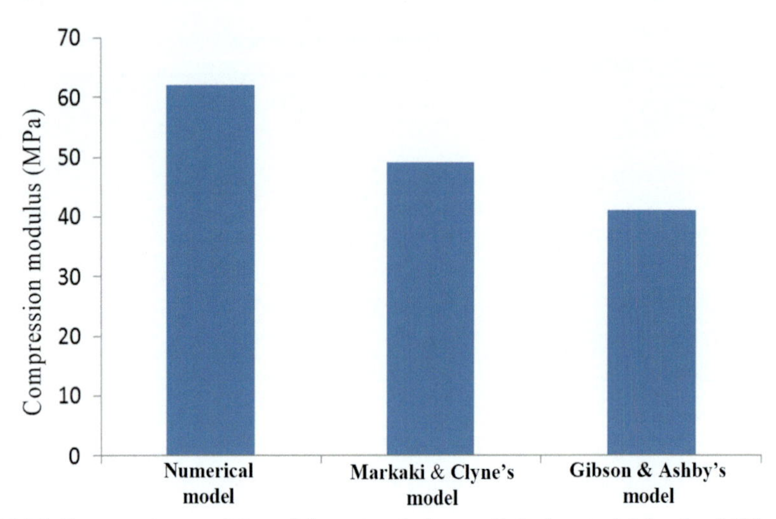

Fig. 12.24 Compression modulus of the entangled cross-linked material for $f = 8.5\%$. Comparison of the values obtained by the numerical model and the analytical models.

where E_f is the Young's modulus of the fibre, L is the average length between junctions, D is the fibre diameter, and f is the volume fraction.

The three values of the compression modulus are of the same order of magnitude. The values of the analytical models are lower than those found numerically; this is probably due to the analytical model, which is developed by determining the overall behaviour of the sample from the behaviour of a single fibre under bend loading and of an idealised fibre network.

12.4.2 Effect of the morphological parameters of the model

12.4.2.1 Effect of the distance between junctions

The average distance between bonded junctions is a parameter strongly influencing the compressive stiffness of the entangled cross-linked material. In this work, four ratios of contact cross-linking are compared: all contacts cross-linked (1/1), 50% cross-linked (1/2), 33% cross-linked (1/3), and 25% cross-linked (1/4). The objective is to look at the impact of the average distance between bonded joints on the stiffness. In practice, when creating the numerical material, the first contact of the first fibre is cross-linked, e.g. in the 1/3 case, the next two are not cross-linked, the fourth is, etc. At the end of this fibre, it is continued with fibre 2, and so on, until all the fibres of the RVE have been considered. At the same time, the effect of the anisotropy of the orientation of the fibres is also tested using four different distributions (Fig. 12.25B). The isotropic distribution is represented by the red curve in Fig. 12.25B. The other curves of Fig. 12.25B represent three anisotropic distributions characterised by a gradual increase in the ratio of fibres having directions quasi-perpendicular to the compression direction (Fig. 12.25B). Fig. 12.25A shows that the average distance between

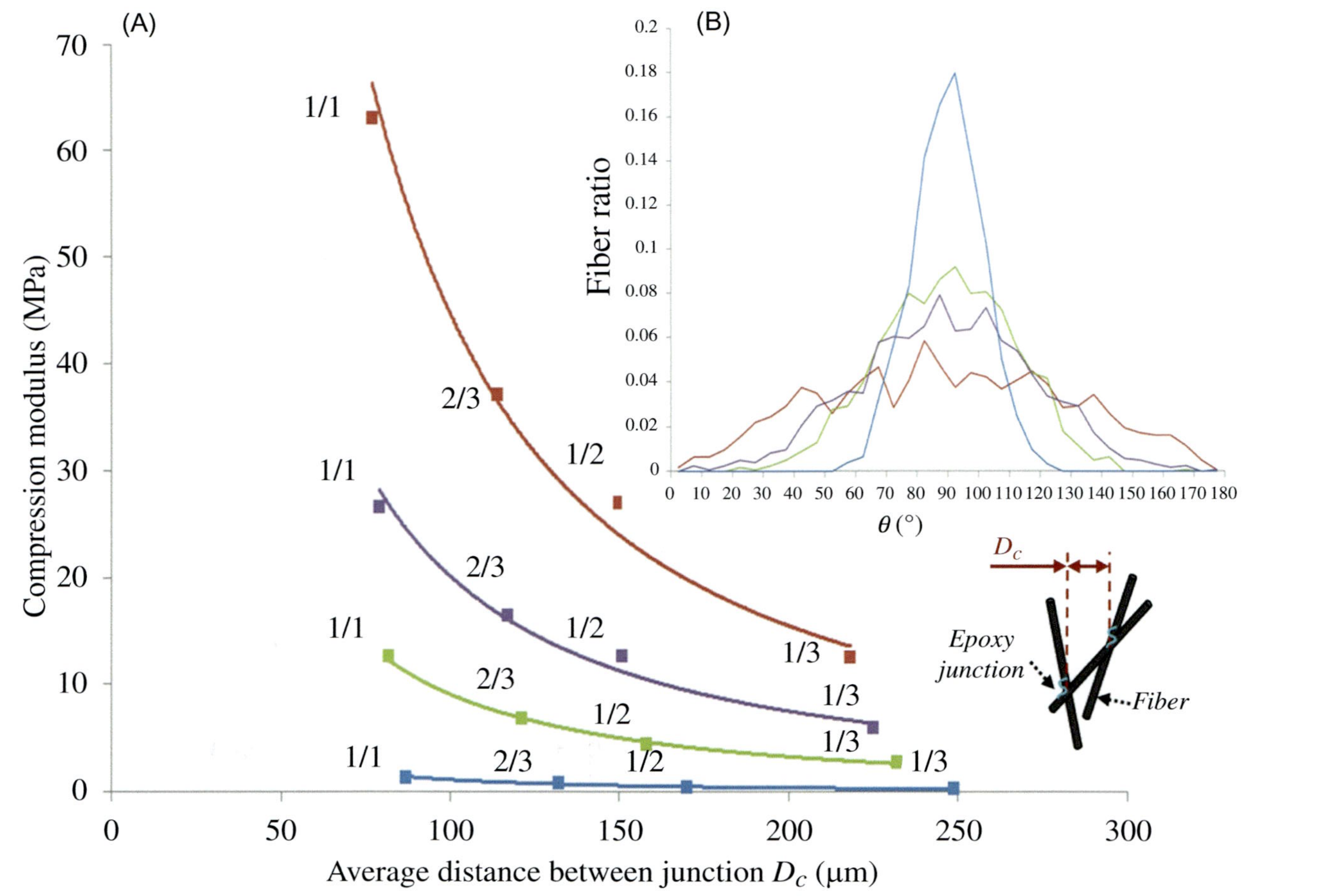

Fig. 12.25 (A) Influence of the average distance between junctions on the compression modulus for (B) different distributions of fibre orientations and different proportions of bonded contacts (1/3, 1/2, 2/3, and 1).

junctions (bonded contacts) induces an important effect on the compression modulus, whatever the distribution of the orientations of fibres is. The effect of the average distance between junctions is almost the same for the four distributions of fibre orientations. The lower the average distance between junctions, the higher the stiffness in compression is. The relationship between the compression modulus and the average distance between junctions is a power function with an exponent of -1.5 for all distributions of fibre orientations. This exponent value of $-3/2$ is different from the value of -4 proposed by Gibson and Ashby (Eq. 5.1). This may be due to the modelling of the junction linking the fibres (edges) of the cells of their model. Indeed, in their approach, these junctions are assumed to be infinitely rigid (embedding). On the contrary, in the proposed model, the epoxy springs have a certain flexibility. The value of -1.5 resulting from our simulations is closer to the value of -2 proposed by Markaki and Clyne (Eq. 5.2). The dependence obtained by these authors is, however, more marked than the one observed in this study.

The average distance between bonded junctions observed with SEM by Mezeix et al. is approximately 120 μm [28,29]. It corresponds roughly to a cross-linking of 2/3 of the contacts. It can be noted that in the case where all the contacts are blocked by the resin, we could then expect a much higher compressive stiffness of the entangled cross-linked material. In the case of an isotropic distribution of fibre orientations, the compression module goes from 37 to 63 MPa when all the contacts are blocked instead of the 2/3 observed experimentally. There is approximately a coefficient of 2 for the compression modulus between the two cases of cross-linking (2/3 and 1/1).

Increasing the number of bonded contacts, which results in a decrease in the average distance between junctions, can be done in two ways. The first entails increasing the volume fraction of fibres, which is not easy to achieve for technical reasons beyond $f = 11\%$. Indeed, it is difficult to manufacture samples with a density higher than 200 kg/m^3 because the entanglement becomes almost impossible to achieve. The second method is to spray more resin, but in this case the density of the sample will increase and the stiffness-to-mass ratio will decrease.

The creation of a greater number of junctions requires a spraying of very fine drops of resin of good fluidity. An improvement in the manufacturing process, and in particular the entanglement and spraying process of the resin, could thus generate significant gains in terms of the rigidity of the entangled cross-linked material. What the simulations show is that a gain of a factor of 5–10 seems possible between the real material and the ideal material.

12.4.2.2 Effect of the fibre diameter

In this section, the effect of diameter on the compressive stiffness of the entangled cross-linked material has been studied. Three fibre diameters (Ø7, Ø8, and Ø9 μm) were tested. Fig. 12.26 shows the evolution of the compression modulus as a function of the average distance between junctions for these three fibre diameters. These three curves are obviously produced for the same volume fraction of 8.5% and for the same junction stiffness ($K_{te} = 0.4$ N/mm and $K_{to} = 3 \times 10^{-4}$ N mm). Obviously, the

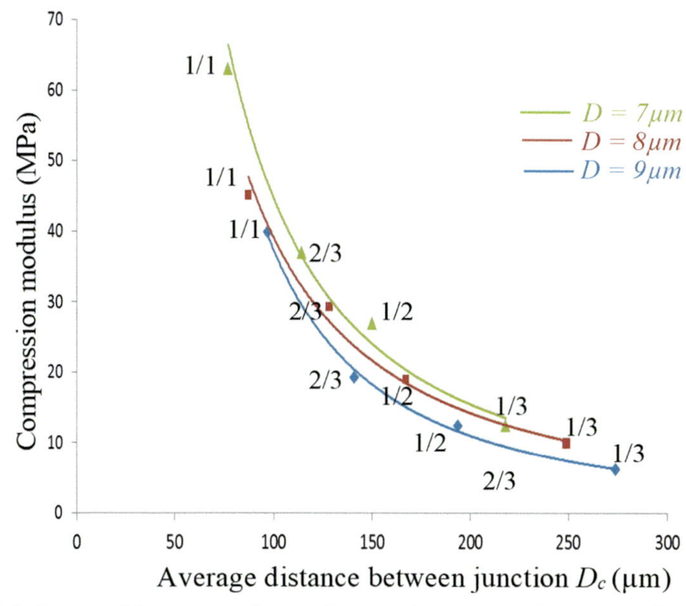

Fig. 12.26 Influence of the average distance between junctions on the compression module for different fibre diameters (Ø7, Ø8, and Ø9 μm).

stiffness in compression increases with the decrease in the average distance between junctions. The relationship between the compression modulus and the average distance between junctions is again characterised by a trend curve in power law with exponent $-3/2$, whatever the diameter of the fibre. The compression modulus increases when the fibre diameter decreases. For the same volume fraction, the increase in the diameter of fibre makes it possible to increase the bending stiffness, but this does not allow for the increase in compressive rigidity of the material because the number of fibres generated also decreases. Indeed, the change in diameter induces a change in the number of fibres generated and, therefore, in the total length of fibres generated (Table 12.2). It can therefore be concluded that for an equal distance

Table 12.2 Details of the three numerical simulations with different fibre diameters.

Fibre diameter (μm)	Number of fibres generated	Total fibre length generated in the RVE (μm)	Number of beam elements
Ø7	2430	2095	62,007
Ø8	1890	1605	42,177
Ø9	1500	1153	28,455

between junctions, the increase in diameter does not improve the mechanical characteristics of the material. However, this increase can be useful if it makes it possible to increase the size of the junctions and therefore their rigidity [28,29]. The influence of the stiffness of the junction will be studied in the next section to see if this is a point of interest.

12.4.2.3 Effect of the tension/torsion stiffness of the junction

In this work, the epoxy junctions are modelled using springs presented in more detail in Section 12.3.2. The stiffness of these junctions is described numerically by only two stiffnesses (a tensile stiffness and a torsional stiffness) to obtain an isotropic stiffness matrix that is independent of the coordinate system as well as the direction of the fibres. The following paragraphs will individually discuss the effects of these two stiffnesses on the compressive stiffness of the entangled cross-linked material.

Fig. 12.27A exhibits the effect of junction tensile stiffness on the compressive stiffness of the entangled cross-linked material. The compression module increases and then tends towards an asymptote. This asymptote is reached more quickly the greater the average distance between junctions (reference I in the figure). For an average distance between junctions equal to 218 μm, the asymptote is reached at a value of $K_{te} = 1.1$ N/mm. In the case of an RVE with a smaller average distance between junctions ($D_c = 78$ μm), the asymptote has not yet been reached for a value of $K_{te} = 2$ N/mm (II).

It can be seen that an increase in the tensile stiffness of the junction would have a greater influence on the samples for small average distances between junctions. In the case of $D_c = 218$ μm, the compressive stiffness is only multiplied by a factor of approximately 1.8 when the tensile stiffness of the spring drops from 0.4 to 2 N/mm. For the case $D_c = 115$ μm, the compression modulus is multiplied by a factor of approximately 2.1 when the tensile stiffness of the spring used is increased from $K_{te} = 0.4$ to 2 N/mm. This factor continues to increase in the last case, which corresponds to the smallest average distance between the junction $D_c = 78$ μm; it is equal to approximately 2.4 for the same increase in K_{te}. Thus, the decrease in the distance between junctions induces an increase in the number of junctions, which explains this greater influence of the properties of the resin ensuring bonding in the case of small distances between junctions.

The effect of tensile stiffness is greater than that of torsional stiffness. This last also depends on the average distance between junctions but always remains weak compared to the effect of the tensile stiffness on the compressive stiffness of the entangled cross-linked material (Fig. 12.27B). As with the tensile stiffness, the effect of the torsional stiffness is greater when the average distance between junctions is small. Table 12.3 presents the increase ratio ($E_{0.3}/E_{0.0003}$) for different average distances between junctions. It can be seen that the ratio varies little.

The increase in the torsional stiffness induces an increase in the compression modulus, which seems to stabilise towards a quasi-asymptote. This is reached more or less quickly depending on the average distance between junctions. It is reached even more quickly as the average distance between junctions is low (reference III in Fig. 12.27B).

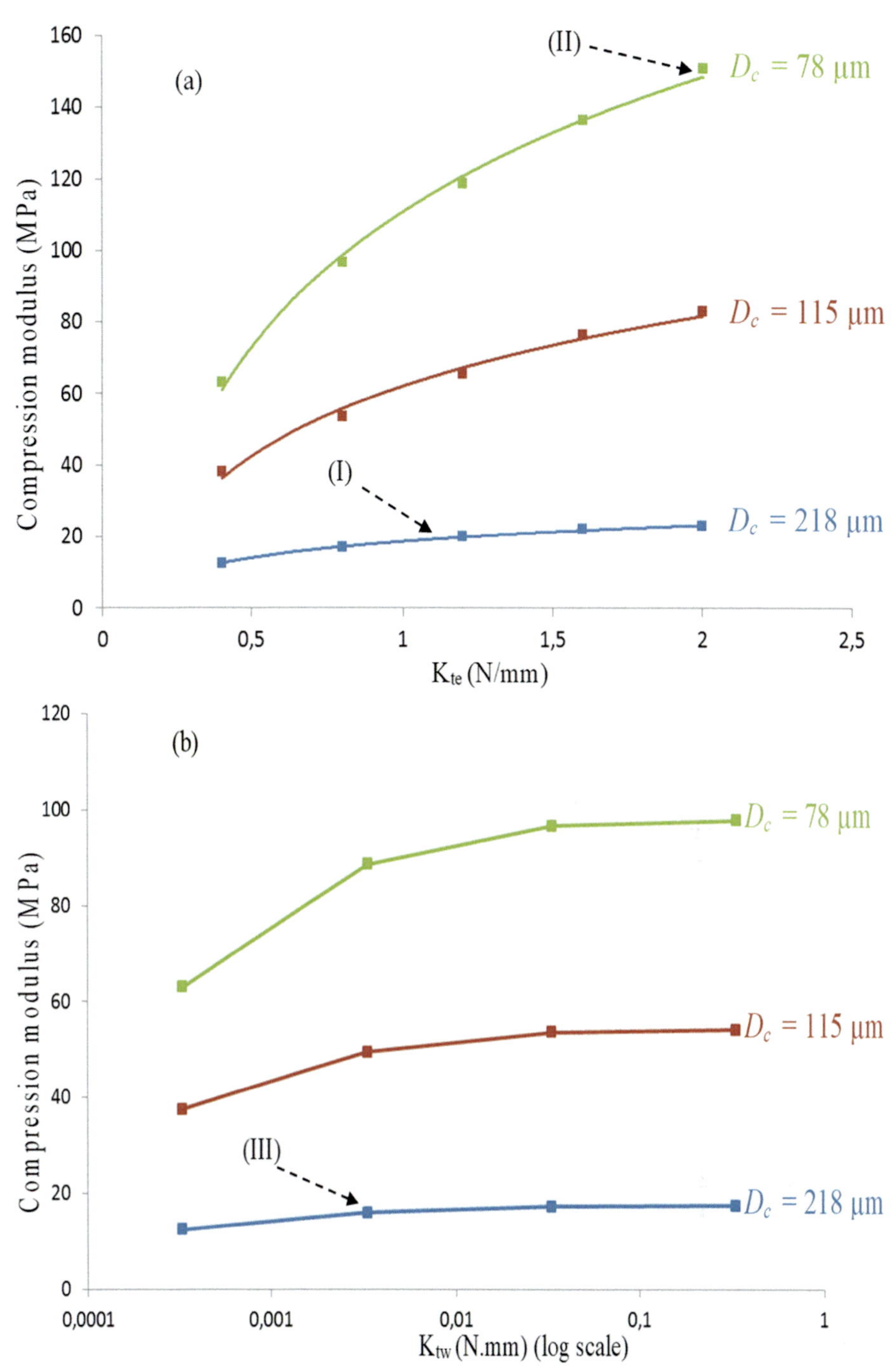

Fig. 12.27 Influence of (A) the tensile rigidity and (B) the torsional rigidity of the epoxy joint on the compression module for different average distances between bonded joints. All RVEs tested have the same configuration with a volume fraction of fibres equal to 8.5% and $D_c = 78$, 115, and 218 µm corresponding, respectively, to bridging of all contacts, 2/3 of contacts, and 1/3 of contacts.

Table 12.3 Evolution of the increase ratio ($E_{0.3}/E_{0.0003}$) for different average distances between bonded junctions ($D_c = 78, 115, 218$ μm).

	Compression modulus E (MPa)		Increase ratio ($E_{0.3}/E_{0.0003}$)
	$K_{to} = 0.0003$ N mm	$K_{to} = 0.3$ N mm	
$D_c = 78$ μm	63	98	1.55
$D_c = 115$ μm	37.5	54	1.44
$D_c = 218$ μm	12.5	17.5	1.93

In the case where $D_c = 218$ μm, the asymptote is reached at only 3×10^{-3} N mm. In the other two cases of mean distance between junctions, it is reached at around 3×10^{-2} N mm.

The increase in the stiffness of the junctions can theoretically be obtained experimentally by modifying the mechanical properties of the resin and/or increasing the size of the sprayed resin droplets. The latter is difficult to control during the spraying phase. One solution to increase the size of the junction could be to play on the surface treatment of the fibres modifying the surface energies and therefore the resin/fibre drop angles.

Another method of increasing the stiffness in joints would be to find a resin with better mechanical characteristics than epoxy to bridge the contacts. This is not immediate either, because epoxy resin is the most commonly used resin in aeronautics due to its good mechanical characteristics.

12.4.2.4 *Effect of the volume fraction*

One of the most important morphological parameters of fibrous materials is the volume fraction of fibres. It must therefore be carefully chosen to optimise the mass-to-rigidity ratio. In this section, the variation in compressive stiffness is presented as a function of the average distance between junctions for different volume fractions (6.5%, 7.5%, and 8.5%).

Fig. 12.28 shows that the compression modulus trend curves as a function of the average distance between junctions are governed by the power law with exponent $-3/2$, regardless of the volume fraction. The compression modulus increases almost linearly with the increase in the volume fraction, which is in agreement with the analytical models of Gibson and Ashby [11] and Markaki and Clyne [13].

12.4.2.5 *Effect of the Young's modulus of fibre*

An entangled cross-linked material can be made from different fibres such as glass, carbon, aramid, or stainless steel. This section will discuss the influence of the nature of the fibre on the compressive stiffness of the developed fibrous material for three distributions of fibre orientations. The red curve of Fig. 12.29B represents an isotropic distribution of fibre orientations, whereas the two other curves present anisotropic

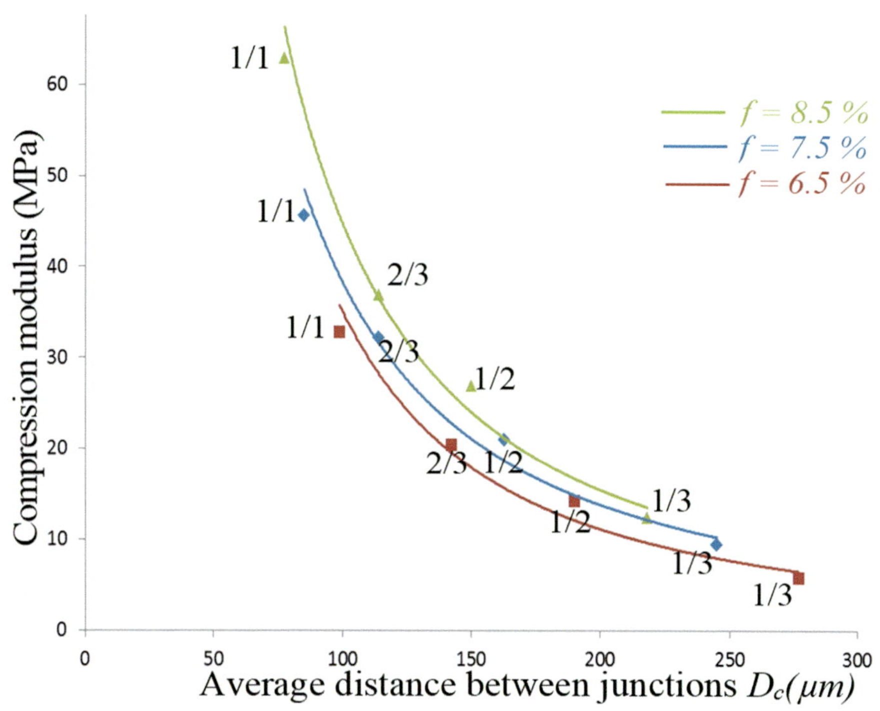

Fig. 12.28 Influence of the volume fraction of fibres and of the average distance between junctions on the compression stiffness.

distributions with more and more fibres perpendicular to the direction of compression. For each distribution, the same morphological parameters are used and only the Young's modulus of the fibre is modified to study its influence on the stiffness in compression (Table 12.4). Fig. 12.29A shows that the fibres with the highest Young's modulus offer the most interesting stiffness in compression. Carbon fibres logically lead to the highest compression modulus whilst aramid gives the lowest compression modulus, as expected.

For a material intended for sandwich core application, the entangled cross-linked material made from carbon fibre seems to be the best choice. However, thanks to its low cost, glass fibre can be a good compromise for applications requiring lower rigidity. Stainless steel fibre can be of interest in applications where electrical conductivity is important. In this study, the fibres are assumed to be linear elastic, but in reality, their behaviour may be more complex. For example, the plasticity of stainless steel fibre could be interesting for dissipating energy [30]. Finally, the manufacture of an entangled cross-linked material from a mixture of fibres of different natures may be advantageous to develop a multifunctional material.

It is seen that the relation between the compression modulus and the Young's modulus is a linear function, which is in agreement with the analytical models of Gibson and Ashby [11] and Markaki and Clyne [13]. The slopes of the three straight lines,

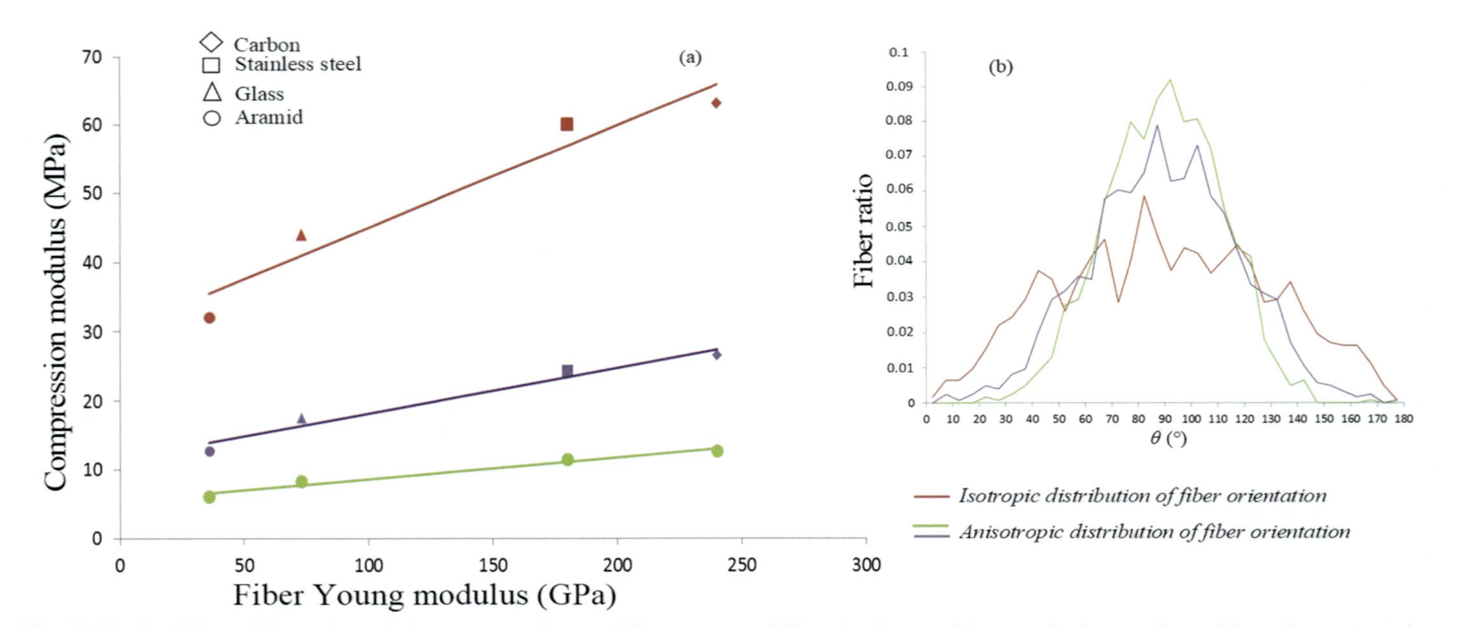

Fig. 12.29 (A) Effect of Young's modulus corresponding to different types of fibres (carbon, stainless steel, glass, and aramid) on the compressive stiffness at a volume fraction of 8.5% for (B) three distributions of fibre orientations (one isotropic case in *red* and two anisotropic cases in *green* and *purple*).

Table 12.4 Young's modulus used for fibres of different natures.

Nature	Young's modulus *E* (GPa)
Carbon	240
Stainless steel	180
Glass	73
Aramid	36

Table 12.5 Slope values for three different distributions of fibre orientations (with respect of Fig. 12.29, the slope unit should be in MPa/GPa, but the percent unit was chosen).

Distribution of fibre orientation		Slope value (%)
Anisotropic distribution	Green curve (Fig. 12.29)	3.1
	Purple curve (Fig. 12.29)	6.6
Isotropic distribution		14.8

corresponding to the three distributions of fibre orientations, are reported in Table 12.5. In addition, the effect of the Young's modulus on the stiffness in compression is greater when the distribution of the orientations of the fibres is isotropic. For example, for a distribution of orientations of the anisotropic fibres, the glass fibre may prove to be an attractive choice in view of its lower cost, whilst obtaining a compressive rigidity a little lower than that obtained for carbon fibre.

12.5 Conclusion

A numerical model of the mechanical behaviour in compression loading of a fibrous material made of carbon fibres cross-linked with spray of epoxy resin has been developed. This model makes it possible to study the effect of some morphological or mechanical parameters.

In a first step, the size of the RVE was evaluated for it to be representative of the used fibrous material. The size of 1 mm^3 was large enough to be representative of the generated fibre network and to avoid too much calculation time. This size can be compared to the fibre diameter and to the distance between junctions; in fact, it is about 100 times the fibre diameter and about 10 times the distance between cross-links.

The model made it possible to study the effect of the fibre diameter and the Young's modulus, of the distance between junctions or of the volume fraction. Of course, the higher the fibre diameter, the Young's modulus of the fibre, or the volume fraction, the higher is the compression Young's modulus of the obtained fibrous material, but only if the same number of fibres is used. Indeed, if the fibre diameter is increased keeping the volume fraction constant, then the fibres number decreased and the Young's modulus increases. At the same time, less is the distance between junctions and higher is this Young's modulus. The obtained value of the Young's modulus using the proposed numerical model with an isotropic fibrous network was compared to the literature and showed similar results [11,13].

One of the most important parameters seems to be the distance between the epoxy junctions. A power function, with an exponent of -1.5, is found between the compressive Young's modulus and the distance between junctions. This parameter is very interesting because it can be changed by only modifying the morphology of the fibre network. Of course, the increasing of the volume fraction of fibres also increases the Young's modulus, but it also induces an increase in the mass of the fibrous material, which is not the case of the modification of the morphology change.

Indeed the increasing of the compressive Young's modulus is an important goal to use the proposed fibrous material as the core material for a sandwich structure. Indeed the currently obtained Young modulus, less than 100 MPa, is yet too low compared to the classical core material used for sandwich structures. For example, the foam PMI 110 currently used for aeronautical applications reaches a Young's modulus of about 160 MPa [31], and the Nomex honeycomb, also currently used in the aeronautical field, reaches 620 MPa of Young's modulus [31]. But with a better driving of the morphological parameters of the proposed fibrous material, it seems possible to reach similar value, at least that of the PMI foam. Nevertheless, the compression behaviour is not the only property to consider for the use of fibrous material as core materials of sandwich structure(s). Other mechanical properties such as shear stiffness of mechanical damping, as well as other properties such as thermal or electrical properties, should be also taken into account.

Moreover, the proposed model seems to show that an important gain is possible if the manufacturing process is better controlled, in particular if the fibre tows are better separated, if the epoxy spray is more homogeneous, or if the fibre orientation is more isotropic. Of course, this control of the manufacturing process is not so obvious because it can induce some other drawbacks, such as damage to fibres or expansion of the epoxy junctions. Indeed, the huge difference between a model and reality is that it is possible to independently change certain parameters of the fibrous material in the model but not in reality. It is, of course, an important quality of the model to be able to highlight the effect of each parameter independently, whilst for the real manufacturing process, some parameters are linked and cannot be changed independently.

Acknowledgement

The authors would like to gratefully acknowledge CALMIP (CALcul en Midi-Pyrénées) (https://www.calmip.univ-toulouse.fr) for access to the HPC resources under the allocation p1026.

References

[1] M. Jebahi, D. André, I. Terreros, I. Iordanoff, Discrete Element Method to Model 3D Continuous Material, 2015, ISBN: 978-1-848-21770-6.

[2] C.M. Van Wyk, Note on the compressibility of wool, Text. Res. J. Inst. 37 (1946) T285–T292.
[3] T. Komori, K. Makishima, Numbers of fiber-to-fiber contacts in general fiber assemblies, Text. Res. J. 47 (13) (1977) 13–17.
[4] O. Kallmes, H. Corte, The Structure of Paper. I. The Statistical geometry of an ideal two dimensional network, TAPPI J. 43 (1960) 737–752.
[5] N. Pan, A modified analysis of the microstructural characteristics of general fiber assemblies, Text. Res. J. 63 (6) (1993) 336–345.
[6] D.H. Lee, G.A. Camaby, S.K. Tandon, Compressional energy of the random fiber assembly. Part II: evaluation, Text. Res. J. 62 (5) (1992) 258–265.
[7] D.H. Lee, G.A. Carnaby, Compressional energy of the random fiber assembly: part I: theory, Text. Res. J. 62 (4) (1992) 15–191.
[8] T. Komori, M. Itoh, A new approach to the theory of the compression of fiber assemblies, Text. Res. J. 61 (7) (1991) 420–428.
[9] T. Komori, M. Itoh, A. Takaku, A model analysis of the compressibility of fiber assemblies, Text. Res. J. 62 (10) (1992) 567–574.
[10] S. Toll, Packing mechanics of fiber reinforcements, Polym. Eng. Sci. 38 (8) (1998) 1337–1350.
[11] L.J. Gibson, M.F. Ashby, Cellular Solids: Structure and Properties, second ed., Cambridge University Press, 1997.
[12] L. Mezeix, C. Bouvet, J. Huez, D. Poquillon, Mechanical behavior of entangled fibers and entangled cross-linked fibers during compression, J. Mater. Sci. 44 (14) (2009) 3652–3661.
[13] A.E. Markaki, T.W. Clyne, Mechanics of thin ultra-light stainless steel sandwich sheet material - part I. stiffness, Acta Mater. 51 (5) (2003) 1341–1350.
[14] N.B. Beil, W.W. Roberts Jr., Modeling and computer simulation of the compressional behavior of fiber assemblies, part I: comparison to van Wyk's theory, Text. Res. J. 72 (4) (2002) 341–351.
[15] N.B. Beil, W.W. Roberts Jr., Modeling and computer simulation of the compressional behavior of fiber assemblies, part II: hysteresis, crimp, and orientation effects, Text. Res. J. 72 (5) (2002) 375–382.
[16] W.W. Roberts Jr., N.B. Beil, Fibrous assemblies: modeling/computer simulation of compressional behavior, Int. J. Cloth. Sci. Technol. 16 (1/2) (2004) 108–118.
[17] D. Durville, Numerical simulation of entangled materials mechanical properties, J. Mater. Sci. 40 (22) (2005) 5941–5948.
[18] D. Rodney, M. Fivel, R. Dendievel, Discrete modeling of the mechanics of entangled materials, Phys. Rev. Lett. 95 (2005), 108004.
[19] C. Barbier, R. Dendievel, D. Rodney, Numerical study of 3-D compressions of entangled materials, Comput. Mater. Sci. 45 (3) (2009) 593–596.
[20] A.I. Abd El-Rahman, C.L. Tucker III, Mechanics of random discontinuous long-fiber thermoplastics. Part I: generation and characterization of initial geometry, ASME J. Appl. Mech. 80 (5) (2013), 051007.
[21] A.I. Abd El-Rahman, C.L. Tucker III, Mechanics of random discontinuous long-fiber thermoplastics. Part II: direct simulation of uniaxial compression, J. Rheol. 57 (2013) 1463, https://doi.org/10.1122/1.4818804.
[22] Y. Zhang, Z. Lu, Z. Yang, D. Zhang, J. Shi, Z. Yuan, Q. Liu, Compression behaviors of carbon-bonded carbon fiber composites: experimental and numerical investigations, Carbon 116 (2017) 398–408.

[23] Y.H. Ma, H.X. Zhu, B. Su, G.K. Hu, R. Perks, The elasto-plastic behavior of three dimensional stochastic fiber networks with cross-linkers, J. Mech. Phys. Solids 110 (2018) 155–172.
[24] A. Shahdin, L. Mezeix, C. Bouvet, J. Morlier, Y. Gourinat, Fabrication and mechanical testing of glass fiber entangled sandwich beams: a comparison with honeycomb and foam sandwich beams, Compos. Struct. 90 (2009) 404–412.
[25] F. Chatti, D. Poquillon, C. Bouvet, G. Michon, Numerical modelling of entangled carbon fibre material under compression, Comput. Mater. Sci. 151 (2018) 14–24.
[26] F. Chatti, C. Bouvet, D. Poquillon, G. Michon, Numerical modelling of shear hysteresis of entangled cross-linked carbon fibers intended for core material, Comput. Mater. Sci. 155 (2018) 350–363.
[27] T. Kanit, S. Forest, I. Galliet, V. Mounoury, D. Jeulin, Determination of the size of the representative volume element for random composites: statistical et numerical approach, Int. J. Solids Struct. 40 (13–14) (2003) 3647–3679.
[28] L. Mezeix, D. Poquillon, C. Bouvet, Entangled cross-linked fibers for an application as core material for sandwich structures part I: experimental investigation, Appl. Compos. Mater. 23 (1) (2015) 71–86.
[29] L. Mezeix, D. Poquillon, C. Bouvet, Entangled cross-linked fibres for an application as core material for sandwich structures—part II: analytical model, Appl. Compos. Mater. (2015) 1–14.
[30] J.P. Masse, L. Salvo, D. Rodney, Y. Bréchet, O. Bouaziz, Influence of relative density on the architecture and mechanical behaviour of a steel metallic wool, Scr. Mater. 54 (7) (2006) 1379–1383.
[31] D. Zenkert, The Handbook of Sandwich Construction, Engineering Materials Advisory Services Ltd, 1997.

Mechanics of interactions of F-actin and vimentin networks

Horacio Lopez-Menendez
Department of Physical-Chemistry, Complutense University of Madrid, Madrid, Spain

13.1 Introduction

The mechanical structure of the cell, called cytoskeleton, plays a key role for many mechanical duties such as mechano-sensing, motility, contraction, division, and extrusion and its alterations are associated to several pathological conditions. It is formed by biopolymeric structures such as F-actin, microtubules, and intermediate filaments (IF) that create networks, which are critical in determining the mechanical properties of the cells [1, 2]. The cell along its journey in many developmental and pathological processes, as during classical cell migration or invasion in cancer metastasis, withstands severe large deformations. The way in which cells are able to sustain this mechanical conditions remains unclear, due to the fact that F-actin and microtubules yield or disassemble under moderate strains. Nevertheless, the vimentin IF hyperelastic network synergistically interacts with other quickly recoverable cytoplasmic components, substantially enhancing the strength, stretchability, resilience, and toughness of the mechanical robust structure, which preserves the cell viability at large deformations [3, 4].

Interestingly the IF are the most diverse of the three major cytoskeletal filaments, being encoded by approximately 70 different genes and its dysregulation can causes a wide range of human diseases including blistering diseases of the skin, cardiomyopathies, lipodystrophy, muscular dystrophies, neuropathy, Crohn's disease, rheumatoid arthritis, HIV, and cancer [5]. Even considering its importance the mechanics of IF and the way in which the IF network interact with others cytoskeleton components remains unclear. In this sense, a recent work by Serres et al. studies the F-actin interactome in mitotic cells, by using super-resolution microscopy, and screened for regulators of cortex architecture and found that IF are recruited to the mitotic cortex. They demonstrated that cortical vimentin recruitment regulates actin network organization and fine-tunes the mechanics required for successful cell division in constrained environments [6].

The in vivo IF can be found from dense amorphous networks to well-organised bundled arrays and all these assemblies are very dynamic, and evolve by non-equilibrium actin polymerisation/depolymerisation and also by the active force such as the generated by the myosin motors. Nevertheless, an interesting way to tackle the complexity of the interactions and couplings between structures is by taking a reductionistic approach using synthetic in vitro networks. In this sense, the previous in vitro

Mechanics of Fibrous Networks. https://doi.org/10.1016/B978-0-12-822207-2.00010-6

studies have reported the mechanics of either single filaments [7, 8], or by hydrogels of filaments made by single biopolymer species [9–11]. These studies will provide a deep understanding of the emergent behaviour of the composite networks, where the copolymerisation of the two networks promotes strong alterations into the assembly dynamics and their mutual steric constrains [12, 13].

Taking a similar experimental set-up, Jensen et al. made a crosslinked F-actin network interpenetrated with a vimentin IF network and used bulk rheology to investigate the mechanics of the composite network at large deformations. They found two well-defined regimes. On the one hand, they showed that copolymerisation with vimentin strengthens F-actin networks when actin crosslinks are abundant, as expected from the overall increase in the amount of polymers in the network. On the other hand, unexpectedly, they found that the mechanical response of the F-actin networks is weakened due to the copolymerisation with vimentin when the F-actin crosslinking density is low compared to the network mesh size. Then the alterations in the network elasticity, the yielding point, can be an emergent response that comes from steric constraints on F-actin by vimentin (IF), promoting a lower degree of F-actin crosslinking in the final network. One interesting point to remark is that for the range of explored concentrations in the experiments, the vimentin network is very flexible and has a small role in the definition of the mechanical properties of the composite network. Nevertheless, it becomes more important for the mechanical properties by setting physical crosslinks or steric constrains over the actin network. In addition, recently was showed by taking a synthetic approach that the interaction of IF with key structural elements within cells can be cross linked by the regulation of ions or PH. Thus, cells may use the presence of different species to precisely modulate their couplings to control the internal mechanical properties [14, 15].

Then, this work deals with the development of a mechanical model capable to explain these interesting observed rheological experiments carried out by Jensen et al. [16] by developing a model defining an effective crosslinked F-actin network, which condenses the alteration of its structure on its main physical variables. In addition, to improve the insight of the experimental studies many theoretical and computational models have been developed. On the one hand, mesoscopic models which introduce microstructural details based on the worm-like chain model and represent an excellent description of rheological experiments [17–22]. On the other hand, several computational studies evaluating a large-scale fibre models have made a substantial progress, considering stochastic dynamics under passive and active situations taking into account the evolution of internal stresses due to the effect of the entanglements and polymerisation dynamic [23–26].

Another way to gain clues about the emerging complexity of composite networks is in the context of hydrogels. More specifically on a very active field of research is on the interpenetrating polymer networks (IPN), which consist of two or more polymeric networks, where at least one of them polymerised and/or crosslinked in the presence of the other. These hydrogels are designed to achieve large deformations and to manifest high toughness, Mullins effect, and necking instabilities [27–29]. These effects emerge due to the fact that the polymeric networks are interlaced on a molecular scale but not covalently bonded to each other.

To better understand the micromechanics of IPN, a constitutive modelling of interpenetrating networks has also been proposed [30, 31].

Inspired by the advances in the mentioned complementary fields, in this work we propose a model into the framework of nonlinear continuum mechanics by using the semiflexible filament described by a worm-like chain following the Blundel–Terentjev formalism [32]. To capture the continuum structure, we homogenise the F-actin network following the three-chain model as was implemented by Meng et al. [20, 33]. On the basis of this model, we introduce the dynamic effect of the crosslinks in order to capture the strengthening–weakening transition manifested by the network [19, 34, 35]. Then, to describe the effects of the interaction between networks we propose an energy term associated to the interaction energy, by using the Landau model for continuous phase transition. This term defines the interaction parameter, which describes the effects of the alteration over the F-actin network due to the interaction [36, 37]. The expressions obtained through the model are compared with experimental data coming from Jensen et al. [16], and finally we discuss the results and future works.

13.2 Formalism of nonlinear elasticity with phase transitions

13.2.1 Free energy

In a first approximation, we consider a Helmholtz free energy which accounts the strain energy functions (SEFs) associated with the elasticity of the crosslinked F-actin network and the IF network made by vimentin, where its mechanical strain deformation is described by the Cauchy–Green tensor **C**. Also, an energy term associated with the interaction between networks which are proportional to the ratio between the concentration of vimentin and the concentration of actin (c), will allow us to define an interaction parameter (Γ)

$$\Psi(\mathbf{C},\Gamma,c) = \Psi_{IF}(\mathbf{C},\Gamma) + \Psi_{actin}(\mathbf{C},\Gamma) + \Psi_{inter}(c,\Gamma) + U(J). \tag{13.1}$$

In the following, we first describe the SEFs without considering the effects of the interaction defined by Γ, which will be defined further for clarity. The first term in the free energy function refers to the IF. They have a long contour length and a low bending stiffness and manifest a much higher flexibility in comparison with the F-actin network. In order to describe the soft mechanics of IF, we consider an isotropic neo-Hookean SEF as

$$\Psi_{IF}(\mathbf{C}) = \frac{c_1}{2}(I_1 - 3), \tag{13.2}$$

where $c_1 > 0$ is a stiffness parameter.

The second term in the free energy function represents the SEF for the crosslinked actin network, based on the worm-like chain model for semiflexible filaments, following the Blundel–Terentjev formalism [32], see Fig. 13.1A, in which the main parameters are the contour length of the filament, L_c, and the persistence length l_p, which is a characteristic length that compares the bending energy with the thermal energy, it represents a measure of the bundle stiffness. In addition, $c_s = l_p/2L_c$ is a dimensionless stiffness parameter reflecting the flexibility of the chain.

The primitive cube for the homogenisation is built with lattice points representing the crosslink sites, and the edges are aligned along the principle directions of deformation tensor $\mathbf{C}$, see Fig. 13.1B. The three chains are linked with their end-to-end vectors along the edges and the equilibrium mesh size ξ. On deformation, the lengths of the perpendicular edges over the lattice point become $\lambda_1\xi$, $\lambda_2\xi$, and $\lambda_3\xi$, respectively. Then, the free energy density of a semiflexible network can be expressed as

$$\Psi_{3c}(\lambda_{i=1,2,3}) = \frac{n}{3}\sum_{i=1,2,3}\psi_{chain}(\lambda_i\xi), \tag{13.3}$$

$$\Psi_{3c} = \frac{nk_BT}{3}\left[\pi^2 c_s\pi c(3 - I_1x^2) + \frac{3 - 2I_1x^2 + I_2x^4}{\pi c_s\pi c(1 - I_1x^2 + I_2x^4 - I_3x^6)}\right] \tag{13.4}$$

with $x = \xi/L_c$

If the stress tensor is expressed as a function of the strain invariants for an incompressible material, where the $I_3 = 1$, it can be expressed as

$$\boldsymbol{\sigma} = 2\left[\left(\frac{\partial\Psi}{\partial I_1} + I_1\frac{\partial\Psi}{\partial I_2}\right)\mathbf{C} - \left(I_1\frac{\partial\Psi}{\partial I_1} + 2I_2\frac{\partial\Psi}{\partial I_2}\right)\frac{\mathbf{I}}{3} - \frac{\partial\Psi}{\partial I_2}\mathbf{C}\cdot\mathbf{C}\right] - p\mathbf{I}. \tag{13.5}$$

The mechanical properties of the network are defined by the interaction between actin and crosslinks. In particular, the network fluidisation is due to the transient chemical crosslinks that are not covalent bonds (with high adhesion energy). In fact their adhesion energy is in the order of tens of k_BT, and have a transient dynamics [38]. In general terms, this kind of gels with chemical crosslinks (proteins as α-actinin) behaves as physical gels [39]. Then, to capture this interaction and the size of the mesh, i.e. the contour length, L_c, we propose, in a similar manner as proposed by Lopez-Menendez et al. [19, 37], the following expression $L_c = L_c^0 + \delta L_c^{cl} P_{ub}$. Where the adhesion is modelled as a two states process in which P_{ub} defines the unbinding probability. From a mesoscopic perspective, the network can fluidise if the crosslinks undergo any transition from a bind, fold, or rigid state toward a state of unbind, unfold, or flexible. L_c^0 represents the contour length when $P_{ub} = 0$ (bind crosslink), and δL_c represents the average increment of the contour length when the unbinding probability is one. Then, this sort of interactions can be modelled as a reversible two-state equilibrium process [37, 40–43].

Moreover, taking into account that the shear velocity is much slower than the internal crosslinks dynamics we can consider the interaction at steady state. Then, it can be expressed as $\frac{P_{ub}}{P_b} = \exp -\frac{(E_b - W_{ext})}{k_BT}$, where the two-state model has the bind state as the preferred low free energy equilibrium state at zero force and the unbind state as the high free energy equilibrium state at zero force. In addition, E_b represents the

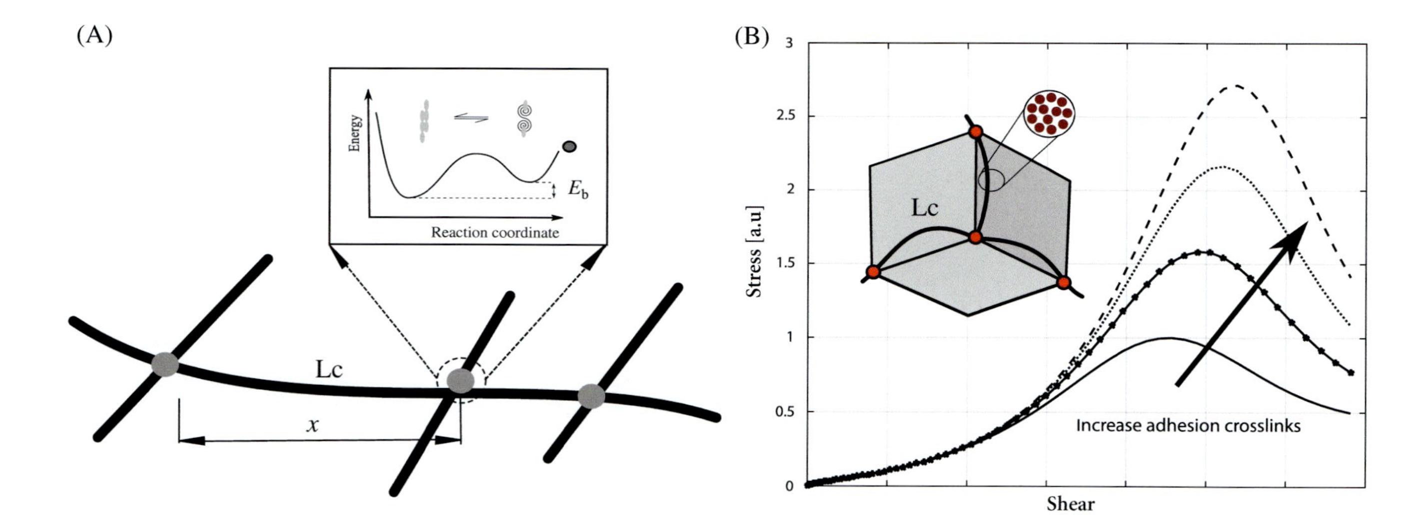

Fig. 13.1 (A) General cartoon showing the organisation of the bundles with crosslinks. (B) The figure details the stress–strain network response for increasing values of γ_0 showing an extension of the solid-like regime.

difference in the free energy between these states. w_{ext} represents the external mechanical work that induces the deformation of the crosslink. Then, we write an expression for the unbinding probability considering the shear strain as the main driving force, by using scaling arguments [39, 44]. To do that, we write $w_{ext} = f \cdot a$, where a is the distance between states in the direction of the reaction coordinate; this is a length scale in the order of the monomer size. The force f can be expressed as $f \sim G\gamma\xi^2$ in which γ is the shear strain, G is the shear modulus which can be estimated as $G \sim \frac{l_p k_B T}{L_c \xi^3}$, and ξ is the network mesh size. In addition, the unbinding transition due to the bundle strain happens in the semiflexible regime when $\xi \sim L_c$. In order to express qualitatively the behaviour of the coupled set of equations under alterations in the prestrain and the adhesion energy of the crosslinks, we evaluate them in the regime of semiflexible response, i.e. $L_c \propto l_p$. Where the coupled set of equations is given by the contour length as

$$L_c(c,\gamma) = L_c^0 + \delta L_c^{cl} P_{ub}, \quad P_{ub} = \frac{1}{1 + \exp[\kappa(\gamma_0 - \gamma)]}, \quad \gamma_0 \sim \left(\frac{E_b \xi^2}{k_B T l_p a}\right), \tag{13.6}$$

where the parameter $\kappa \sim \frac{l_p a}{\xi^2}$ is proportional to the sharpness of the transition between states and γ_0 is the characteristic strain, which is proportional to the adhesion energy and defines the point at which the probability of unbinding is 0.5. Then, if $\gamma_0 \ll \gamma$, the network is easy to be remodelled showing a fluid-like behaviour. On the contrary, if $\gamma_0 \gg \gamma$, the crosslinks are stable and the probability of transition is low; consequently the network behaves as a solid-like structure. Moreover, we can clearly identify that the characteristic strain γ_0, scales proportionally with the adhesion energy E_b, with the mesh size ξ and increases when the bundle stiffness l_p, becomes smaller.

Then, rewriting Eq. (13.5) considering that the incompressibility is satisfied automatically and the remaining of the invariants are: $I_1 = I_2 = 3 + \gamma^2$ we obtain the shear stress as

$$\sigma_{xz}(\gamma) = \frac{2}{3} n k_B T \gamma x^2 \left[\frac{(1 - x^2)}{c\pi[1 - (2 + \gamma^2)x^2 + x^4]^2} - c\pi^2\right]. \tag{13.7}$$

In order to illustrate the general trend of the model, Fig. 13.1B shows the response when γ_0 increases the initial stiffness of the network remains unaltered while the yielding stress and strain increase, extending the solid-like regime. This implies that as γ_0 increases the crosslinks become more stable.

13.2.2 Interaction between F-actin with vimentin: Softening and strengthening

Regarding the interaction between networks, experimentally it was observed that, on the one hand, for very low concentration, the changes are almost negligible and, on the other hand, once a certain value is overpassed, the effects become more relevant until reaching some asymptotic value.

This effect can be interpreted by means of the phase transition formalism. In this sense, we propose to use an interaction energy Ψ_{int} by means of the Landau functional that couples the effects with the networks [36, 37]. This energy is written in terms of an interaction parameter defined $\Gamma = \Gamma(c)$, where c represents the ratio between the concentrations of vimentin and actin. As our aim is to determine when the effect of the vimentin (IF) becomes relevant on the mechanical response, we focus on the concentration of c when it is near the critical point and the interaction parameter Γ assumes a very small value. This allows us to expand the free energy in even powers of Γ and retain only the lowest order terms for the Landau energy term, associated with the remodelling.

Then, we rewrite the Helmholtz free energy as follows:

$$\Psi(\mathbf{C},\Gamma) = \frac{\alpha}{2}\Gamma^2 + \frac{\beta}{4}\Gamma^4 + \Psi_{actin}(\mathbf{C},\Gamma) + \Psi_{IF}(\mathbf{C}). \tag{13.8}$$

In which the first two terms define the Landau energy that corresponds to the interaction parameter; the third term interprets the strain energy for the network without any coupling as a function of the isochoric Cauchy strain tensor and the interaction parameter. The behaviour from the equilibrium position (minimum) of the interaction energy changes at $\alpha = 0$, which is associated with the critical point at $c = c_{cr}$. It allows to choose $m\hat{c}$ as α, where m is a positive constant and $\hat{c} = (c - c_{cr})/c_{cr}$ is the deviation of the concentration ratio from the critical point normalised by c_{cr}, which is defined as a reduced concentration ratio. Then, the simplest selection is $\alpha = m\hat{c}$, for which $\alpha > 0$ above the critical point and $\alpha < 0$ below. The dependence of β with $\hat{c}$ does not affect qualitatively the behaviour of the free energy in the vicinity of the critical point and, therefore, we take β as a constant. Then, by deriving the free energy with respect to the interaction parameter Γ, we obtain the equilibrium condition as $\frac{\partial \Psi}{\partial \Gamma} \approx 2\alpha\Gamma + 4\beta\Gamma^3 = 0$. Thus, the equilibrium value of remodelling, Γ is

$$\Gamma \approx \pm\left(\frac{-\alpha}{2\beta}\right)^{1/2} = \begin{cases} \pm\left[\dfrac{m(c - c_{cr})}{2\beta c_{cr}}\right]^{1/2} & \forall \quad c > c_{cr} \\ 0 & \forall \quad c < c_{cr}. \end{cases} \tag{13.9}$$

The interaction parameter is cancelled when the concentration ratio $c \approx c_{cr}$, above the critical value scale as $\Gamma \sim (c - c_{cr})^{1/2}$. For values of c below the critical point, the level of interaction is zero.

This effect leads us to think that the formation of the cytoskeleton scaffolding elements can drive to a broad phase diagram for the cellular mechanical properties. In this sense, Fig. 13.2 condenses our interpretation of the process. We consider that the effects of strengthening and weakening can be considered as the action of two ratios of concentrations (which could also be described as chemical potentials), one is defined as $c = \frac{[vimentin]}{[F-actin]}$, and the other as $\chi = \frac{[actin]}{[crosslinks]}$. The first concentration ratio, c defines the intensity of the interaction and above a critical value, c_{cr}, it occurs a phase transition in which the coupling between networks becomes dominant. The second

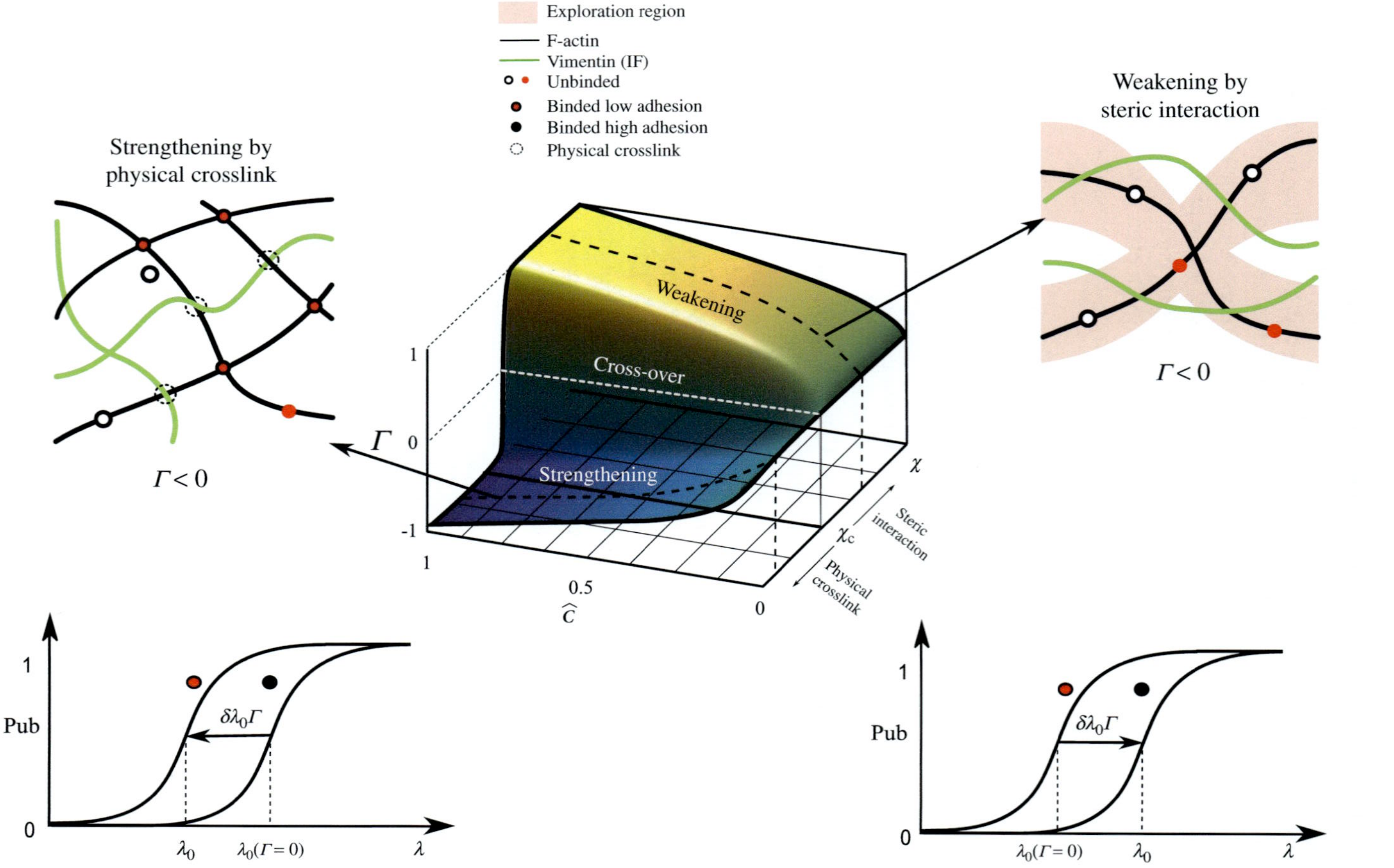

Fig. 13.2 (*Central*) Phase diagram for the interaction parameter showing the weakening and the strengthening effects. (*Right*) The strengthening effect promoted by the formation of physical crosslinks, which reduce the contour length. (*Left*) The weakening effect is promoted by the steric interaction that disturbs the formation of chemical crosslinks, which increase the contour length.

concentration ratio controls the sort of the resulting interaction. For $\chi = \chi_c$ exists a crossover between the two regimes.

Then, once the interaction parameter has been defined, next we will describe the internal variables that encode the interplay between the two networks and how they are driven by the interaction parameter $\Gamma(c)$. We consider the following hypothesis.

Interaction induces strengthening

(i) A rise in the concentration of IF enhances the deformation of the physical crosslinks with the F-actin networks and on average it reduces the contour length and the degree of fluctuations of the actin. Then, the L_c is reduced, the ratio r/L_c tends to one and the composite network manifests a rise in the stress. It is important to point out that as the IF filaments are very flexible, the increment in the density of physical crosslinks due to the interaction with F-actin does not produce a relevant change over the stress sustained by the IF. Consequently, to simplify the model, we neglect the effect of the physical crosslinks over the IF and only focus on the role of physical crosslinks over the F-actin, as can be observed in Fig. 13.2 (left-top). Then finally, the effective contour length L_c due to the alterations of vimentin is

$$L_c(\Gamma) = L_c^0 - \delta L_c^\Gamma \Gamma + \delta L_c^{cl} P_{ub}, \tag{13.10}$$

where $\delta L_c^\Gamma \Gamma$ represents the reduction of the contour length due to vimentin interaction, as was outlined previously, and L_c^0 represents the contour length of the mesh without vimentin. The second term $\delta L_c^\Gamma \Gamma$ represents the effective reduction in the length associated with the formation of the physical crosslinks promoted by the vimentin, see Fig. 13.2 (left-bottom). (ii) The effective network represents a network build-up by two kinds of transient crosslinks. On the one hand, the chemical interactions given by the biological crosslinks; on the other hand, the physical crosslinks due to the interaction between F-actin with vimentin. Then, we expect that the effective yielding strain γ_0, which is proportional to the mixture between physical and chemical crosslinks, will be smaller as it represents a lower effective adhesion energy. Hence, according to the proposed model the alterations promoted by the increment of the density of IF in the F-actin network are: a decrease in L_c, and in γ_0. Then, we write it as a combination of the parameters associated in a network without IF plus a perturbation associated with the interaction parameter Γ, as follows:

$$\gamma_0(\Gamma) = \bar{\gamma}_0 - \delta\gamma_0 \Gamma. \tag{13.11}$$

Interaction induces weakening

(i) Unexpectedly, at high actin concentrations, the additional polymer results in a weaker composite network with a lower elasticity and yield stress. When the actin concentration increases and the concentrations of crosslinks and vimentin are the same as in the strengthening experiments, the ratio $\chi = \frac{[crosslinks]}{[actin]}$ is lowered.

Then, a rise in the mesh size promotes a rise in the flexibility of the bundle, which is $\propto (\frac{L_c}{lp})$, at the same time increases the explored region by the bundle due to the thermal fluctuations, this effect is illustrated in Fig. 13.2 (right-top) as a pink contour representing the fluctuations tube.

Then, the probability of bond formation becomes scarce. Furthermore, when the networks are copolymerised with this ratio $\chi = \frac{[crosslinks]}{[actin]}$ and with the range of concentrations of vimentin, the interaction disturbs the crosslinking process. It provides an additional steric constraint imposed by vimentin IF, which results in a loss of F-actin crosslinking.

Then, according to that, we write it as a combination of the parameters associated in a network without IF plus a perturbation dependent on the interaction parameter Γ, as follows:

$$L_c(\Gamma) = L_c^0 + \delta L_c^\Gamma \Gamma + \delta L_c^{cl} P_{ub}, \tag{13.12}$$

where L_c^0 represents the contour length of the mesh without vimentin and $\delta L_c^\Gamma \Gamma$ the effective increase in the length due to the steric interaction promoted by the vimentin. (ii) The rise of the internal stress is propagated to the chemical crosslinks lowering the characteristic strain γ_0, as shown in Fig. 13.2 where the red and black dots describe the effect of the prestress over the P_{ub} of the chemical crosslinks [45, 46]. Then, according to the proposed model, the changes induced by the increment of the density of IF in the F-actin network are encoded as an increase in γ_0.

$$\gamma_0(\Gamma) = \bar{\gamma}_0 + \delta\gamma_0 \Gamma. \tag{13.13}$$

13.3 Results

The proposed theory is used to depict the experiments carried out by Jensen et al. [16] on copolymerised F-actin/vimentin network. We evaluate the proposed model for the set of parameters identified by means of nonlinear least-square fit with experiments of monotonic shear tests, in a regime of large deformation, as is reported in Ref. [16]. Subsequently, solving the following coupled set of equations we can obtain the stress–strain relation for the different analysed networks:

$$\gamma_0(c) = \bar{\gamma}_0 \pm \delta\gamma_0 \left[\frac{m(c_{cr} - c)}{2\beta c_{cr}} \right]^{1/2}, \tag{13.14}$$

where $\pm$, as was explained previously, will depend on the actin concentration. Next, the contour length is

$$L_c(c, \gamma) = L_c^0 \pm \delta L_c^\Gamma \left[\frac{m(c_{cr} - c)}{2\beta c_{cr}} \right]^{1/2} + \frac{\delta L_c^{cl}}{1 + \exp[\kappa(\gamma_0 - \gamma)]}. \tag{13.15}$$

Next, the introduced alterations on L_c by the ratio of concentrations c and the strain of the network structure imply the need to updated mesh size for the reference configuration as

$$x(L_c) = (1+\epsilon)\left(1 - \frac{2L_c(c,\gamma)}{l_p \pi^{\frac{3}{2}}}\right)^{1/2}. \tag{13.16}$$

Finally, the Cauchy shear stress considering the neo-Hookean contribution, plus the nonlinear response of the effective semiflexible network, defined in the Eq. (13.7), is expressed as

$$\sigma_{xz}(\gamma) = c_1\gamma + \frac{2}{3}nk_BT\gamma x^2\left[\frac{(1-x^2)}{c\pi[1-(2+\gamma^2)x^2+x^4]^2} - c\pi^2\right]. \tag{13.17}$$

The parameters of the model can be divided in two types: (i) rigid-worm-like chain parameters L_c^0, l_p, δL_c^{Γ}, δL_c^{cl}, and ϵ which are of the order of magnitude of the values used to describe the experiments of in vitro F-actin networks and to keep on the regime of semiflexible entropic elasticity [19, 33, 47]. (ii) The parameters associated with the remodelling dynamics of the crosslinks κ and γ_0, and the parameters that describe the interaction parameter $\Gamma(c)$. These parameters encode the transitions to induce the fluidisation of the network and represent an indirect measure of the adhesion force of crosslinks. These values were identified in order to fit the experimental data.

13.3.1 Strengthening phase

The formation of physical crosslinks encodes the strengthening phase and it is made in a network with constant concentration of F-actin of 6μM and a concentration of vimentin in the range of: 0, 0.3, 1.5, and 3μM.

Next, the simulated model by considering the following scaled material parameters as: $\frac{\delta L_c^{\Gamma}}{L_c^0} = 0.4$; $\frac{\delta L_c^{cl}}{L_c^0 + \delta L_c^{\Gamma}} = 0.5$; $\frac{l_p}{L_c^0} = 0.8$; $\frac{\gamma_0}{\gamma^{\max}} = 0.9$; $\frac{\delta\gamma_0}{\gamma^{\max}} = 0.4$; $\frac{c_1}{\sigma^0} = 0.2$; $\kappa = 30$; $\frac{m}{2\beta} = 0.5$; and $\epsilon = 0.03$. Then, according with these parameters, we see, in Fig. 13.3A, the experimental results associated with the strengthening phase, the increment of the $\sigma_{\max}$, and the reduction of the γ_c according with the rise of the concentration of vimentin (IF). In addition, Fig. 13.3B describes the modulus $K = \frac{d\sigma}{d\gamma}$ in which can be observed the increment in the value of $G_0 \approx K_{\gamma=0}$ following the rise in the concentration of vimentin. Finally, in Fig. 13.3C, we observe the interaction parameter $\Gamma(c)$ as a function of the ratio of concentrations c where the points express the associated values of the ratio $c = \frac{[vimentin]}{[F-actin]}$ and Γ that allow the description of the experiments.

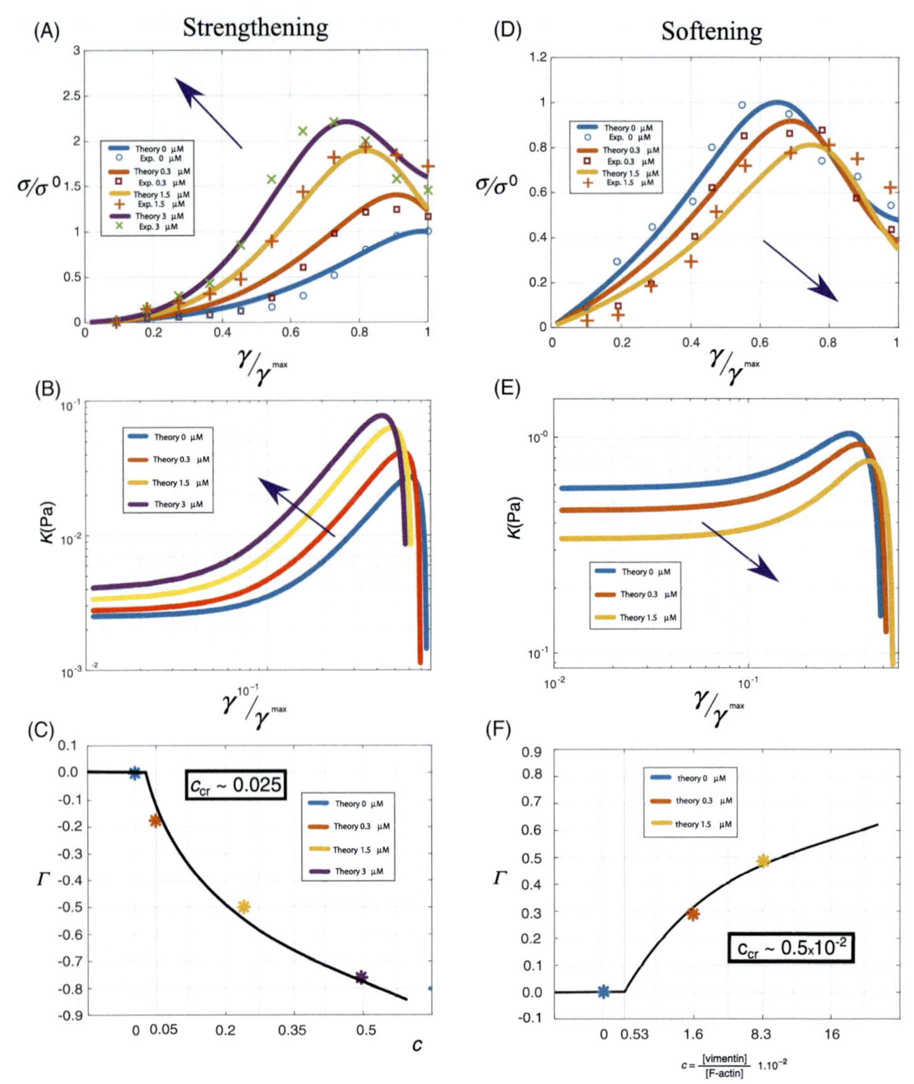

Fig. 13.3 *Strengthening effect by physical crosslinks.* (A) Stress–strain (shear) curves showing the nonlinear elastic effects. It shows a good agreement between the theory and the experimental measurements. (B) Effect of initial strengthening is illustrated by $K = \frac{d\sigma}{d\gamma}$; the *blue arrow* points the direction of the strengthening increase. (C) Interaction parameter $\Gamma(c)$ as a function of the concentrations ratio c. *Weakening effect by steric interactions.* (D) Stress–strain (shear) curves showing the nonlinear elastic effects. It can be observed a good agreement between the model predictions and the experimental measurements. (E) The effect of the initial weakening is illustrated by $K = \frac{d\sigma}{d\gamma}$; the *blue arrow* points the direction of the stiffening decrease. (F) Interaction parameter $\Gamma(c)$ as a function of the concentrations ratio c.

13.3.2 Weakening phase

In the following, we describe the results provided by the model to capture the experimentally reported emergent softening phase into the composite networks F-actin/vimentin, which promotes a steric interaction that blocks the formation of crosslinks. The studied concentration of F-actin kept constant at 18μM and the vimentin encompasses in the range: 0, 0.3, 1.5, and 3μM.

Then, the simulated model considers the following scaled material parameters as: $\frac{\delta L_c^{\Gamma}}{L_c^0}=0.4$; $\frac{\delta L_c^{cl}}{L_c^0+\delta L_c^{\Gamma}}=0.6$; $\frac{l_p}{L_c^0}=1.2$; $\frac{\overline{\gamma_0}}{\gamma^{\max}}=0.68$; $\frac{\delta\gamma_0}{\gamma^{\max}}=0.4$; $\frac{c_1}{\sigma^0}=0.2$; $\kappa=60$; $\frac{m}{2\beta}=0.45$; and $\epsilon=0.02$. With this parameters, the model describes the general trend of the experimental results, associated with the weakening as well as the reduction of the $\sigma_{\max}$ and the increment of the γ_c when the concentration of vimentin (IF) increases. In Fig. 13.3D, we plot the model predictions and the experimental measurements from Jensen et al. for the stress–strain curve of the composite actin–vimentin network. Fig. 13.3E shows the modulus $K=\frac{d\sigma}{d\gamma}$ in which is observed the value of $G_0 \approx K_{\gamma=0}$ decreases with the concentration of vimentin. This effect is explained by the alterations over the mesh size on the effective network, its changes, due to the strain γ and the interaction parameter $\Gamma(c)$, over the contour length L_c with respect the contour length for a mesh without vimentin L_c^0. Finally, Fig. 13.3F depicts the positive functional form of the interaction parameter $\Gamma(c)$ as a function of the ratio of concentrations c. The points express the associated values of the ratio $c=\frac{[vimentin]}{[F-actin]}$ and Γ the described stress–strain curves.

13.4 Discussion and conclusions

In summary, our work left from two previous studies on composite biopolymer networks of F-actin/vimentin. On the one hand, rheological measurements which showed that the composite networks always induce strain strengthening in comparison with the single F-actin network and, on the other hand, from the experiments of Jensen et al. which demonstrated that those networks can drive either the mechanical strengthening or weakening, during the copolymerisation of the two semiflexible species.

From that point on, we developed our theory into the framework of nonlinear continuum mechanics, where we define a free energy functional considering the role of the entropic-elastic for semiflexible networks with weak crosslinks and also, in order to describe the interaction parameter, we include an energetic term which allows the coupling between the two networks.

Unexpectedly, our phenomenological approach provides a very simple and useful constitutive model, which successfully reproduces the experimental observations and can capture the two described mechanisms of strengthening and softening just as a change in the sign of the interaction parameter $\Gamma(c)$. Then, it can be readily implemented into a field theory and also used to calculate the behaviour of a composite network of actin/vimentin under complex loading conditions. In addition, from a general perspective the described coupling between χ and Γ could has a functional form as $\sim \tanh(\chi - \chi_c)$ (see Fig. 13.2, central). Thus, it allows the change of the sign in the interaction parameter depending on whether the value is above or below the crossover χ_c.

The model presented here defines an effective crosslinked F-actin network that incorporates the actin/vimentin interactions that drive the microstructural remodelling effects via the alterations of the contour length L_c, and the characteristic stretch γ_0. Our methodology can be thought as a balance between microstructural and phenomenological phase transition formalism proposed by Landau. It provides a description to incorporate the remodelling exerted by vimentin without the associated details of the steric interaction. Nevertheless, the model provides a useful way to describe experiments by defying better metrics considering the nonlinear elasticity or to predict the susceptibility of the emergent network to concentrations changes.

Acknowledgement

The author thanks to Prof. Eugene Terentjev from the University of Cambridge for his valuable feedback.

References

[1] J. Howard, Mechanics of Motor Proteins and the Cytoskeleton, Sinauer Associates, Sunderland, MA, 2001.

[2] R. Phillips, J. Kondev, J. Theriot, Physical Biology of the Cell, Garland Science, 2009.

[3] E. Latorre, S. Kale, L. Casares, M. Gómez-González, M. Uroz, L. Valon, R.V. Nair, E. Garreta, N. Montserrat, A.D. Campo, Active superelasticity in three-dimensional epithelia of controlled shape, Nature 563 (7730) (2018) 203–208.

[4] J. Hu, Y. Li, Y. Hao, T. Zheng, S.K. Gupta, G.A. Parada, H. Wu, S. Lin, S. Wang, X. Zhao, High stretchability, strength, and toughness of living cells enabled by hyperelastic vimentin intermediate filaments, Proc. Natl. Acad. Sci. 116 (35) (2019) 17175–17180.

[5] F. Danielsson, M.K. Peterson, H.C. Araújo, F. Lautenschläger, A.K.B. Gad, Vimentin diversity in health and disease, Cells 7 (10) (2018) 147.

[6] M.P. Serres, M. Samwer, B.A.T. Quang, G. Lavoie, U. Perera, D. Görlich, G. Charras, M. Petronczki, P.P. Roux, E.K. Paluch, F-actin interactome reveals vimentin as a key regulator of actin organization and cell mechanics in mitosis, Dev. Cell 52 (2) (2020) 210–222.

[7] F. Gittes, B. Mickey, J. Nettleton, J. Howard, Flexural rigidity of microtubules and actin filaments measured from thermal fluctuations in shape, J. Cell Biol. 120 (4) (1993) 923–934.

[8] N. Mücke, L. Kreplak, R. Kirmse, T. Wedig, H. Herrmann, U. Aebi, J. Langowski, Assessing the flexibility of intermediate filaments by atomic force microscopy, J. Mol. Biol. 335 (5) (2004) 1241–1250.

[9] P.A. Janmey, U. Euteneuer, P. Traub, M. Schliwa, Viscoelastic properties of vimentin compared with other filamentous biopolymer networks, J. Cell Biol. 113 (1) (1991) 155–160.

[10] M.L. Gardel, J.H. Shin, F.C. MacKintosh, L. Mahadevan, P. Matsudaira, D.A. Weitz, Elastic behavior of cross-linked and bundled actin networks, Science 304 (5675) (2004) 1301–1305.

[11] M.L. Gardel, K.E. Kasza, C.P. Brangwynne, J. Liu, D.A. Weitz, Mechanical response of cytoskeletal networks, Methods Cell Biol. 89 (08) (2008) 487–519.
[12] V. Pelletier, N. Gal, P. Fournier, M.L. Kilfoil, Microrheology of microtubule solutions and actin-microtubule composite networks, Phys. Rev. Lett. 102 (18) (2009) 188303.
[13] J. Kayser, H. Grabmayr, M. Harasim, H. Herrmann, A.R. Bausch, Assembly kinetics determine the structure of keratin networks, Soft Matter 8 (34) (2012) 8873–8879.
[14] H. Wu, Y. Shen, D. Wang, H. Herrmann, R.D. Goldman, D.A. Weitz, Effect of divalent cations on the structure and mechanics of vimentin intermediate filaments, Biophys. J. 119 (1) (2020) 55–64.
[15] A.V. Schepers, C. Lorenz, S. Köster, Tuning intermediate filament mechanics by variation of pH and ion charges, Nanoscale 12 (28) (2020) 15236–15245.
[16] M.H. Jensen, E.J. Morris, R.D. Goldman, D.A. Weitz, Emergent properties of composite semiflexible biopolymer networks, BioArchitecture 4 (4–5) (2014) 138–143.
[17] R.C. Picu, Mechanics of random fiber networks—a review, Soft Matter 7 (15) (2011) 6768–6785.
[18] C.P. Broedersz, F.C. MacKintosh, Modeling semiflexible polymer networks, Rev. Mod. Phys. 86 (3) (2014) 995.
[19] H. López-Menéndez, J.F. Rodríguez, Microstructural model for cyclic hardening in F-actin networks crosslinked by α-actinin, J. Mech. Phys. Solids 91 (2016) 28–39.
[20] F. Meng, E.M. Terentjev, Theory of semiflexible filaments and networks, Polymers 9 (2) (2017) 52.
[21] F.J. Vernerey, Transient response of nonlinear polymer networks: a kinetic theory, J. Mech. Phys. Solids 115 (2018) 230–247.
[22] M.R. Buche, M.N. Silberstein, Statistical mechanical constitutive theory of polymer networks: the inextricable links between distribution, behavior, and ensemble, Phys. Rev. E 102 (1) (2020) 012501.
[23] T. Kim, W. Hwang, H. Lee, R.D. Kamm, Computational analysis of viscoelastic properties of crosslinked actin networks, PLoS Comput. Biol. 5 (7) (2009) e1000439.
[24] C. Borau, T. Kim, T. Bidone, J.M. García-Aznar, R.D. Kamm, Dynamic mechanisms of cell rigidity sensing: insights from a computational model of actomyosin networks, PLoS One 7 (11) (2012) e49174.
[25] A.S. Abhilash, P.K. Purohit, S.P. Joshi, Stochastic rate-dependent elasticity and failure of soft fibrous networks, Soft Matter 8 (26) (2012) 7004–7016.
[26] T. Su, P.K. Purohit, Semiflexible filament networks viewed as fluctuating beam-frames, Soft Matter 8 (17) (2012) 4664–4674.
[27] T. Nakajima, T. Kurokawa, S. Ahmed, W.-L. Wu, J.P. Gong, Characterization of internal fracture process of double network hydrogels under uniaxial elongation, Soft Matter 9 (6) (2013) 1955–1966.
[28] X. Zhao, Multi-scale multi-mechanism design of tough hydrogels: building dissipation into stretchy networks, Soft Matter 10 (5) (2014) 672–687.
[29] E. Ducrot, Y. Chen, M. Bulters, R. Sijbesma, C. Creton, Toughening elastomers with sacrificial bonds and watching them break, Science 344 (6180) (2014) 186–189.
[30] Z. Suo, J. Zhu, Dielectric elastomers of interpenetrating networks, Appl. Phys. Lett. 95 (23) (2009) 232909.
[31] X. Zhao, A theory for large deformation and damage of interpenetrating polymer networks, J. Mech. Phys. Solids 60 (2) (2012) 319–332.
[32] J.R. Blundell, E.M. Terentjev, Stretching semiflexible filaments and their networks, Macromolecules 42 (14) (2009) 5388–5394.

[33] F. Meng, E.M. Terentjev, Nonlinear elasticity of semiflexible filament networks, Soft Matter 12 (32) (2016) 6749–6756.
[34] H. Lopez-Menendez, J.F. Rodriguez, Towards the understanding of cytoskeleton fluidisation-solidification regulation, Biomech. Model. Mechanobiol. 16 (4) (2017) 1159–1169.
[35] H. Lopez-Menendez, L. Gonzalez-Torres, A theory to describe emergent properties of composite F-actin and vimentin networks, J. Mech. Phys. Solids 127 (2019) 208–220.
[36] H. Nishimori, G. Ortiz, Elements of Phase Transitions and Critical Phenomena, OUP, Oxford, 2010.
[37] H. Lopez-Menendez, J. D'Alessandro, Unjamming and nematic flocks in endothelial monolayers during angiogenesis: theoretical and experimental analysis, J. Mech. Phys. Solids 125 (2019) 74–88.
[38] J.M. Ferrer, H. Lee, J. Chen, B. Pelz, F. Nakamura, R.D. Kamm, M.J. Lang, Measuring molecular rupture forces between single actin filaments and actin-binding proteins, Proc. Natl. Acad. Sci. 105 (27) (2008) 9221–9226.
[39] P.-G. De Gennes, Scaling Concepts in Polymer Physics, Cornell University Press, 1979.
[40] A. Brown, R. Litvinov, D. Discher, P. Purohit, J. Weisel, Multiscale mechanics of fibrin polymer: gel stretching with protein unfolding and loss of water, Science 325 (5941) (2009) 741–744.
[41] O. Lieleg, M.M.A.E. Claessens, A.R. Bausch, Structure and dynamics of cross-linked actin networks, Soft Matter 6 (2) (2010) 218–225.
[42] P. Purohit, R. Litvinov, A. Brown, D. Discher, J. Weisel, Protein unfolding accounts for the unusual mechanical behavior of fibrin networks, Acta Biomater. 7 (6) (2011) 2374–2383.
[43] H. Lopez-Menendez, A mesoscopic theory to describe the flexibility regulation in f-actin networks: an approach of phase transitions with nonlinear elasticity, J. Mech. Behav. Biomed. Mater. 101 (2020) 103432.
[44] G. Bell, Models for the specific adhesion of cells to cells, Science 200 (4342) (1978) 618–627.
[45] O. Lieleg, K. Schmoller, M. Claessens, A. Bausch, Cytoskeletal polymer networks: viscoelastic properties are determined by the microscopic interaction potential of cross-links, Biophys. J. 96 (11) (2009) 4725–4732.
[46] O. Lieleg, J. Kayser, G. Brambilla, L. Cipelletti, A. Bausch, Slow dynamics and internal stress relaxation in bundled cytoskeletal networks, Nat. Mater. 10 (3) (2011) 236–242.
[47] J. Palmer, M. Boyce, Constitutive modeling of the stress-strain behavior of F-actin filament networks, Acta Biomater. 4 (3) (2008) 597–612.

Effect of interfibre bonding on mechanical behaviour of electrospun fibrous mats

14

Mir Karim Razavi Aghjeh[a,b] *and Mir Jalil Razavi*[c]
[a]Department of Macromolecular Science and Engineering, Case Western Reserve University, Cleveland, OH, United States, [b]Faculty of Polymer Engineering, Institute of Polymeric Materials, Sahand University of Technology, Tabriz, Iran, [c]Department of Mechanical Engineering, State University of New York at Binghamton, Binghamton, NY, United States

14.1 Introduction

Electrospinning, as a simple and versatile technique, is widely employed in the fabrication of micro-, nano-, and even smaller than 1 nm fibrous products [1–4]. A variety of organic (polymeric) and inorganic (ceramic) materials and their composites have been used to produce fibrous mats via this method [2,5,6]. The most significant features of the electrospun fibrous mats are their large surface-to-volume ratio, high porosity, and interconnected pore structures, which originate with the production method [2]. Those features have led them to be used in a wide range of applications including but not limited to energy storage systems [7], electrodes [8], fuel cells [9], sensors [4], water purification [10], air filtration [11], heterogeneous catalysis [5], super-absorbent media [12], textiles [13], encapsulation and delivery of drugs [14], tissue engineering [15], tissue-engineering scaffolds [16], and medical disposables [17].

Because of the complexity of the production process and diversity of involved parameters, characterisation of the mechanical properties of electrospun fibrous mats has been a difficult task [18]. Beyond the intrinsic properties of the bulk materials and porosity, the mechanical properties of the electrospun mats are influentially affected by the small-scale architecture of the electrospun mats, which include fibre diameter, the orientation of the fibres along the mat, and the interaction between the individual fibres at the crossing points (interfibre bonding) [19–22]. Depending on the properties of the used materials, the process parameters, and possible post-treatment, the interaction between the individual fibres can alter from weak Van der Waals interactions to strong welding at the crossing points [19,23,24]. Different studies have shown that the interfibre bonding is a major damage mechanism and plays a key role in determining the mechanical properties of the electrospun fibrous mats [7,18,19,25–30]. Before focusing exclusively on the interfibre bonding and its effect on the mechanical properties of electrospun fibrous mats, a brief overview is presented to introduce the most

Mechanics of Fibrous Networks. https://doi.org/10.1016/B978-0-12-822207-2.00007-6

important parameters affecting the mechanical properties of electrospun mats. In general, the mechanical properties of electrospun fibrous mats are dependent on the meso- and microstructure features of the mats including fibre diameter [31–36], alignment of the fibres along the mat [37–40], porosity and interconnectivity of the mat [31,38], porosity [41–43], crystallinity [13], and interaction between the adjacent fibres [19–22,25,44,45]. These characteristics are in turn controlled by the intrinsic properties of the base materials, the electrospinning process parameters as well as possible post-treatments. In electrospun fibrous mats, the morphology of the fibres, particularly the fibre diameter, is affected by a variety of parameters ranging from the solution properties and electrospinning process conditions to ambient atmosphere conditions [31,32,46–48].

The results of experimental studies have shown that the Young's modulus and the tensile strength of the electrospun fibrous mats are enhanced by a reduction in average fibre diameter in which, the extent of increase is considerable when the fibre diameter falls below 1 μm (Fig. 14.1). These were mainly attributed to the higher molecular orientation and crystallinity (in semi-crystalline polymers) of the thinner fibres. While the strain at break of the single fibres increases with the decrement of fibre diameter, it usually passes through a maximum for the mats (Fig. 14.1) [31,32]. However, because of the complexity of the process and diversity of the involved parameters, the results of different researches may not necessarily show the same trend [31,32,38]. For instance, efforts to alter the fibre diameter may lead to changes in the porosity of the mats as well, which is independently in the favour of lower tensile modulus and strength [38]. Therefore, it would be valuable if the effect of fibre diameter on the mechanical properties of the fibrous mats is considered at constant porosity. The alignment of the fibres during electrospinning is another key parameter that controls the mechanical properties of electrospun mats. The mats composed of aligned fibres exhibit favourable mechanical behaviour in aligned direction relative to their randomly deposited counterparts [6,49–52]. Generally, alignment of the fibres increases the tensile modulus and strength while decreasing the elongation at break (draw ratio) in the orientation (tensile) direction and vice versa [38].

In addition to contributions of fibre diameter and fibre alignment, the interaction between the adjacent fibres (interfibre interaction) plays a key role in determining the mechanical properties as well as the integrity and free standing of the electrospun mats. Relatively poor mechanical properties of electrospun fibrous mats originate from their nonwoven structure and insufficient interfibre bonding [53]. Because electrospun nonwoven mats are collected layer by layer, the interaction between the fibres at cross points is inherently weak, which may in turn lead to delamination of as-spun mats under tension [28]. This naturally leads to unfavourable mechanical properties as well as structural instability. Therefore, to take the advantages of the electrospun mats, many attempts have been made to improve interfibre adhesion by adjusting the process variables and by post-process treatments or employing both the methods, simultaneously. In general, improvement of the interfibre interaction at the cross points leads to enhanced mechanical strength as well as better free standing of the fibrous mats. The following section discusses the methods that have been employed to date to enhance interfibre bonding and the effect of interfibre bonding on the mechanical properties of the electrospun mats.

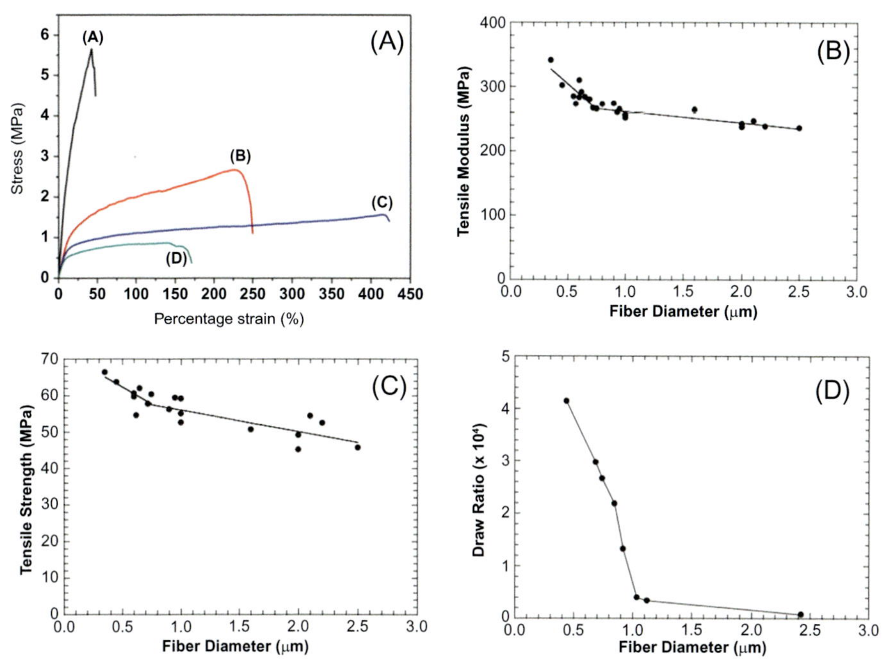

Fig. 14.1 (A) Typical stress–strain curves of electrospun poly(caprolactone) (PCL) mats with various average fibre diameters; (a) 0.1, (b) 0.8, (c) 1.9, and (d) 3.4 μm. (B) Tensile modulus, (C) tensile strength, and (D) draw ratio of fibres vs fibre diameter for poly(caprolactone) (PCL) fibres.

(A) Reprinted with permission from H.H. Kim, M.J. Kim, S.J. Ryu, C.S. Ki, Y.H. Park, Effect of fiber diameter on surface morphology, mechanical property, and cell behavior of electrospun poly(ε-caprolactone) mat, Fibers Polym. 17(7) (2016) 1033–1042, https://doi.org/10.1007/s12221-016-6350-x. (B–D) Reprinted with permission from S.-C. Wong, A. Baji, S. Leng, Effect of fiber diameter on tensile properties of electrospun poly(ε-caprolactone), Polymer 49(21) (2008) 4713–4722, https://doi.org/10.1016/j.polymer.2008.08.022.

14.2 Effect of interfibre bonding on mechanical properties of electrospun mats and methods to enhance the interfibre bonding

Because of the lack of interaction between the fibres and, hence, the unfavourable mechanical performances of as-spun fibrous mats in most cases, efforts have been made to increase the strength of interfibre bonding using different in-process and post-process methods and strategies. Among different methods, thermal bonding (or thermal welding) [7,13,19,26,28,39,54], vapour- or solvent-induced bonding (or welding) [2,38,55–58], and chemical or crosslinking bonding (or welding) [53,59–61] are most frequently employed to alter the interaction between the adjacent fibres of the electrospun mats. Physical methods such as fibre charging [62] and hydrogen bonding [63] as well as additive fillers [64] have also been employed to improve the interfibre bonding.

14.2.1 Thermal bonding

Electrospun fibrous mats can be welded at the cross points of the adjacent fibres via thermal treatment above the glass transition temperature (T_g) and below the melting temperature (T_m) of the base polymer to preserve the fibrous structure of the mats [13,19,26]. In the case of semi-crystalline polymers, the thermal treatment is also carried out at temperatures above the T_m of the base polymer [28] or above the T_m of a component in the mixture [14,16,39].

Electrospun nanofibres of poly(etherimide) (PEI) were post-treated at different temperatures ranging from 80°C to 240°C to evaluate the interfibre bonding improvement [19]. The results showed that the thermal treatment below T_g of PEI (220°C) did not increase interfibre bonding, while treatment at temperatures between T_g and T_m (≥330°C) of PEI led to the fusion of nanofibres at the intersections and hence to significantly enhanced stiffness and ultimate strength, while decreased stretchability (Fig. 14.2).

It should be considered that regardless of the effect of thermal treatment on interfibre adhesion, the mechanical strength of the electrospun mats can be affected by thermal annealing as a consequence of the alteration in molecular features and the crystalline structure of the individual fibres. Pan and Lim [13] indicated that the thermal annealing increased the Young's modulus of poly(lactic acid) (PLA) nanofibres mainly because of the increase in crystallinity and transformation of the morphology from a purely fibrillar structure to a mixture of fibrillar and nano-granular.

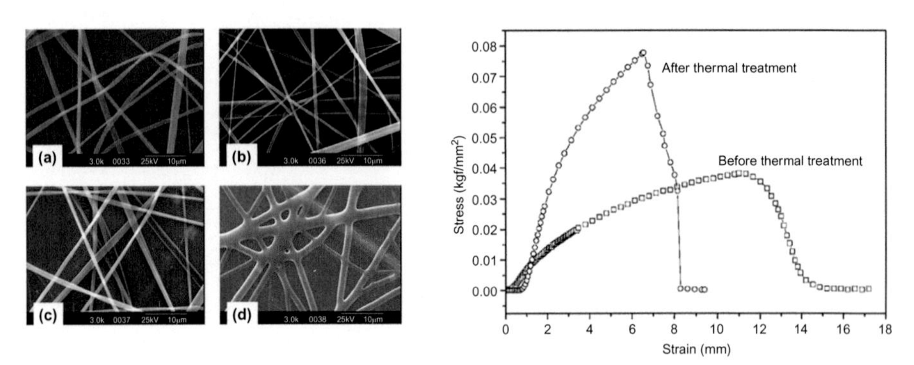

Fig. 14.2 (Left) Scanning electron microscopy (SEM) images of electrospun poly(etherimide) (PEI) fibres after thermal treatment at 80°C (a), 150°C (b), 220°C (c), and 240°C (d). (Right) Stress–strain curves of the PEI nanofibre web before the thermal treatment and the PEI nanofibre web treated at 240°C.
Reprinted with permission from S.-S. Choi, S.G. Lee, C.W. Joo, S.S. Im, S.H. Kim, Formation of interfiber bonding in electrospun poly(etherimide) nanofiber web, J. Mater. Sci. 39(4) (2004) 1511–1513, https://doi.org/10.1023/B:JMSC.0000013931.84760.b0.

The effects of thermal treatment on the structure and mechanical properties of electrospun chitosan-gelatin (CG) nanofibre membranes were studied by Wang et al. [26]. The results showed that the significant improvement of mechanical properties (Young's modulus, tensile strength, and fracture energy) was achieved for the samples that underwent the thermal annealing at temperatures above T_g of CG. Using different analyses, it was demonstrated that the interfibre bonding and the intermolecular bonding (electrostatic interaction and covalent bonds) make contributions to improve mechanical properties following thermal annealing at proper temperatures. To avoid the effect of interfibre bonding as well as to determine the contribution of intermolecular bonding, the effect of annealing treatment on mechanical properties of cast CG membranes was studied. The extent of improvement in mechanical properties for electrospun membranes was greater than that of cast membranes indicating that the main source of improvement is thermally enhanced interfibre bonding at the intersections rather than intermolecular bonding (Fig. 14.3).

Because of the restricted chain mobility of the semi-crystalline polymers below T_m, the thermal welding of the individual fibres inside the fibrous mats may be more efficient at temperatures above T_m. You et al. [28] thermally bonded the PLA electrospun fibres at 180°C (T_m of the used PLA was 173.2°C) for different lengths of time. For short times, the thermal treatment led to welded nanofibres at the intersections and to a slight increase in the fibres' diameters but maintained the fibre geometry. At longer treatment times, the nanofibres completely fused at the cross points and the mats shrunk significantly, leading to the formation of a film-like morphology. The Young's modulus and, interestingly, the elongation at break of the treated mats were larger than

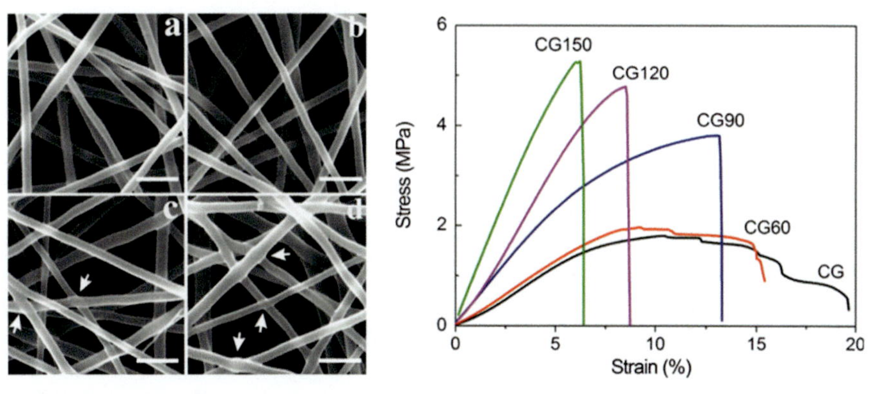

Fig. 14.3 (Left) Scanning electron microscopy (SEM) images of chitosan-gelatin (CG) nanofibres untreated (a), and thermally annealed at 60°C (b), 90°C (c), and 150°C (d). Scale bar for each image is 1 μm. *Arrows* in images c and d show the welded points of the fibres. (Right) Stress–strain curves of chitosan-gelatin (CG) nanofibre membranes at different annealing temperatures.

Reprinted with permission from Z. Wang, et al., Effect of thermal annealing on mechanical properties of polyelectrolyte complex nanofiber membranes, Fibers Polym. 15(7) (2014) 1406–1413, https://doi.org/10.1007/s12221-014-1406-2.

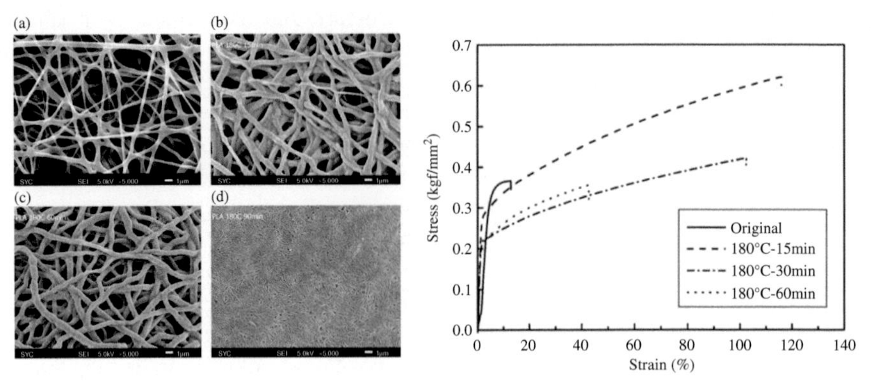

Fig. 14.4 (Left) Scanning electron microscopy (SEM) images of poly(lactic acid) (PLA) nanofibrous mat before and after thermal treatment at 180°C; (a) as-spun, (b) treated for 15 min, (c) treated for 60 min, and (d) treated for 90 min. (Right) Stress–strain curves of poly(lactic acid) (PLA) nanofibrous mat before and after thermal treatment.
Reprinted with permission from Y. You, S. Won Lee, S. Jin Lee, W.H. Park, Thermal interfiber bonding of electrospun poly(l-lactic acid) nanofibers, Mater. Lett. 60(11) (2006) 1331–1333, https://doi.org/10.1016/j.matlet.2005.11.022.

that of untreated counterpart. While the modulus enhancement was expected, the increase in elongation at break was attributed to shrinkage of the mats during thermal treatment (Fig. 14.4).

As mentioned before, thermal treatment is a simple and versatile method to increase interfibre bonding in electrospun mats. However, finding the optimum conditions at which thermal welding is achieved while preserving the fibrous structure of the mat is a challenge, particularly for the polymers with relatively high T_m [2,15]. Moreover, this treatment is usually associated with alteration in the structural properties such as porosity and pore size, which are crucial characteristics in applications such as membranes [5]. An efficient strategy to overcome that drawback is using a bicomponent system associated with thermal annealing at temperatures between the T_m of two components. While the component with a higher T_m preserves the structure of the mat, the low melting point component adheres the fibres at the cross points [16,39]. The poly(caprolactone)/poly(ethylene terephthalate) (PCL/PET) blends of different ratios were electrospun to nanofibrous mats followed by thermal treatment (thermal sintering) at 100°C to improve the mechanical properties and stability of the resulted capsules [14].

To improve the interfibre bonding and the adhesion between the fibres, Kancheva et al. [16] prepared PLA/PCL nanofibrous mats via both electrospinning and electrospraying routes, followed by thermal treatment of the mats at the melting temperature of PCL (about 60°C), well below the melting temperature of PLA (about 160°C) to preserve the integrity of the fibrous mats. They used various strategies including the electrospinning of a single solution, simultaneous electrospinning of two separate solutions, and simultaneous electrospinning and electrospraying of two separate solutions to prepare the nanofibrous mats of PLA/PCL. The results showed that the thermal treatment enabled the welding of the fibres via melting of the PCL fibres or particles and thus enhanced the mechanical properties of the mats, the level of which was dependent on the production strategy of the mats. The most improvement in mechanical properties in conjunction with most durable heat-treated fibrous mats was obtained in the case of electrospinning of PLA/PCL single solution. In another work [39], poly(3-hydroxybutyrate) (PHB) and poly(caprolactone) (PCL) were electrospun from separate jets to create hybrid mats and then thermally heated at 80°C for 15 min. The thermal treatment near the melting temperature of PCL (60°C) led to the well-sealed and adhered fibres while preserving the fibrous structure, thus enhancing the mechanical properties of the mats [39].

Electrospinning has widely been used to fabricate fibrous mats with material that is the precursor of carbon nanofibres (CNFs) [7,54,65,66]. Generally, the results have shown that starting with well-bonded fibrous mats leads to mechanically strong and electrically conductive CNFs after thermostabilisation and carbonisation treatments. For instance, Niu et al. [66] showed that using side-by-side polyaniline (PAN) and poly(vinylpyrrolidone) (PVP) nanofibre precursors led to strongly interbonded carbon nanofibres and, therefore, larger electrochemical capacitances than if the nanofibres were made solely of PAN or PAN/PVP blend. To improve the interfibre bonding of the lignin/PAN precursor and the mechanical properties of the resulting CNFs, Ding et al. [54] used butyrated lignin (B-lignin). The electrospun nanofibres of lignin/PAN and B-lignin/PAN were subjected to oxidative thermostabilisation and carbonisation at 200°C and 1000°C, respectively, to obtain CNFs. The results showed that the strong interfibre bonding of B-lignin/PAN fibres during the thermostabilisation stage, which originated from the enhanced thermal mobility of the modified lignin molecules, led to the fused carbon fibres at the intersections and significant improvement in mechanical properties of the resulted carbon fibres after carbonisation. Wang et al. [7] studied the effect of polyethylene oxide (PEO) concentration on different properties of the electrospun nanofibres of lignin-PEO blends as well as the resulted electrospun carbon nanofibres via thermostabilisation and carbonisation (Fig. 14.5). They showed that the high proportions of PEO ($\geq$10%) resulted in fusion among the adjacent carbon fibres (as shown in Fig. 14.5F), which were ascribed to the strong melting behaviour of PEO (mp 66–70°C) during the heat treatment. They stated that the separated carbon fibres (as shown in Fig. 14.5D and E), the interconnected fibre webs can provide shorter and more continuous pathways for electron transportation, and thus reduce the electrochemical resistance of the electrode. The electrical conductivity of the nonfused carbon fibres and fused carbon fibres was 7.34 and 10.53 S/cm, respectively.

The effect of a low melting point polymer additive (PEO) on morphology and mechanical properties of electrospun lignin-based CNFs, as high-performance anode materials for lithium-ion batteries, was studied [6]. The electrospun fibrous mats of lignin/PEO blends and pristine lignin were subjected to carbonisation and thermal annealing with urea and the results showed that the presence of PEO enhanced the interfibre bonding leading to improved mechanical and electrochemical properties of the resulted CNFs. The response mechanism of the lignin content in lignocellulosic fibres to their interfibre bonding properties by using fibre deformation behaviour

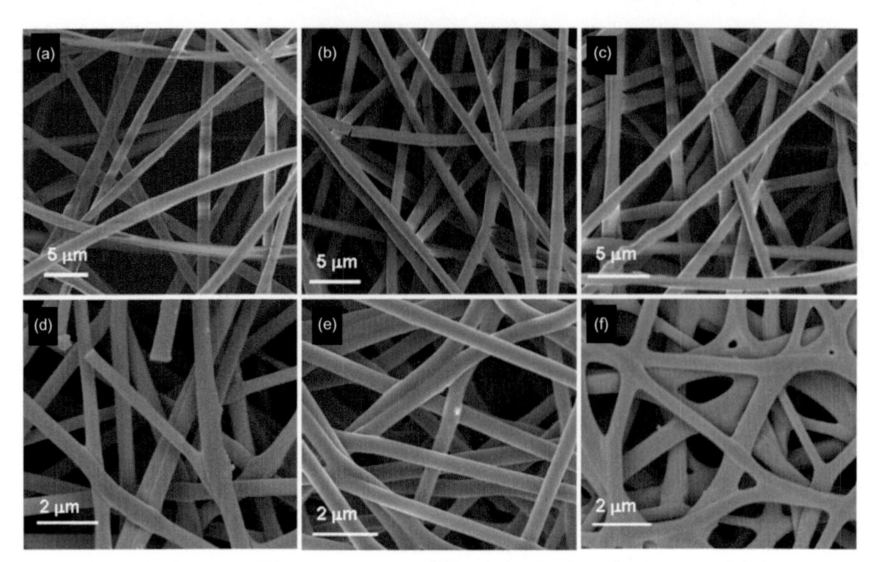

Fig. 14.5 FESEM images of electrospun nanofibres from lignin-PEO blends with lignin to PEO ratio of (A) 97:3, (B) 95:5, and (C) 90:10. FESEM images of electrospun carbon nanofibres derived from lignin-PEO blends with lignin to PEO ratio of (D) 97:3, (E) 95:5, and (F) 90:10. Reprinted with permission from S.-X. Wang, L. Yang, L.P. Stubbs, X. Li, C. He, Lignin-derived fused electrospun carbon fibrous mats as high performance anode materials for lithium ion batteries, ACS Appl. Mater. Interfaces 5(23) (2013) 12275–12282, https://doi.org/10.1021/am4043867.

theory was investigated by Li et al. [65]. They showed that a decrease in lignin content enlarged the bonded area and strengthened the bonding while increasing the deformability of lignocellulosic fibres. Consequently, a more compact fibre network was formed, showing a significant decrease in the bulk property. The dependency between physical strength and bulk properties of the fibre network was justified by the wet fibre deformation behaviour, which was influenced significantly by the lignin content in lignocellulosic fibres.

Solvent-induced thermal treatment is another strategy to enhance interfibre bonding. Wang et al. [27] optimised the mechanical stability of polyethersulphone (PES) electrospun nanofibrous membranes (ENMs) via a thermal treatment method, in which the evaporation of the residual solvent in as-spun mats facilitated the welding of the fibres. Thermal treatment was applied at 190°C, above the solvent boiling temperature (dichloromethane (DCM) 153°C) and below T_g of PES (225°C). The treated materials showed significant improvement in mechanical properties (Fig. 14.6).

The welding of nanofibres at their intersections, induced by thermal treatment, usually increases the Young's modulus, mostly at the cost of decreased elongation at break of the nanofibrous mats. Using other strengthening strategies along with the thermal treatment may lead to simultaneous improvement of toughness and stiffness. Namsaeng et al. [67] achieved enhancement of both modulus and elongation at break

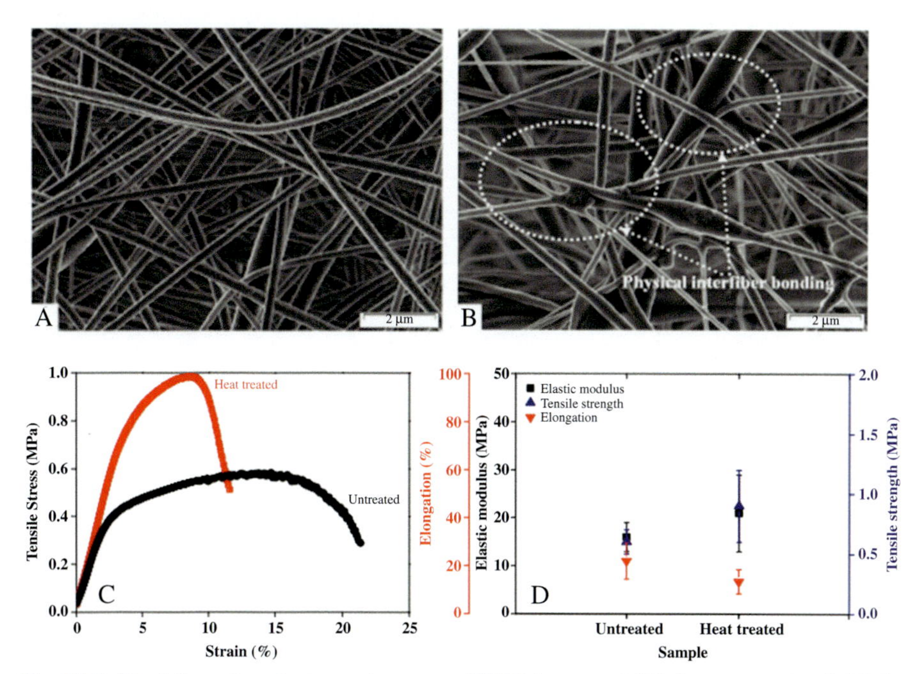

Fig. 14.6 (Top) Scanning electron microscopy (SEM) images of (A) the untreated and (B) the heat-treated nanofibres. (Bottom) The results of the tensile test (C) stress–strain curves and (D) the tensile properties including elastic modulus, tensile strength, and elongation for untreated and heat-treated nanofibres.
Reprinted with permission from S. Homaeigohar, J. Koll, E.T. Lilleodden, M. Elbahri, The solvent induced interfiber adhesion and its influence on the mechanical and filtration properties of polyethersulfone electrospun nanofibrous microfiltration membranes, Sep. Purif. Technol. 98 (2012) 456–463, https://doi.org/10.1016/j.seppur.2012.06.027.

for as-spun PAN–poly(vinyl chloride) (PAN-PVC) mats by thermal welding of the fibres at their cross points synergistically in combination with multiwalled carbon nanotube (MWCNT) reinforcement, to be used in water filtration applications (Fig. 14.7).

To further enhance the interfibre adhesion, the electrospun PLA nanofibres were incorporated with MWNT [68]. Thermal bonding improved the interfibre bonds of the electrospun fibres. The modulus increased at lower bonding temperatures and reduced at higher bonding temperatures. Cold crystallisation of PLA nanofibres, activated at the temperatures near the melting temperature of PLA, sharply increased the tensile strength as well as the Young's modulus. The electrical conductivity was best when the fibres were bonded close to their melting points, representative of the importance of fibre–fibre bonds in electrical connectivity. The fusion of the fibres at the junctions occurred at the temperatures above T_g and below T_m of the PLA. Further increase in treatment temperature to above T_m of the PLA led to highly bonded flattened fibres and finally to film formation (Fig. 14.8) [68].

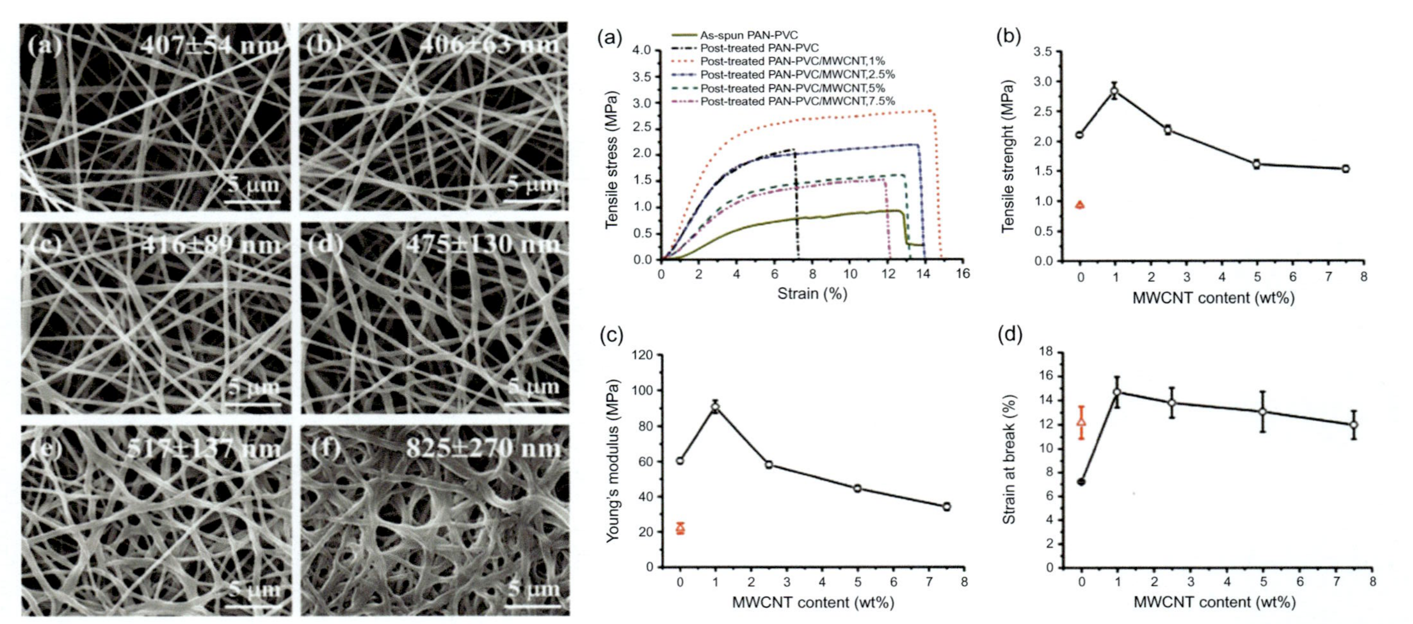

Fig. 14.7 (Left) Scanning electron microscopy (SEM) images of PAN–PVC/MWCNT, 1% composite mats: (a) without solvent-vapour treatment and with solvent exposure times of (b) 6 h, (c) 9 h, (d) 15 h, (e) 24 h, and (f) 30 h. (Right) The mechanical properties of as-spun PAN–PVC, post-treated PAN–PVC, and post-treated PAN–PVC/MWCNT composite mats with various MWCNT loadings of 1.0, 2.5, 5.0, and 7.5 wt%: (a) representative stress–strain curves, (b) tensile strength, (c) Young's modulus, and (d) strain at break. In (b)–(d), the mechanical properties of as-spun PAN–PVC mats are shown as red triangles for comparative purposes.

Reprinted with permission from J. Namsaeng, W. Punyodom, P. Worajittiphon, Synergistic effect of welding electrospun fibers and MWCNT reinforcement on strength enhancement of PAN–PVC non-woven mats for water filtration, Chem. Eng. Sci. 193 (2019) 230–242, https://doi.org/10.1016/j.ces.2018.09.019.

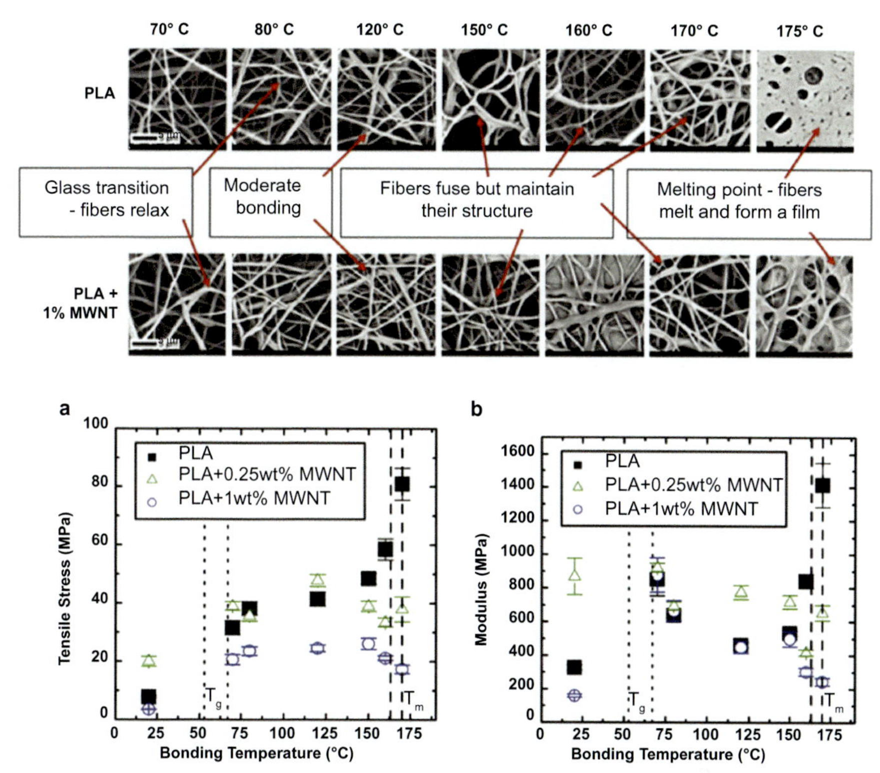

Fig. 14.8 (Top) Scanning electron microscopy (SEM) images of PLA and PLA/1% MWNT nanofibre morphology as a function of bonding temperature. (Bottom) (a) Tensile strength and (b) tensile modulus as a function of bonding temperature for PLA *(filled squares)*, PLA-0.25 wt % MWNT *(open triangles)*, and PLA-1 wt% MWNT *(open circles)*. The T_g and T_m regions indicate the range of observed T_g and T_m values for different sample types.
Reprinted with permission from S. Ramaswamy, L.I. Clarke, R.E. Gorga, Morphological, mechanical, and electrical properties as a function of thermal bonding in electrospun nanocomposites, Polymer 52(14) (2011) 3183–3189, https://doi.org/10.1016/j.polymer.2011.05.023.

Thermal and solvent-vapour post-treatment strategies, the latter of which will be discussed below, to increase the interfibre bonding of the nanofibres at their cross points, are usually associated with undesired structural alterations such as whole nanofibres' welding or dimensional shrinkage. Therefore, selective welding of the nanofibrous mats at the intersection of the fibres, without or with little structural changes away from the cross points, can be an interesting challenge. Photothermal heating has been used to weld a variety of materials from polymers, polymer blends, and composites and metals via doping the materials with plasmonic nanoparticles or light-absorbing dyes [69–72]. Wu et al. [20] used the concept of *photothermal welding* to weld or over-weld the PCL nanofibres by doping the nanofibres with various amounts of a light-absorbing dye (indocyanine green (ICG)) followed by laser irradiation at different irradiances and exposure time. The results showed that at low dye concentration and laser irradiance, the nanofibres were selectively welded at their junctions. This was attributed to higher temperature rise at the cross points because of the larger surface area at the cross points, which in turn led the cross points to reach their melting points ahead of other regions (Fig. 14.9). An increase in dye concentration or laser irradiance as well as a long exposure led to over-welded fibres and even the formation of transparent films. All the welded and over-welded mats showed improved Young's modulus, tensile strength, and elongation at break than that of the pristine mats (Fig. 14.9).

14.2.2 Solvent and vapour bonding

The solvent evaporation rate can alter the merging and fusion of the fibres at the cross points [73,74]. Raghavan and Coffin [55] demonstrated that the fusion of fibres at their intersections can be controlled by adjusting the humidity in the surrounding environment of the electrospinning process. They showed that the bonding of the electrospun polyethylene oxide (PEO) fibres increased with humidity, as a consequence of a decreased evaporation rate. Further increases in the humidity led to flattened networks. Nonetheless, the results for other polymers have shown diverse effects of ambient humidity on fibre bonding and, hence, on mechanical properties. The effect of relative humidity (RH) of the spinning environment on mechanical properties of PAN and polysulphone (PSU) electrospun nanofibrous mats were studied by Huang et al. [56]. The results indicated that the mechanical strength of the mats increased with RH at low RHs and decreased sharply after passing through a maximum at high RHs. The electrospun fibres obtained at low RH had better mechanical strength than those obtained at high RH. This was mainly attributed to the phase separation of the solution leading to the formation of skin layers, which hindered the interfibre bonding in the resulting mat. These diversities originate from the different solubility of polymers in used solvents as well as different phase separation rate of solution for different polymers.

Electrospun nanofibrous membranes (ENMs) of PAN and PSU were exposed to dimethylformamide (DMF) vapour for different times to find the evaporation period that best improved interfibre bonding and the mechanical properties. The results showed that under the optimum exposure time, PAN and PSU ENMs underwent the fusion of adjacent fibres at the intersections leading to greatly improved mechanical properties with negligible change in membrane flux performance. They also showed that vapour treating of the fibres in an impermeable substrate resulted in stronger interfibre bonding and higher improvement in mechanical strength for PAN and PSU fibre mats [58].

The effect of solvent-induced fibre bonding along with the effect of other controlling parameters on mechanical properties of electrospun PLA fibrous mats were studied by Selatile et al. [38]. Using solvents with different volatilities (DCM and DMF) as well as their mixture, they controlled the bonding strength of the fibres at the intersections. The results showed that using a low-volatile solvent led to enhanced fibre–fibre bonding and, therefore, to better mechanical properties of the mats.

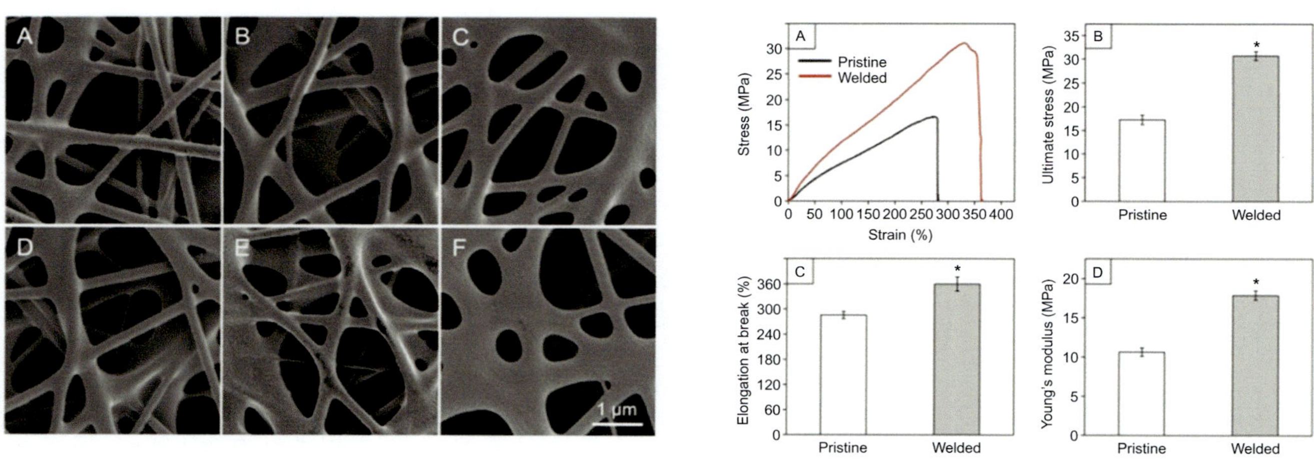

Fig. 14.9 (Left) Scanning electron microscopy (SEM) images showing the morphology of ICG-doped PCL nanofibres electrospun from solutions containing ICG at concentrations of (a, d) 1%, (b, e) 2%, and (c, f) 5% after exposure to an 808-nm diode laser at irradiances of 0.2 (a–c) and 0.4 $W cm^{-2}$ (d–f) for 2 s. (Right) (a) Stress–strain curves, (b) ultimate stress, (c) elongation at break, and (d) Young's modulus of mats of the ICG-doped PCL nanofibres (electrospun from a solution containing 1% ICG) without and with welding at the cross points by exposure to the laser at an irradiance of 0.2 $W cm^{-2}$ for 2 s.

Reprinted with permission from T. Wu, H. Li, J. Xue, X. Mo, Y. Xia, Photothermal welding, melting, and patterned expansion of nonwoven mats of polymer nanofibers for biomedical and printing applications, Angew. Chem. Int. Ed. 58(46) (2019) 16416–16421, https://doi.org/10.1002/anie.201907876.

However, the fibre diameter and fibre alignment played the main roles in determining the mechanical properties.

In some applications such as thin-film nanofibrous composite (TFNC) membranes, a nanofibrous mid-layer scaffold is created on a non-woven microporous support. Adhesion between the nanofibres and the non-woven support as well as interfibre bonding between the nanofibres are crucial parameters in determining the performance and durability of such layered structures [57]. That study showed the adhesion strength between the electrospun nanofibres of PES and non-woven PET microporous support increased with an increase in the ratio of DMF to *N*-methyl-2-pyrrolidone (DMF), PES concentration, flow rate, and the presence of PEO as a primer-coating layer of the non-woven PET mat. Moreover, the adhesion between the PES nanofibre and the PET non-woven support increased with RH possibly because of the skin formation and decreased after passing through a maximum threshold because of PES precipitation, showing the competitive affecting parameters on the interfibre bonding [57].

Li et al. [2] introduced a simple and versatile methodology to selectively weld the electrospun nanofibres of PCL at the intersections to improve the interfibre bonding and the mechanical properties of the electrospun nanofibrous mats. The welding was achieved by contacting the nonwoven mat with the vapour of DCM, as a proper solvent, in certain partial pressure and exposure times (Fig. 14.10). The results interestingly showed that the treated electrospun nanofibrous mats at the optimum conditions exhibited significantly increased Young's modulus, tensile strength, and elongation at break compared with pristine nanofibre mat (Fig. 14.10). It is worth noting that several past efforts to increase the elastic modulus and tensile strength of the nanofibre mats, via improvement of the interfibre bonding, have been associated with remarkably reduced elongation at break. This achievement i.e. simultaneous stiffness and toughness improvements were attributed to the targeted welding of the electrospun nanofibres at the intersections without altering the packing of polymer chains and severe structural destruction. They successfully employed the same methodology to weld poly(D,L-lactic-*co*-glycolic acid) (PLGA) nanofibres and polystyrene (PS) microfibres, and showed the generality of the method.

The solvent-vapour bonding methodology was employed to bond the polyimide (PI) nano- and microfibres (PI-N and MFA) in nanofibre-derived aerogels (NFAs) using DCM. Controlled welding of the adjacent nano- and microfibres at their intersections led to the formation of super elastic aerogels with a recoverable ultimate strain of 80%, with acceptable structural retaining (Fig. 14.11) [75].

14.2.3 *Chemical and* crosslinking bonding

Depending on the type of the base material, the thermal treatment process may lead to crosslinking of the fibrous materials at the cross points instead of fusion or simply welding. Ghorani et al. [61] studied the effect of heat and alkaline treatment on mechanical properties of cellulose acetate (CA) electrospun webs and showed that the treated webs exhibited high tensile strength and low elongation at break compared to the untreated webs (Fig. 14.12). The results were attributed to the crosslinking

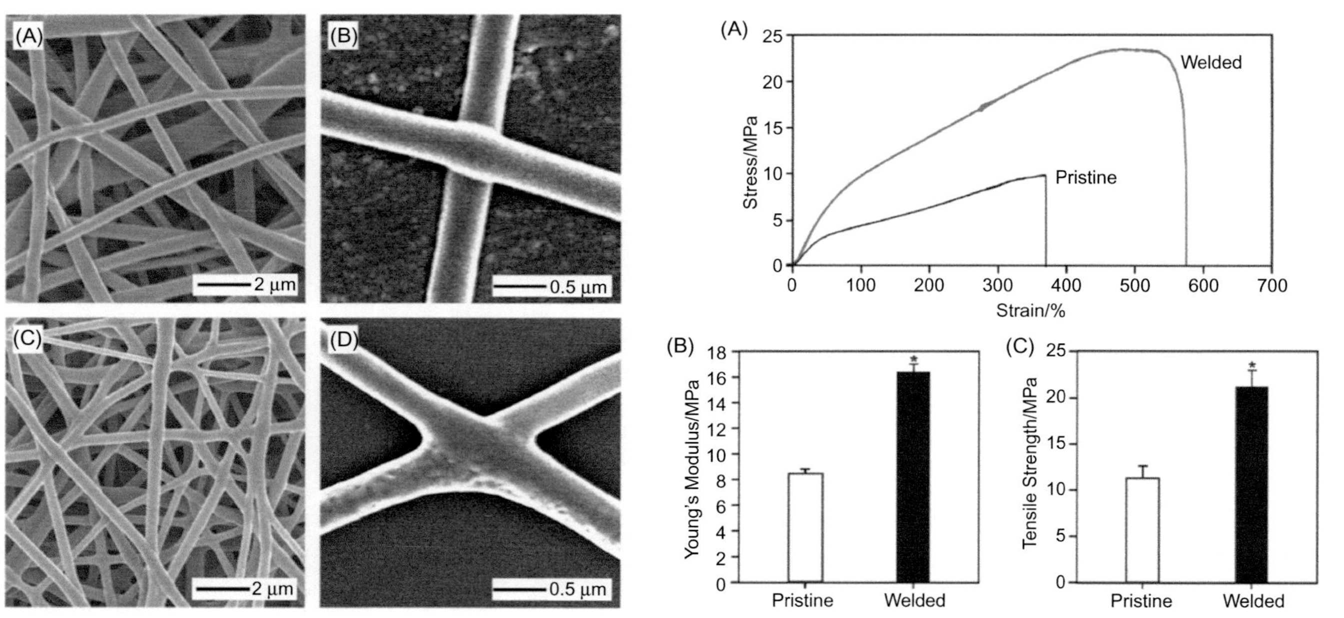

Fig. 14.10 (Left) Scanning electron microscopy (SEM) images of the nonwoven mats of electrospun PCL nanofibres after exposure to the vapour generated from 25 μL of DCM in a closed vial for different periods. (a, b) 30 min and (c, d) 60 min. (Right) Tensile mechanical assessment of electrospun PCL nanofibre mats before and after the treatment with 25 μL DCM vapour for 60 min. (a) Stress–strain curves. (b) Young's modulus. (c) Tensile strength. $n = 5$ for each test, * indicates a significant difference between the two types of samples ($P < .05$).

Reprinted with permission from H. Li, C. Zhu, J. Xue, Q. Ke, Y. Xia, Enhancing the mechanical properties of electrospun nanofiber mats through controllable welding at the cross points, Macromol. Rapid Commun. 38(9) (2017) 1600723, https://doi.org/10.1002/marc.201600723.

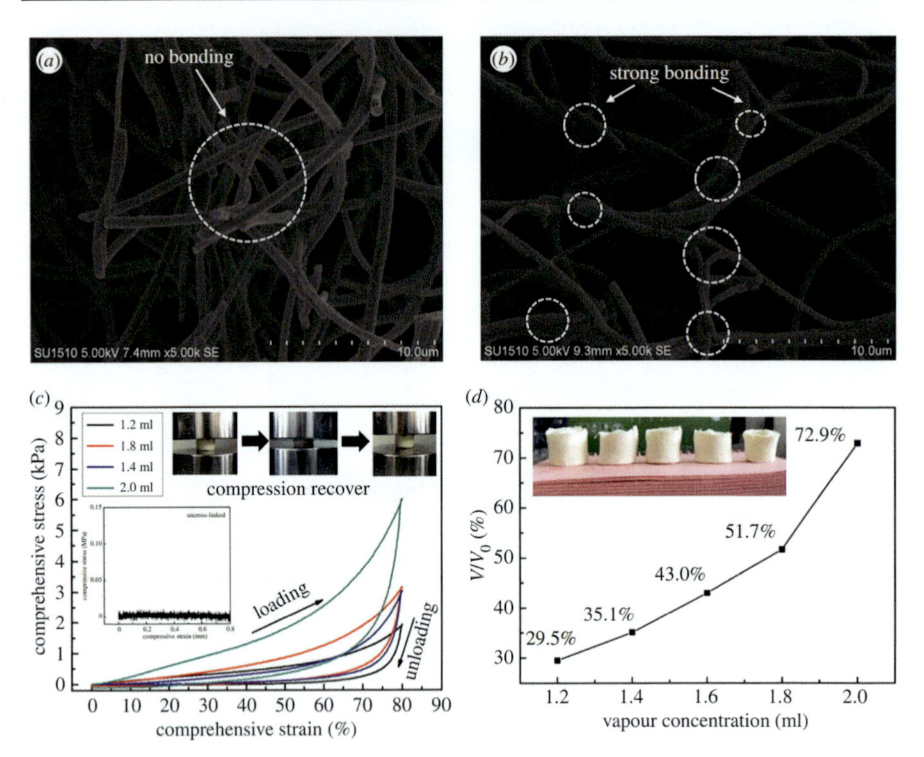

Fig. 14.11 Scanning electron microscopy (SEM) images showing the PI-N/MFA$_0$ (A) before and (B) after treatment by the DCM vapour generated from 1.4 mL DCM. (C) Compressive strain–stress curves and (D) volume shrinkage behaviours of the PI-N/MFA$_0$ before and after treatment by the DCM vapour coming from different volumes of DCM solutions: 1.2, 1.4, 1.8, and 2.0 mL.
Reprinted with permission from Y. Shen, D. Li, B. Deng, Q. Liu, H. Liu, T. Wu, Robust polyimide nano/microfibre aerogels welded by solvent-vapour for environmental applications, R. Soc. Open Sci. 6(8) (2019) 190596, https://doi.org/10.1098/rsos.190596.

between the CA fibres after treatment, leading to interfacial bonding between fibres at the cross-over points, without significant change in the fibre morphology.

Interfibre fusion and welding can be induced as side effects of the other treatment methods. Wang et al. [60] showed that crosslinking in glutaraldehyde (GTA) vapour led to significantly enhanced thermal and mechanical properties of the electrospun starch nanofibres. Morphological characterisation clearly showed that the crosslinked starch nanofibrous mat maintained the fibrous structure while the interfibre welding had taken place (Fig. 14.13).

Zhao et al. [53] used poly(ethylene glycol) dimethacrylate (PEGDMA) to crosslink the poly(vinylidene fluoride-*co*-hexafluoropropylene) (PVDF–HFP) membranes. The PVDF–HFP/PEGDMA solution was electrospun, dried, and thermally treated at 125°C or 130°C for the crosslinking of PEGDMA. Morphological studies showed that in a spatial range of PVDF–HFP to PEGDMA ratios, PEGDMA-induced crosslinking

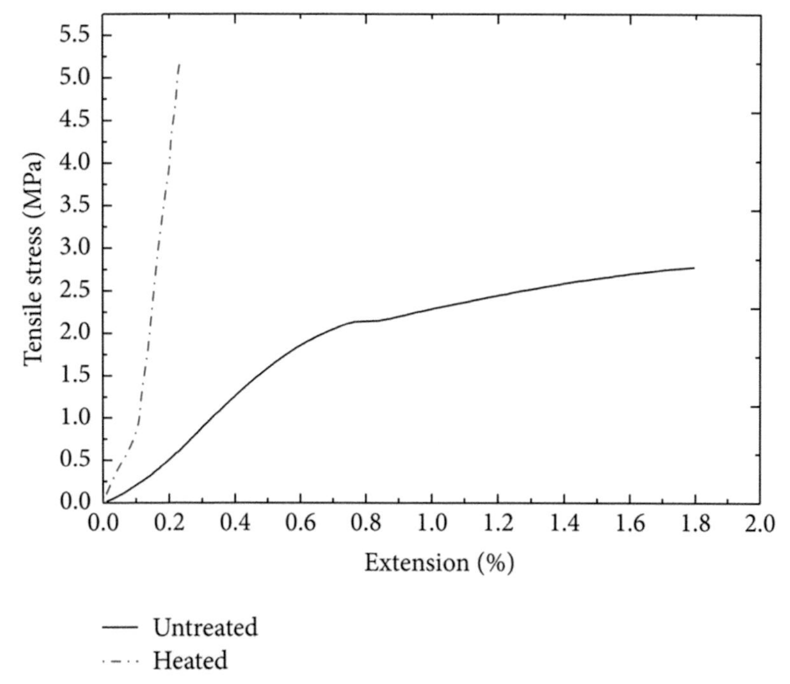

Fig. 14.12 Tensile stress–strain curve of the untreated CA fibre web and the heat-treated CA webs. Crosshead speed was 10 mm min^{-1}, web thickness: 150 μm, sample width: 6.35 mm. Reprinted with permission from B. Ghorani, S.J. Russell, P. Goswami, Controlled morphology and mechanical characterisation of electrospun cellulose acetate fibre webs, Int. J. Polym. Sci. 2013 (2013) 1–12, https://doi.org/10.1155/2013/256161.

of the nanofibres was accompanied by welding of the nanofibres at the cross points. This led to an appreciable increase in tensile modulus and strength and a little declined elongation at break (Fig. 14.14) [53].

The interfibre bonding can also be improved via a chemical reaction between the fibres at cross points. Merlini et al. [59] coated the surface of the electrospun PVDF fibrous mat with polyaniline (PAN) via in situ oxidative polymerisation of aniline (ANI) inside. After coating with PAN, the Young's modulus of PVDF mat enhanced significantly mainly because of the adhered PAN on the PVDF fibres, which increased the interfibre bonding. However, the elongation at break was reduced because of the brittle nature of the PAN component.

14.2.4 Physical bonding (hydrogen bonding and fibre charging)

It has well been known that operation of hydrogen bonding between the chains of a polymer such as polyamides or between the chains of different polymers in a blend greatly affects the mechanical properties of the polymer or the blend. Thus, the mechanical properties of the electrospun single fibres of such polymers would be

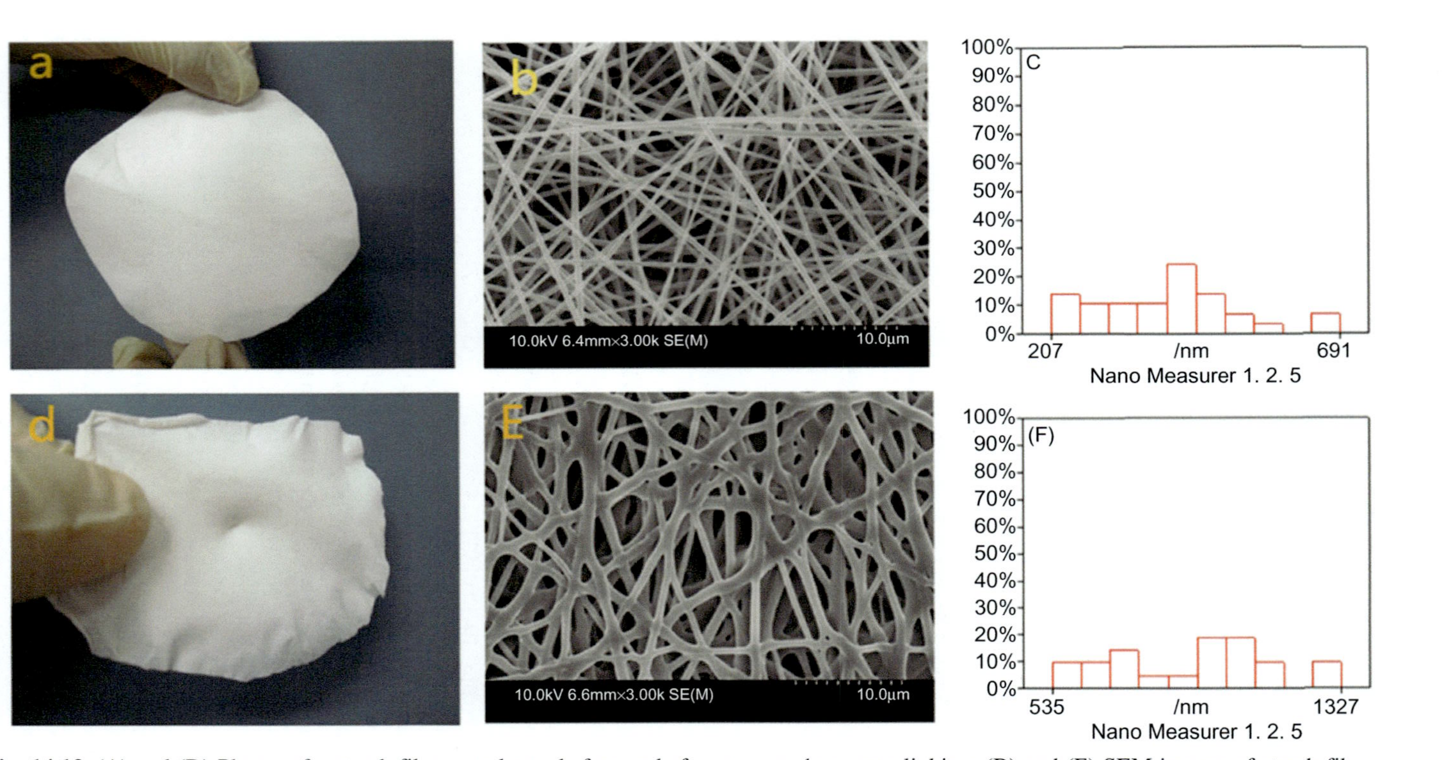

Fig. 14.13 (A) and (D) Photos of a starch fibre membrane before and after vapour phase crosslinking; (B) and (E) SEM images of starch fibre membrane before and after vapour phase crosslinking; and (C) and (F) fibre diameter distribution of un-crosslinked and crosslinked starch fibre membrane.

Reprinted with permission from W. Wang, et al., Effect of vapor-phase glutaraldehyde crosslinking on electrospun starch fibers, Carbohydr. Polym. 140 (2016) 356–361, https://doi.org/10.1016/j.carbpol.2015.12.061.

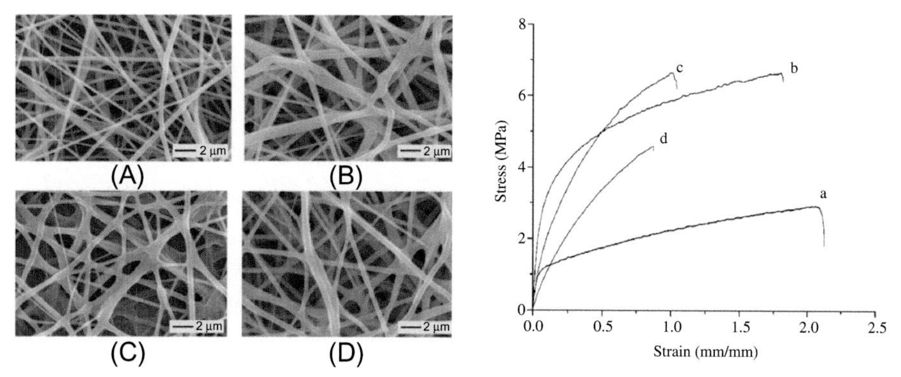

Fig. 14.14 (Left) Scanning electron microscopy (SEM) images of the electrospun PVDF–HFP/PEGDMA fibres prepared from solutions with different PVDF–HFP/PEGDMA ratios after heat treatment at 130°C for 2 h: (a) 4/1, (b) 4/2, (c) 4/3, and (d) 4/4 (PVDF–HFP concentration = 0.16 g/mL, voltage = 15 kV, distance = 15 cm, flow rate = 0.47 mL/h, DMF/acetone = 6/4 v/v). (Right) Typical stress–strain tensile curves of the electrospun PVDF–HFP/PEGDMA membranes with different mass ratios after heat treatment at 130°C for 2 h: (a) 4/0, (b) 4/1, (c) 4/2, and (d) 4/3.
Reprinted with permission from L. Zhao, H. Zhang, X. Li, J. Zhao, C. Zhao, X. Yuan, Modification of electrospun poly(vinylidene fluoride-*co*-hexafluoropropylene) membranes through the introduction of poly(ethylene glycol) dimethacrylate, J. Appl. Polym. Sci. 111(6) (2009) 3104–3112, https://doi.org/10.1002/app.29374.

affected by the existence of hydrogen bonding. On the other hand, hydrogen bonds between the adjacent fibres can develop at their intersections leading to strong interfibre bonding. This would enhance the mechanical properties of the resulting fibrous mats. Chen et al. [63] showed that *co*-electrospinning of poly(arylene sulphide sulphone) (PASS) with a semi-aromatic nylon, poly(m-xylene adipamide) (MXD6), led to significant enhancement in mechanical properties of the blend electrospun membrane compared to neat PASS membranes. It was understood that the hydrogen bonding between PASS and MXD6 chains in single fibres as well as the hydrogen bonding between the crossing fibres at the intersections were the main source of the improvement of mechanical properties. The development of hydrogen bonding at the intersections of the adjacent fibres could facilitate the stress transfer during tensile testing.

The introduction of a negative charge on cellulosic fibres (inner charge or surface charge) significantly enhances the physico-mechanical properties of fibres and fibre-based materials. Zhao et al. [62] studied the effect of 2,2,6,6-tetramethyl-piperidine-1-oxyl radical (TEMPO)-mediated oxidation and carboxymethyl cellulose (CMC) attachment, as inner charging and surface charging treatments, respectively, on the mechanical properties of cellulosic fibre-based materials. They showed that introducing a high negative charge to cellulosic fibre can effectively increase the tensile strength with just a slight reduction in fibre network integrity. They also indicated that the increased fibre inner charge, mostly contributed to the improved interfibre bonding of the fibres in the network, while the fibre surface charge, directly affected the formation of hydrogen bonding. Both the surface charge and inner charge were shown to increase the tensile strength and apparent density.

14.2.5 Interfibre bonding induced by rigid fillers (composites)

The effect of NaCl concentration on the morphology and mechanical properties of poly(vinyl alcohol) (PVA) electrospun nanofibrous mats was studied by Shi et al. [64]. The results showed that the mean diameter of nanofibres decreased and then increased after passing through a minimum with the NaCl concentration. The composite PVA/NaCl mats showed greatly improved yield stress, modulus, and tensile strength, at the cost of decreased strain at break compared to PVA mats. It was suggested and confirmed via morphological characterisation that the creation of a much more point-bonding structure (interfibre bonding) in the composite mats was the main reason for the observed mechanical properties (Fig. 14.15).

It was shown that the interfibre bonding properties of the mats composed of lignocellulosic fibres are strongly affected by the wettability of the fibre surface [76].

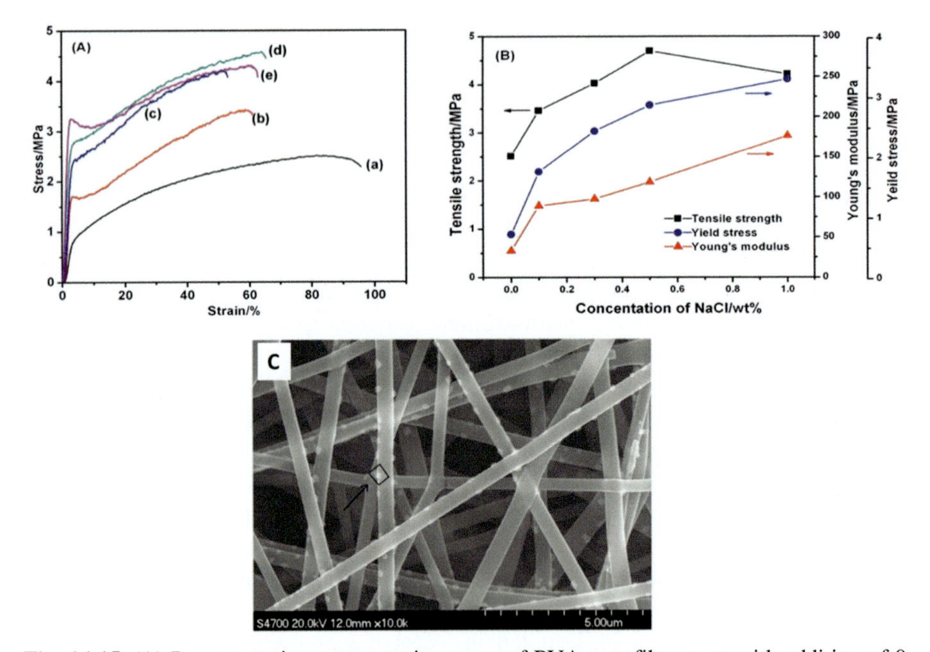

Fig. 14.15 (A) Representative stress–strain curves of PVA nanofibre mats with addition of 0 (a), 0.1 (b), 0.3 (c), 0.5 (d), and 1.0 (e) wt% NaCl. (B) Plot of average tensile strength and Young's modulus. (C) SEM image of the PVA nanofibre with the addition of 1 wt% NaCl. Reprinted with permission from Y. Shi, Y. Zhao, X. Li, D. Yan, D. Cao, Z. Fu, Enhancement of the mechanical properties and thermostability of poly(vinyl alcohol) nanofibers by the incorporation of sodium chloride, J. Appl. Polym. Sci. 135(13) (2018) 45981, https://doi.org/10.1002/app.45981.

Although this method has been applied for non-electrospun fibres, the results can be inspiring for electrospun mats. The effect of surface wettability on the bonding properties of mechanically refined fibres was studied by Xie et al. [76]. It was evaluated by the surface free energy, surface lignin, and surface charge values. The results showed that during the fibre mechanical treatment, the fibre surface charge increased, and the surface lignin decreased, leading to the improvement of the fibre surface free energy. As a result, the bonding strength index tripled without significant loss in bulk properties.

The dependency of the mechanical response to the interaction between fibres is also true for the other types of fibrous mats such as cellulose nanopaper. Cellulose nanopaper shows superior mechanical properties thanks to the strong, stiff nanofibres, and the enhanced interfibre interactions [77]. The reported elastic modulus of cellulose nanopapers is around 12 GPa [78,79], while it is around 3–4 GPa for conventional papers [80].

14.3 Mechanical simulation of interfibre bonding in fibrous mats

As discussed in the previous sections, to date, many experimental studies have been conducted to characterise the material of constituent fibres to regulate and tune the mechanical properties of electrospun fibrous mats [1]. Nevertheless, it is not an easy task to experimentally link the behaviour of bulk material to the characteristics of constituent fibres [81]. Contrary to continuum materials, complex constituent material properties and the microstructural heterogeneity hamper the accurate characterisation of fibrous mats [25]. This statement is true when it comes to studying the interfibre bonding mechanics and the effect of interfibre bonding on the mechanical properties of the electrospun mats. Because, single fibres usually show a non-linear stress–strain behaviour in finite strains of stretching, bending, rotation, and the failure of interfibre bonds take place at the early stages of deformation under external stimuli [82–84]. As a result, the stress distribution in fibrous mats is nonuniform and mainly localised, resulting in their distinctive behaviour from continuum materials. Therefore, models predicting the mechanical behaviour of these classes of materials should consider their discrete nature by taking into account the non-linear behaviour of constituent fibres as well as associated bonding between crossing fibres.

The discrete simulation method can explicitly link the micromechanics of constituent elements to the bulk material behaviour of the mat. Various multiscale modelling techniques have been developed to correlate the behaviour of the bulk material to the microstructure architecture [25,81,85–89]. Among different multiscale modelling methods, the finite element method (FEM) has successfully been used to gain insight into the effect of the mechanical behaviour of single fibres on the mechanical performance of fibrous mats [25,40,85,90–95]. The results of these studies confirm that constitutive-relation and finite element models can satisfactorily elucidate the non-linear mechanical behaviours of the mats. These modelling attempts, in agreement with experimental studies, show that the microstructure of fibrous mat (fibre diameter, fibre alignment, porosity, etc.) and the mechanical properties of single fibres have a significant effect on the elastic modulus of the fibrous mats. However, in most of the above-mentioned studies, the bonding between the intersecting fibres was not considered in the models. Ignoring interfibre bonding simplifies models but can non-negligibly reduce the accuracy of the results, particularly in the well-bonded (welded) fibres.

More recently, the interfibre bonding, as one of the effective and non-negligible factors, has been considered in the mechanical models. Wei et al. [25] in a molecular dynamics (MD) model included fusion bonding and Van der Waals interactions at the nanoscale. The results indicated that the tensile strength of the mat is very dependent on the interfibre fusion. The strength of the mat increased as the number of bonding points increased, while the fracture energy was reduced by the over-fusion of the fibres. For more accurate simulations, multiscale finite element models have been developed to predict the mechanical performance of fibrous mats by simultaneously taking the characteristics of constituent fibres and their interfibre bonding into account [18,29,96]. Shahsavari et al. [97] showed that the elastic modulus is very sensitive to the crosslinking density in the affine limit but is independent of this parameter in the non-affine regime. The latter is because increasing the bond density reduces the fibre segment length causing the fibres to deform axially rather than bend. Therefore, the effectiveness of interfibre bonding is dependent on the other parameters such as the density of fibres.

The analysis of the representative volume element (RVE) of mats allows the control and measurement of parameters that are hardly reachable through experiments, specifically the density of fibres and the location of their cross points. The reliability of these models relies on the accuracy of assumptions that are made according to the existing experimental data. In discrete network models (DNMs), individual fibres are considered as finite elements in the form of truss or beam elements, which can possess straight or curved forms. These simulations with great flexibility can find the effect of the determinant factors such as porosity, the morphology of fibres, the orientation of fibres, and interfibre interactions. Therefore, DNMs are effective tools to investigate the mechanics of fibrous mats in a multiscale media [25,85,98–102].

The results of the recent simulation work by Chavoshnejad et al. [103] showed that in a fixed porosity, higher bonding densities lead to the affine deformation of the mat, resulting in the loss of bonding effectiveness. Fig. 14.16 shows their developed finite element model to study the effect of interfibre bonding on the elastic modulus and rupture of the electrospun polymethylmethacrylate (PMMA) fibrous mats.

The effect of the interfibre bonding on the mechanical properties and failure of a PMMA mat from their results is depicted in Fig. 14.17. The bonding percentage was defined as the percentage ratio of the number of bonded cross points to the total number of cross points in the model. The effect of interfibre bonding on the mechanical behaviour of the mat is obvious by comparing the lowest-percentage bonded case with the fully bonded one. For the case with 100% bonding, the elastic modulus of the mat is considerably larger than the cases with low percentages of bonding. Interfibre bonding increases the stiffness of the mat [96] as discussed in the previous section. Fig. 14.17 clearly shows that at 50% porosity, increasing the bonding percentage from

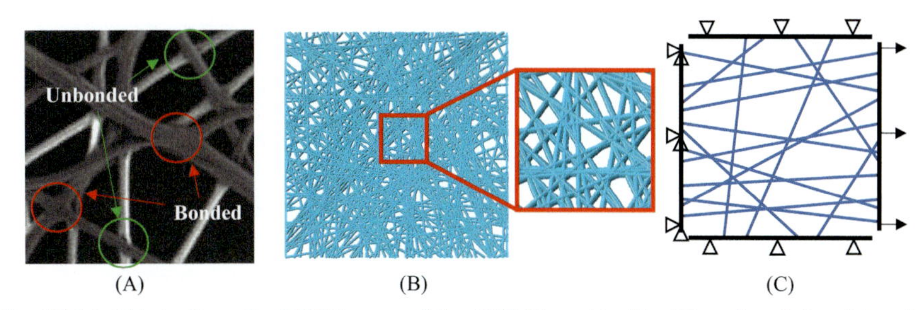

Fig. 14.16 (A) A slice of a SEM image of the PMMA mat to identify unbonded and bonded cross points. (B) Finite element model with random location and orientation of the fibres for the case with 25% porosity. (C) Boundary condition of the finite element model.
Reprinted with permission from P. Chavoshnejad, O. Alsmairat, C. Ke, Effect of interfiber bonding on the rupture of electrospun fibrous mats, J. Phys. Appl. Phys. 54, 025302.

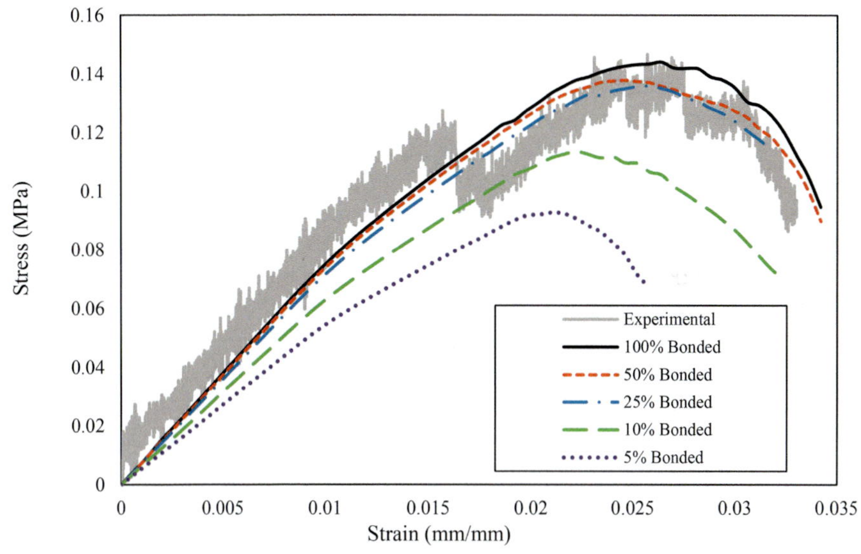

Fig. 14.17 Stress–strain curves of the mat with 0%, 10%, 25%, 50%, and 100% bonding at the cross points. The porosity of all models is 50%.
Reprinted with permission from P. Chavoshnejad, O. Alsmairat, C. Ke, Effect of interfiber bonding on the rupture of electrospun fibrous mats, J. Phys. Appl. Phys. 54, 025302.

5% to 10% significantly increases the elastic modulus, while increasing the bonding percentage from 50% to 100% has a negligible effect.

The authors of Ref. [96] showed that at any predefined porosity, increasing the percentage of bonding produces many smaller segments (Fig. 14.18). Segment length was defined as the distance between two bonded cross points. Small segments intrinsically prefer to be stretched and rotated rather than bent and, consequently, the mat tends to undergo the affine deformation. As it is clear, at 25% porosity, there are many more

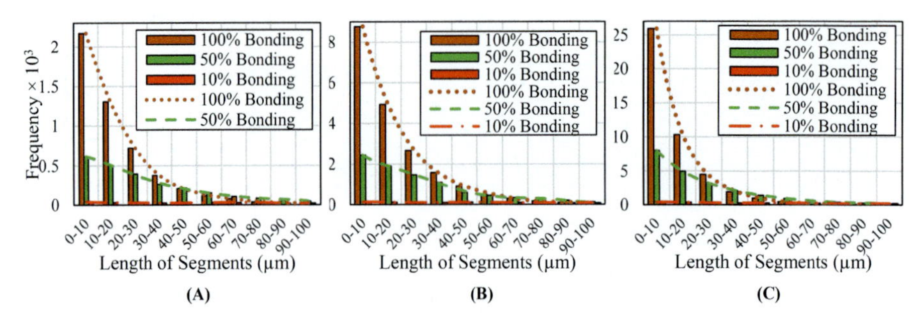

Fig. 14.18 Histogram of length of segments of fibres for cases with (A) 75% porosity, (B) 50% porosity, and (C) 25% porosity.
Reprinted with permission from P. Chavoshnejad, O. Alsmairat, C. Ke, Effect of interfiber bonding on the rupture of electrospun fibrous mats, J. Phys. Appl. Phys. 54, 025302.

small segments than at 50% and 75% porosities. The low density of interfibre bonding in a predefined porosity produces scatter locations for the stress concentration, while a high density of interfibre bonding removes stress concentrations and shifts the behaviour of the mat towards isotropic material behaviour.

The dependency of the elastic modulus of fibrous mats to the interfibre bonding density has been discussed in other research studies [29,96]. Using a three-dimensional (3D) finite element fibrous network model, Goutianos et al. [29] found that in the case of fixed fibre properties, the elastic modulus and strength are mainly controlled by the density and strength of the interfibre bonding. The elastic modulus and strength of the fibrous mat are increased by the increase of the interfibre bond density as well as the strength of interfibre bonding. However, the mechanical property of interfibre bonds and their density have a slight effect on the failure strain. In this study, the deposition of the fibres was modelled by adding fibre as 3D (Timoshenko) beam elements with circular cross sections one by one on a flat surface. Fig. 14.19 shows how a greater number of bonds alter the mechanical properties of the mat in different fibre volume fractions. An increase in bond density increases the elastic modulus and strength of the mat. This study showed that a 10% increase in the fibre volume fraction (and consequently the bond density) results in a ~22% increase in the mat strength. Interestingly, the increase in the fibre volume fraction (and bond density) has a minor effect on the strain to failure. In their study, a 3.2-fold increase in the bond density resulted in an increase of the elastic modulus of the mat around 1.8–2.2 times depending on the bond strength. In another study, Chavoshnejad and Razavi [96] showed although the bonding increases the stiffness of the mat for all the ranges of porosities, it is more effective at low porosity ranges.

Different approaches have been used to model the interfibre bonding between crossing fibres in the fibrous mats. However, the modelling of interfibre bonding does not have a unique and established method, and every research group has their own strategy to simulate the effect of this parameter on the mechanical behaviour of the mats. Fig. 14.20A–H summarises the employed strategies in different studies to model the interfibre bonding in the process of simulation. Kumar et al. [104] generated an

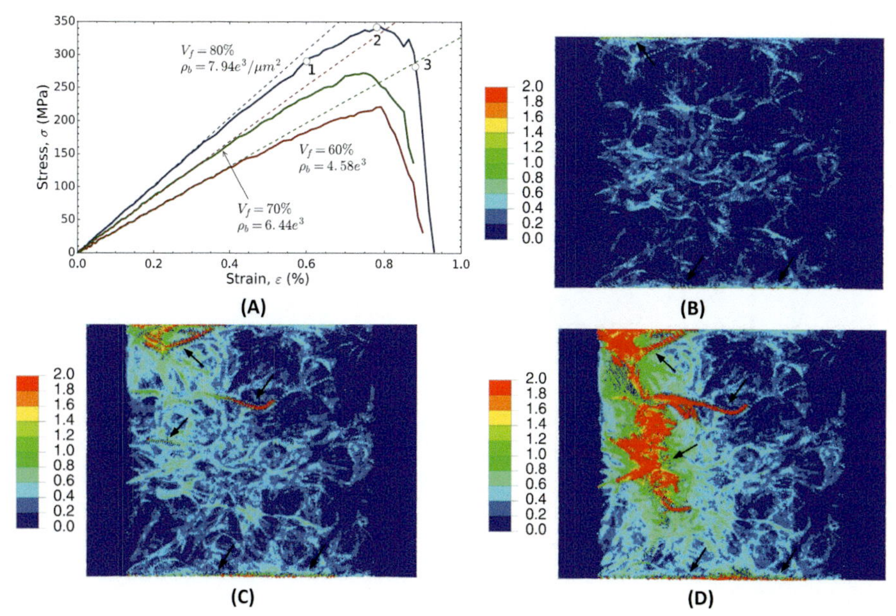

Fig. 14.19 (A) Fibre network stress-strain curves for three different fibre volume fractions, 60%, 70%, and 80% respectively. ρ_b shows the bond density. Bond density is defined as the number of inter-fibre bonds in the network divided by the surface of the fibre network. (B)–(D) show contour plots of the inter-fibre bond state for increasing applied displacement. The contours correspond to points 1–3 of A.
Reprinted with permission from S. Goutianos, R. Mao, T. Peijs, Effect of inter-fibre bonding on the fracture of fibrous networks with strong interactions, Int. J. Solids Struct. 136–137 (2018) 271–278, https://doi.org/10.1016/j.ijsolstr.2017.12.020.

RVE by controlling the arrangement and diameter of the fibres as well as the porosity of the mat. The fibres were marginally fused at their crossing points. They showed that both the Young's modulus and the toughness of the polystyrene mats of aligned fibres are greater than mats consisting of random fibres (Fig. 14.20A). Torgnysdotter et al. [105] developed a finite element model to illustrate the influence of the fibre and the contact region properties on the stress-strain behaviour and strength of the bond between two crossing fibres (Fig. 14.20B). They showed that the degree of contact between fibres is the most controlling parameter in determining the maximum stress and the fibre–fibre joint strength.

The tie constraint concept, which assumes no debonding or slippage at the cross points, was used to model the perfect bonding in several finite element-based studies [96,103,106] (Fig. 14.20C). Zhang et al. [106] developed a finite element RVE to study the large deformation mechanics of random nonwoven networks, benefiting the tie constraint concept. The results showed that the porosity of the network, the fibre aspect ratio, and the fibre curliness have considerable effects on the network stiffness and strength. In another study [29], the bonds between the fibres were modelled by a one-dimensional (1D) non-linear spring with damage possibility

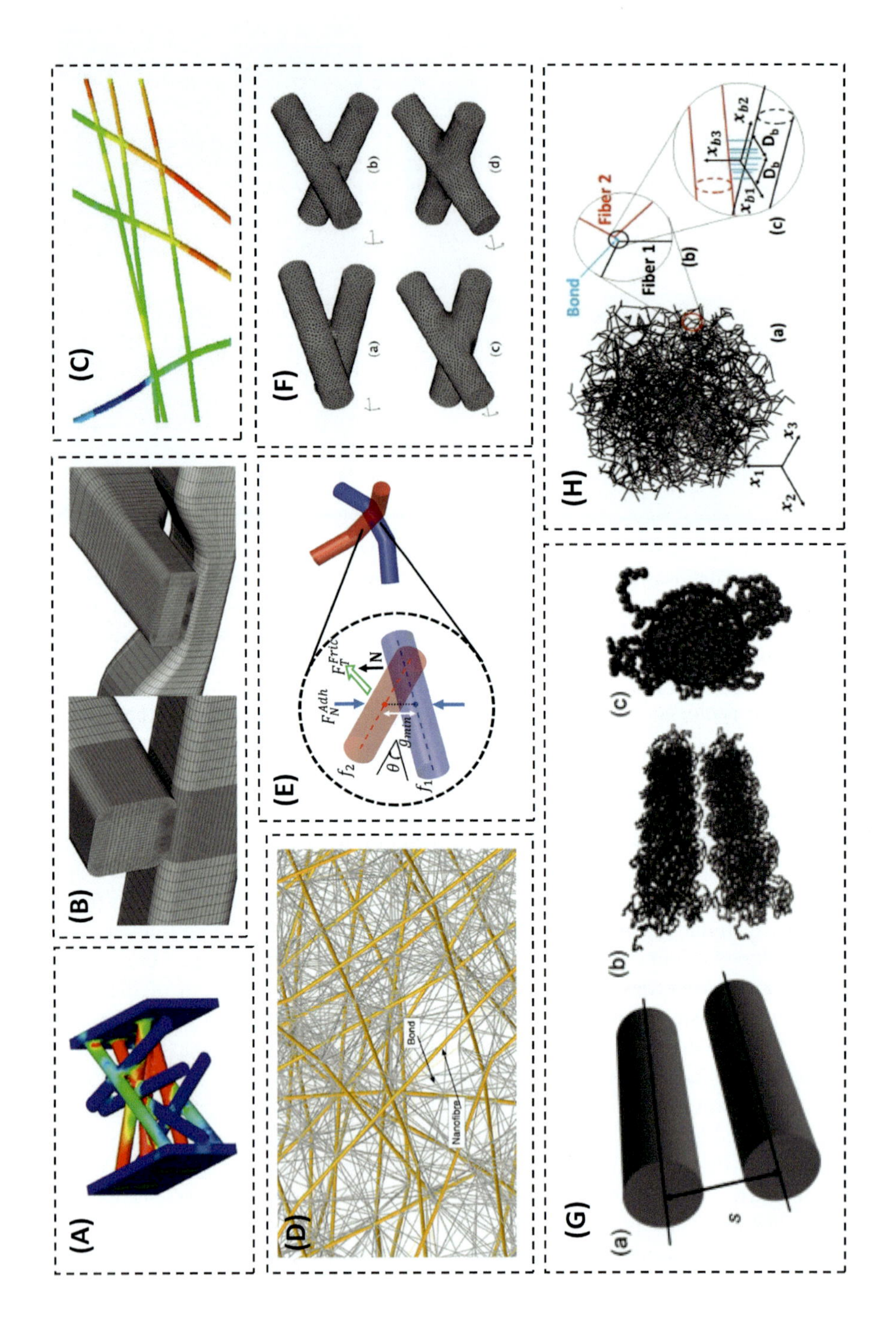

Fig. 14.20 See figure legend on opposite page

(Fig. 14.20D). Negi and Picu [107] developed a non-bonded model for modelling the interaction among fibres. Unlike the permanent bonding, the developed non-bonded interaction by considering friction and adhesion acts only at inter-fibre contacts during the duration of the contact (Fig. 14.20E). Berhan et al. [108] studied the geometry and response of 3D interconnects between cylindrical elements that abundantly can be seen in fibrous mats. By developing a finite element model for fused bonds, they showed the local stress at the intersections is very dependent on the degree of intersection (d.o.i.) of the joint (Fig. 14.20F). d.o.i. is defined as the ratio of the diameter of the cylinder (e.g. diameter of the fibre) to the width of the joint. Small d.o.i. results in great local stress. Therefore, to improve the bond performance of the fused fibrous networks, the d.o.i. of the joints should be kept high. As discussed in Section 14.2, the d.o.i. of the non-woven mats can be controlled by adjusting the thermal treatment time and temperature and the fibre contact force. Buell et al. [44] employed molecular dynamics (MD) and molecular statics (MS)-based simulations to study the interfibre interactions between prototypical polymeric fibres of 4.6 nm diameter (Fig. 14.20G). Deogekar and Picu [109] modelled the bonds as uncoupled springs with translational and rotational stiffness using the connector element CONN3D2 in Abaqus (Fig. 14.20H). They observed that the strength of a fibrous material is linearly

Fig. 14.20 Interfibre bonding modelling in a few selected simulation studies. (A) Fibres are marginally fused at their crossing area section in an RVE. (B) Modelling fibre–fibre interaction by a finite element model. (C) Tie constraint for modelling perfectly bonding. (D) Bonds between fibres are modelled by one-dimensional non-linear spring with damage. (E) Non-bonded interaction by considering friction and adhesion. (F) Finite element modelling of interfibre bonding with different degrees of intersection (d.o.i.). (G) Molecular dynamics (MD) and molecular statics (MS) simulations to study the interfibre interactions. (H) Modelling of bonding by the number of fibrils uniformly distributed over the contact area of size $D_b \times D_b$. Reprinted with permission from S. Goutianos, R. Mao, T. Peijs, Effect of inter-fibre bonding on the fracture of fibrous networks with strong interactions, Int. J. Solids Struct. 136–137 (2018) 271–278, https://doi.org/10.1016/j.ijsolstr.2017.12.020; S. Buell, G.C. Rutledge, K.J.V. Vliet, Predicting polymer nanofiber interactions via molecular simulations, ACS Appl. Mater. Interfaces, 2(4) (2010) 1164–1172, https://doi.org/10.1021/am1000135; P. Chavoshnejad, M.J. Razavi, Effect of the interfiber bonding on the mechanical behavior of electrospun fibrous mats, Sci. Rep. 10(1) (2020) 7709, https://doi.org/10.1038/s41598-020-64735-5; P. Kumar, R. Vasita, Understanding the relation between structural and mechanical properties of electrospun fiber mesh through uniaxial tensile testing, J. Appl. Polym. Sci. 134(26) 2017, https://doi.org/10.1002/app.45012; A. Torgnysdotter, A. Kulachenko, P. Gradin, L. Wågberg, Fiber/fiber crosses: finite element modeling and comparison with experiment, J. Compos. Mater. 41(13) (2007) 1603–1618, https://doi.org/10.1177/0021998306069873; V. Negi, R.C. Picu, Mechanical behavior of nonwoven non-crosslinked fibrous mats with adhesion and friction, Soft Matter 15(29) (2019) 5951–5964, https://doi.org/10.1039/C9SM00658C; L. Berhan, A.M. Sastry, On modeling bonds in fused, porous networks: 3D simulations of fibrous–particulate joints, J. Compos. Mater. 37(8) (2003) 715–740, https://doi.org/10.1177/002199803029725; S. Deogekar, R.C. Picu, On the strength of random fiber networks, J. Mech. Phys. Solids 116 (2018) 1–16, https://doi.org/10.1016/j.jmps.2018.03.026.

proportional to its fibre density and the strength of the interfibre bonds. Using a finite element model for a pre-cracked mat, Theng and Chung [110] showed that reducing the fibre density or percentage of bonding increase the toughness of the fibrous network. Increasing the bonding percentage increases the degree of non-affine deformation and results in a more compliant response at the crack-tip region. That study showed that the effect of the cross-link percentage (bonding percentage), in a predefined porosity, diminishes enormously after a cross-link percentage threshold of 30%. Negi et al. [107] developed a model to study the mechanical behaviour of nonwoven fibrous mats with adhesion and friction without crosslinking. The results of the study indicated that in the absence of interfibre bonding the response of the mat to tensile loading has three regimes: a short elastic regime where there is no sliding at the contacts of the fibres, a sliding regime characterised by strain hardening, and a fast stiffening regime at larger strains.

So far, it has been discussed that the interfibre bonding affects the mechanical behaviour of the fibrous mats. However, it should be emphasised that not only the density of the bonding but also the mechanical strength of the bonds can alter the mechanics of the non-woven mats [18,29]. Meanwhile, accurate characterisation and measurement of the mechanical properties of the bonds are not straightforward mainly because of the nano- and microscales of the constituent fibres. Therefore, the constitutive behaviour of the bonds has not been often considered in simulations. The mechanical properties of a bond have been considered in a few studies employing the tensile testing or atomic force microscopy (AFM) [105,111,112]. The mechanical strength of the fibrous mats was found to increase with increasing bond strength [55]. The damage takes place by either the interfibre bond fracture or the fibre breakage [113]. The interfibre bond damage usually starts at small strains and affects mechanical strength and elasticity degradation. Bond fracture greatly affects the strength and damage progression in a fibre network structure [18].

The effect of the interfibre debonding and damage of the interfibre bonding can be considered by providing a constitutive behaviour. In this strategy, the debonding and damage of the interfibre bonding plays a role in the overall mechanical behaviour of the mat. Bond damage was found to be the main damage mechanism in many non-woven products such as paper, geosynthetic materials, and electrospun mats [29,82,109,114,115]. However, the bond strength is rarely evaluated because of the lack of an effective characterisation technique. The values of the bond strength of various non-woven materials such as paper have great variance [111,116]. Various experimental techniques such as acoustic emission detection, in situ SEM, and in situ computed tomography have been used to capture the bond fracture under deformation [82,114,117]. The results of a two-dimensional (2D) model developed by Hägglund et al. [118] showed that bond fracture occurs if the energy stored in the bond exceeds a certain threshold value. Bond damage in the non-woven mats is observed starting at strains as small as 2.5% nominal strain [82,114].

A few constitutive models tried to model bond damage by introducing a damage variable [18,29]. However, this variable did not have a direct relationship to bond properties because of the lack of known bond properties [119]. Chen and Silberstein [18] presented a combined experimental and computational approach to study the bond strength in non-woven mats. The bond properties were measured by matching the results predicted by finite element modelling to the experimental data. They employed the Ortiz's interfacial cohesive law to describe the bond behaviour, using a four-parameter bi-linear interface law with normal stiffness, k, shear stiffness, βk, separation at the start of damage, d_1, and separation at total loss of the bond stiffness, d_2, was used. Interfacial effective displacement is defined as $\Delta_{\text{eff}} = \sqrt{\Delta_n{}^2 + (\beta \Delta_s{}^2)}$, where Δ_n is normal displacement and Δ_s is shear displacement. The bond behaviour is linearly elastic when $\Delta_{\text{eff}} \leq d_1$, and the bond elastic modulus linearly degrades when $d_1 \leq \Delta_{\text{eff}} \leq d_2$. The load displacement of a bond under the normal separation is shown in Fig. 14.21A. Bond strengths in pure normal and pure shear deformation modes are calculated as $\sigma_{\text{normal}} = kd_1$ and $\sigma_{\text{shear}} = \beta kd_1$, respectively. The bonds were modelled by a connector element, which constitutively defines the interaction between two bonding nodes. For a commercial non-woven polypropylene mat composed of fibres with diameter 40–60 μm, the normal and shear strength were obtained $1.3 \pm 0.3 \times 10^2$ and $1.0 \pm 0.2 \times 10^2$ MPa, respectively.

The effect of bond strength parameters on force-displacement behaviour and the percentage of the broken bonds of the mat are shown in Fig. 14.21B and C. It is clear there is a considerable difference in the mechanical behaviour of the no-bonded model and the models consisting of rigid or even damageable bonds. A similar method and results have been presented by Goutianos et al. [29]. They described the response of the bonds by a 1D cohesive law acting in the direction of the bond. This method defines the bond as a damageable non-linear spring. They argued that the strength of the mat is mainly controlled by the interfibre bond strength and secondarily by the interfibre bond fracture energy.

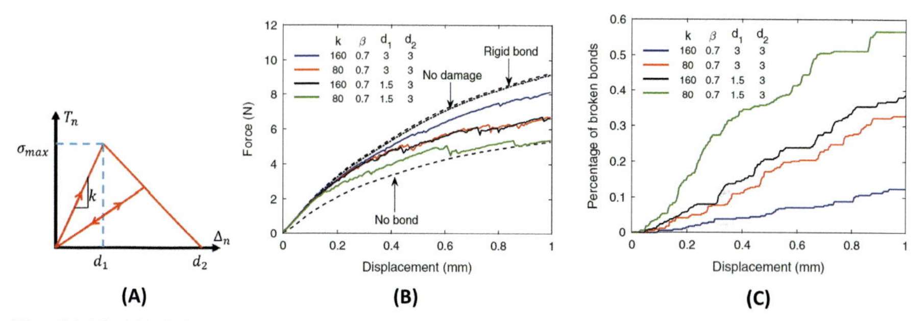

Fig. 14.21 (A) Schematic illustration of bi-linear, under tension, traction-separation law simulating the mechanical response of inter-fibre bonds. Line with double arrows indicates the unloading-reloading path for this bond model. (B) Load displacement curves for parameter study of k and d_1. (C) Corresponding damage progression curves.

Reprinted with permission from N. Chen, M.N. Silberstein, Determination of bond strengths in non-woven fabrics: a combined experimental and computational approach, Exp. Mech. 58(2) (2018) 343–355, https://doi.org/10.1007/s11340-017-0346-3.

14.4 Summary and future perspective

In this chapter, after a brief introduction to the electrospinning process, the materials used, and the effect of dominant parameters e.g. porosity, fibre diameter, fibre alignment, and the interfibre interaction on the mechanical properties of the electrospun mats were presented. Among these parameters, special attention was paid to the effect of interfibre bonding at the cross points of the individual fibres on determining the mechanical performances as well as the structural integrity of the electrospun fibrous mats. The damage at the interfibre bonds is the main failure mechanism of the electrospun fibrous mats and is mainly activated at the early stages of the deformation. Considering the inherently weak interactions at the intersections of the fibres, the methods to improve the interfibre bonding e.g. thermal (or photothermal) treatment bonding, solvent- and vapour-induced bonding, chemical or crosslinking bonding, physical bonding, as well as the results of different studies were reviewed. Moreover, the effectiveness, advantages, and disadvantages of these either in-process or post-process methods, on the improvement of mechanical properties as well as morphological futures of the electrospun fibrous mats were discussed. It was also discussed that discrete computational models can accurately link the mechanical properties of a mat to the mechanical properties of the constituent microstructure. Considering the determining role of the interfibre bonding on mechanical properties as well as the structural integrity of the electrospun mats from one side and with the increasing need of industry for high-performance electrospun mats from the other side, the goals of future studies, the research works, and innovations in the context of interfibre bonding can be defined as follows:

- Efforts to develop new methods and strategies as well as to improve the efficiency of the current methods in enhancing of the interfibre bonding in electrospun fibrous mats, particularly the in-process and solvent-less methods to make the process more cost effective and environmentally friendly.
- Attempts shall also be made to employ new methodologies and improve the developed processes to bond the fibres more selectively at their cross points without alteration in morphological characteristics and structural integrity of the electrospun mats.
- Development and application of modern methods as well as improving existing methods in evaluating the mechanical properties of single electrospun fibres and the mechanics of the cross points of the fibres.
- Developing efficient computational models to mimic and predict the mechanical performance of fibrous mats according to their microstructure mechanical properties.

References

[1] J. Xue, T. Wu, Y. Dai, Y. Xia, Electrospinning and electrospun nanofibers: methods, materials, and applications, Chem. Rev. 119 (8) (2019) 5298–5415, https://doi.org/10.1021/acs.chemrev.8b00593.

[2] H. Li, C. Zhu, J. Xue, Q. Ke, Y. Xia, Enhancing the mechanical properties of electrospun nanofiber mats through controllable welding at the cross points, Macromol. Rapid Commun. 38 (9) (2017) 1600723, https://doi.org/10.1002/marc.201600723.

[3] S. Jian, et al., Nanofibers with diameter below one nanometer from electrospinning, RSC Adv. 8 (9) (2018) 4794–4802, https://doi.org/10.1039/C7RA13444D.
[4] Y. Li, M.A. Abedalwafa, L. Tang, D. Li, L. Wang, Electrospun nanofibers for sensors, in: Electrospinning: Nanofabrication and Applications, Elsevier, 2019, pp. 571–601, https://doi.org/10.1016/B978-0-323-51270-1.00018-2.
[5] J. Xue, J. Xie, W. Liu, Y. Xia, Electrospun nanofibers: new concepts, materials, and applications, Acc. Chem. Res. 50 (8) (2017) 1976–1987, https://doi.org/10.1021/acs.accounts.7b00218.
[6] D. Li, Y. Wang, Y. Xia, Electrospinning of polymeric and ceramic nanofibers as uniaxially aligned arrays, Nano Lett. 3 (8) (2003) 1167–1171, https://doi.org/10.1021/nl0344256.
[7] S.-X. Wang, L. Yang, L.P. Stubbs, X. Li, C. He, Lignin-derived fused electrospun carbon fibrous mats as high performance anode materials for lithium ion batteries, ACS Appl. Mater. Interfaces 5 (23) (2013) 12275–12282, https://doi.org/10.1021/am4043867.
[8] C. Kim, K.-S. Yang, W.-J. Lee, The use of carbon nanofiber electrodes prepared by electrospinning for electrochemical supercapacitors, Electrochem. Solid St. 7 (11) (2004) A397, https://doi.org/10.1149/1.1801631.
[9] C. Kim, S.-H. Park, W.-J. Lee, K.-S. Yang, Characteristics of supercapaitor electrodes of PBI-based carbon nanofiber web prepared by electrospinning, Electrochim. Acta 50 (2–3) (2004) 877–881, https://doi.org/10.1016/j.electacta.2004.02.071.
[10] K. Yoon, B.S. Hsiao, B. Chu, Functional nanofibers for environmental applications, J. Mater. Chem. 18 (44) (2008) 5326, https://doi.org/10.1039/b804128h.
[11] T. Grafe, K. Graham, Polymeric nanofibers and nanofiber webs: a new class of nonwovens, Int. Nonwovens J. os-12 (1) (2003), https://doi.org/10.1177/1558925003os-1200113, 1558925003os–12.
[12] R. Aminyan, S. Bazgir, Fabrication and characterization of nanofibrous polyacrylic acid superabsorbent using gas-assisted electrospinning technique, React. Funct. Polym. 141 (2019) 133–144, https://doi.org/10.1016/j.reactfunctpolym.2019.05.012.
[13] E.P.S. Tan, C.T. Lim, Effects of annealing on the structural and mechanical properties of electrospun polymeric nanofibres, Nanotechnology 17 (10) (2006) 2649–2654, https://doi.org/10.1088/0957-4484/17/10/034.
[14] F.J. Chaparro, et al., Sintered electrospun poly(ε-caprolactone)–poly(ethylene terephthalate) for drug delivery, J. Appl. Polym. Sci. 136 (26) (2019) 47731, https://doi.org/10.1002/app.47731.
[15] A. Morel, Understanding and Tailoring the Multiscale Architectural and Mechanical Properties of Electrospun Membranes for Tissue Engineering Applications, ETH Zurich, 2019, https://doi.org/10.3929/ETHZ-B-000404888.
[16] M. Kancheva, A. Toncheva, N. Manolova, I. Rashkov, Enhancing the mechanical properties of electrospun polyester mats by heat treatment, Express Polym. Lett. 9 (1) (2015) 49–65, https://doi.org/10.3144/expresspolymlett.2015.6.
[17] F.E. Ahmed, B.S. Lalia, R. Hashaikeh, A review on electrospinning for membrane fabrication: challenges and applications, Desalination 356 (2015) 15–30, https://doi.org/10.1016/j.desal.2014.09.033.
[18] N. Chen, M.N. Silberstein, Determination of bond strengths in non-woven fabrics: a combined experimental and computational approach, Exp. Mech. 58 (2) (2018) 343–355,-https://doi.org/10.1007/s11340-017-0346-3.
[19] S.-S. Choi, S.G. Lee, C.W. Joo, S.S. Im, S.H. Kim, Formation of interfiber bonding in electrospun poly(etherimide) nanofiber web, J. Mater. Sci. 39 (4) (2004) 1511–1513,-https://doi.org/10.1023/B:JMSC.0000013931.84760.b0.

[20] T. Wu, H. Li, J. Xue, X. Mo, Y. Xia, Photothermal welding, melting, and patterned expansion of nonwoven mats of polymer nanofibers for biomedical and printing applications, Angew. Chem. Int. Ed. 58 (46) (2019) 16416–16421, https://doi.org/10.1002/anie.201907876.
[21] D. Cao, Z. Fu, C. Li, Heat and compression molded electrospun poly(l-lactide) membranes: preparation and characterization, Mater. Sci. Eng. B 176 (12) (2011) 900–905, https://doi.org/10.1016/j.mseb.2011.05.015.
[22] K. Lee, B. Lee, C. Kim, H. Kim, K. Kim, C. Nah, Stress-strain behavior of the electrospun thermoplastic polyurethane elastomer fiber mats, Macromol. Res. 13 (5) (2005) 441–445, https://doi.org/10.1007/BF03218478.
[23] Q. Shi, K.-T. Wan, S.-C. Wong, P. Chen, T.A. Blackledge, Do electrospun polymer fibers stick? Langmuir 26 (17) (2010) 14188–14193, https://doi.org/10.1021/la1022328.
[24] U. Stachewicz, R.J. Bailey, W. Wang, A.H. Barber, Size dependent mechanical properties of electrospun polymer fibers from a composite structure, Polymer 53 (22) (2012) 5132–5137, https://doi.org/10.1016/j.polymer.2012.08.064.
[25] X. Wei, Z. Xia, S.C. Wong, A. Baji, Modelling of mechanical properties of electrospun nanofibre network, Int. J. Exp. Comput. Biomech. 1 (1) (2009) 45, https://doi.org/10.1504/IJECB.2009.022858.
[26] Z. Wang, et al., Effect of thermal annealing on mechanical properties of polyelectrolyte complex nanofiber membranes, Fibers Polym. 15 (7) (2014) 1406–1413, https://doi.org/10.1007/s12221-014-1406-2.
[27] S. Homaeigohar, J. Koll, E.T. Lilleodden, M. Elbahri, The solvent induced interfiber adhesion and its influence on the mechanical and filtration properties of polyethersulfone electrospun nanofibrous microfiltration membranes, Sep. Purif. Technol. 98 (2012) 456–463, https://doi.org/10.1016/j.seppur.2012.06.027.
[28] Y. You, S. Won Lee, S. Jin Lee, W.H. Park, Thermal interfiber bonding of electrospun poly(l-lactic acid) nanofibers, Mater. Lett. 60 (11) (2006) 1331–1333, https://doi.org/10.1016/j.matlet.2005.11.022.
[29] S. Goutianos, R. Mao, T. Peijs, Effect of inter-fibre bonding on the fracture of fibrous networks with strong interactions, Int. J. Solids Struct. 136–137 (2018) 271–278, https://doi.org/10.1016/j.ijsolstr.2017.12.020.
[30] M. Zündel, E. Mazza, A.E. Ehret, A 2.5D approach to the mechanics of electrospun fibre mats, Soft Matter 13 (37) (2017) 6407–6421, https://doi.org/10.1039/C7SM01241A.
[31] S.-C. Wong, A. Baji, S. Leng, Effect of fiber diameter on tensile properties of electrospun poly(ε-caprolactone), Polymer 49 (21) (2008) 4713–4722, https://doi.org/10.1016/j.polymer.2008.08.022.
[32] H.H. Kim, M.J. Kim, S.J. Ryu, C.S. Ki, Y.H. Park, Effect of fiber diameter on surface morphology, mechanical property, and cell behavior of electrospun poly(ε-caprolactone) mat, Fibers Polym. 17 (7) (2016) 1033–1042, https://doi.org/10.1007/s12221-016-6350-x.
[33] C.-L. Pai, M.C. Boyce, G.C. Rutledge, Mechanical properties of individual electrospun PA 6(3)T fibers and their variation with fiber diameter, Polymer 52 (10) (2011) 2295–2301, https://doi.org/10.1016/j.polymer.2011.03.041.
[34] F. Croisier, et al., Mechanical testing of electrospun PCL fibers, Acta Biomater. 8 (1) (2012) 218–224, https://doi.org/10.1016/j.actbio.2011.08.015.
[35] C.R. Carlisle, C. Coulais, M. Guthold, The mechanical stress–strain properties of single electrospun collagen type I nanofibers, Acta Biomater. 6 (8) (2010) 2997–3003, https://doi.org/10.1016/j.actbio.2010.02.050.

[36] M. Naraghi, S.N. Arshad, I. Chasiotis, Molecular orientation and mechanical property size effects in electrospun polyacrylonitrile nanofibers, Polymer 52 (7) (2011) 1612–1618, https://doi.org/10.1016/j.polymer.2011.02.013.
[37] A. Baji, Y.-W. Mai, S.-C. Wong, M. Abtahi, P. Chen, Electrospinning of polymer nanofibers: effects on oriented morphology, structures and tensile properties, Compos. Sci. Technol. 70 (5) (2010) 703–718, https://doi.org/10.1016/j.compscitech.2010.01.010.
[38] M.K. Selatile, S.S. Ray, V. Ojijo, R. Sadiku, Correlations between fibre diameter, physical parameters, and the mechanical properties of randomly oriented biobased polylactide nanofibres, Fibers Polym. 20 (1) (2019) 100–112, https://doi.org/10.1007/s12221-019-8262-z.
[39] I. Borisova, O. Stoilova, N. Manolova, I. Rashkov, Modulating the mechanical properties of electrospun PHB/PCL materials by using different types of collectors and heat sealing, Polymers 12 (3) (2020) 693, https://doi.org/10.3390/polym12030693.
[40] T. Stylianopoulos, et al., Tensile mechanical properties and hydraulic permeabilities of electrospun cellulose acetate fiber meshes, J. Biomed. Mater. Res. B Appl. Biomater. 100B (8) (2012) 2222–2230, https://doi.org/10.1002/jbm.b.32791.
[41] L.T.(.S.). Choong, Z. Khan, G.C. Rutledge, Permeability of electrospun fiber mats under hydraulic flow, J. Membr. Sci. 451 (2014) 111–116, https://doi.org/10.1016/j.memsci.2013.09.051.
[42] W.W. Sampson, A multiplanar model for the pore radius distribution in isotropic near-planar stochastic fibre networks, J. Mater. Sci. 38 (8) (2003) 1617–1622, https://doi.org/10.1023/A:1023298820390.
[43] R. Bagherzadeh, M. Latifi, S.S. Najar, L. Kong, Experimental verification of theoretical prediction of fiber to fiber contacts in electrospun multilayer nano-microfibrous assemblies: effect of fiber diameter and network porosity, J. Ind. Text. 43 (4) (2014) 483–495,-https://doi.org/10.1177/1528083712463400.
[44] S. Buell, G.C. Rutledge, K.J.V. Vliet, Predicting polymer nanofiber interactions via molecular simulations, ACS Appl. Mater. Interfaces 2 (4) (2010) 1164–1172, https://doi.org/10.1021/am1000135.
[45] X.-F. Wu, Y.A. Dzenis, Adhesive contact in filaments, J. Phys. Appl. Phys. 40 (14) (2007) 4276–4280, https://doi.org/10.1088/0022-3727/40/14/026.
[46] V. Beachley, X. Wen, Effect of electrospinning parameters on the nanofiber diameter and length, Mater. Sci. Eng. C 29 (3) (2009) 663–668, https://doi.org/10.1016/j.msec.2008.10.037.
[47] C.J. Thompson, G.G. Chase, A.L. Yarin, D.H. Reneker, Effects of parameters on nanofiber diameter determined from electrospinning model, Polymer 48 (23) (2007) 6913–6922, https://doi.org/10.1016/j.polymer.2007.09.017.
[48] S.-H. Tan, R. Inai, M. Kotaki, S. Ramakrishna, Systematic parameter study for ultra-fine fiber fabrication via electrospinning process, Polymer 46 (16) (2005) 6128–6134, https://doi.org/10.1016/j.polymer.2005.05.068.
[49] J. Lee, Y. Deng, Increased mechanical properties of aligned and isotropic electrospun PVA nanofiber webs by cellulose nanowhisker reinforcement, Macromol. Res. 20 (1) (2012) 76–83, https://doi.org/10.1007/s13233-012-0008-3.
[50] F. Chen, Y. Su, X. Mo, C. He, H. Wang, Y. Ikada, Biocompatibility, alignment degree and mechanical properties of an electrospun chitosan–P(LLA-CL) fibrous scaffold, J. Biomater. Sci. Polym. Ed. 20 (14) (2009) 2117–2128, https://doi.org/10.1163/156856208X400492.

[51] S. Park, K. Park, H. Yoon, J. Son, T. Min, G. Kim, Apparatus for preparing electrospun nanofibers: designing an electrospinning process for nanofiber fabrication: apparatus for preparing electrospun nanofibers, Polym. Int. 56 (11) (2007) 1361–1366, https://doi.org/10.1002/pi.2345.
[52] W.E. Teo, S. Ramakrishna, A review on electrospinning design and nanofibre assemblies, Nanotechnology 17 (14) (2006) R89–R106, https://doi.org/10.1088/0957-4484/17/14/R01.
[53] L. Zhao, H. Zhang, X. Li, J. Zhao, C. Zhao, X. Yuan, Modification of electrospun poly(vinylidene fluoride-*co*-hexafluoropropylene) membranes through the introduction of poly(ethylene glycol) dimethacrylate, J. Appl. Polym. Sci. 111 (6) (2009) 3104–3112,-https://doi.org/10.1002/app.29374.
[54] R. Ding, Fabrication and Characterization of Novel Polymer-Matrix Nanocomposites and Their Constituents, Doctor of Philosophy, Iowa State University, Digital Repository, Ames, 2016, https://doi.org/10.31274/etd-180810-4721.
[55] B.K. Raghavan, D.W. Coffin, Control of inter-fiber fusing for nanofiber webs via electrospinning, J. Eng. Fibers Fabr. 6 (4) (2011), https://doi.org/10.1177/155892501100600401, 155892501100600.
[56] L. Huang, N.-N. Bui, S.S. Manickam, J.R. McCutcheon, Controlling electrospun nanofiber morphology and mechanical properties using humidity, J. Polym. Sci. B Polym. Phys. 49 (24) (2011) 1734–1744, https://doi.org/10.1002/polb.22371.
[57] Z. Tang, et al., Design and fabrication of electrospun polyethersulfone nanofibrous scaffold for high-flux nanofiltration membranes: PES electrospun nanofibrous scaffold for TFNC NF, J. Polym. Sci. B Polym. Phys. 47 (22) (2009) 2288–2300, https://doi.org/10.1002/polb.21831.
[58] L. Huang, S.S. Manickam, J.R. McCutcheon, Increasing strength of electrospun nanofiber membranes for water filtration using solvent vapor, J. Membr. Sci. 436 (2013) 213–220, https://doi.org/10.1016/j.memsci.2012.12.037.
[59] C. Merlini, et al., Electrically conductive polyaniline-coated electrospun poly(vinylidene fluoride) mats, Front. Mater. 2 (2015), https://doi.org/10.3389/fmats.2015.00014.
[60] W. Wang, et al., Effect of vapor-phase glutaraldehyde crosslinking on electrospun starch fibers, Carbohydr. Polym. 140 (2016) 356–361, https://doi.org/10.1016/j.carbpol.2015.12.061.
[61] B. Ghorani, S.J. Russell, P. Goswami, Controlled morphology and mechanical characterisation of electrospun cellulose acetate fibre webs, Int. J. Polym. Sci. 2013 (2013) 1–12, https://doi.org/10.1155/2013/256161.
[62] C. Zhao, H. Zhang, X. Zeng, H. Li, D. Sun, Enhancing the inter-fiber bonding properties of cellulosic fibers by increasing different fiber charges, Cellulose 23 (3) (2016) 1617–1628, https://doi.org/10.1007/s10570-016-0941-y.
[63] L. Chen, et al., Enhanced mechanical properties of poly(arylene sulfide sulfone) membrane by co-electrospinning with poly(m-xylene adipamide), Chin. J. Polym. Sci. 38 (1) (2020) 63–71, https://doi.org/10.1007/s10118-019-2297-x.
[64] Y. Shi, Y. Zhao, X. Li, D. Yan, D. Cao, Z. Fu, Enhancement of the mechanical properties and thermostability of poly(vinyl alcohol) nanofibers by the incorporation of sodium chloride, J. Appl. Polym. Sci. 135 (13) (2018) 45981, https://doi.org/10.1002/app.45981.
[65] Z. Li, H. Zhang, X. Wang, F. Zhang, X. Li, Further understanding the response mechanism of lignin content to bonding properties of lignocellulosic fibers by their deformation behavior, RSC Adv. 6 (110) (2016) 109211–109217, https://doi.org/10.1039/C6RA22457A.

[66] H. Niu, X. Wang, T. Lin, Carbon nanofibers with inter-bonded fibrous structure for supercapacitor application, in: 2012 World Congr. Adv. Civ. Environ. Mater. Res. ACEM' 12, 2012.
[67] J. Namsaeng, W. Punyodom, P. Worajittiphon, Synergistic effect of welding electrospun fibers and MWCNT reinforcement on strength enhancement of PAN–PVC non-woven mats for water filtration, Chem. Eng. Sci. 193 (2019) 230–242, https://doi.org/10.1016/j.ces.2018.09.019.
[68] S. Ramaswamy, L.I. Clarke, R.E. Gorga, Morphological, mechanical, and electrical properties as a function of thermal bonding in electrospun nanocomposites, Polymer 52 (14) (2011) 3183–3189, https://doi.org/10.1016/j.polymer.2011.05.023.
[69] R. Wang, et al., Plasma-induced nanowelding of a copper nanowire network and its application in transparent electrodes and stretchable conductors, Nano Res. 9 (7) (2016) 2138–2148, https://doi.org/10.1007/s12274-016-1103-0.
[70] E.C. Garnett, et al., Self-limited plasmonic welding of silver nanowire junctions, Nat. Mater. 11 (3) (2012) 3, https://doi.org/10.1038/nmat3238.
[71] Y. Liao, et al., Carbon nanotube-templated polyaniline nanofibers: synthesis, flash welding and ultrafiltration membranes, Nanoscale 5 (9) (2013) 3856, https://doi.org/10.1039/c3nr00441d.
[72] Q. Qian, J. Wang, F. Yan, Y. Wang, A photo-annealing approach for building functional polymer layers on paper, Angew. Chem. 126 (17) (2014) 4554–4557, https://doi.org/10.1002/ange.201310714.
[73] C.-M. Hsu, S. Shivkumar, Nano-sized beads and porous fiber constructs of poly(ε-caprolactone) produced by electrospinning, J. Mater. Sci. 39 (9) (2004) 3003–3013, https://doi.org/10.1023/B:JMSC.0000025826.36080.cf.
[74] J. Lannutti, D. Reneker, T. Ma, D. Tomasko, D. Farson, Electrospinning for tissue engineering scaffolds, Mater. Sci. Eng. C 27 (3) (2007) 504–509, https://doi.org/10.1016/j.msec.2006.05.019.
[75] Y. Shen, D. Li, B. Deng, Q. Liu, H. Liu, T. Wu, Robust polyimide nano/microfibre aerogels welded by solvent-vapour for environmental applications, R. Soc. Open Sci. 6 (8) (2019) 190596, https://doi.org/10.1098/rsos.190596.
[76] J. Xie, et al., Role of a 'surface wettability switch' in inter-fiber bonding properties, RSC Adv. 8 (6) (2018) 3081–3089, https://doi.org/10.1039/C7RA12307H.
[77] M. Henriksson, L.A. Berglund, P. Isaksson, T. Lindström, T. Nishino, Cellulose nanopaper structures of high toughness, Biomacromolecules 9 (6) (2008) 1579–1585, https://doi.org/10.1021/bm800038n.
[78] S.-J. Chun, S.-Y. Lee, G.-H. Doh, S. Lee, J.H. Kim, Preparation of ultrastrength nanopapers using cellulose nanofibrils, J. Ind. Eng. Chem. 17 (3) (2011) 521–526, https://doi.org/10.1016/j.jiec.2010.10.022.
[79] A. Retegi, et al., Bacterial cellulose films with controlled microstructure–mechanical property relationships, Cellulose 17 (3) (2010) 661–669, https://doi.org/10.1007/s10570-009-9389-7.
[80] R. Mao, et al., Comparison of fracture properties of cellulose nanopaper, printing paper and buckypaper, J. Mater. Sci. 52 (16) (2017) 9508–9519, https://doi.org/10.1007/s10853-017-1108-4.
[81] S. Mohammadzadehmoghadam, Y. Dong, I.J. Davies, Modeling electrospun nanofibers: an overview from theoretical, empirical, and numerical approaches, Int. J. Polym. Mater. Polym. Biomater. 65 (17) (2016) 901–915, https://doi.org/10.1080/00914037.2016.1180617.

[82] A. Ridruejo, C. González, J. LLorca, Micromechanisms of deformation and fracture of polypropylene nonwoven fabrics, Int. J. Solids Struct. 48 (1) (2011) 153–162, https://doi.org/10.1016/j.ijsolstr.2010.09.013.
[83] M.N. Silberstein, C.-L. Pai, G.C. Rutledge, M.C. Boyce, Elastic–plastic behavior of non-woven fibrous mats, J. Mech. Phys. Solids 60 (2) (2012) 295–318, https://doi.org/10.1016/j.jmps.2011.10.007.
[84] Y. Chen, A. Ridruejo, C. González, J. Llorca, T. Siegmund, Notch effect in failure of fiberglass non-woven materials, Int. J. Solids Struct. 96 (2016) 254–264, https://doi.org/10.1016/j.ijsolstr.2016.06.004.
[85] T. Stylianopoulos, C.A. Bashur, A.S. Goldstein, S.A. Guelcher, V.H. Barocas, Computational predictions of the tensile properties of electrospun fibre meshes: effect of fibre diameter and fibre orientation, J. Mech. Behav. Biomed. Mater. 1 (4) (2008) 326–335,- https://doi.org/10.1016/j.jmbbm.2008.01.003.
[86] C.-L. Pai, M.C. Boyce, G.C. Rutledge, On the importance of fiber curvature to the elastic moduli of electrospun nonwoven fiber meshes, Polymer 52 (26) (2011) 6126–6133,- https://doi.org/10.1016/j.polymer.2011.10.055.
[87] M.S. Rizvi, P. Kumar, D.S. Katti, A. Pal, Mathematical model of mechanical behavior of micro/nanofibrous materials designed for extracellular matrix substitutes, Acta Biomater. 8 (11) (2012) 4111–4122, https://doi.org/10.1016/j.actbio.2012.07.025.
[88] M.S. Rizvi, A. Pal, Statistical model for the mechanical behavior of the tissue engineering non-woven fibrous matrices under large deformation, J. Mech. Behav. Biomed. Mater. 37 (2014) 235–250, https://doi.org/10.1016/j.jmbbm.2014.05.026.
[89] E. Vatankhah, D. Semnani, M.P. Prabhakaran, M. Tadayon, S. Razavi, S. Ramakrishna, Artificial neural network for modeling the elastic modulus of electrospun polycaprolactone/gelatin scaffolds, Acta Biomater. 10 (2) (2014) 709–721, https://doi.org/10.1016/j.actbio.2013.09.015.
[90] K. Polak-Krasna, A. Georgiadis, P. Heikkila, Mechanical characterisation and modelling of electrospun materials for biomedical applications, in: 2015 IEEE International Symposium on Medical Measurements and Applications (MeMeA) Proceedings, Torino, Italy, May, 2015, pp. 507–511, https://doi.org/10.1109/MeMeA.2015.7145256.
[91] Y. Yin, J. Xiong, Finite element analysis of electrospun nanofibrous mats under biaxial tension, Nanomaterials 8 (5) (2018) 348, https://doi.org/10.3390/nano8050348.
[92] Y. Yin, Z. Pan, J. Xiong, A tensile constitutive relationship and a finite element model of electrospun nanofibrous mats, Nanomaterials 8 (1) (2018) 29, https://doi.org/10.3390/nano8010029.
[93] C.T. Koh, C.Y. Low, Y.B. Yusof, Structure-property relationship of bio-inspired fibrous materials, Procedia Comput. Sci. 76 (2015) 411–416, https://doi.org/10.1016/j.procs.2015.12.278.
[94] F. Farukh, E. Demirci, B. Sabuncuoglu, M. Acar, B. Pourdeyhimi, V.V. Silberschmidt, Numerical analysis of progressive damage in nonwoven fibrous networks under tension, Int. J. Solids Struct. 51 (9) (2014) 1670–1685, https://doi.org/10.1016/j.ijsolstr.2014.01.015.
[95] T.P. Driscoll, N.L. Nerurkar, N.T. Jacobs, D.M. Elliott, R.L. Mauck, Fiber angle and aspect ratio influence the shear mechanics of oriented electrospun nanofibrous scaffolds, J. Mech. Behav. Biomed. Mater. 4 (8) (2011) 1627–1636, https://doi.org/10.1016/j.jmbbm.2011.03.022.
[96] P. Chavoshnejad, M.J. Razavi, Effect of the interfiber bonding on the mechanical behavior of electrospun fibrous mats, Sci. Rep. 10 (1) (2020) 7709, https://doi.org/10.1038/s41598-020-64735-5.

[97] A.S. Shahsavari, R.C. Picu, Elasticity of sparsely cross-linked random fibre networks, Philos. Mag. Lett. 93 (6) (2013) 356–361, https://doi.org/10.1080/09500839.2013.783241.
[98] G. Argento, M. Simonet, C.W.J. Oomens, F.P.T. Baaijens, Multi-scale mechanical characterization of scaffolds for heart valve tissue engineering, J. Biomech. 45 (16) (2012) 2893–2898, https://doi.org/10.1016/j.jbiomech.2012.07.037.
[99] A. Agic, B. Mijovic, Mechanical properties of electrospun carbon nanotube composites, J. Text. Inst. 97 (5) (2006) 419–427, https://doi.org/10.1533/joti.2006.0264.
[100] B. Mijovic, Bio-inspired electrospun fibre structures—numerical model, J. Fiber Bioeng. Inform. 6 (1) (2013) 23–32, https://doi.org/10.3993/jfbi03201302.
[101] A. D'Amore, et al., From single fiber to macro-level mechanics: a structural finite-element model for elastomeric fibrous biomaterials, J. Mech. Behav. Biomed. Mater. 39 (2014) 146–161, https://doi.org/10.1016/j.jmbbm.2014.07.016.
[102] Y. Liu, Y. Dzenis, Explicit 3D finite-element model of continuous nanofibre networks, Micro Nano Lett. 11 (11) (2016) 727–730, https://doi.org/10.1049/mnl.2016.0147.
[103] P. Chavoshnejad, O. Alsmairat, C. Ke, Effect of interfiber bonding on the rupture of electrospun fibrous mats, J. Phys. Appl. Phys. 54 (2021), https://doi.org/10.1088/1361-6463/abba95, 025302.
[104] P. Kumar, R. Vasita, Understanding the relation between structural and mechanical properties of electrospun fiber mesh through uniaxial tensile testing, J. Appl. Polym. Sci. 134 (26) (2017), https://doi.org/10.1002/app.45012.
[105] A. Torgnysdotter, A. Kulachenko, P. Gradin, L. Wågberg, Fiber/fiber crosses: finite element modeling and comparison with experiment, J. Compos. Mater. 41 (13) (2007) 1603–1618, https://doi.org/10.1177/0021998306069873.
[106] M. Zhang, Y. Chen, F. Chiang, P.I. Gouma, L. Wang, Modeling the large deformation and microstructure evolution of nonwoven polymer fiber networks, J. Appl. Mech. 86 (1) (2019) 011010, https://doi.org/10.1115/1.4041677.
[107] V. Negi, R.C. Picu, Mechanical behavior of nonwoven non-crosslinked fibrous mats with adhesion and friction, Soft Matter 15 (29) (2019) 5951–5964, https://doi.org/10.1039/C9SM00658C.
[108] L. Berhan, A.M. Sastry, On modeling bonds in fused, porous networks: 3D simulations of fibrous–particulate joints, J. Compos. Mater. 37 (8) (2003) 715–740, https://doi.org/10.1177/002199803029725.
[109] S. Deogekar, R.C. Picu, On the strength of random fiber networks, J. Mech. Phys. Solids 116 (2018) 1–16, https://doi.org/10.1016/j.jmps.2018.03.026.
[110] K.C. Theng, S.M. Chung, Effects of microstructure architecture on the fracture of fibrous materials, Int. J. Integr. Eng. 12 (1) (2020). Accessed 07 August 2020 [Online]. Available from: *https://publisher.uthm.edu.my/ojs/index.php/ijie/article/view/4294*.
[111] F.J. Schmied, C. Teichert, L. Kappel, U. Hirn, R. Schennach, Joint strength measurements of individual fiber-fiber bonds: an atomic force microscopy based method, Rev. Sci. Instrum. 83 (7) (2012) 073902, https://doi.org/10.1063/1.4731010.
[112] A. Kulachenko, T. Uesaka, Direct simulations of fiber network deformation and failure, Mech. Mater. 51 (2012) 1–14, https://doi.org/10.1016/j.mechmat.2012.03.010.
[113] N. Chen, M.N. Silberstein, A micromechanics-based damage model for non-woven fiber networks, Int. J. Solids Struct. 160 (2019) 18–31, https://doi.org/10.1016/j.ijsolstr.2018.10.009.
[114] N. Chen, M.K.A. Koker, S. Uzun, M.N. Silberstein, In-situ X-ray study of the deformation mechanisms of non-woven polypropylene, Int. J. Solids Struct. 97–98 (2016) 200–208, https://doi.org/10.1016/j.ijsolstr.2016.07.028.

[115] P. Isaksson, P.A. Gradin, A. Kulachenko, The onset and progression of damage in isotropic paper sheets, Int. J. Solids Struct. 43 (3–4) (2006) 713–726, https://doi.org/10.1016/j.ijsolstr.2005.04.035.

[116] F.J. Schmied, C. Teichert, L. Kappel, U. Hirn, W. Bauer, R. Schennach, What holds paper together: nanometre scale exploration of bonding between paper fibres, Sci. Rep. 3 (1) (2013) 2432, https://doi.org/10.1038/srep02432.

[117] P. Isaksson, R. Hägglund, Evolution of bond fractures in a randomly distributed fiber network, Int. J. Solids Struct. 44 (18–19) (2007) 6135–6147, https://doi.org/10.1016/j.ijsolstr.2007.02.013.

[118] R. Hägglund, P. Isaksson, On the coupling between macroscopic material degradation and interfiber bond fracture in an idealized fiber network, Int. J. Solids Struct. 45 (3–4) (2008) 868–878, https://doi.org/10.1016/j.ijsolstr.2007.09.011.

[119] A. Ridruejo, C. González, J. LLorca, A constitutive model for the in-plane mechanical behavior of nonwoven fabrics, Int. J. Solids Struct. 49 (17) (2012) 2215–2229, https://doi.org/10.1016/j.ijsolstr.2012.04.014.

Index

Note: Page numbers followed by *f* indicate figures and *t* indicate tables.

D

E

F

G

H

U

V

W

X

Y

Printed in the United States
by Baker & Taylor Publisher Services